DIESEL ENGINE OPERATION AND MAINTENANCE

The Construction, Operation, Maintenance, and Repair of Modern Diesel Engines

V. L. MALEEV, M.E., DR. A.M.

Fellow ASME, Member SAE
Professor Emeritus of Mechanical Engineering
Oklahoma Agricultural and Mechanical College
Formerly Senior Mechanical Engineer (Diesel)
U.S.N. Engineering Experiment Station, Annapolis
Formerly Chief Engineer, Western-Enterprise Engine Co.

McGRAW-HILL BOOK COMPANY, INC.

New York   Toronto   London

1954

DIESEL ENGINE OPERATION AND MAINTENANCE

Copyright, 1954, by the McGraw-Hill Book Company, Inc. Printed in the United States of America. All rights reserved. This book, or parts thereof, may not be reproduced in any form without permission of the publishers.

Library of Congress Catalog Card Number: 53-12431

THE MAPLE PRESS COMPANY, YORK PA.

PREFACE

Each new diesel engine sold or installed represents an avenue whereby skilled men, versed in the operation and repair of diesel engines, are able to find profitable and continuous employment. To become a diesel-engine man requires practical experience, but to become a really qualified diesel-engine man requires the backing-up of this experience by a clear understanding of the principles underlying the operation and maintenance of diesel engines.

To meet this demand for an understanding of diesel engines and their construction, operation, maintenance, and repair and to help those who wish to become fully qualified operators, this volume has been prepared, presenting all information in a practical way.

The engine is divided into its main parts, and each part is described in respect to its construction, operation, and maintenance. The troubles which are likely to occur are brought out, and the best methods of adjustment and repair are given and illustrated.

The questions at the end of each chapter have two objects. First, they call the attention of the reader to the main topics discussed in the chapter, and second, they help him master the contents of the chapter. If the reader can write down the correct answers to all these questions, he understands and knows the chapter. If the answers to some questions are incorrect or not clear, he should find the corresponding pages in the text, read them carefully, and thus become familiar with the topics he missed the first time.

Thus by reading the book, answering the questions, and checking back to material of which he is not certain, the student will become thoroughly conversant with the principles behind the operation and maintenance of diesel engines and will be prepared to take his place among those men in great demand—skilled diesel-engine men.

V. L. MALEEV

CONTENTS

Preface iii

PART I. INTRODUCTION

Chapter 1. General Description and Construction 3
  General. Diesel-engine Characteristics. Engine Classification. Engine Parts. Materials. Cylinder Arrangement. Engine Designation.

Chapter 2. Diesel-engine Principles 16
  Four-stroke-cycle Events. Timing of Events. Compression. Combustion. Two-stroke-cycle Events. Scavenging Methods. Supercharging. Piston Speed. Speed Factor.

PART II. DIESEL-ENGINE COMPONENTS

Chapter 3. Frames, Cylinders, and Heads 31
  Framework. Cylinders. Cylinder Heads.

Chapter 4. Running Gear 41
  Pistons. Piston Pins. Piston Rings. Connecting Rods. Crosshead. Crankshafts. Counterweights.

Chapter 5. Bearings. 65
  Definitions. Bearing Classification. Bearing Loads. Operating Conditions. Bearing Construction. Bearings for Reciprocating Motion. Crosshead Guides. Bearings with Rolling Contact.

Chapter 6. Flywheels 74
  Flywheel Action. Uniformity of Rotation. Flywheel Rim Markings. Rim Velocity. Balancing. Flywheel Knock. Construction Features.

Chapter 7. Valve Gear 83
  Definitions. Cams. Rocker Arms. Valves. Replaceable Parts. Valve Springs. Valve Lash and Adjustment.

Chapter 8. Fuel Injection 98
  Methods. Air Injection. Mechanical Injection. Common-rail System. Jerk-pump System. Jerk Pumps. Fuel Nozzles. Unit Injector. Distributor System. Precombustion Chambers.

Chapter 9. Governors 118
  Engine Loads. Functions of a Governor. Governor Characteristics. Governor Types. Mechanical Governors. Relay-type Governors. Load-limit Governors. Overspeed Governors and Trips.

CONTENTS

PART III. FUNDAMENTALS OF DIESEL ENGINE WORK

CHAPTER 10. FUELS AND COMBUSTION 141
Diesel Fuels. Properties of Fuel Oils. Ignition. Combustion. Heat Values. Combustion in Diesel Engines.

CHAPTER 11. FUEL SYSTEM 155
Fuel System. Strainers and Filters. Centrifuging. Fuel Transfer Pumps. Tanks. Fuel Lines. Fuel Metering.

CHAPTER 12. LUBRICATION AND LUBRICANTS 174
Objects of Lubrication. Principles of Lubrication. Wedge Action Lubrication. Ball-bearing Lubrication. Lubricants. Lubricating-oil Characteristics. Additive Oils. Greases. Filtration. Centrifuging.

CHAPTER 13. ENGINE LUBRICATION 196
Importance of Lubrication. Bearing Lubrication. Diesel-engine Lubrication. Oil Contamination and Dilution. Used-oil Analysis. Oil Circulation. Oil Consumption. Friction Surfaces. Pistons and Cylinders. Rotating Parts. Nonrotating Parts. Air Compressor. Oil Filtering. Oil Coolers. Pressure Regulators. Pumps. Temperature Measurement. Safety Devices.

CHAPTER 14. ENGINE COOLING 218
Necessity of Cooling. Heat Transfer. Liquid Cooling. Construction Features. Water Circulation. Recooling of Water. Direct Air Cooling. Cooling Equipment. Cooling Controls.

CHAPTER 15. AIR-INTAKE SYSTEM 244
Air-intake System. Air Cleaners. Intake Silencers. Intake Headers and Piping. Supercharging. Scavenging of Two-stroke Engines.

CHAPTER 16. EXHAUST SYSTEM 261
Exhaust in Four-stroke Engines. Exhaust System. Exhaust Manifold. Exhaust Pipe. Mufflers. Tail Pipe. Exhaust-gas Turbines. Recovery of Waste Heat. Exhaust and Scavenging in Two-stroke Engines. Exhaust Temperatures.

CHAPTER 17. STARTING AND REVERSING METHODS. 284
Requirements. Starting Methods. Auxiliary Gasoline Engine. Air Starting. Cam-operated Valves. Air-operated Valves. Starting Position. Cartridge Starter. Low-temperature Starting. Reversing.

CHAPTER 18. VIBRATIONS AND BALANCING 307
Causes of Vibration. Engine Vibrations. Unbalanced Engine Forces. Crankshaft Balance. Balancers. Vibration Absorbers. Torsional Vibration. Torsional-vibration Dampers and Absorbers.

PART IV. ENGINE INSTALLATION AND OPERATION

CHAPTER 19. ENGINE INSTALLATION. 325
Selection of Engine. Foundations. Vibrations. Excavation for the Foundation. Leveling the Engine. Grouting the Bed. Aligning the Shaft Extension. Protection of Parts.

CONTENTS vii

CHAPTER 20. OPERATING A DIESEL ENGINE 337
Before Starting. Starting. Warming Up. Running. Stopping the Engine. Types of Trouble. Failure to Start. Failure to Come Up to Speed. Failure to Develop Full Power. Irregular Engine Speed. Engine Overspeeds. Engine Stops Suddenly. Smoky Exhaust. Abnormal Cylinder Pressure and Exhaust Temperature. Incorrect Coolingwater Temperature. Excessive Piston Cooling-oil Temperature. Overheating of the Engine. Engine Is Noisy. Vibration. Starting-air Pipe Hot. Gumming Up of Pistons and Rings. Carbon Deposits on Fuel Injector and Exhaust Valve. Water in Crankcase. General Procedure.

CHAPTER 21. ENGINE PERFORMANCE. 356
Indicator Diagrams. Indicators. Offset Diagrams. Draw Cards. Firing Diagrams. Intake and Exhaust. Pressure-Time Diagrams. Mean Indicated Pressure. Indicated Horsepower. Brake Horsepower. Engine Efficiencies. Limitations of Performance.

CHAPTER 22. RATING AND TESTING 375
Power Rating. Torque. Performance Curves. Testing. Brake Horsepower. Electric Loading. Fuel Measurements. Exhaust Temperature.

PART V. MAINTENANCE AND REPAIRS

CHAPTER 23. MAINTENANCE RECORDS AND DATA 395
Importance of Maintenance. Qualifications of Operators. Combustion. Inspection Schedules. Maintenance Log Sheets. Maintenance Requirements. Compression and Combustion Pressures. Details of Maintenance. Safety Precautions. Average Clearances.

CHAPTER 24. MAINTENANCE AND REPAIRS 411
Cylinder Heads. Cylinders and Liners. Pistons. Bearings. Valves. Fuel Pumps. Mechanically Operated Fuel Valves. Hydraulic Nozzles. Camshaft Gears. Governor. Lubricating-oil System. Air Compressor.

CHAPTER 25. ENGINE OVERHAULING. 429
Definitions. Dismantling. Cylinder Liners. Pistons. Bedplate. Connecting Rods. Crankshaft. Main Bearings. Fitting Rebabbitted Bearings. Checking Bearing Alignment. Precision Bearings. Checking Final Bearing Fit. Tramming.

CHAPTER 26. SPECIAL REPAIRS AND SALVAGING 456
General Considerations. Repair and Salvage Methods. Welding. Brazing. Cold-welding. Shrink Members. Surface Welding. Metal Spraying. Electroplating. Repair Examples. Cylinder Jackets. Crankcases and Bedplates. Cylinder Liners. Pistons. Piston Pins. Valves. Crankshafts.

APPENDIX. GLOSSARY OF TERMS USED IN DIESEL ENGINEERING. . . . 475
LIST OF VISUAL AIDS 493
INDEX. 501

A WORD ABOUT DIESEL ENGINES

The diesel engine is slightly over fifty years old. This is quite young compared with the steam engine, which is well over a hundred years old, and even gasoline engines appeared about 35 years before the diesel engine. At the time of the introduction of the first diesel engines on the market, combustion engines operating on cheap fuel oil already existed both in Europe and in this country. De La Vergn and Mietz & Weiss were two American concerns building oil engines in comparatively large numbers as far back as 1893. De La Vergn continues to build oil engines and has become one of the leading diesel-engine manufacturers. At present it is a division of the Baldwin Locomotive Works.

In this country the diesel-engine industry did not actually start until 1912, when the original patents of Rudolf Diesel expired.

The first years of the diesel-engine development in this country showed very slow progress in spite of the promising fuel economy and dependability of this new type of prime mover. At first the engine use was limited chiefly to stationary power-plant applications. From 1916 on the diesel engine began to come into its own, and by 1920 diesels were being used extensively in pipe-line pumping service. This was the beginning of the general use of diesel engines in the oil field in many types of applications.

The application of diesel engines to marine service started about 1923 and gradually became very popular. The installation of diesel engines in ocean-going passenger liners began in 1924. Today they are found also in cargo ships, fishing boats, tugboats, dredges, ferries, submarines, and a variety of other craft.

The first diesel-powered bus made its appearance in 1925, almost simultaneously with the first diesel-powered locomotive. The first diesel-powered truck was introduced in 1929. The full significance of diesel power for the railroads was not demonstrated until 1934, when the first main-line passenger diesel locomotive made its record-breaking run between Chicago and Denver. The number of diesel engines in stationary power plants, chiefly for generation of electric current, was growing more steadily, though not in such a spectacular way.

In 1936, diesel engines in daily use totaled 6 million horsepower. The rate of new installations at that time was 1 million horsepower per year. By 1947 this country had installations totaling between 44 and 45 million horsepower, exclusive of some 35 to 45 million horsepower in military and naval installations. A conservative estimate of new diesel-engine installations, on the basis of the present trends, is 10 million horsepower per year—truly a tremendous industry with a big future! As time goes on, we may expect to find the diesel engine taking an ever-increasing part in industries of all kinds.

PART 1

INTRODUCTION

CHAPTER 1

GENERAL DESCRIPTION AND CONSTRUCTION

1-1. General. Diesel engines are a special type of *internal-combustion engine*. As the name indicates, the internal-combustion engine is a heat engine in which the chemical energy of combustion is released inside the engine cylinder, while in the other group of heat engines—steam engines and steam turbines—the energy developed during combustion of a fuel is transmitted first to steam and only through steam does its work in the engine or turbine. However, since no engines with external combustion exist, except the latest development, the gas turbine, which in every respect is in a class by itself, there is at present a tendency to call all heat engines operated directly by combustion gases simply *combustion engines*. This shorter name will be used in the present text.

It may also be of interest to note that in Germany, the country of its birth, the diesel engine is always called, simply, a combustion engine, as are all other so-called *internal-combustion engines*.

There are several reasons why diesel engines not only are competing with other heat engines but in many instances are dominating the field. The class of service is the main factor in many cases. One of the outstanding applications of diesel engines is transportation, on land and water, in trucks, railcars, locomotives, boats, and ships. In many small and medium-sized stationary plants, on farms and in small industrial enterprises, simplicity and low cost of operation are determining the adoption of oil engines in preference to steam engines or electrical motors. In large power plants, used for the production of electric current or propulsion of ships, the fuel economy decides the choice in favor of diesel engines.

1-2. Diesel-engine Characteristics. The main characteristic of diesel engines which distinguishes them from other combustion engines is the method of igniting the fuel. In a diesel engine the fuel is injected into the cylinder, which contains highly compressed air. During compression of the air in the engine cylinder the temperature of the air goes up so that when the fuel, in the form of a fine spray, comes in contact with this hot air, it ignites, and no other external means of ignition is required. For this reason diesel engines are also called *compression-ignition engines*.

Another important diesel-engine characteristic is that the engine produces a torque which is more or less independent of the speed, as the amount of air taken into the cylinder during each suction stroke of the piston is little affected by the speed of the engine. The amount of fuel that can be burnt in the cylinder with each suction stroke and the useful effort developed by the action of the piston are, therefore, almost constant.

Equally important is the fact that a diesel engine has a higher thermal efficiency than other heat engines, uses less fuel for the same power delivered, and in addition uses a fuel which is cheaper than gasoline.

Naturally, there are also some disadvantages as compared with gasoline engines: (1) slightly greater weight for the same horsepower, (2) in high-speed engines, a certain roughness in operation, particularly at light loads, and (3) a considerably greater initial cost.

1-3. Engine Classification. Diesel engines may be divided into several groups, in each of which one of the following features is characteristic: operating cycle, method of charging the cylinder, and general design. The last classification, in turn, includes the number and position of cylinders, the method of injecting and burning the fuel, the speed, etc. These classifications often overlap, engines in the same class in regard to one feature belonging to different classes in regard to another.

Operating Cycles. Diesel engines may be divided into those operating on a *constant-pressure cycle* and those operating on a *combination* cycle. Engines with combustion taking place at constant pressure are large low-speed air-injection engines. A combination, or dual-combustion, cycle is a cycle in which part of the fuel burns at constant volume, as in a gasoline engine, and part burns at an approximately constant pressure. In engines operating on the combination cycle, the pressure first increases to a peak during the first part of the combustion, then stays more or less constant and, as the piston moves farther away from the dead center, begins to drop toward the end of the combustion process; this cycle is typical for medium and high-speed airless-injection engines.

Methods of Charging. Diesel engines may be divided into *four-stroke* and *two-stroke engines.* In four-stroke engines, during two strokes of the piston, or one revolution of the crankshaft, the piston and cylinder act as a pump which removes products of combustion of the preceding cycle and fills the cylinder with a fresh air charge. In two-stroke engines the cylinder is scavenged and filled with a fresh air charge by slightly compressed air furnished by some external pump or blower. Four-stroke engines in turn may be divided into engines with *natural aspiration* and *supercharged* engines. In the former the fresh charge is drawn in by the vacuum created when the piston moves away from the combustion space. In supercharged engines the charge is admitted into the cylinder at a higher than atmospheric pressure. This high air pressure is produced by a pump or blower similar to the ones used in two-stroke engines.

General Design. All engines may be divided into single- and double-acting engines. Double-acting design is used only for large engines. Other classifications under this head are: horizontal, vertical, in-line, V-type, radial, and opposed-cylinder and opposed-piston engines, meaning engines with horizontal, vertical, parallel, inclined, and star-shaped center lines of the cylinders. Also single- and multicylinder engines, with two, three, four, six, and sometimes as many as 24 cylinders.

Method of Fuel Injection. In the original low-speed diesel engines the fuel is injected into the cylinder by a blast of high compressed air, hence the name of *air-injection engines*. Air-injection equipment is too heavy and complicated for small-bore, high-speed engines, which use various types of airless, or mechanical, injection. At present mechanical injection is being used for all types and sizes of diesel engines.

Speed. Classification of engines according to their speed as low-, medium-, and high-speed engines has for its reason the fact that the speed factor influences the design of engines, their maintenance, and life, as explained in Sec. 2-9.

1-4. Engine Parts. An understanding of the operation or purpose of the various parts is necessary for a complete understanding of the whole engine. Each part or unit has its own special function to perform and in conjunction with other parts makes up a diesel engine. A person who wants to operate, repair, or otherwise service diesel engines must be able to recognize the different parts by sight and know what their particular functions are. A knowledge of the engine parts will be gradually acquired, first by reading attentively the following sections and later by looking up in the Glossary at the end of the book every term not sufficiently familiar from a reading of the corresponding chapters.

Diesel engines vary in outer appearance, size, number and arrangement of cylinders, and details of construction. However, they all have the same main parts which, though they may look different, all perform the same functions. Each diesel engine has only a few main working parts; the auxiliary parts are necessary to hold the working parts together or to assist the main working parts in their performance. The main working parts are (*a*) the cylinder; (*b*) the cylinder head, usually holding inlet and exhaust valves; (*c*) the piston; (*d*) the connecting rod; (*e*) the crankshaft; (*f*) the crankshaft or main and connecting-rod bearings; and (*g*) the fuel pump and fuel nozzle.

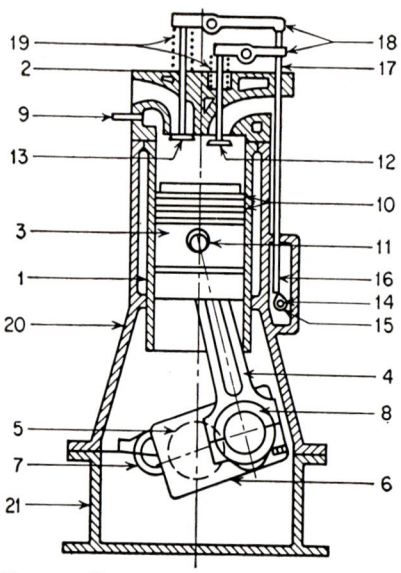

Fig. 1-1. Cross section of a typical diesel engine: 1, cylinder liner; 2, cylinder head; 3, piston; 4, connecting rod; 5, crankshaft; 6, crank web; 7, main bearing; 8, crank-pin and bearing; 9, fuel nozzle; 10, piston rings; 11, piston pin and bearing; 12, intake valve; 13, exhaust valve; 14, camshaft; 15, cam; 16, cam follower; 17, push rod; 18, rocker arms; 19, valve springs; 20, cylinder block or crankcase; 21, bedplate.

Figure 1-1 shows schematically a typical four-stroke diesel engine, in which the working and main subordinate parts have been keyed to the legend.

Cylinder. The heart of the engine is the cylinder, where the fuel is burned and the power developed. The inside of the cylinder is formed by the *liner*, or *sleeve*.

The inside diameter of the cylinder is called the *bore*.

The *cylinder head* closes one end of the cylinder and often contains the valves through which air and fuel are admitted and the exhaust gases discharged.

Piston. The other end of the working space of the cylinder is closed by the piston that transmits to the crankshaft the power developed by the burning of the fuel. *Piston rings* lubricated with engine oil produce a gastight seal between the piston and the cylinder liner. The distance that the piston travels from one end of the cylinder to the other is called the *stroke*.

Connecting Rod. One end, called the *small end* of the connecting rod, is attached to the *wrist pin* or *piston pin* located in the piston. The other or *big end* has a bearing for the crankpin. The connecting rod changes and transmits the reciprocating motion of the piston to the continuously rotating crankpin during the working stroke and vice versa during the other strokes.

GENERAL DESCRIPTION AND CONSTRUCTION

Crankshaft. The crankshaft turns under the action of the piston through the connecting rod and crankpin located between crank webs, or cheeks, and transmits the work from the piston to the driven shaft. The parts of the crankshafts supported by and rotating in the main bearings are called the *journals*.

A *flywheel* of sufficient weight is fastened to the crankshaft and stores kinetic energy during the power stroke and returns it during the other strokes. The flywheel helps to start the engine and also serves to make the rotation of the crankshaft more or less uniform.

A *camshaft* driven from the crankshaft by a chain drive or by timing gears operates the *intake* and *exhaust valves* through *cams, cam followers, push rods,* and *rocker arms*. Valve springs serve to close the valves.

A *crankcase* serves to hold together the cylinder, piston, and crankshaft, to protect all moving parts and their bearings, and to provide a reservoir for lubricating oil. It is called a *cylinder block* if the cylinder liners are inserted into it. The lower part of the crankcase is called a *bedplate*.

Fuel is delivered into the combustion space by an injection system consisting of a *pump, fuel line,* and *injector,* also called *fuel-injection nozzle* or *spray nozzle*.

1-5. Materials. Most diesel engine parts are of cast iron. However, in order to obtain a lighter engine, certain parts, such as pistons and cylinder heads, may be made of light aluminum alloys. Cylinder blocks, crankcases, and bedplates are often fabricated of welded steel plates and shapes. Moving parts subjected to relatively great stress or wear, such as crankshafts and connecting rods, are made of steel. However, connecting rods of high-speed engines may be made of forged aluminum in order to decrease the weight and inertia forces of reciprocating parts.

Crankshafts until recently were forged from steel. Lately crankshafts cast of special iron alloys are being used in engines of all sizes and speeds. One reason is that it is much easier to give the complicated shape required by the theoretical design to a cast crankshaft than to a forged one. Another reason is a lower manufacturing cost, particularly for small, mass-produced engines.

Bearing shells are made of cast iron, bronze, or steel for rigidity, lined with various low-friction alloys, such as babbitt or copper-lead, to reduce friction between the bearing and rotating shaft.

1-6. Cylinder Arrangement. *Cylinders-in-line.* This is the simplest arrangement, with all cylinders parallel, *in line,* as in Fig. 1-2. This construction is used normally for engines having up to eight cylinders. In-line engines usually have vertical cylinders. However, engines with horizontal cylinders are used for buses. Such an engine is basically a vertical engine laid on its side to reduce its height.

V Arrangement. If an engine has more than eight cylinders, it is difficult to make a sufficiently rigid frame and crankshaft with an in-line arrangement. The V arrangement (Fig. 1-3a) with two connecting rods attached to each crankpin, allows the engine length to be cut in half and thus makes it much more rigid, with a stiffer crankshaft. This is the most common arrangement for engines with eight to sixteen cylinders. Cylinders lying in one plane are called a *bank;* the angle a between the two banks varies from 30 to 120 deg, the more common angle being between 40 and 75 deg.

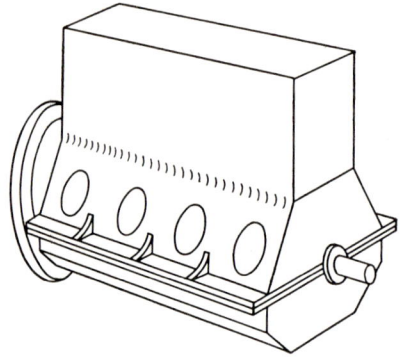

Fig. 1-2. In-line engine.

Radial engines have all cylinders in one plane with the center lines at equal angles and only one crank to which are attached the connecting rods. Such engines are built with five, seven, nine, and eleven cylinders.

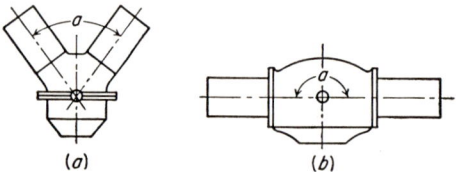

Fig. 1-3. V-type and flat engines.

The *flat* engine (Fig. 1-3b) is a V engine with an angle of 180 deg between banks. This arrangement is used for buses and trucks.

Multiple-engine Units. The weight per horsepower, called *specific engine weight*, increases with an increase of the engine size, its bore and stroke. In order to obtain an engine with a considerably higher power output without increasing its specific weight, two and four complete engines, having six or eight cylinders each, are combined in one unit by connecting each engine to the main drive shaft s (Fig. 1-4a and b) by means of clutches and roller chains or clutches and gears.

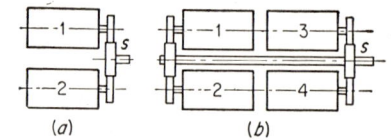

Fig. 1-4. Multiple-unit engines.

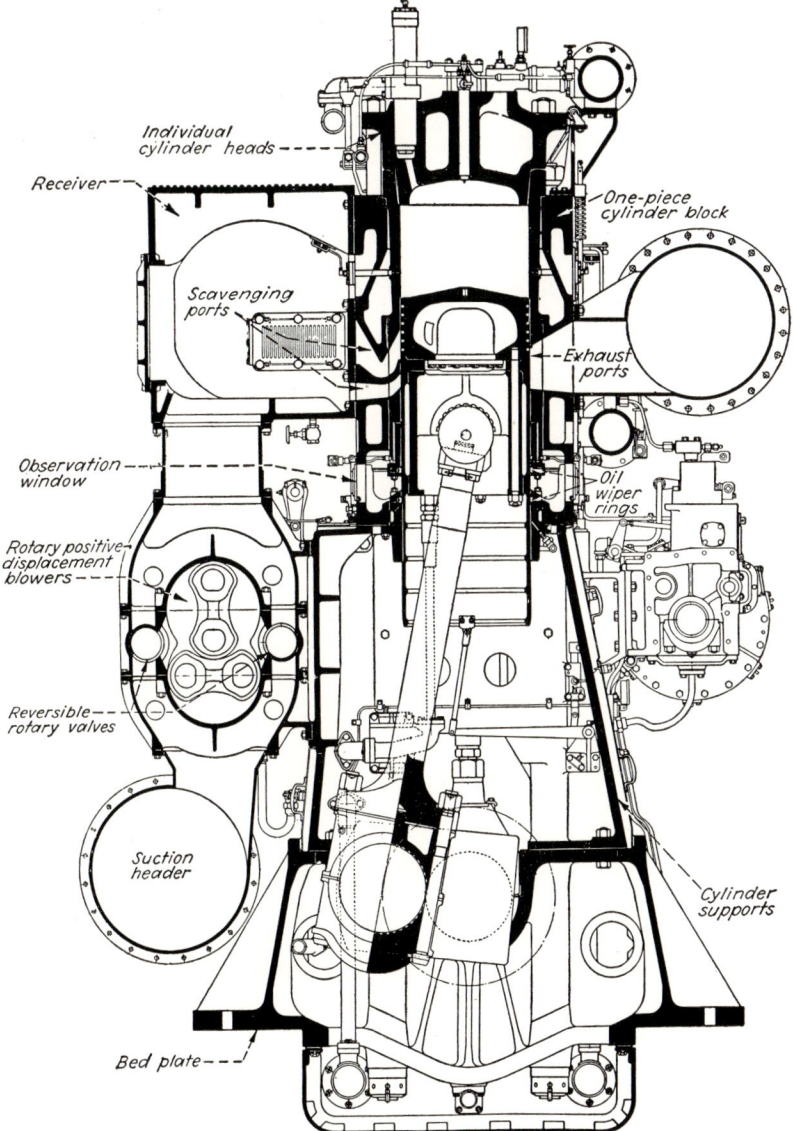

Fig. 1-5. Cross section of a 20½ by 27½ in. two-stroke engine. *Courtesy of Nordberg Manufacturing Company.*

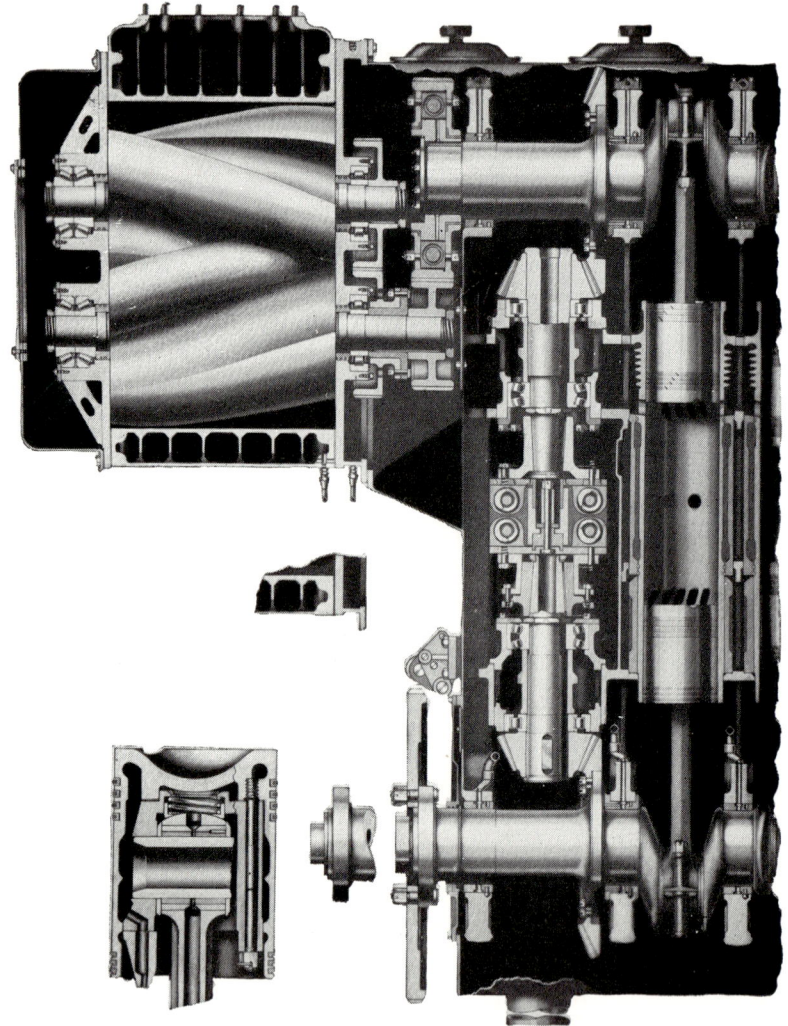

Fig. 1-6. Longitudinal section of an opposed-piston diesel engine. *Courtesy of Fairbanks, Morse & Co.*

Opposed-piston Engines. Engines having two pistons per cylinder driving two crankshafts (Fig. 1-6) are used in marine and railroad service. The design presents many advantages from the standpoint of combustion of the fuel, balancing of the reciprocating masses, engine maintenance, and accessibility.

Representative Engine Types. Figure 1-5 shows a cross section of a large, comparatively low-speed two-stroke diesel engine, as used in

Fig. 1-7. Cross section of an opposed-piston diesel engine. *Courtesy of Fairbanks, Morse & Co.*

power plants to drive electric generators. The engine develops 340 bhp (brake horsepower) per cylinder at 240 rpm and is built in units with three to ten cylinders, 1,000 to 3,400 bhp per engine.

Figure 1-6 shows a longitudinal view and Fig. 1-7 a cross section of an opposed-piston two-stroke engine used mostly for marine installations and locomotives. The largest American engine of this type develops in ten cylinders of 8⅛-in. bore and 10-in. × 2 stroke up to

2,000 bhp at 750 rpm, which is a rather high speed for an engine of such size.

Figure 1-8 shows a medium-speed industrial engine of the four-stroke type and Fig. 1-9 a cross section of a small industrial two-

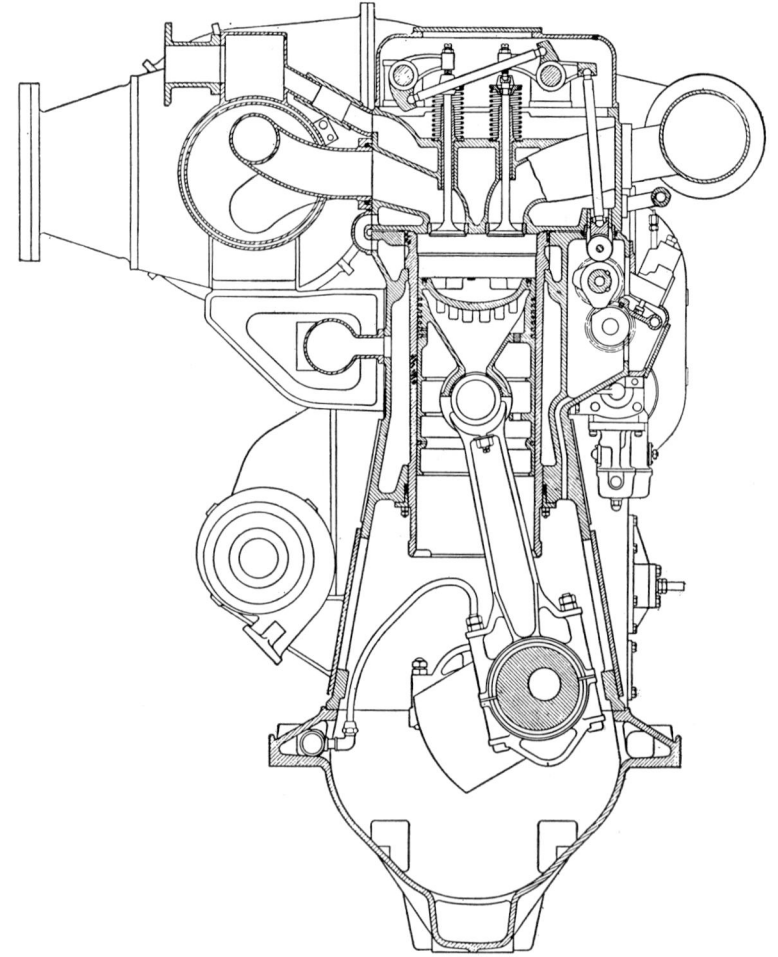

Fig. 1-8. Cross section of a 13 by 16-in. supercharged stationary diesel engine. *Courtesy of The Cooper-Bessemer Corporation.*

stroke engine. Figure 1-10 shows a cross section and Fig. 1-11 a longitudinal section of an automotive-type two-stroke engine used extensively in trucks, buses, and tractors.

1-7. Engine Designation. Diesel engines are, like automotive engines, designated by their bore, stroke, number of cylinders, and

GENERAL DESCRIPTION AND CONSTRUCTION 13

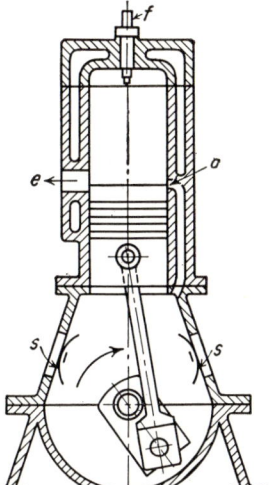

Fig. 1-9. Crankcase-scavenge engine. The openings in the crankcase with plate valves s, s admit air during the upward stroke of the piston. During the downward stroke the air in the crankcase is compressed. When the piston uncovers the exhaust port e, part of the exhaust gases escapes through e and the rest is scavenged by the air from the crankcase which is admitted through port a uncovered by the piston. The fuel nozzle is indicated by f.

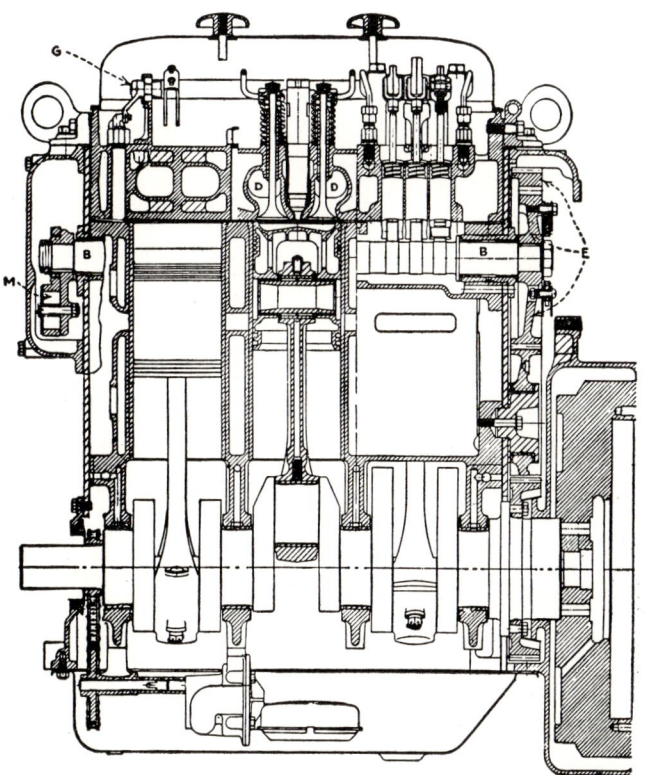

Fig. 1-10. Longitudinal section of a GM two-stroke truck diesel engine.

speed, if the engine operates at a fixed speed, in the order named. Thus a $4\frac{3}{4}$-in. × 6-in. × 6 cyl × 800 rpm or often simply $4\frac{3}{4}$ × 6 × 6 × 800 designation means an engine with a $4\frac{3}{4}$-in. bore, 6-in. stroke, 6 cylinders, rated speed 800 rpm. In the designation used in the United States Navy the number of cylinders is given first and then the bore and stroke.

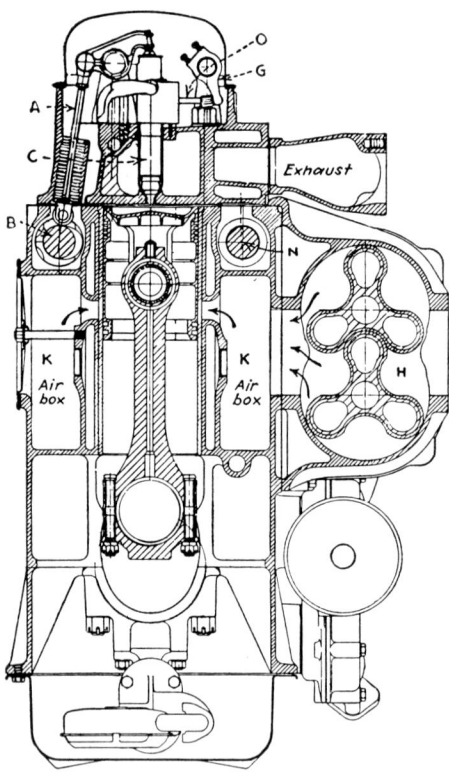

FIG. 1-11. Cross section of a GM two-stroke truck diesel engine. The camshaft B is driven by gears E (Fig. 1-10), which also drive the blower H and the balance shaft N. Camshaft B operates the exhaust valves through push rods A and also the fuel injectors C. The blower H (Fig. 1-11) has two three-lobe rotors and delivers slightly compressed scavenge air to the air box K, from which it is admitted to the cylinders when a piston uncovers the ports near bottom dead center, as indicated by the arrows.

Another designation is by *piston displacement* or, for short, by *displacement*. The engine displacement is computed as the product of the piston area, piston stroke, and the number of cylinders. In English units, displacement is given in cubic inches. Engine speed does not enter into the determination of engine displacement. Sometimes the displacement is given as the displacement of one cylinder and the number of cylinders. Thus, the $4\frac{3}{4}$ × 6 × 6 × 800 engine may be specified either as a 638-cu in. or as a 106.3 × 6-cu in. engine.

GENERAL DESCRIPTION AND CONSTRUCTION

QUESTIONS

1. What is the main difference between an internal-combustion engine and a steam engine or steam turbine?
2. Enumerate the advantages of diesel engines over other heat engines.
3. What is the main characteristic of a diesel engine that distinguishes it from other combustion engines?
4. Why is a diesel engine often called a compression-ignition engine?
5. How does the torque produced by a diesel engine change with the speed of the engine?
6. How does the thermal efficiency of a diesel engine compare with the thermal efficiency of other combustion engines?
7. What are the disadvantages of diesel engines as compared with gasoline engines?
8. What are the bases of classification of diesel engines?
9. State what methods of supplying diesel engines with a fresh air charge are in use.
10. What is a diesel engine with natural aspiration?
11. What is a supercharged diesel engine?
12. Enumerate the different classifications of diesel engines in respect to their general design.
13. What are the two main methods of introducing the fuel into the combustion space of a diesel engine?
14. What is the advantage of knowing the purpose and operation of each diesel-engine part?
15. What is a glossary?
16. Enumerate the working parts present in every diesel engine.
17. What does the term *bore* mean?
18. What is the cylinder head of an engine?
19. What does the term *piston stroke* or *engine stroke* mean?
20. What is the function of a connecting rod?
21. Which end of a connecting rod is the small end and which one the big end?
22. What is the function of a crankshaft?
23. Draw a sketch of a crankshaft and give the names of its main parts.
24. State the functions of a flywheel.
25. How is a camshaft connected with the crankshaft?
26. How does a camshaft operate the intake and exhaust valves?
27. What is the function of a crankcase?
28. What is a cylinder block?
29. Of what material is each of the main parts of a diesel engine made?
30. Of what main parts does a fuel-injection system consist?
31. Enumerate and illustrate by sketches the most common diesel-engine cylinder arrangements.
32. What is a cylinder bank? In what engines is this term used?
33. Draw sketches of multiple-engine units.
34. What is the main object of making multiple-engine units?
35. How are combustion engines designated?
36. How are diesel engines designated in the United States Navy?
37. What is the displacement of an engine?
38. What are the two methods of designation by displacement that are used for multicylinder engines?

CHAPTER 2

DIESEL ENGINE PRINCIPLES

2-1. Four-stroke-cycle Events. *Cycle.* A sequence of events which recur regularly and in the same order is called a *cycle*. The following events form the cycle in a diesel engine:
 1. Filling of the engine cylinder with fresh air.
 2. Compression of the air charge which raises its pressure and temperature so that when the fuel is injected, it ignites readily and burns efficiently.
 3. Combustion of the fuel and expansion of the hot gases.
 4. Emptying the products of combustion from the cylinder.

When these four events are completed, the cycle is repeated. When each of these four events requires a separate stroke of the piston, the cycle is called a *four-stroke* cycle.

Dead Centers. The positions of the piston when the piston is closest to the cylinder head and farthest away from it are called *top* and *bottom dead center*, respectively, or shorter, *top* and *bottom center* and are indicated as t.c. and b.c. The reason for these designations is that at these positions the center line of the crankpin is in the same plane with the center lines of the piston pin and the shaft journals and the piston cannot be moved by gas pressure. The moving force must come from the rotating crankpin acting through the connecting rod.

Main Events. The four main events are shown schematically in Fig. 2-1. During the first, or *suction stroke*, (Fig. 2-1a) the piston moves downward, pulled by connecting rod r, the lower end of which is moved by crank c. The piston, moving away from the cylinder head, creates a vacuum in the cylinder, and outside air is drawn or sucked into the cylinder through the intake valve i that opens about the beginning of the suction stroke and stays open until the piston reaches bottom center.

When the piston has passed bottom center, the second, or *compression, stroke* begins (Fig. 2-1b): the intake valve is closed and the piston, pushed upward by the crank and connecting rod, compresses the air in the cylinder and raises its temperature. Shortly before the piston reaches top center, liquid fuel, in the form of a finely atomized spray, is admitted gradually into the hot air in the cylinder. The fuel ignites and burns during the first part of the working stroke, thus raising the pressure in the cylinder. During this third stroke, called *working*, or *power*, *stroke* (Fig. 2-1c), the hot gases push the piston downward or forward. The gases expand from the increasing cylinder volume and through the connecting rod and crank transmit the developed energy to the rotating crankshaft.

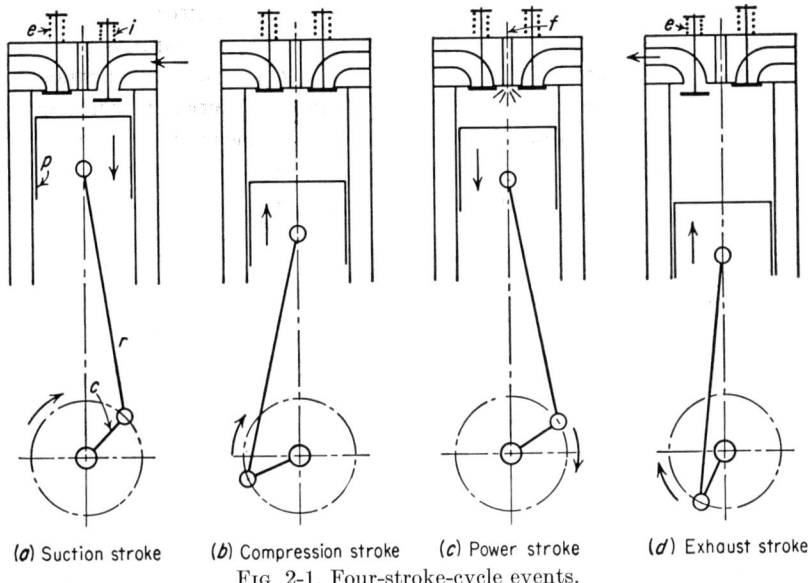

(*a*) Suction stroke   (*b*) Compression stroke   (*c*) Power stroke   (*d*) Exhaust stroke
Fig. 2-1. Four-stroke-cycle events.

Shortly before the piston reaches bottom center, exhaust valve e opens (Fig. 2-1d) and the hot products of combustion that are still under high pressure begin to escape through the exhaust ports to the outside. During the fourth, or *exhaust*, *stroke* the piston moves upward, pushed by the crank and connecting rod, expelling the remaining products of combustion.

Near top center the exhaust valve is closed, the intake valve is opened, and the cycle starts over. As may be seen, the four strokes require two revolutions of the crankshaft. Thus in a four-stroke engine one power stroke is obtained for every two crankshaft revolutions, or the number of power impulses per minute is one-half the rpm rating.

2-2. Timing of Events. Actually the dividing points between the four main events do not coincide with the beginning and end of the corresponding strokes. The differences are smaller in low-speed engines and increase as the engine speed increases. The intake valve begins to open before top center, by some 10 to 25 deg of crank travel. This anticipation allows the valve to be appreciably open at top center when the piston starts the suction stroke. The intake valve is closed from 25 to 45 deg after bottom center. The fuel injection starts from 7 to 26 deg before top center. The end of fuel injection depends upon the load of the engine. In order to release the exhaust-gas pressure before the piston starts the return stroke, the exhaust valve begins to open 30 to 60 deg *before* b.c. and closes 10 to 20 degrees *after* t.c.

2-3. Compression. There are two purposes in compressing the air charge during the second or compression stroke: first, to increase the thermal or over-all efficiency of the engine by increasing the density of the charge and thus obtaining higher temperatures during combustion; this applies to all internal-combustion engines, both of the spark-ignition and compression-ignition type. Second, to raise the temperature of the air charge so much that, when the finely atomized fuel is injected into it, the fuel will ignite and begin to burn without needing an outside source of ignition such as the spark plug used in gasoline engines.

The *compression ratio* of a combustion engine is the ratio of the volume V_1, cubic inch, of the gases in the cylinder with the piston at bottom center to the volume V_2 of the gases with the piston at top center. The compression ratio is designated by r:

$$r = \frac{V_1}{V_2} \qquad (2\text{-}1)$$

Volume V_2 is called the *compression volume* or *combustion space;* volume V_1 is equal to the sum of the piston displacement and the compression volume.

The usual compression ratios of diesel engines are from about 12:1 to about 19:1. With a compression ratio lower than 12:1 there is the danger that the temperature of the compressed air may not be high enough to ensure ignition of the fuel when the engine is started "cold." The high limit is set by practical considerations. Theoretically, an increase of the compression ratio of an engine raises its thermal efficiency and lowers its fuel consumption. However, an increase of the compression ratio increases the maximum gas pressures and the combustion temperatures. This causes increased stresses and pressures in various engine parts and higher friction losses. It requires stronger and heavier engine parts and increases the weight of the whole engine. Higher pressures and temperatures also increase the wear of the engine and thus reduce its life and reliability. Therefore each engine type has a limit above which it is not advisable to raise the compression.

2-4. Combustion. There are two distinct methods of burning the fuel in an engine cylinder: (1) at *constant volume* and (2) at *constant pressure*. Combustion at constant volume means that during combustion the volume does not change and all the heat energy developed by the fuel goes into an increase of the gas temperature and pressure. In an engine it means that combustion proceeds at such a high rate that the piston has no time to move during combustion. Such a combustion is obtained when the piston is at top center. The advantage of this method of fuel combustion is a high thermal efficiency. Its disadvantage is a very sudden pressure increase and resulting noisiness of the engine. Such combustion is more or less approached in spark-ignition gasoline engines.

Combustion at constant pressure means that during combustion the temperature increases at a rate such that the resulting increase of pressure is just enough to counteract the influence of the increasing volume due to the piston motion, and the pressure does not change. The heat energy developed by the fuel goes partly into an increase of the gas temperature and partly into performing outside work. In an engine with a constant-pressure combustion, the fuel is burned gradually so that the pressure attained at the end of the compression stroke is maintained during most of the combustion event. Such combustion was used in the original low-speed air-injection diesel engines. Its advantage is a smoothly running engine producing a more even torque because of the extended combustion pressure. However, it is not suitable to high-speed oil engines.

Modern high-speed diesel engines operate on a cycle which is a combination of the above two methods, also called *dual-combustion* cycle; part of the fuel is burned rapidly, almost at a constant volume near the dead center, the rest is burned while the piston begins to move away from the top center. However, the high pressure does not remain constant, but usually first increases and then decreases. In general this cycle resembles the constant-volume combustion cycle more than the cycle of the original diesel engines. Its advantage is a high efficiency and a low fuel consumption. Its drawback is the difficulty of preventing rough and noisy operation of the engine.

2-5. Two-stroke-cycle Events. A two-stroke cycle is completed in two strokes, or one revolution of the crankshaft, whereas a four-stroke cycle requires two revolutions. The main difference between two-stroke and four-stroke engines is in the method of removing the burned gases and filling the cylinder with a fresh charge of air. In a four-stroke engine these operations are performed by the engine piston during the exhaust and suction strokes. In a two-stroke engine these operations are performed near bottom dead center by means of a separate air pump or blower.

The compression, combustion, and expansion events do not differ from those of a four-stroke engine. The removal of the used gases and the filling of the cylinder with a fresh charge takes place as follows: When the piston has travelled 80 to 85 per cent of the expansion stroke,

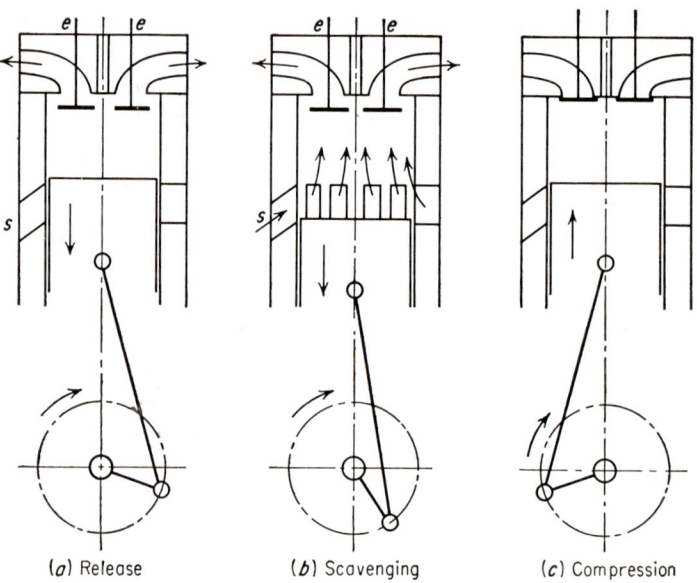

FIG. 2-2. Scavenging of a two-stroke-cycle engine.

exhaust valves e, e (Fig. 2-2a) are opened, the exhaust gases are released and begin to escape from the cylinder and the pressure in the cylinder begins to drop. The piston continues to move toward the bottom center and soon uncovers ports s, s through which slightly compressed air begins to enter the cylinder. This air has a slightly higher pressure than the hot gases in the cylinder, and thus pushes them out through valves e, e (Fig. 2-2b) to the outside atmosphere. This operation is called *scavenging*, the air admitted is called *scavenge air* and the ports through which the air is admitted are called *scavenge ports*. About the time when the piston on its upward stroke closes ports s, s, the exhaust valves e, e are also closed (Fig. 2-2c) and the compression stroke begins.

The advantage of two-stroke operation is the elimination of the two charging strokes required in four-stroke operation. Thus the cylinder delivers one power stroke for every revolution of the engine as compared with one power stroke for every two revolutions in a four-stroke-cycle engine. If all other conditions, such as bore, stroke, speed, and mean effective gas pressures are equal, a two-stroke engine will develop twice the power of a four-stroke engine. This means also that a two-stroke engine, roughly speaking, weighs one-half as much as a four-stroke engine of the same power, and produces a more even torque.

However, it must be noted that this is true only for engines having the same mean effective pressure. Thus a two-stroke engine with crankcase scavenging has a low mean effective pressure and therefore develops less power than a comparable four-stroke engine. On the other hand, a supercharged four-stroke engine can develop the same or even a greater power than a two-stroke engine of the same displacement.

These advantages are very important in ships and locomotives and therefore two-stroke engines are used in these installations much more than four-stroke engines, particularly in larger power units. A disadvantage of all two-stroke engines is high temperatures of the piston and cylinder head caused by the fact that combustion occurs at every revolution.

2-6. Scavenging Methods. Figure 2-2 illustrates only one of several methods of cylinder scavenging. In some engines the exhaust gases are let out through ports, uncovered by the piston the same as the scavenge ports s, s (Fig. 2-2). Depending upon the location of the exhaust ports in respect to the scavenge ports, there exist two basically different methods of scavenging: *cross-flow* scavenging (Fig. 2-3) and *loop*, or *return-flow*, scavenging (Fig. 2-4).

Cross-flow Scavenging. With this method the piston uncovers first the exhaust ports e, e and releases the pressure; going down further the piston uncovers the scavenge ports s, s and begins to admit slightly compressed air whose stream is directed mainly upward, as indicated by the arrows, and thus pushes out the exhaust gases through ports e, e. Having passed dead center the piston closes first the scavenge ports and soon afterward the exhaust ports. The fact that the exhaust ports are closed after the scavenge ports permits some of the air charge to escape from the cylinder. This is a disadvantage of this scavenge scheme. However, it has also a decided advantage—simplicity of construction and maintenance, absence of valves which must be kept tight.

Some large, low-speed engines use the cross-flow scavenging scheme improved by the addition of special check valves located near the scavenge ports. In this case, the scavenge ports are made of the same height or even slightly higher than the exhaust ports, as shown in Fig. 1-5. Thus the scavenge ports are uncovered by the piston simultaneously with or slightly before the exhaust ports; however, the check valves prevent exhaust gases from going into the scavenge-air receiver. As soon as the pressure in the cylinder drops below the pressure in the air receiver, the pressure in the latter opens the check valves and

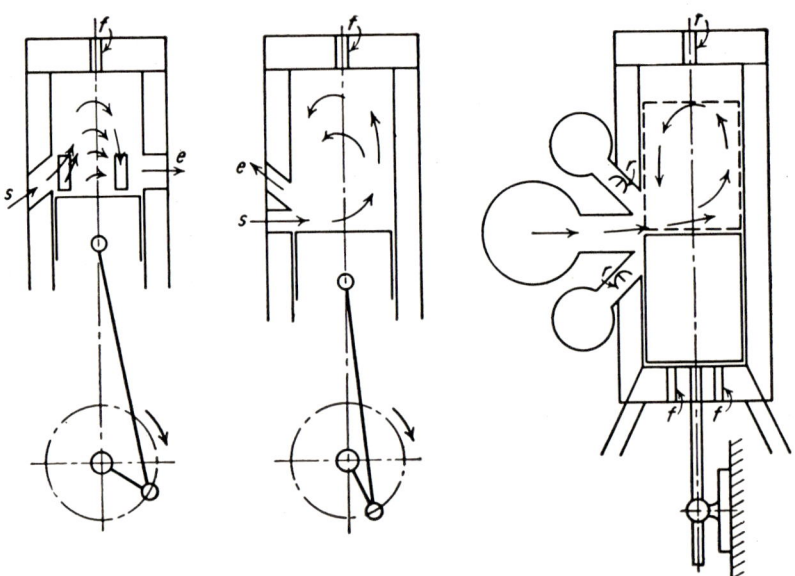

Fig. 2-3. Cross-flow scavenging.

Fig. 2-4. Loop- or return-flow scavenging.

Fig. 2-5. Return-flow scavenging in a double-acting engine.

admission of scavenge air begins. Scavenging continues until both scavenge and exhaust ports are closed by the piston. This scheme gives an increased scavenge efficiency, resulting in a higher mean effective pressure at a very nominal cost both of the valves and of their maintenance.

Loop scavenging (Fig. 2-4) is similar to cross flow in the sequence of the port opening. However, the direction of air flow is different, as indicated by the arrows. Its advantage is that the bulky scavenge-air and exhaust-gas receivers are located on one side of the cylinders, thus giving a better accessibility. This scheme is particularly suitable for double-acting engines, as with them the operation of exhaust valves (Fig. 2-2) for the lower combustion space becomes very complicated. When used for double-acting engines (Fig. 2-5), the scheme is improved by the introduction of rotary exhaust valves r, r. During the release of the exhaust gases valve r is open but is being closed when the piston covers the scavenge ports on the return stroke. By this arrangement for escape of air charge during the beginning of the compression stroke, when the exhaust ports are covered by the piston, the rotary valve is opened and made ready for the next cycle. As may be seen from Fig. 2-5, the length of the piston is made exactly equal to the length

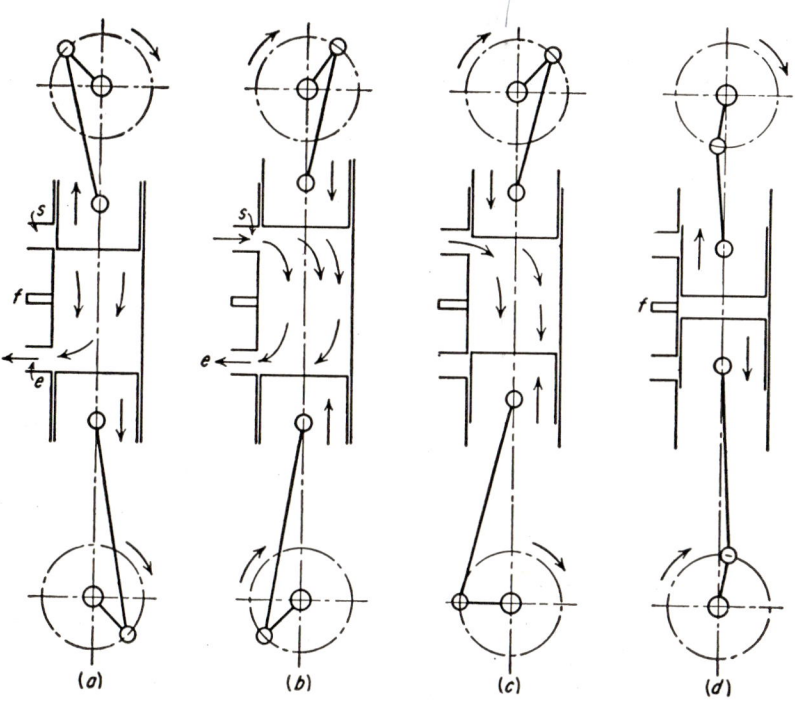

FIG. 2-6. Opposed-piston operation.

of the stroke in order to control the exhaust and scavenge events alternately by the upper and lower edges of the piston.

The *opposed-piston scheme* is shown in Fig. 2-6. The lower piston controls the exhaust ports, the upper one the scavenge ports. In order to obtain the necessary preliminary release of the exhaust gases by uncovering the exhaust ports e ahead of the scavenge ports s, the crank of the lower crankshaft is advanced in respect to the crank of the upper crankshaft, that is, it *leads* the upper crank by some 10 to 15 deg. In this way the exhaust ports are opened first (Fig. 2-6a); when the pressure is sufficiently reduced, the scavenge ports are uncovered (Fig. 2-6b) and scavenging proceeds. After the exhaust ports are closed, additional admission of air takes place (Fig. 2-6c) until the scavenge ports are also covered and compression of the air charge takes place. Slightly before the pistons reach the point at which they are closest together, fuel is injected, ignites, and burns while the expansion stroke starts (Fig. 2-6d). The rotation of the lower and upper crankshaft is transmitted to the lower main crankshaft by means of an intermediate vertical shaft and two pairs of bevel gears.

The advantages of this scheme are (1) efficient scavenging of the cylinder and hence greater power developed; (2) absence of valves and valve-operating gears; (3) absence of cylinder heads which, because of their complicated shape, are a source of trouble in engine operation; and (4) good accessibility for inspection and repairs of most parts.

The two scavenge schemes (Figs. 2-2 and 2-6) are also classified as *uniflow* scavenging. In both cases the exhaust gases and scavenging air are flowing in the same direction, with less chance for formation of turbulences which are unavoidable with cross- and return-flow scavenging.

2-7. Supercharging. Supercharging has for its object an increase in the power which an engine of given piston displacement and speed can develop. In a diesel engine the power is developed by burning of fuel, and if it is desired to increase the power, more fuel must be burned and therefore more air must be available, since each pound of fuel requires a certain amount of air. Other conditions being the same, a given volume, or space, will hold a greater weight of air, if the air pressure is increased. Thus supercharging is obtained by a higher pressure of the air charge in the cylinder at the beginning of the compression stroke.

In order to increase the air pressure in four-stroke engines, the air charge is not sucked into the cylinder, or as it is said, is not admitted by *natural aspiration* by the receding piston, but is pushed in by a higher pressure created by a separate air pump or blower.

There are three types of blowers in use: (1) the reciprocating piston pump, similar to an air compressor; (2) the rotating positive-displacement blower of the Roots type, and (3) the centrifugal high-speed blower, similar to the centrifugal pump, usually driven by a gas turbine that utilizes the kinetic energy of the exhaust gases.

When a supercharger is applied to a four-stroke engine, the main change required in the engine design is a change in timing of the intake and exhaust valves. The intake-valve opening time is advanced and the exhaust-valve closing is retarded. The two valves are designed to stay open simultaneously for about 50 to 100 deg, the selection depending upon the normal engine speed. This simultaneous opening is called *overlapping*. The advantage gained by a large overlap is a better scavenging of the combustion space. Tests have shown that an overlap of 40 to 50 deg increases the power output of an engine from about 5 per cent—if the supercharging is very small, sufficient only to eliminate the vacuum in the cylinder during the suction stroke—to 8 per cent with a supercharger pressure of 12 in. of mercury, as compared with an overlap of 10 to 20 deg commonly used in unsupercharged engines. The total power gain due to supercharging varies from 20 to 50 per cent, depending upon the supercharger pressure, which in present diesel engines varies from 5 to about 12 in. of mercury.

It should be noted that, simultaneously with an increase of the mean effective pressure, supercharging increases also the maximum or firing pressure and the maximum temperatures. On the other hand, the fuel consumption per horsepower-hour usually decreases with supercharging because, as a result of an increase of air turbulence, a better mixing of the fuel and air charge and hence a better combustion of the fuel takes place and also because the mechanical efficiency of the engine increases—from the fact that the output is being increased more than the mechanical losses.

Two-stroke engines usually already have a blower to obtain scavenge air and their supercharging is obtained simply by increasing the amount and pressure of scavenge air. In addition, a slight change of the exhaust and scavenge timing is made in order to retain more scavenge air at the beginning of the compression stroke.

2-8. Piston Speed. While the velocity of rotation of the crankpin may be considered uniform, piston travel is not: at dead center the piston is standing still—its velocity is zero; as the piston begins to move, its velocity gradually increases and reaches a maximum about the middle of the stroke; from there on the piston velocity begins to decrease, and at the opposite dead center the piston comes again to a stop. Thus the piston velocity, or *speed*, as it is usually called, varies all the time. For many calculations it is necessary to know the average or mean piston speed, that is, the constant speed with which the piston would need to move to cover the same distance in the same time as it covers with the variable speed. This mean speed is usually referred to simply as the *piston speed* of the engine. It is customary to measure piston speed in feet per minute. The distance traveled by the piston in one minute is equal to two strokes made per revolution times the number of revolutions per minute n and represents the mean piston speed c. If the length of the stroke is l in., dividing by 12 to convert inches to feet gives

$$c = \frac{ln}{6} \tag{2-2}$$

2-9. Speed Factor. Engines are often divided into several classes of speed performance. Some are classified as low-speed engines, some as medium-speed, and some as high-speed engines. However, unless a definite yardstick is used, the designations remain vague. There have been attempts to use either engine speed, revolutions per minute, or piston speed, feet per minute, as a measure of speed performance, but neither of these two methods can give meaningful indications. The reason rotative speed as such is not suitable as a speed characteristic is that it does not take into consideration the size of the engine. A $3\frac{1}{2} \times 4\frac{1}{2} \times 6$ engine operating at 900 rpm is not a high- but only a medium-speed engine, as an engine of this size may be found operating at speeds up to 2,000 rpm and higher. On the other hand, $8\frac{1}{2} \times 10\frac{1}{2}$ diesel engines usually operate at speeds not exceeding 750 rpm and even at this lower speed have many features in common with high-speed engines and an $8\frac{1}{2} \times 10\frac{1}{2} \times 900$ engine would be decidedly a high-speed engine.

The same is true, only in reverse, in respect to piston speeds. In a large engine a relatively high piston speed, 1,800 ft per min or more, may be obtained with a relatively low rotational speed; in a small high-speed engine the piston speed is not high.

A good speed characteristic, called *speed factor* and designated c_s, is obtained as a product of revolutions per minute and piston speed. For the sake of obtaining smaller, more easily remembered figures, the product is divided by 100,000. Thus

$$c_s = \frac{nc}{100,000} \tag{2-3}$$

The speed factors for various existing diesel engines lie between the limits of 1 and slightly above 81. According to these data all engines may be divided into four classes, in each class the high limit being obtained by multiplying the low limit by 3:

1. Low-speed engines with a speed factor 1 to 3.
2. Medium speed engines, with a speed factor 3 to 9.
3. High-speed engines with a speed factor 9 to 27.
4. Super-high-speed engines with a speed factor 27 to 81 or higher.

Practical Meaning. Knowledge of the speed group to which engine belongs is of great value to the engine operator: the higher the speed classification of an engine, the more the operator should try to keep the engine in its best running condition and to observe every detail given in the manufacturer's instruction book, and the more careful he should be when inspecting or overhauling the engine.

QUESTIONS

1. What is a cycle?
2. Enumerate the events that form a cycle in a diesel engine.
3. What is a four-stroke cycle?
4. What is dead center in an engine with reciprocating motion?
5. What is the reason for applying the term *dead center* to certain positions of the piston?
6. Explain by sketches the sequence of main events in a four-stroke cycle engine.
7. State the usual timing of a diesel-engine event in terms of crank travel in reference to the two dead centers.
8. What are the two purposes in compressing the air charge before injecting the fuel?
9. What is the compression ratio of a combustion engine?
10. What are the usual compression ratios of diesel engines?
11. What are the two distinct methods of burning the fuel in a combustion-engine cylinder?
12. What cycle, in respect to burning of the fuel, is used in modern high-speed diesel engines?
13. What are the advantages and drawbacks of the combination cycle as compared with the constant-pressure combustion cycle?
14. What is the main difference between a two-stroke and a four-stroke engine?
15. What is scavenging in a two-stroke engine?
16. Explain the operation of a simple-port two-stroke engine.
17. What are the theoretical advantages of a two-stroke operation?
18. What are the main practical disadvantages of two-stroke operation?
19. Enumerate and illustrate by diagrammatic sketches the various methods of scavenging two-stroke engines at present in use.
20. What is the main advantage of a cross-flow scavenging scheme with special check valves near the scavenge-air ports?
21. What is the main advantage of the loop-scavenging scheme? For what engines is it particularly suitable?
22. Draw a diagrammatic sketch of an opposed-piston two-stroke engine and explain the operation of the engine.
23. Enumerate the advantages of the opposed-piston scheme.
24. Draw diagrammatic sketches of the two differents cavenge schemes, both classified as uniflow scavenging.
25. What is the main advantage of uniflow scavenging?
26. What does *supercharging* a combustion engine mean?
27. What is the main object of supercharging?
28. Referring to a four-stroke combustion engine, what is *natural aspiration?*
29. What are the three types of blowers used for supercharging diesel engines?
30. What is called *overlapping* in reference to the valve timing of a four-stroke engine? How is it expressed?

31. What are the amounts of overlapping used in an unsupercharged engine and a supercharged one?
32. What is the object of a large overlap?
33. How does supercharging affect the over-all efficiency of a diesel engine?
34. What is the piston speed of an engine?
35. In what units is piston speed expressed?
36. What is the expression for calculating the piston speed of an engine?
37. What is called *speed factor?*
38. What is the expression for calculating the speed factor?
39. Enumerate the four classes into which engines may be divided according to their speed factors and indicate the range of speed factors for each class.
40. What is the practical meaning of the speed factor or speed classification?

PART 2

DIESEL-ENGINE COMPONENTS

CHAPTER 3

FRAMES, CYLINDERS, AND HEADS

3-1. Framework. The framework of a large vertical diesel engine usually consists of a bedplate, crankcase, and cylinder jackets. In smaller medium-speed engines the cylinder jackets are often cast as one piece, or a block, and sometimes the crankcase and cylinder block are also a one-piece casting or weldment. High-speed diesel engines,

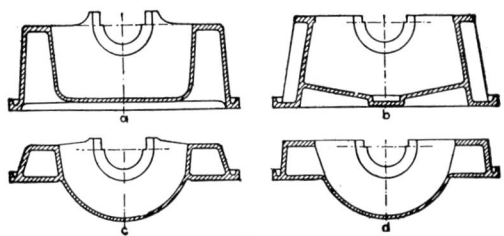

Fig. 3-1. Types of bedplates for vertical engines.

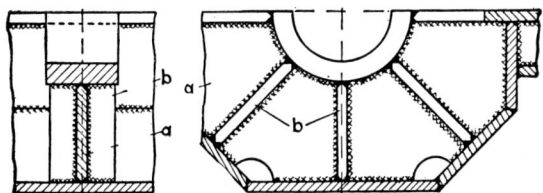

Fig. 3-2. Welded bedplate and mainbearing.

in the design of the frame and cylinder, follow the accepted forms of automotive gasoline-engine construction, except that many have removable cylinder liners.

Bedplates. Cross sections and proportions of typical cast-iron bedplates for medium-sized and large engines are shown in Fig. 3-1: Fig. 3-1a shows a section of a bedplate for an A-frame stationary engine; Fig. 3-1b shows a section of a bedplate for a stationary engine with a continuous enclosed crankcase; Fig. 3-1c and d shows the corresponding sections for marine engines, in which it is desirable to have a low center of gravity of the engine.

In a lightweight welded steel bedplate (Fig. 3-2) the question of rigidity is of particular importance. While the center support a is sufficient for strength, ribs b, b are added for rigidity.

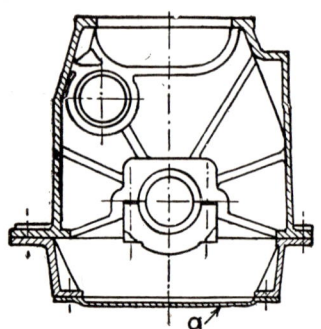

FIG. 3-3. Automotive-type crankcase.

Regardless of the amount of rigidity incorporated in the construction of the bedplate, a long bedplate of a medium- or large-bore engine will warp or sag if not supported at a sufficient number of points.

Automotive engines usually are built with the bedplate combined with the crankcase. The crossties containing the main bearings are inverted and the bearing caps hold the crankshaft (Fig. 3-3). An oil pan of sheet metal or a light aluminum casting closes the crankcase from the bottom.

Frames and Crankcases. The A-type frame, so called from its transverse section (Fig. 3-4), is the earliest type of diesel-engine frame and is still used with some low-speed medium-size engines. Each cylinder stands on its own legs, independent of the other cylinders. The legs of the column are cast in one piece with the cylinder jacket into which the liner if fitted.

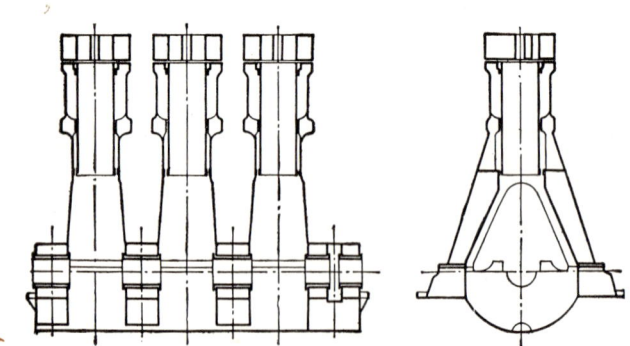

FIG. 3-4. A-type diesel-engine frame.

A typical B, or box, frame, often called a *crankcase*, is shown in Fig. 3-5; it is used in many medium-sized and large engines. In some engines the box construction of the crankcase transmits the forces acting between cylinder and bedplate. A better construction is to provide steel rods for this purpose (Fig. 3-6). Here the crankcase is simply a spacer between cylinder and bedplate. In large engines the crankcase may be divided into two or more pieces, sometimes a separate piece under each cylinder, and the separate pieces bolted together by flanges, thus forming a continuous box of great rigidity.

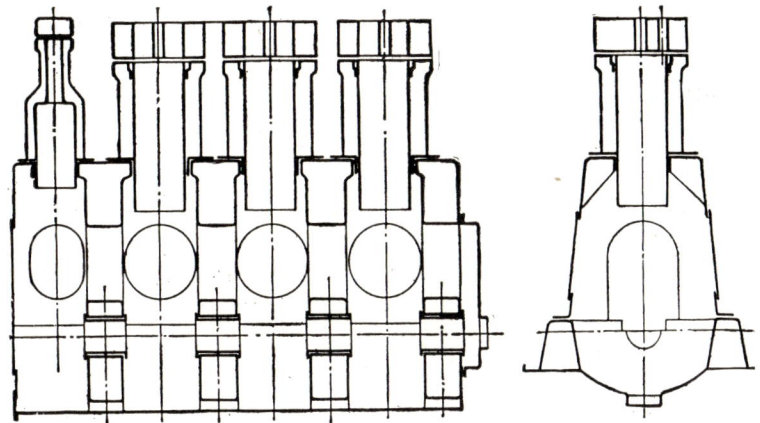

Fig. 3-5. B-type frame of an air-injection diesel engine.

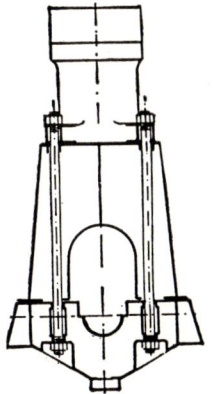
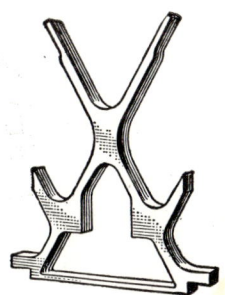

Fig. 3-6. Crankcase with tie bolts.   Fig. 3-7. Flame-cut part of a crankcase.

Welded crankcases are used chiefly to reduce engine weight. Figures 3-7 to 3-9 illustrate the building of a welded one-piece bedplate, crankcase, and cylinder block for a twelve-cylinder V-type 1,000-hp oil engine. Figure 3-10 shows a similar frame which contains several drop forged pieces in order to strengthen the construction without making it heavier.

3-2. Cylinders. Only a few makes of large engine use separate cylinders as shown in Fig. 3-5 and 3-6. The majority of multicylinder engines have the cylinders cast in block, even in large engines; usually each cylinder is fastened in a separate compartment with cross bracing between the compartments. Some engines use cylinders cast with the water-jacket walls as one piece. However, the majority of diesel builders and users prefer removable cylinder liners for the reason that certain conditions, such as inferior or dirty fuel oil, or dust and sand in the intake air, may cause rather rapid wear of the cylinder surface. This will give an increased clearance between the piston and cylinder

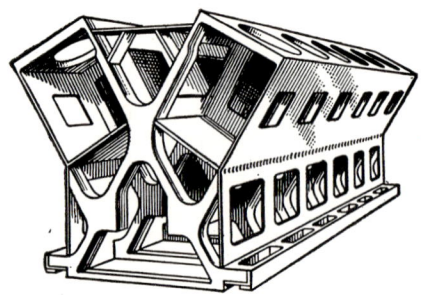

Fig. 3-8. Crankcase in process of fabricating by welding.

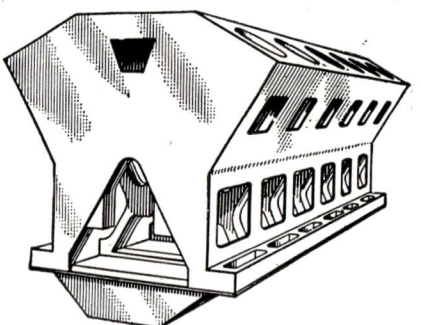

Fig. 3-9. Finished welded crankcase.

and cause gas blow-by, loss of compression, and excessive contamination of the lubricating oil. As a result either the cylinder must be rebored and new, oversize pistons installed or the whole cylinder must be exchanged for a new one. If the cylinders are cast in block, the whole block must be replaced. This is an expensive procedure. To reduce this expense, general practice is to use removable liners. Another advantage of removable liners is that they can be made of better wear-resisting cast iron without appreciably increasing the cost of the engine.

Cylinder liners are made of two types, the wet type, when the outside surface of the liner is in direct contact with water, as in Fig. 3-11, and the dry type, pressed into the cylinder proper (Fig. 3-12). While in some instances it is more difficult to remove a dry liner, it is much simpler to replace it, since there is no danger of water leakage into the combustion space or the crankcase. Its disadvantage is a decrease of heat conduction through the composite wall.

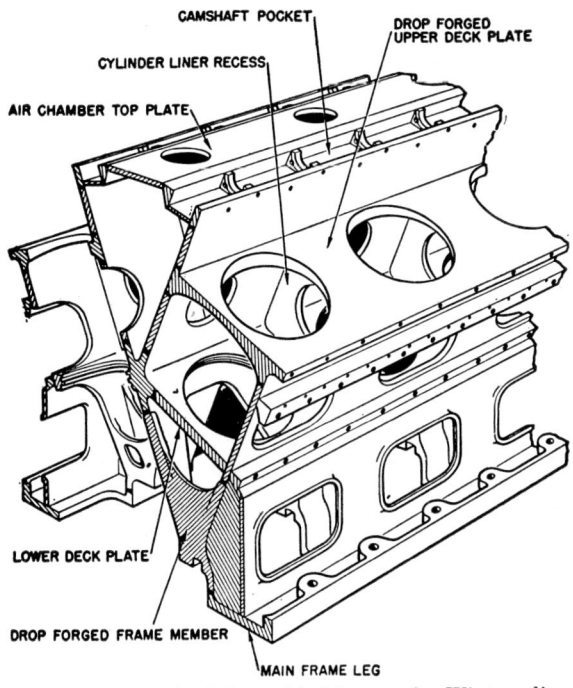

Fig. 3-10. Sectional sketch of the welded frame of a Winton diesel engine.

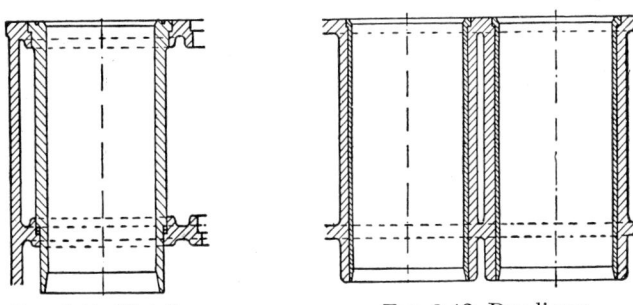

Fig. 3-11. Wet liner. Fig. 3-12. Dry liners.

In some two-stroke engines the liner is combined with the water jacket (Fig. 3-13) to prevent the possibility of water leaking into the cylinder through a joint.

Honed Liners. The first advance beyond the finishing of liner bores on a lathe or boring mill was reaming. Reaming was an improvement over boring and is still adhered to for large-bore liners. Grinding, used by some engine builders, is not very good, as it leaves in the surface particles of abrasive which prevent the wearing in of the liner and cause rapid wear of piston rings and pistons.

The best method is honing. Honing, which was first adopted by builders of automobile engines, has been applied to the finishing of diesel liners after boring, especially for bores of less than 10 in. The honing machine gives a reciprocating and circular motion to the hone head. The oil stones, or hones, rapidly give the bored liner a mirror-like finish. The tolerance may be held close by constant checking of the hone set-up.

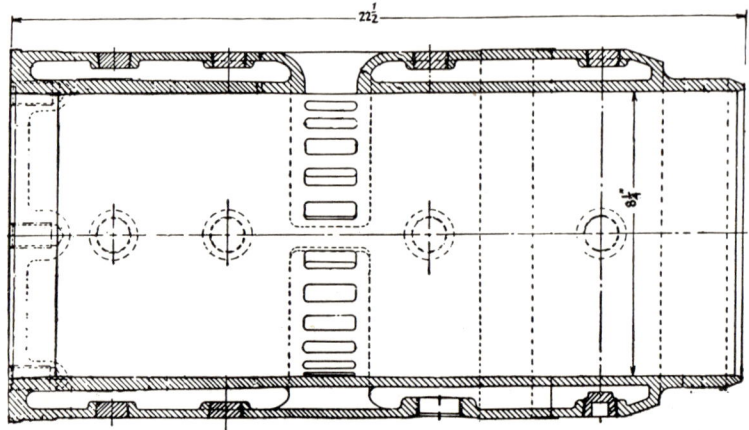

Fig. 3-13. Liners with integral water jacket.

Advantage of Honing. A cylinder surface must be smooth, because it must maintain an unbroken thin oil film. In order to maintain an unbroken oil film, the cylinder must be straight and round and the piston and rings moving within the cylinder must expand and contract sufficiently uniformly so as not to produce high-pressure areas that will break through the oil film. This factor must be controlled by accurate machine work.

A perfectly smooth cylinder surface will maintain a thin oil film. The smoother the surface, the thinner the oil film can be and still prevent metal-to-metal contact. This statement can best be illustrated by placing a drop of oil between two pieces of plate glass. The surfaces of glass are so smooth that the oil film can be squeezed down so thin that it is hardly noticeable on the surface of the glass, yet the two pieces of glass will not contact each other.

In order to prevent wear, an oil film must be maintained between the piston and rings and the cylinder surface. If the cylinder is rough, the oil film must be thick enough to cover up all the little high points on the rough surface, or excessive wear will result. Even if it were possible to maintain a sufficiently thick oil film to prevent wear between rough surfaces, minimum oil consumption could not be attained.

While laboratory equipment for determining smoothness of cylinders is not available in the field or in the usual service station, there is a very simple test for checking the smoothness of cylinders: just take a copper penny and rub the corner of it against the surface of the cylinder; if the cylinder is smooth enough and does not produce a bright spot on the edge of the penny, the finish is commercially satisfactory; if the surface is rough enough to cut the edge of the penny and leave a mark, the cylinder is too rough and will result in excessive wear to the pistons, the rings, and the cylinder itself.

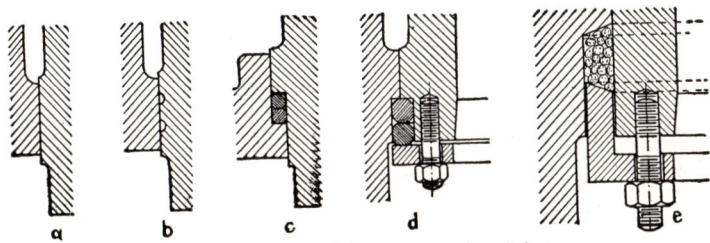

Fig. 3-14. Types of liner expansion joints.

Expansion. In order to prevent water leaks at the lower, free end of a wet liner when it moves from heat expansion, some kind of expansion joint must be used. Figure 3-14 shows some typical expansion joints: *a* is simply a close fit of the liner to the bore in the jacket with no provisions to stop a leak if one should develop; *b* has grooves on the liner which supposedly act as a *labyrinth gland*—so called because the grooves in a smooth bushing present an additional assistance to the flow of a fluid through the clearance between a cylindrical part and its guiding bushing; *c* and *d* have rubber rings slightly compressed between flat or conical surfaces; and, finally, *e* is a regular stuffing box with a gland such as those used in very large engines. Natural-rubber rings are satisfactory for temperatures up to 225 F; synthetic-rubber rings can be used up to 350 F.

Liners of double-acting two-stroke engines can be made to expand inwardly (Fig. 3-15).

Distorted Liners. If a cylinder liner is measured after the cylinder head has been removed, in many instances the micrometer will not show the same diameter at all points. The reason is that tightening up the cylinder-head bolts distorts the liner. The piston at once proceeds to wear the distorted liner to a true circle; but, when the head is removed, the true circle of the liner distorts as soon as it is freed from the head constraint and the cylinder appears out of round.

For this reason, a liner, during honing, should be held in the frame by a dummy head, so that actual working pressures and consequent frame distortion are produced. The liner will then be true when in operating condition.

Life of Liners. Liners of all diesels in time become inefficient because of wear or scoring. Research indicates that the wear of a cylinder liner, regardless of engine size, is approximately 0.001 in. per 1,000 hr of operation in low-speed engines. This means that an engine running 8,000 hr yearly will show 0.008 in. wear in this time. An engine with a 16-in. bore will operate satisfactorily even if the clearance—the difference between the piston diameter and diameter of the cylinder liner—is as great as 0.050 in. If the original clearance is 0.020 in., the permissible wear is $0.050 - 0.020 = 0.030$ in. With a wear rate of 0.001 in. per 1,000 hr of operation—an average figure for low-speed engines—such a liner would have a life of

$$1,000(0.030/0.001) = 30,000 \text{ hr}$$

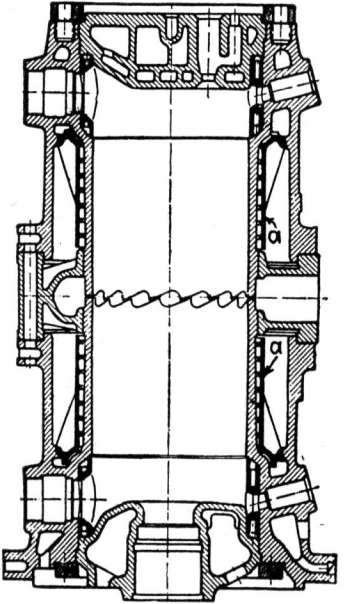

FIG. 3-15. M.A.N. diesel-engine cylinder, 33½ by 42 in. 160 rpm.

or about four years of constant day-and-night operation. This is about the normal life of such a liner.

In high-speed diesels, the wear will be greater, approximately 0.003 in. per 1,000 hr, or 0.024 in. in a year of 8,000 operating hours. Consider now an engine with a 4-in. bore: the original clearance will be around 0.005 in., and the maximum clearance, before blow-by occurs, around 0.015 in. The permissible wear will then be

$$0.015 - 0.005 = 0.010 \text{ in.}$$

At a rate of 0.024 in. per year, a liner will need to be replaced in $0.010/0.024 = 0.4$ year, or about five months. Experience shows that on an average a high-speed engine operating on such a schedule actually will require replacement in this time. Fortunately, however, the 4-in. liner costs but little compared to the cost of a 16-in. liner, and its frequent replacement is not so objectionable.

Chrome-plating. The life of a liner can be materially increased by chrome-plating the working surface. A cylinder liner of an engine of medium size and speed with a properly chrome-plated bore will have a wear of 0.001 in. in about 4,000 hr. With a 0.008-in.-thick chrome surface the liner will give about 32,000 hr of service before it will need replating. The cost of replating is only a small fraction of that of putting in a new liner. The life of chrome-plated cylinder liners of high-speed engines may be assumed to be about one-half that of medium-speed engines or about 16,000 hr or two years, compared with 5 months for bare cast-iron liners.

Another way to reduce liner wear is to use chrome-plated piston rings, usually in the second and third grooves from the top. The smooth, hard chrome surface has a good affinity for lubricating oil and the rate of wear of the liner is reduced by at least one-half, compared with the wear by ordinary cast-iron piston rings.

3-3. Cylinder Heads. *Materials.* The cylinder heads of small and medium-sized engines are usually made of close-grained cast iron, Meehanite, or nickel cast iron. Heads of large engines are made of low-carbon cast steel or welded of steel plates and drop forgings.

Rigidity. The shape of the cylinder head itself with the cooling-water space, reinforced openings for the valves or valve cages, and ports leading to the intake and exhaust manifolds, ensures great rigidity. Rigidity is necessary to keep the joint between cylinder and head tight and to prevent the gasket under the cylinder from being blown out by the high combustion pressure. Because of this rigidity a small number of hold-down studs may be used. In some designs the cylinder head is made square and is held by only four studs, one at each corner; such a design facilitates the arrangement of valves and ports. However, a more common design is to have the cylinder head round or octagonal with six or eight studs.

The smaller the number of studs, the larger is the stud diameter and the smaller is the danger of overstressing the studs by excessive tightening of the nuts.

Cleaning. The high temperatures to which the inside cylinder-head wall is exposed have a tendency to precipitate salts from the cooling water and thus to produce scale. Scale is a very poor heat conductor and therefore a wall on which scale is being deposited transmits less and less heat to the cooling water and finally becomes overheated. Overheating lowers the strength of the head material and at the same time creates additional stresses and thus may easily cause cracking of the cylinder head.

To prevent such conditions, cylinder heads have hand holes or plugs which permit the cleaning out of scale.

QUESTIONS

1. Enumerate the parts that make up the framework of a large vertical diesel engine.
2. What is the difference in the framework construction of large low-speed, smaller medium-speed, and high-speed diesel engines?
3. Draw a sketch with the necessary views and sections of a typical cast-iron bedplate for medium-sized and large engines.
4. Draw a sketch with the necessary views and sections of a typical welded bedplate.
5. What is the main difference in the design of bedplates for a stationary and an automotive-type diesel engine?
6. Draw a sketch of a crankcase for a medium-sized diesel engine.
7. What is the object of providing steel rods between the bedplate and cylinders of medium-sized and large engines? Explain by means of a sketch.
8. What is the object in making crankcases welded of steel plates and forgings?
9. What are the types of cylinder design used in multicylinder engines?
10. What are the two main advantages of removable cylinder liners?
11. What are the two types of removable liners in use with diesel engines?
12. What are the different methods used for finishing the inside working surface of cylinders and cylinder liners?
13. What is the main advantage of having cylinder liners?
14. What simple test may be used to determine whether the cylinder is sufficiently smooth?
15. What is the purpose of expansion joints on the lower end of wet liners?
16. Draw sketches of typical expansion joints of cylinder liners.
17. What distorts the cylinder liner during engine assembly?
18. What procedure should be used when the liner is being honed to keep the bore of the liner perfectly round when it is in place in the engine?
19. What is the approximate rate of wear in low- and high-speed engine liners?
20. What is the expected life of cast-iron liners in low-speed engines? In high-speed engines?
21. What are the object and effect of chrome-plating the working surface of cylinder liners?
22. What is the effect of using chrome-plated piston rings?
23. Of what materials are cylinder heads of small and medium-sized engines usually made?
24. Of what materials are cylinder heads of large diesel engines made?
25. What effect does the rigidity of cylinder heads have upon the number and size of hold-down studs? What is the advantage of large studs?
26. What attention should be given to the water space in cylinder heads?
27. What is the effect of scale on the internal stresses in the cylinder-head material, and if scale is not removed periodically, what may be the consequences?

CHAPTER 4

RUNNING GEAR

Under the classification running gear are included pistons, connecting rods, and crankshafts.

4-1. Pistons. There exist two types of pistons—those used in connection with crossheads, mostly in double-acting engines, and those in which the piston itself performs the function of a crosshead. Pistons of the first type, when used in double-acting engines, must be cooled by circulation of water or oil, and hence are called *barrel pistons*. Pistons of the second type are called *trunk pistons* and are used in all single-acting engines.

All pistons have to carry pressures varying from a slight vacuum during the suction stroke to peak pressures of 1,000 to 1,200 psi, with the resulting fluctuation in temperature and heat expansion; they have to withstand high bearing pressures from side thrust against the cylinder liners; they often have insufficient lubrication; and they have to resist wear on the outer cylindrical surface and in the ring grooves from the pressure and sliding action of the compression rings.

Pistons used in double-acting engines are built up of several sections and are closed at both ends because both ends of the cylinder are used for combustion chambers. They are cooled by water or oil entering and leaving through the hollow piston rod.

Trunk Pistons. The *functions* of a trunk piston are the following:

1. To transmit the gas pressure to the crankshaft.
2. To take the side pressures due to angularity of the connecting rod.
3. To seal the inside of the cylinder from the crankcase.
4. To dissipate heat absorbed by the piston top during combustion and the early part of the expansion stroke.

Materials used in making trunk pistons are, in the order of their importance, cast iron, cast aluminum, forged aluminum, cast steel, and forged steel.

Cast iron is an excellent material; its main drawback is that it gives a slightly heavier piston than aluminum. However, with proper design the difference is only about 10 to 20 per cent. Properly designed and machined cast-iron pistons may produce even less cylinder-liner wear than aluminum ones, especially if they are tin-plated.

Cast aluminum alloy gives better heat dissipation and lighter weight but costs considerably more than cast iron. The strength is about the same as that of cast iron.

Forged aluminum pistons are stronger and still lighter. They are used for aircraft engines and heavy-duty high-speed compression-ignition oil engines.

Alloy cast-steel pistons are used in some automotive engines and require liners of great surface hardness. The same is true of welded

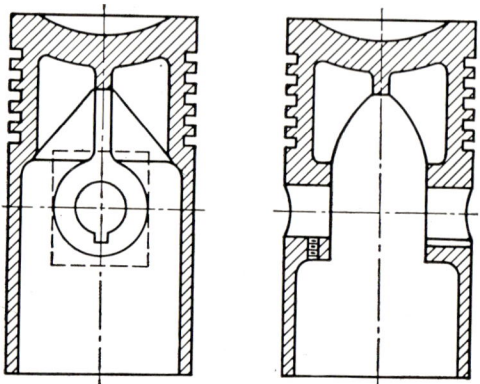

FIG. 4-1. Trunk piston of a medium-size diesel engine.

steel pistons used in some high-speed diesels and forged-steel pistons used in some aircraft engines.

Design Features. Some pistons are cooled by water or oil circulated through baffles within the piston, but generally trunk pistons are cooled merely by contact with the watercooled cylinder walls and by lubricating oil impinging on ribs under the crown of the piston. Pistons of high-speed engines are made as light as possible because of the effect of the mass on the bearing loads. For several years the tendency was to use aluminum pistons. Recently the trend has changed to pistons of cast iron with thin walls cooled by lubricating oil circulated through a hollow crown. The cast iron has one important advantage —it gives a piston with the same coefficient of heat expansion as the cylinder liner. This permits the use of smaller piston-to-liner clearances when the engine is cold without the danger of seizure when it is operated under a heavy load.

RUNNING GEAR 43

Piston Head. The thickness of the head or crown is controlled by two requirements: strength and heat dissipation.

In order to provide a sufficient cross-sectional area for the heat flow, the crown often is made thicker than is necessary for strength.

Typical Constructions. A typical cast-iron piston of a diesel engine of medium size and speed is shown in Fig. 4-1. The piston has a fixed piston pin and ribs that take part of the gas load acting on the

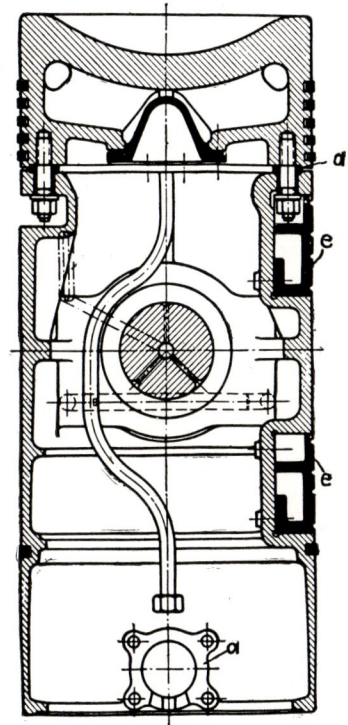

Fig. 4-2. Oil-cooled diesel-engine piston. Scale 1:13.

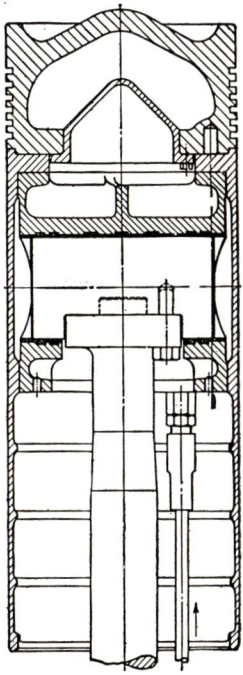

Fig. 4-3. Three-piece piston of large Nordberg diesel engine.

crown, but, even more, help it to dissipate heat absorbed from the hot gases.

In large oil engines the upper part of the piston is sometimes made as a separate piece fastened by studs a to the lower part (Figs. 4-2 and 4-3). The object of such a construction is to use for the head a material of greater strength, sometimes cast steel.

In cast-iron pistons with a steel crown or upper part the diameter of the latter is made such that it will not touch the cylinder walls even at the highest temperature. Another advantage of this construction is the lower cost of exchanging the top of the piston if it develops cracks from the high-temperature stresses or from being burned by the fuel hitting it.

Figure 4-4 shows an aluminum piston with a cast-iron top. A typical piston of a truck diesel is shown in Fig. 4-5.

The length of the piston determines the side pressure resulting from the angularity of the motion of the connecting rod. A longer piston runs more quietly and has less slap, as the increase of its length decreases the angle of lateral movement from α_1 to α_2 (Fig. 4-6) for the given clearance c between piston and cylinder necessary to take care of expansion. Figure 4-3 shows such a piston which is so long that it requires an extra long connecting rod, $l/r = 5.25$, to clear the piston skirt.

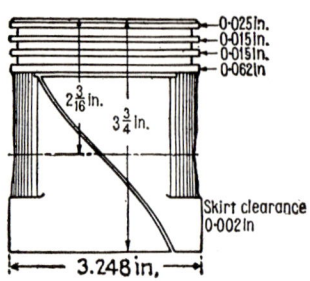

Fig. 4-4. Split-skirt piston.

Another means of decreasing the lateral movement and thus to avoid piston slap is the use of separate adjustable glide pieces e, e (Fig. 4-2). However, this construction is used only in very large engines.

To reduce the weight of the reciprocating parts, pistons of high-speed engines are made considerably shorter. This results in increased side pressures and slightly greater wear of the pistons, but on the other

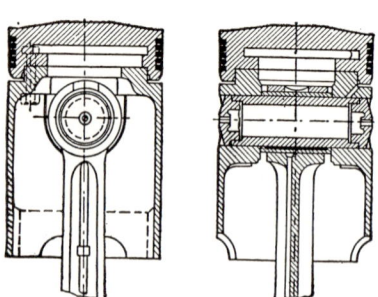

Fig. 4-5. Beardmore composite piston. Scale 1:10.

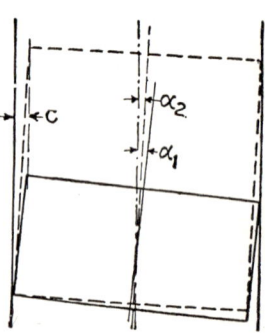

Fig. 4-6. Influence of piston length.

hand the smaller weight reduces the wear of the crankshaft bearings.

Friction. The friction between piston and cylinder is reduced by using proper materials, smooth finish of the surfaces, piston rings with minimum pressure, reasonably free pistons, good lubrication, and reasonably high temperatures of the oil.

The pistons are usually ground. The pressure of the piston rings necessary to seal the wearing surface is small, about 5 to 6 psi in low-speed engines and decreasing down to 3.5 psi in high-speed engines. Gas pressures penetrating behind the rings increase the wall pressure and friction losses considerably. This is the reason why many engine builders prefer simple snap rings to special composite rings in which a greater sealing capacity is obtained through gas pressure under the ring.

Because the side pressure on the pistons is small, there is not much danger of destruction of the oil film, if its temperature does not exceed 190 to 200 F. However, in some engines the oil-film temperature at high loads goes up much higher and a so-called *lacquering* of the piston skirt may then occur.

To take care of the large expansion coefficient of aluminum and at the same time to avoid loose pistons at low temperatures, aluminum-alloy pistons are often made with split skirts (Fig. 4-4) or sometimes with steel rings cast into the aluminum to give rigidity and to force the aluminum skirt to expand as though it were of steel.

Piston Clearance. Clearance between piston and liner varies with the engine design and piston diameter. A general rule is to have a clearance of at least 0.005 in. between the skirt of the piston and the liner for cast-iron pistons up to $4\frac{1}{2}$ in. diameter. The top section is tapered, so that the clearance at the top of the piston is greater. These are *cold* measurements, and when the engine becomes hot, the top clearance is about the same as the cold skirt clearance. The manufacturer of a particular engine should be consulted on these clearances.

With aluminum pistons the clearance must be about twice as large as with cast-iron pistons.

4-2. Piston Pins. All of the load developed in the cylinder passes through the piston pin, also called *wrist pin*. It is the only link between the piston and the connecting rod. Wrist pins that are supported at both ends by bronze bushings in the piston bosses, with the connecting rod swinging on a bronze bushing or needle bearing at the center of the pin, are termed *full-floating* (Figs. 4-5 and 4-8). In larger engines the wrist pins are sometimes locked in the piston at both ends (Fig. 4-1). Such pins are termed *fixed pins*. This design has the disadvantage that all swinging movement is confined to the connecting-rod bearing and there is some danger that its wear may not be uniform. *Semifloating* pins are pins clamped in the small end of the connecting rod, as in automobile practice.

Piston-pin bearings operate under rather severe conditions: in addition to the great loading from the piston pressures, there is the handicap of less efficient lubrication because the swinging motion does not help to form an oil film as much as the rotary motion of a journal. The wrist pin is made of a steel alloy of great strength to carry the load, and it must have a fine-finish hardened surface to obtain good bearing action.

In order to obtain uniform distribution of pressure between piston and cylinder, the piston pin should be placed in the center of the piston part carrying the load. This would be halfway between the edge of the last ring groove and the end of the piston. However, to shorten the height of the engine the piston pin is usually moved nearer to the top (Fig. 4-1), sometimes about halfway between the two ends (Fig. 4-2), and sometimes even nearer to the top end (Fig. 4-5). In vertical engines placing the piston pin nearer to the top allows the engine to run more quietly.

The high pressures between the piston pin and its bearings and the pounding action of the very fast pressure rise require the use of bushings of special high-grade bronze, and even these show fast wear, as do the stationary bearings in the piston bosses.

In high-speed oil engines the piston pins are usually of the full-floating type, with bronze bushings in the connecting rod and in the piston bosses and with either bronze or aluminum plugs (Fig. 4-5) or spring retainers (Fig. 4-8) to prevent the pin from touching and scoring the cylinder liner.

In trunk pistons in which the pressures act always in one direction, as in two-stroke engines or in low-speed engines with comparatively light pistons, the bearing pressures on the piston pin may be lowered by the construction shown in Fig. 4-3. This piston is for a two-stroke engine and is made up of three pieces—the head, the barrel, and a center piece in the shape of a hollow casting which rests on an oversized piston pin. The head is a steel forging; the wristpin is bolted to the connecting rod and increases the bearing area so as to lower the pressure to about 1,250 psi. This pressure also permits the use of a babbitt bearing surface. In this construction the working pressures are transmitted directly to the connecting rod and the barrel is not strained or distorted by wristpin bosses.

A similar design for a four-stroke medium-speed engine is shown in Fig. 4-7. The piston pin is bolted to the connecting rod and works in a bushing pressed into the piston. The piston head is cooled by oil admitted through the drilled connecting rod and discharged into the crankcase.

In pistons with fixed wrist pins, the barrel is usually slightly relieved around the piston pin, as shown by dotted lines in Fig. 4-1; it is still more relieved in the pistons shown in Figs. 4-4 and 4-7.

The methods of fitting the piston pin to the piston and holding it in place may be seen in the different drawings.

In automotive engines the piston pin is usually clamped in the connecting rod but sometimes is made floating both in the piston bosses

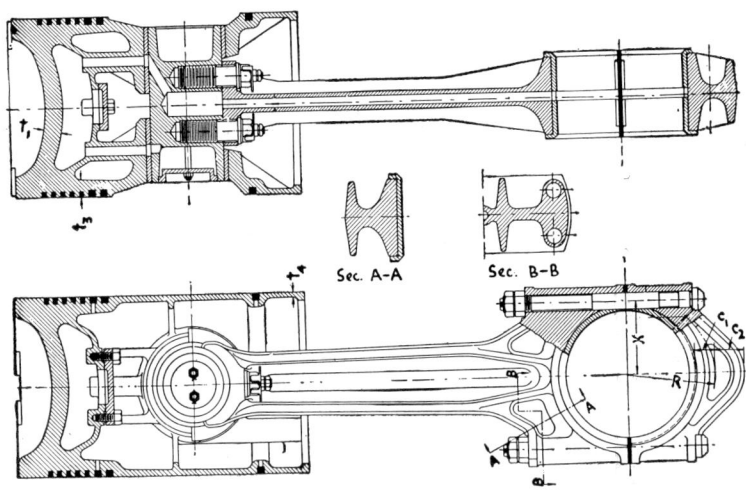

Fig. 4-7. Piston and connecting rod of a Cooper-Bessemer diesel engine.

and in a bushing that is inserted into the end of the connecting rod, as shown in Fig. 4-5.

For larger engines the piston pin is made of mild steel, case-hardened and ground to final size; for small high-speed engines it is made of nickel or vanadium steel, hardened and ground.

Lubrication of the piston pin is provided either by forcing oil under pressure from the hollow crankshaft and up the rifle-drilled connecting rod (Fig. 4-7) or by collecting the oil from the cylinder surface and leading it to the hollow wrist pin and from there to the bushing (Fig. 4-2).

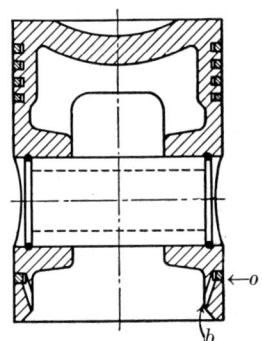

Fig. 4-8. Piston with full-floating pin. *Courtesy of International Harvester Company.*

4-3. Piston Rings. At the top of the piston are inserted several *compression rings* which (1) seal the space between the piston and liner, preventing the high-pressure combustion gases or the air charge from escaping down the liner during the compression stroke; (2) transmit heat from the piston to the water-cooled cylinder liner; and (3) absorb a certain part of the fluctuations of the piston side thrust.

The *oil-scraper* or *oil-control rings*, represented by ring *o* in Fig. 4-8, usually are located at the bottom end of the piston with holes (*b*) to drain the oil. However, in larger engines one of the oil-control rings may be located above the piston pin. Small engines use one, larger ones two or sometimes three oil-control rings to a piston. The purpose of oil rings is to scrape off most of the lubricating oil splashed upward by the crankshaft and connecting rod in order to reduce the amount of oil carried upward and burnt in the combustion chamber. At the same time they must allow sufficient lubricating oil to be carried to the upper part of the liner during the upstroke to give proper lubrication for the piston and the compression rings. Double-acting pistons have no oil-scraper rings, since no oil is splashed on the liner.

Compression rings are machined from gray-iron castings. Some types have special facings, such as a bronze insert (Fig. 4-9*a*) or a

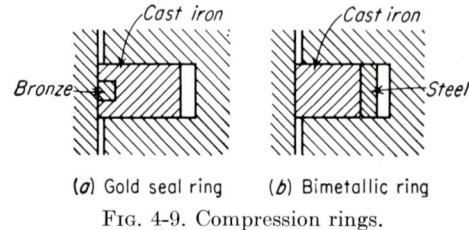

(*a*) Gold seal ring   (*b*) Bimetallic ring

Fig. 4-9. Compression rings.

treated surface, to facilitate seating in to the liner. To expedite wearing in or seating of the ring face, some rings have a slight angle, $\frac{1}{2}$ to 1 deg, to the face so that at first the contact area is very small and wear is rather fast, decreasing later. Such rings give a good, gastight seal.

Since the Second World War chrome plating of the outer, wearing surface of the upper compression ring by a special process called *hard-chroming* has become popular because it seems to reduce the wear both of the piston ring and of the liner.

The type of compression ring most widely used has a rectangular cross section. The diameter of the rings is slightly larger than the cylinder bore, and part of the ring is cut away to permit it to go into the cylinder. The difference in diameters produces a pressure against the liner wall. The pressure of the upper rings is increased by the action of the gases. The combustion gases or compression air enters behind the ring through the vertical clearance which always exists between a ring and its groove and forces the ring against the cylinder liner. Some piston-ring manufacturers advocate the use of rings peened from the inside.

Some engines have compression rings with the bottom or both bottom and top beveled, making the ring thinner at the inside than at the outside diameter (Fig. 4-10). The groove in the piston is machined to the same shape. The gas pressure acting on the top wall, owing to the beveled bottom surface, produces an additional force pressing the ring against the cylinder wall and helping to seal it. On the other hand, at each reversal of the side thrust of the piston the ring slides

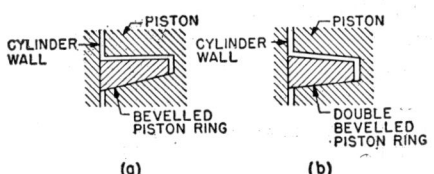

FIG. 4-10. Beveled piston rings.

slightly into the groove, is pressed against the upper groove wall, crushes the carbon which is deposited on it, and keeps the ring from sticking. Some engines use bimetal rings (Fig. 4-9b), in which the cast-iron wearing face is brazed to an inner steel ring to obtain increased strength and reduce the probability of ring fracture.

Oil pumping is prevented by rounding off the upper edges of the piston rings (Fig. 4-11a). The last ring is made as an oil scraper with a slight bevel (Fig. 4-11b) by beveling off the lower groove edge and drilling small holes toward

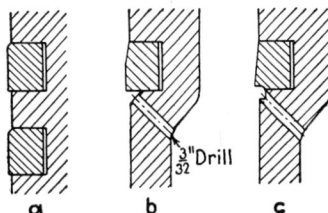

FIG. 4-11. Methods of preventing oil pumping.

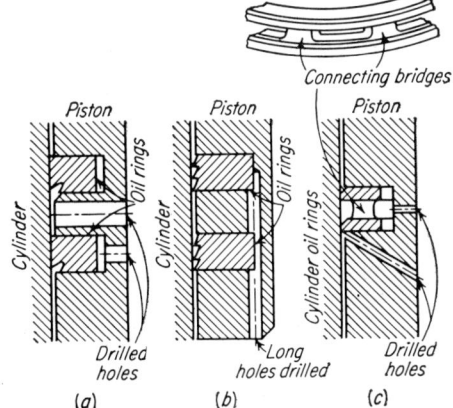

FIG. 4-12. Oil-control piston rings.

the inside of the piston to assist in the removal of excess oil. Where lubricating-oil consumption is excessive, a more effective scraper ring with an undercut may be used (Fig. 4-11c). Piston rings with a high specific pressure act in a way similar to scraper rings, but their drawback is increased cylinder wear.

The oil-control rings have a narrow face designed to provide a higher unit wall pressure and are undercut to give a scraping edge. In some designs the ring has one, in others two narrow scraping edges and the piston has rows of holes drilled for draining the oil through the bottom of the ring grooves, the lands between the grooves, or both (Fig. 4-12a, b, and c). The oil scraped by the ring must be drained off immediately, otherwise it will build up a pressure which will force the ring back into its groove and stop the scraping action. It is important that the drainage from the piston grooves be complete. Inadequate drainage means faulty scraping, higher lubricating-oil consumption, and a darker color of the engine exhaust gases. Spring-steel expanders are

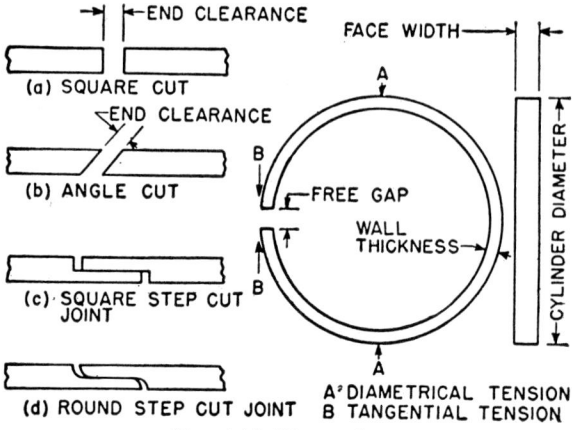

Fig. 4-13. Piston rings.

sometimes used behind the rings to increase the wall pressure and improve the scraping action.

Piston-ring Clearance. A piston ring should have no more clearance between its sides and grooves than is needed to permit movement. This calls for a side clearance of from 0.002 to 0.0025 in. in small-bore engines (bores of about 4 to 5 in.). Engines of an 11- to 12-in. bore should have clearances of from 0.003 to 0.005 in. A piston ring when slipped into the cylinder bore should have a clearance at the ends, that is, should have a *gap clearance* of from 0.015 to 0.020 in. in small engines and up to 0.070 in. in large-bore engines.

The gap between the ends of the compression rings, when they are inserted cold in the cylinder, must be sufficiently large so that when the ring expands with the maximum piston temperature the ends will not be pressed together and buckle the ring. The way the ends are cut varies. Most rings have the ends cut square (Fig. 4-13a). A design which makes gas blow-by more difficult has the ends cut at a 45-deg angle (Fig. 4-13b). There are several designs of step-cut—also called *lap-joint*—rings (Fig. 4-13c and d). However, experience shows that little is gained by this more complicated shape. Narrow piston rings in two-stroke engines are likely to catch the ring ends in the ports over which the rings slide because of the ring flexibility. To prevent this the ends are notched and a pin is installed in the piston groove to hold the ring ends always in line with a bridge between the ports.

4-4. Connecting Rods. Connecting rods used in diesel engines are similar to those used in automobiles (Fig. 4-14). They have an eye at the small end for the piston-pin bearing, a long shank, and a big-end opening which is split to take the crankpin bearing shells. The rods are forged of a high-strength alloy steel. Most connecting rods are rifle-drilled from the big end to the eye for oil flow from the crankshaft to the piston pin. Often the rod shank is H-shaped for maximum strength with minimum weight. Types of rods include (1) the normal

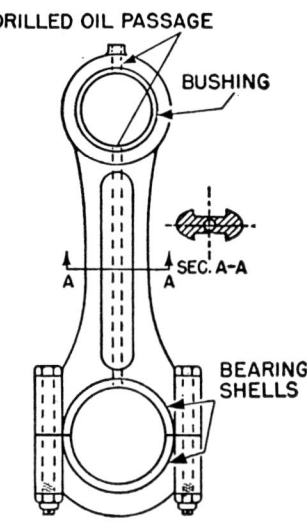

Fig. 4-14. Standard connecting rod.

shape (Fig. 4-14), used with only one cylinder to a crankpin or two cylinders offset so the rods can operate side by side; (2) fork-and-blade rods in V-type engines (Fig. 4-15), in which the big end of one rod has

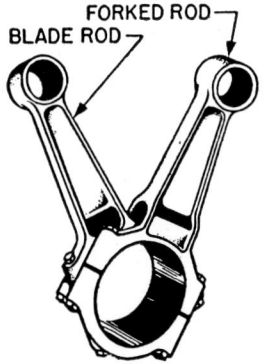

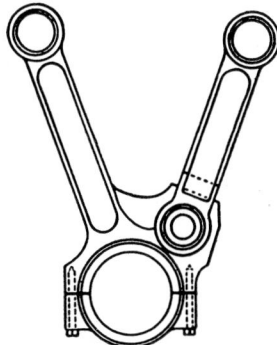

Fig. 4-15. Fork-and-blade connecting rod.

Fig. 4-16. Articulated connecting rod.

the normal shape while the rod of the piston in the opposite bank is widened and split into a fork shape straddling the first rod; (3) articulated connecting rods of V-type engines (Fig. 4-16), in which one rod, the *master rod*, resembles the conventional connecting rod but has a projection off the shank with an eye to which the rod for the piston in the opposite bank is attached. This second rod is called an *articulated* or *link rod*.

Connecting rods of medium-sized and large diesel engines are made with adjustable bearings (Fig. 4-17). Connecting rods of double-

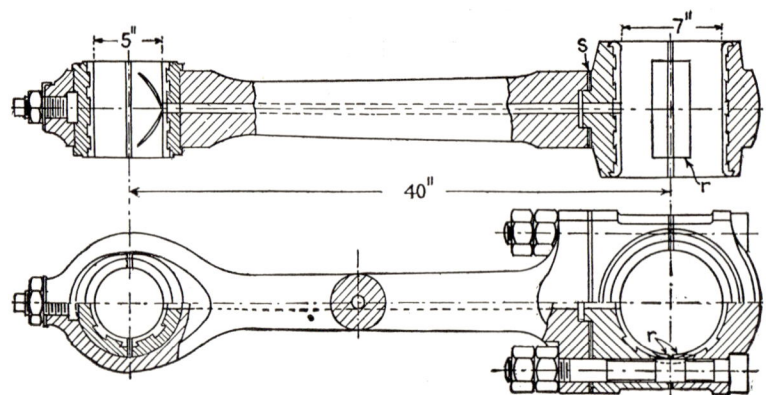

Fig. 4-17. Connecting rod with adjustable bearings.

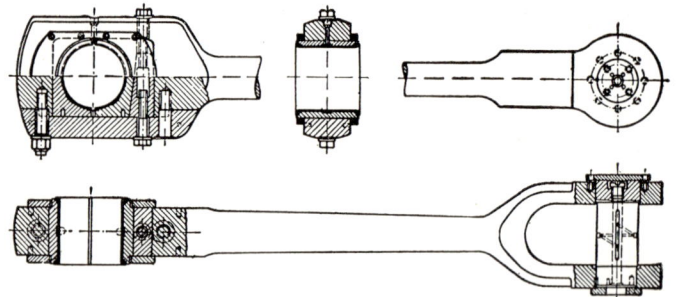

Fig. 4-18. Connecting rod for a large horizontal engine.

acting engines have a forked small end with the wrist pin fastened in it and swinging in the bearing fastened to the crosshead (Fig. 4-18). Another type of a forked small end for a crosshead engine is shown in Fig. 4-19.

Angularity. At dead centers the center line of the connecting rod coincides with the center line of the cylinder. As the crank turns away from dead center, the big end of the connecting rod moves with it while the small end moves along the cylinder center line, the connecting-rod center line thus beginning to form an angle with the cylinder center line. This angle reaches a maximum when the connecting rod and crank are at right angles to each other; after that the angle begins to decrease. The shorter the length of the connecting rod l in respect to the crank radius r, the smaller the ratio l/r and the greater this maximum angle. This swinging of the connecting rod is called *angularity*. If the connecting rod were infinitely long, it would remain parallel to the cylinder center line at all positions of the crank and there would be no angularity.

The angularity affects the piston motion, making it asymmetrical in respect to the two dead centers—faster near the outer, or top, dead center and slower near the inner, or bottom, one. This affects the timing of the engine. The events near the outer dead center, such as the opening of the intake valve or the closing of the exhaust valve, require a smaller crank-angle motion for a given piston travel than the same events near the inner dead center. The angularity also produces a side pressure of the trunk piston against the cylinder walls or of the crosshead against the guide surface. This pressure and the wear caused by it decrease with a decrease of the angularity, or with an increase of the ratio l/r. However, an increase of l/r increases the over-all height of the engine. Common values of l/r are between 4 and 5.

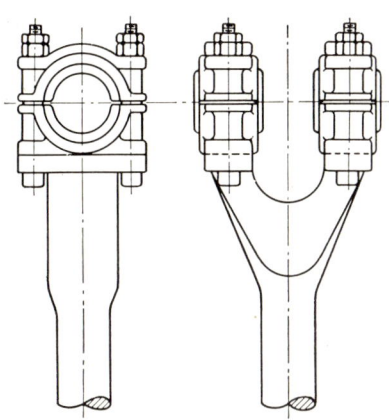

FIG. 4-19. Forked crosshead end of connecting rod.

Crankpin Bearings. Except in very small engines, the big end of the connecting rod has removable precision bearing shells, usually of the same type as those in the main bearings of the particular engine. These bearing shells consist of relatively thin steel, bronze, or brass shells with a lining of bearing metal, which is generally $\frac{1}{32}$ in. or less in thickness. The bearing metal may be one of several kinds which have proved satisfactory: tin- or lead-base babbitt, copper-lead, cadmium-silver, etc. Grooving is kept to the minimum, and wedge-type lubrication is used to the fullest extent, as explained in Chap. 6.

Crankpin-bearing Bolts. The crankpin-bearing cap is fastened to the connecting rod by two or four bolts. In four-stroke engines these bolts are designed to take the maximum inertia load with a small stress based on the inertia and centrifugal loads and referred to the area of the bolts at the bottom of the threads. It is difficult to avoid failure of these bolts in trunk-piston engines when piston seizure occurs.

When the engine is running, the bolts are not stressed equally, owing to journal friction; the pull in bolt 1 (Fig. 4-20) is greater than in bolt 2. With crosshead engines the small-end bolts have to carry the inertia load due to the piston, piston rod, and crosshead and also friction forces acting on the piston and crosshead. As these forces are rather indeterminate, the usual practice is to allow about 30 per cent less stress on these bolts than on the big-end bolts.

Connecting-rod cap bolts of two-stroke engines are made of the same size as those for four-stroke engines, since there is always a possibility of loss of compression resulting in heavy inertia loads.

Crankpin-bearing bolts may be subject to considerable repeated stresses in the case of torsional vibration of the crankshaft, particularly if the crankpin bearing is separate from the connecting rod, as in

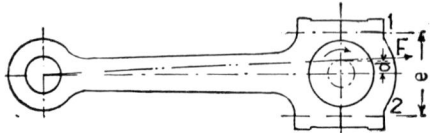

Fig. 4-20. Friction action on big-end bolts.

Fig. 4-17. Spigoting the bearing into the foot of the rod reduces the stress but may not be sufficient to avoid failure. For this reason some engine manufacturers advise changing the bolts at regular intervals, once every 12 to 18 months, as the cost of bolts is slight compared with the consequences of bolt failure while the engine is running.

A better practice is to inspect the bolts periodically after a certain number of hours of engine operation, say, 3,000 to 5,000, by taking them out and painting them with chalk dissolved in alcohol. If a crack has developed, the oil from the crack will turn the chalk over the crack yellow. Such a crack is the beginning of a progressive fracture, and the bolt must be exchanged. Magnaflux inspection is also very good; however, the necessary equipment is available only in larger industrial places.

Piston Rods. In engines with a crosshead, the piston rod connects the barrel piston to the crosshead and is a hollow steel rod. Its motion is straight reciprocation. Within the rod are contained the passages for the oil or water for cooling the piston. If the rod is used with a double-acting engine, the diameter of the rod end that is connected to the crosshead must be not larger than the diameter of the main rod length so that the end can be inserted through the stuffing box. The rod is generally assembled together with the piston, often with the piston parts clamping the end of the rod. It is installed through the stuffing box in the lower head and fastened to the crosshead, usually by a screwed connection (Fig. 4-21).

4-5. Crosshead. The crosshead transmits the load from the piston through the piston rod to the connecting rod. Figure 4-21 shows a crosshead for a double-acting direct-reversible diesel engine; the wrist pin requires a marine-type forked big-end connecting rod, as shown in

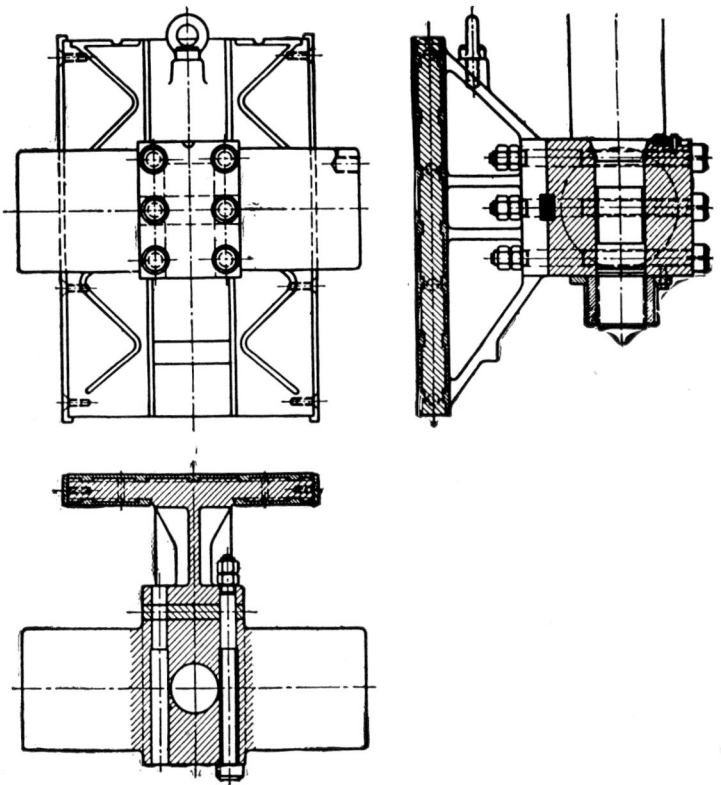

FIG. 4-21. Crosshead of a direct-reversible marine diesel engine, scale 1:14.

Fig. 4-19. The design now used in naval engines mostly includes a ball-and-socket joint with the piston-rod end, to take care of any unavoidable slight misalignment, and a bearing for the wrist pin fastened in the forked end of the connecting rod (Fig. 4-22).

The crosshead, long oil-cooled piston, and long piston rod considerably increase the weight of the reciprocating masses of double-acting engines. At the speed of 720 rpm used in naval engines of this type the loads on all bearings—main, crankpin, and wrist-pin—become rather large owing to the forces of inertia and require good and abundant lubrication of these bearings.

4-6. Crankshafts. The forces acting on a diesel-engine crankshaft are rather high because of the high peak gas pressures and the inertia forces of the moving parts. The main requirements of a crankshaft are mechanical strength and rigidity, both lengthwise and crosswise. An important factor is the number of bearings supporting the shaft. The greater the number of bearings for a given number of cranks, the more rigid the crankshaft. Crankshafts of diesel engines generally have one bearing between each pair of cranks and one bearing on each shaft end.

In V-type engines each crankpin serves two cylinders, and in radial engines one crankpin serves all cylinders—three, five, seven, or nine.

Firing Order. To secure uniform rotation in multicylinder engines, the power impulses should be equally spaced. This consideration, together with balance requirements, determines the conventional arrangements of crankshaft throws and firing orders, as shown in Fig. 4-23. The cylinders are numbered starting either from the flywheel end or, as in automobile engines and some large oil engines, from the forward end, counting toward the flywheel. Seen from the flywheel, most engines rotate counterclockwise.

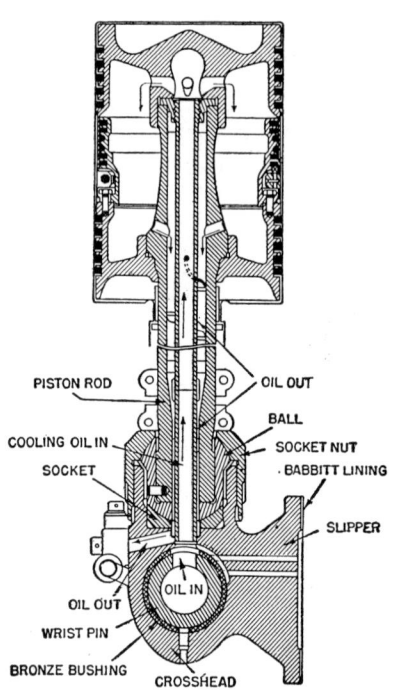

Fig. 4-22. Oil-cooled piston, piston rod, and crosshead of a double-acting diesel engine.

Materials. Crankshafts of stationary engines are usually made of open-hearth steel. The crankshafts of marine oil engines are made of steel with a smaller carbon content. Automotive crankshafts are usually made of chrome-nickel steels which, after heat treating, have a high tensile strength and elastic limit. Crankshafts commonly are forged either from open-hearth or high-strength alloy steels. However, advances in the understanding of the stress distribution are changing the situation. It has been found that the removal of metal at certain sections will redistribute the stresses and result in a stronger shaft. Casting permits easier application of such improved shapes and as a result high-strength cast-iron alloys are used in several modern diesel engines.

The use of cast-iron alloys for crankshafts is steadily increasing, particularly for large engines. Casting of the crankshaft permits a theoretically desirable but complicated shape with a minimum of machining and at the smallest cost. In addition, cast iron has the advantage of a considerably smaller sensitivity to notch effect than forged steel. Figure 4-24 shows such a crankshaft of a small diesel engine. Many forged-steel shafts are surface-hardened by electric

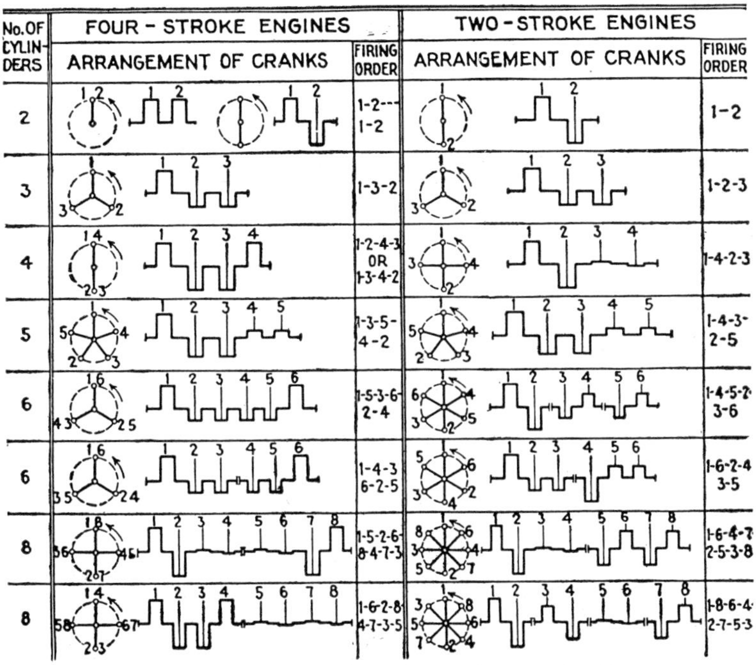

No. OF CYLIN-DERS	FOUR - STROKE ENGINES		TWO - STROKE ENGINES	
	ARRANGEMENT OF CRANKS	FIRING ORDER	ARRANGEMENT OF CRANKS	FIRING ORDER
2		1-2-- 1-2		1-2
3		1-3-2		1-2-3
4		1-2-4-3 OR 1-3-4-2		1-4-2-3
5		1-3-5-4-2		1-4-3-2-5
6		1-5-3-6-2-4		1-4-5-2-3-6
6		1-4-3-6-2-5		1-6-2-4-3-5
8		1-5-2-6-8-4-7-3		1-6-4-7-2-5-3-8
8		1-6-2-8-4-7-3-5		1-8-6-4-2-7-5-3

Fig. 4-23. Arrangements of cranks and firing-order diagram.

heating of the surface only, after which it is quenched with sprays of water.

Lubrication. The shafts are drilled through the crank webs to admit pressure lubrication from the main bearings to the crankpins and wrist pins. In some engines the oil is carried further and is used for cooling the piston crowns.

Manufacturing. The crankshaft is one of the most important parts of an engine and great care is taken in its manufacture. Small crankshafts are drop-forged, larger shafts are forged and machined to shape. Cast-iron crankshafts are cast in permanent molds which ensure a maximum accuracy and a minimum of machining. When a crankshaft is machined, special precautions are taken to avoid springing; the shaft must be well supported between the centers. After turning, the journals and crankpins are ground to exact size. Finally the crankshafts are balanced as explained in Sec. 18-4: large shafts of low-speed engines are balanced statically; crankshafts of high-speed engines are balanced dynamically on special balancing machines.

Grinding. Most crankshafts are ground at the journals and crankpins. In some cases grinding is followed by hand lapping with emery cloth. Some engine builders claim that a good lathe-turned journal is

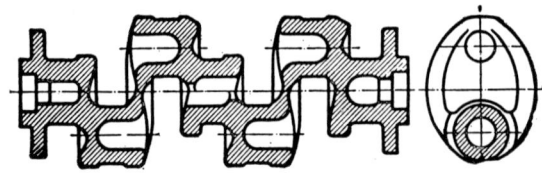

FIG. 4-24. Cast-iron crankshaft of a small high-speed diesel engine. *Courtesy of Sulzer Bros., Ltd.*

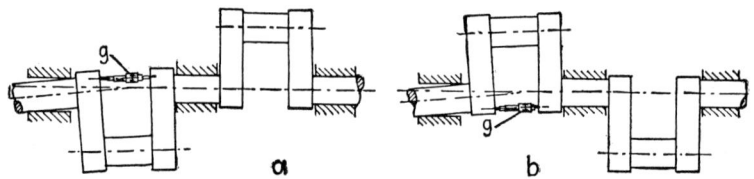

FIG. 4-25. Checking alignment with a micrometer.

better, on the basis that grinding fills the pores of the journal with fine metallic particles and thus prevents them from acting as minute oil reservoirs.

Endurance Failure. It is clear from the appearance of the crack or complete fracture that most crankshaft failures are due to so-called *progressive fracture* either from repeated bending or from reversed torsional stresses. Stresses repeated many times result in the final failure of the stressed part, even though the stresses do not reach the elastic limit of the material. The maximum stress that can be applied indefinitely without causing failure is called the *endurance limit*. For open-hearth steel used for crankshafts the endurance limit for bending is about 32,000 psi. For chrome-nickel and other alloy steels used for automotive crankshafts the endurance limit is about 75,000 psi.

Repeated bending stresses exceeding the endurance limit may be produced in a crankshaft from one or more bearings being lower or higher than the rest. The easiest way to detect such a bearing is by using a micrometer gage g and checking the distance between the cheeks in several positions of the crank, as shown in exaggerated form in Fig. 4-25. The procedure to be followed is explained in Sec. 25-10. Nitriding of the journals increases the endurance strength up to 50 per cent, whereas chrome plating and surface hardening, particularly by high-frequency induction heating and quenching, reduces the endurance strength up to 40 per cent.

Repeated dangerous torsional stresses may occur in multithrow crankshafts from torsional vibration at critical speeds. In single-throw and two-throw crankshafts these stresses may occur from counterweights cast as part of the rim of the flywheel instead of being fastened to the crank cheeks. In a multithrow crankshaft, with a heavy flywheel on one end and the power take-off on the other end, the torsional stresses may exceed the endurance limit at other than the critical speeds.

Torsional vibrations are very dangerous because they act as sudden loads that are repeated with a high frequency, several hundred times per minute or even more. The dangerous feature of a sudden load is that the stress created by it is twice that produced by the same load acting steadily or applied gradually.

Heat Expansion. The expansion of crankshafts both from the heat radiated by the pistons and from heat of friction in the bearings must be taken care of by allowing sufficient clearances between the crank cheeks and bearing shells. This elongation may be considerable.

In order to prevent excessive lengthwise play, the crankshaft is usually fixed in one of the middle bearings. The distance between the cheeks of this journal fits the bearing shells, and all the other bearings have the necessary axial clearance.

Crankshaft Dimensions. While dimensions of crankshafts are chosen by the engine designer, it is always well to make sure that the shaft of an engine to be purchased meets standard requirements.

Lloyd's rules, observed in marine engines and in some stationary diesel engines, require that the crankshaft diameter be determined by the formula

$$d = \sqrt[3]{D^2(AS + BL)} \tag{4-1}$$

where d = shaft diameter, in.
D = cylinder bore, in.
L = distance between bearing centers, in.
S = piston stroke, in.
A = constant (see Table 4-1)
B = constant (see Table 4-1)

Table 4-1. Constants in Lloyd's Formula (4-1)

Number of cylinders		With compressor driven by overhung crank		Without compressor	
Two-stroke	Four-stroke	A	B	A	B
1	1, 2, 3	0.086	0.038	0.089	0.037
2	4	0.093	0.037	0.099	0.036
3	5, 6	0.103	0.035	0.111	0.035
4	8	0.120	0.034	0.131	0.033

Some designers proportion crank webs according to the formula

$$a = \sqrt{\frac{3Fl}{4bs}} \qquad (4\text{-}2)$$

where a = thickness of the web in the direction of the crankpin and shaft length, in.
F = load on the crankpin, lb
l = one-half the distance between web centers, in.
b = breadth or width of the web, in.
s = allowable fiber stress, 7,500 psi for forged steel

Some British designers recommend that, using the cylinder bore as unity, the crankshaft dimensions be taken as follows:

Diameter of shaft and crankpin.................... 0.525–0.54
Length of main journals........................... 0.75 –0.80
Length of crankpin................................ 0.525–0.54
Thickness of web.................................. 0.32 minim.
Width of web...................................... 0.80 –0.92

4-7. Counterweights. Where an engine has only one or two cylinders it is inherently unbalanced. The resulting vibration of the frame can be reduced by adding counterweights of proper mass and shape.

Engines with three and more cylinders are balanced at least in regard to the most important unbalance forces, the so-called *first harmonics.*

Adding counterweights to crankshafts of multicylinder engines takes care of unbalanced couples and has for its chief object a more uniform loading and wear of the main bearings.

Construction Details. In small engines and some special constructions, as in Fig. 4-26, the counterweights are made integral with the shaft. In low- and medium-speed engines, cast-iron counterweights are fastened to the shaft by studs (Fig. 4-27). To prevent movement of the counterweights on the shaft they are machined to fit over the cheeks, but the recess is made slightly smaller than the corresponding projection on the crankshaft. The counterweights are heated before assembling, to provide, after they are in place, a light shrink fit.

It is very important that the counterweights be fastened to the cranks absolutely solidly. When the engine is running, the counterweights are being pulled away from the cranks by the centrifugal force

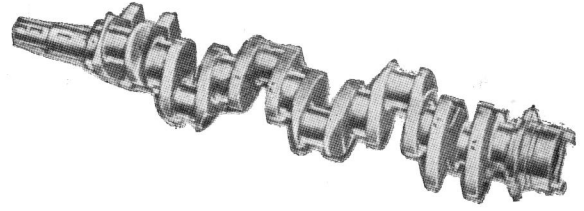

Fig. 4-26. Crankshaft of a six-cylinder in-line diesel engine.

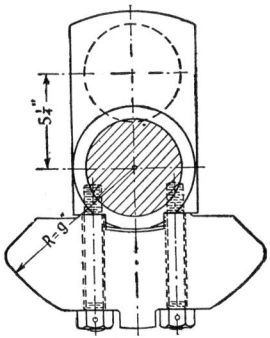

Fig. 4-27. Counterweight of a medium-size diesel engine. (*Courtesy of Cooper-Bessemer Corporation*).

and at the same time, under the action of the force of gravity, have a tendency to shift back and forth when moving from one side of the shaft to the other side. This tendency is facilitated by the fact that the centrifugal force decreases the pressure between the counterweight and the surface of the crank upon which the former rests. If there is any clearance between the registering edges and the crank (Fig. 4-27), these variable forces may start to move the counterweights, at first very little, but gradually they will increase the clearance by pounding until there will develop a danger of breaking the hold-down studs, or unscrewing the nuts.

Another method of fastening counterweights, suitable for high-speed engines, is shown in Fig. 4-28. A combination of both methods is used in Winton diesel engines (Fig. 4-29).

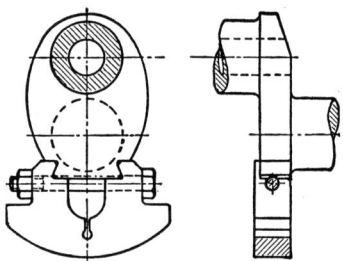
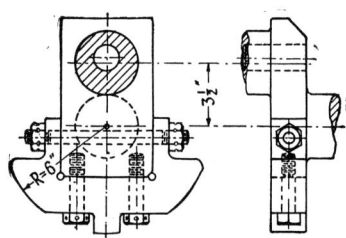

Fig. 4-28. Counterweight for a high-speed engine.

Fig. 4-29. Counterweight for a high-speed diesel engine.

QUESTIONS

1. What engine parts are included in the classification of running gear?
2. What is the difference between a barrel piston and a trunk piston?
3. What material is used for pistons of slow, heavy-duty engines? Of higher-speed engines?
4. How are diesel-engine pistons prevented from overheating?
5. What are the advantages and drawbacks of aluminum pistons?
6. What are the functions of a trunk piston?
7. Enumerate the materials used in making trunk pistons, in the order of their importance.
8. What procedure is used to reduce cylinder-liner wear by cast-iron pistons?
9. What is the basis for determining the length of a trunk piston?
10. How are pistons finished on the outside surface?
11. What is the necessary pressure of the piston rings in low-speed engines? In high-speed engines?
12. What is the maximum permissible temperature of the oil film between the piston and the cylinder liner?
13. What clearance between piston skirt and liner is recommended for small-bore engines, up to about $4\frac{1}{2}$ in.?
14. What engine part is called a *wrist pin?*
15. What is the difference, if any, between a *wrist pin* and a *piston pin?*
16. In what three types are wrist pins grouped in accordance with how they are linked with the piston and connecting rod?
17. Explain why the working conditions of piston-pin bearings are more severe than those of the main bearings.
18. What is the purpose of bronze or aluminum plugs at the ends of hollow wrist pins?
19. Show a sketch of a piston construction that permits reducing the pressure on the piston pin.
20. What is the purpose of relieving the piston barrel in the region of the wrist-pin bosses when using a fixed pin?
21. How is the piston pin lubricated?
22. What are the functions of piston rings?

RUNNING GEAR

23. What is the purpose of oil-scraper rings?

24. How many oil-scraper rings are used in a piston and where are they usually located?

25. Of what material are compression rings made?

26. What forces press the piston rings against the cylinder wall?

27. What is the object of beveling the bottom or both top and bottom of piston rings?

28. What measures may be taken to prevent oil pumping?

29. What side clearance is recommended for piston rings?

30. What should be the end clearance in piston rings of small-bore engines? Of large-bore engines?

31. What prevents piston-ring ends of two-stroke engines with wide ports from being caught in the ports?

32. Give a sketch of a conventional diesel connecting rod and label the parts of the rod.

33. Enumerate and illustrate by sketches the three types of connecting rod used in diesel engines.

34. What is a master rod? What is an articulated, or link, rod? In what engines are these rods used?

35. What is angularity in the motion of the connecting rod?

36. For an engine with a given stroke, when will the angularity be greater—with a short or a long connecting rod?

37. What is the effect of the angularity of the connecting rod upon the timing of an engine?

38. What is the effect of the angularity of the connecting rod upon the side pressure of the trunk piston against the cylinder wall?

39. Give a sketch showing the big end of a connecting rod and indicate what kind of bearing shells are commonly used at present.

40. Why are the bolts holding the bearing cup in the large end of the connecting rod not stressed equally? On what side are they stressed more?

41. What may subject crankpin-bearing bolts to considerable repeated stresses and cause their failure?

42. How is the danger of failure of crankpin-bearing bolts reduced and how can it be practically eliminated?

43. In what engines are piston rods used?

44. Why are piston rods made hollow?

45. What is a crosshead? Draw a sketch with sufficient views to give a clear picture of the crosshead shape.

46. What type of connection between the piston rod and crosshead is used in naval engines? Explain its advantage and give a sketch.

47. What are the main requirements of a diesel crankshaft?

48. Of what influence is the number of bearings supporting a shaft on the shaft characteristic?

49. How many bearings are there in (a) a six-throw crankshaft of a heavy-duty diesel engine, (b) a four-throw crankshaft of an automotive diesel engine?

50. What determines the arrangements of crankshaft throws in diesel engines?

51. Draw diagrammatical sketches of conventional crankshafts indicating the firing order for (a) a four-cylinder in-line two-stroke diesel engine, (b) a five-cylinder four-stroke engine, (c) a six-cylinder four-stroke engine.

DIESEL-ENGINE COMPONENTS

52. What are the materials and methods used in manufacturing diesel-engine crankshafts?

53. What are the advantages of making a crankshaft from cast-iron alloys?

54. How does lubrication of main journals, crankpins, and wrist pins affect the machining of crankshafts?

55. What is the sequence of main operations in manufacturing a forged-steel diesel-engine crankshaft?

56. What are the two most usual causes of failure of crankshafts?

57. What is the endurance limit of a material?

58. What is the elastic limit of a material?

59. What may cause heavy bending stresses in crankshafts?

60. What is the influence of nitriding, chrome plating, and surface hardening upon the endurance strength of forged-steel crankshafts?

61. What may cause dangerous torsional stresses in a diesel crankshaft?

62. Explain why torsional vibrations are particularly dangerous.

63. What causes diesel crankshafts to expand in operation?

64. What precautions are taken to take care of the unavoidable expansion of diesel crankshafts?

65. What is a counterweight in a crankshaft?

66. What is the main purpose of installing counterweights on crankshafts of multicylinder engines that do not need balancing?

67. Show by sketches different types of counterweights and how they are fastened to the crankshaft.

CHAPTER 5

BEARINGS

5-1. Definitions. The term *bearing* is applied to machine parts which transmit forces or loads from moving to stationary parts and thus support the moving parts. The surfaces in contact under pressure are called *bearing surfaces*.

Bearings are important parts in an oil engine and other machinery connected to it. Proper understanding of the functioning of various bearings is important to an engine operator who wants to maintain their satisfactory performance and prevent their failure and the resulting stoppage of the engine.

5-2. Bearing Classification. All bearings may be divided into two main groups, depending upon the type of motion of parts that they support:

1. *Bearings for rotary motion* which, in turn, may be subdivided into (a) journal bearings in which the main load is normal to the axis of rotation and (b) thrust bearings in which the load acts along the axis of rotation.

2. *Bearings for reciprocating motion* which, in turn, may be subdivided into (a) bearings for linear motion, such as a cylinder or cylinder liner supporting the moving piston, and (b) bearings for rocking or oscillating motion, such as piston-pin bushings supporting the piston pin.

In respect to their construction bearings may be divided into bearings (a) with a sliding contact, or plain bearings, and (b) with a rolling contact in which either steel balls or steel rollers are interposed between the working surfaces.

In respect to the type of loading, journal bearings may be divided into two groups: (a) bearings with a steady load, such as bearings of electric generators and motors, centrifugal pumps, or various blowers and fans; and (b) bearings with a variable or fluctuating load, such as main, crankpin, wrist-pin, or camshaft bearings in diesel engines.

Bearings with a steady load are often called *power bearings* in order to differentiate them from *engine bearings* which operate under fluctuating loads. While both bearing groups do not differ in respect to their construction, the operating conditions of the engine bearings are much more severe than those of the power bearings.

5-3. Bearing Loads. Engine-bearing loads are a combination of two kinds of force—of gas pressures acting upon the crown of the engine piston and forces of inertia created by reciprocating and rotating masses acting upon the main journals of the crankshaft. The variation of the gas pressure may be obtained from the so-called *indicator diagram*, as will be explained in Chap. 21. For low- and medium-speed engines the indicator diagram can be taken from the engine cylinder itself by means of an indicator; for high-speed engines a theoretical indicator diagram must be used, constructed from such data as compression ratio, exponents of the compression and expansion lines, maximum pressure which can be measured even in high-speed engines with a sufficient accuracy, and brake load.

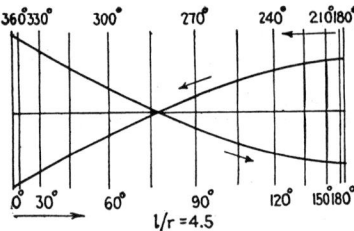

Fig. 5-1. Inertia curves of reciprocating masses of a diesel engine.

The forces of inertia of the reciprocating masses change as the piston moves, and their direction is reversed each time the piston passes dead center. The curve, which represents the force of inertia for any position of the piston, is constructed from such data as weight of the reciprocating parts, ratio of the connecting rod to the crank radius, and engine speed. Figure 5-1 shows a sample of an inertia diagram for a diesel engine.

The force of inertia of the rotating masses is represented by the so-called *centrifugal force*. This force does not change while the engine speed remains constant, but its direction changes all the time as the crank rotates.

All forces of inertia change as the square of the engine speed. As the inertia of reciprocating parts changes also with the piston speed, or position, as shown on Fig. 5-1, inertia forces are added to the gas-pressure load at some points of the stroke and subtracted at others. As a result, with a multicylinder engine there may be engine speeds at which the combined bearing loads, taking into account both reciprocating and centrifugal forces of inertia, will reach maximum and may decrease at both a lower and a higher engine speed.

Bearing Pressure. A factor by which bearing loads are measured is the bearing pressure p, determined as the bearing load F, pounds,

BEARINGS

divided by the product of diameter d and length l, inches, of the bearing:

$$p = \frac{F}{dl} \tag{5-1}$$

Bearing pressure p, as any other pressure, is expressed in psi. However, the actual pressure between a journal and its bearing with proper lubrication varies around the contact surfaces. Thus the value p computed by Eq. (5-1) is only an average figure and must be regarded simply as a useful characteristic of a bearing.

5-4. Operating Conditions. Under operating conditions of bearings one understands bearing pressure, rotative speed, temperature and viscosity of the lubricating oil, and the degree of contamination of the oil.

In respect to operating conditions the two bearing groups, power bearings and engine bearings, are quite different.

Power bearings as a rule operate under rather favorable conditions:

1. Since the loads are kept relatively moderate, the bearing pressures remain rather low, not over 150 to 200 psi.

2. The maximum lubricating-oil temperature seldom exceeds 160 F.

3. It is easy to prevent the oil from being contaminated by dust or other sediments from the atmosphere.

4. It is possible to use an oil with a viscosity that suits the load and speed conditions.

Contrary to this, diesel-engine bearings operate under much less favorable conditions:

1. The load fluctuates—the maximum bearing pressures may reach 3,500 psi and in some instances 5,000 psi; mean bearing pressures seldom are below 600 psi and often exceed 850 psi.

2. The lubricating-oil temperature is high: in large high-output engines it is seldom below 180 F, in smaller high-speed engines it runs as high as 200 to 220 F and in some instances 240 F.

3. The oil is being gradually contaminated with products of oxidation, sludge formed by water condensed from the products of combustion, carbon formed by incomplete combustion of lubricating oil and fuel, and worn-off metal particles; in addition, it may be diluted by fuel oil leaking from the injection system.

4. The same oil is used for lubrication of all bearings, gears, cams, and pistons and at best is only a compromise in respect to viscosity and other characteristics.

5-5. Bearing Construction. Journal bearings with sliding contact are often called *plain bearings*. Generally they are made in two halves, bored so that when they are brought together around the journal, they form a true circle. In some cases, the bearing may be made in the shape of a bushing pressed into a pedestal. To replace worn bearing surfaces, bearings ordinarily are equipped with shells which can easily be exchanged. Large power bearings and main bearings of low-speed engines have cast-iron or cast-steel shells with babbitt lining and *shims* of different thickness inserted between the two halves (Fig. 5-2). The shims serve to adjust the clearance between the journal and bearing surfaces, which has a tendency to increase gradually as a result of bearing wear. Figure 5-3 shows such a bearing construction used

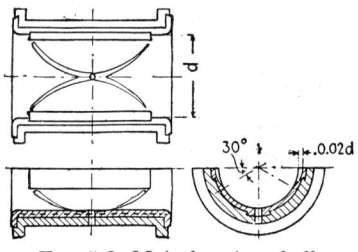

FIG. 5-2. Main-bearing shell.

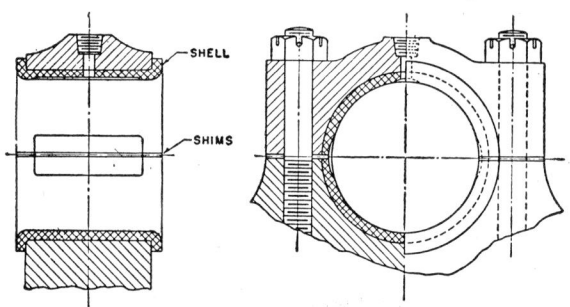

FIG. 5-3. Plain journal bearing.

as a main engine bearing or an outboard bearing for a generator. A similar bearing construction with babbitted shells and shims is used for both the small and large ends of connecting rods of medium-sized and large diesel engines, as shown in Fig. 4-17. High-speed diesel engines employ *precision bearings*, for both main bearings and connecting rods. The main difference between ordinary bearing shells and a pair of precision bearing shells is that the latter are machined to very close tolerances

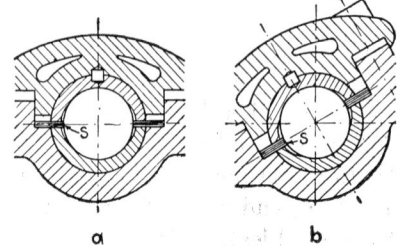

FIG. 5-4. Arrangement of bearing shells and caps.

and are interchangeable without hand scraping of the bearing surfaces. Precision bearing shells, being better supported due to more accurate workmanship, are made much thinner with a very thin babbitt lining.

BEARINGS

Main Bearings. Horizontal Engines. In medium-sized low-speed engines, round shells with horizontal (Fig. 5-4a) or inclined (Fig. 5-4b) caps are used. The wear can be taken up by means of shims.

Ring lubrication for main bearings, used extensively some years ago, is now used only for small- and medium-sized horizontal engines and

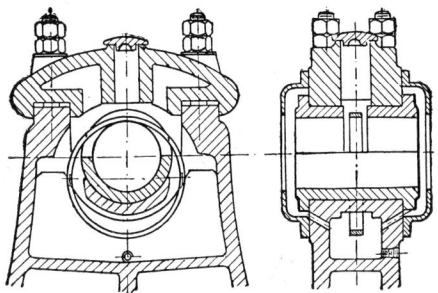

Fig. 5-5. Oil-ring bearing.

for crankcase-scavenging two-stroke engines, which do not lend themselves as well to pressure lubrication. An oil-ring outboard bearing for a low-speed engine is shown in Fig. 5-5.

Vertical Engines. Most vertical oil engines, regardless of size, are equipped with pressure-feed lubrication. The oil is admitted either

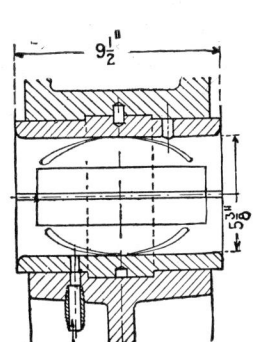

Fig. 5-6. Pressure lubricated main bearing.

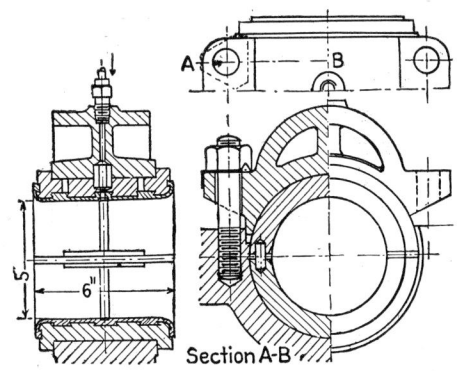

Fig. 5-7. Pressure-lubricated bearing of a marine diesel engine.

from underneath (Fig. 5-6) or through the cap (Fig. 5-7). Theoretically the latter construction is better, as the oil is delivered at the point of minimum pressure; the disadvantage is in the more complicated oil piping and in the necessity of breaking the pipe connection every time the bearing shell must be taken out.

In single-acting vertical engines the wear takes place at the bottom of the lower main-bearing shells. The wear is taken up by reducing the thickness of the shims between the shells.

Bearing Shells. In low- and medium-speed engines of older design the shells are made of brass or bronze in smaller engines, of cast-iron in medium-sized engines, and of steel in large engines. Cast-iron and steel shells are always lined with babbitt and brass and bronze shells are usually so.

The chief requirements for good bearing shells are (1) good fitting to both bedplate and journal; (2) proper quality of babbitt—a sufficient hardness to stand the working pressures and toughness to prevent cracking; and (3) good bond between the babbitt and the backing metal. This is obtained by tinning the shells before babbitting and inserting swallowtail keys if a heavy babbitt layer is used.

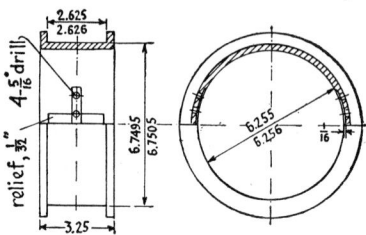

FIG. 5-8. Precision bearing shells for a diesel engine.

Round shells are ground from the outside to exact dimensions in order to insure good contact with the bed. A good fit helps considerably the scraping in of the shells, prevents cracking of babbitt, and provides better transfer of the friction heat.

In high-speed engines only precision-type bearings (Fig. 5-8) are used; these are made of a thin bronze or steel backing with a very thin white-metal or copper-lead lining. These shells are precision-machined and do not require fitting to the journals. When used in larger engines, some fitting of the bearing shells is necessary, but only to the bedplate. Precision bearings do not have any take-up; when a shell is worn, both halves are replaced by a new set.

5-6. Bearings for Reciprocating Motion. The two places where a reciprocating motion is encountered in single-acting engines are the piston pin and the piston itself. As will be explained in Chap. 12, a reversal of motion is a great handicap to proper lubrication.

Piston-pin bearings usually are made in the form of bronze bushings pressed into the piston bosses, sometimes with narrow, slightly helical oil grooves cut through the full length of the bushing to help the distribution of oil. Similar bushings are used in the small end of the connecting rod of many engines. These bushings may be either of the *pressed-in* or of the *full-floating* type. The maximum bearing pressures run very high, up to 6,000 psi, but such high pressures exist only during a very small fraction of a second and therefore are not able to squeeze out the protective oil film entirely. Also, the surface speed is very low. The mean bearing pressure is about one-sixth of the maximum pressure. Needle-roller bearings are used in some engines, but the latest trend is back to plain bronze bushings using improved bronze alloys.

Cylinders and Pistons. The side-thrust pressure exerted by the gas pressure due to the angularity of the connecting rod is rather small, the maximum value usually being under 100 psi. The lubrication of the piston and cylinder is more than required in all medium- and high-speed engines owing to the splash action of the cranks and other rotating parts. As mentioned before, special care must be taken to remove excess oil from the cylinder walls. Pistons of low-speed engines are lubricated by special positive-feed oilers.

5-7. Crosshead Guides. In double-acting engines the side thrust is taken by the crosshead slipper and its guides. The pressure is small, usually under 100 psi; the lubrication does not present any difficulties, and there is no danger of overlubrication as with trunk pistons of single-acting engines. Figure 4-22 shows the method of admitting oil under pressure to the oil grooves of the slipper and also the use of babbitt lining. The crosshead guide is made of cast iron.

5-8. Bearings with Rolling Contact. *Advantages.* Properly designed and precision-manufactured bearings with rolling contact have the following advantages over plain bearings with sliding contact: (1) accurate alignment is kept over long periods of time; (2) heavy momentary overloads do not cause failure; (3) power loss caused by friction is small; (4) since coefficient of friction is practically independent of speed, they are particularly suitable for variable-speed operation; (5) they have a very low starting resistance; and (6) their lubrication is simple and requires little attention.

Because the power loss is small, ball and roller bearings are often called *antifriction* bearings.

Disadvantages. Both ball bearings and roller bearings require precision-made outer and inner races which cannot be split into halves and therefore can be used only on straight shafts of constant diameter or of diameters that decrease toward the end of the shaft. Therefore they cannot be used on crankshafts with several cranks. However, the recent development of split ball bearings in this country promises to broaden the field of application of ball bearings, at least in diameters up to 4 in. Another disadvantage is in that their load capacity decreases with an increase of speed and, finally, that their cost, in large sizes, is higher than that of plain bearings.

Ball Bearings. Since extreme precision in manufacturing, which is so important for their satisfactory operation, is obtained more easily with ball bearings than with roller bearings, ball bearings are more widely used. Ball bearings support the load on a series of hardened steel balls and for simplicity it is said sometimes that ball bearings have point contacts. Actually, however, because of elastic deformation, they perform with very small contact areas. Ball bearings are built to standard sizes and are interchangeable, whether made by the same manufacturer or not. A ball bearing consists of four main elements: (1) an *inner* ring or *race* grooved on its outer surface, (2) an *outer race* grooved on its inner surface, (3) steel *balls*, and (4) a ball *retainer* or *cage* for spacing the balls so that they do not touch one another, thus reducing loss of power, wear, and noise.

Roller Bearings. Roller bearings are a development of ball bearings; they have cylindrical, conical, or barrel-shaped rollers instead of balls and many of them are interchangeable with standard ball bearings, *i.e.*, have the same outside dimensions. Their load capacity, for the same outside dimensions, is greater than that of ball bearings because of a larger contact area. However, they are more sensitive to misalignment. Bearings with conical rollers, often called *tapered rollers*, and correspondingly shaped races can be adjusted for axial clearance, and are usually known as Timken bearings. Roller bearings using relatively long rollers of small diameter, 2 to 4 in. or about $5/64$ to $5/32$ in., are called *needle bearings* and are used mostly as wrist-pin bearings in some diesel engines. Needle bearings do not use cages for spacing their rollers.

Friction. The total friction in ball and roller bearings is made up, in addition to rolling friction, of sliding friction caused by rubbing against retainers and in the bearing seals, of differences in speeds of other parts, and of manufacturing inaccuracies.

The apparent coefficient of friction of deep-groove ball bearings varies from 0.0014 to 0.0025; self-aligning ball bearings have a somewhat smaller coefficient of friction. The coefficient of friction of cylindrical roller bearings is also very low, that of tapered roller bearings slightly higher. The coefficient of friction of needle bearings is about three to four times higher because the needles are used without retainers and have a tendency to lose their parallelism.

QUESTIONS

1. What is a *bearing*?
2. Into what two main groups, depending upon the type of motion of parts they support, may all bearings be divided?
3. Into what two groups may rotary bearings be divided?
4. Into what two groups may reciprocating bearings be divided?
5. Into what two groups, in respect to their main construction feature, may all bearings be classified?

BEARINGS

6. How do journal bearings differ in respect to the type of loading?
7. What bearings are called *power bearings?*
8. What is the load characteristic of engine bearings?
9. What forces or loads are acting simultaneously on engine bearings?
10. What is the name given to the force of inertia of rotating masses?
11. What relation exists between the forces of inertia of reciprocating parts and the engine speed?
12. What is meant by the term *bearing pressure?*
13. What are the operating conditions of power bearings?
14. What are the operating conditions of diesel-engine bearings?
15. What is a plain bearing? Illustrate the description by a sketch.
16. What are precision bearing shells? Explain with sketches and state in what engines they are used.
17. What are the differences between an ordinary and a precision journal bearing?
18. How is a worn precision bearing taken up?
19. State what diesel engines are at present equipped with ring lubrication of main bearings.
20. Enumerate the chief essentials of good bearing shells.
21. When are so-called *swallow-tail keys* put in bearing shells and what is their purpose?
22. What is the object of grinding bearing shells from the outside?
23. What bearing shells are used in high-speed engines?
24. What are in a diesel engine the two places where reciprocating-motion bearings are used?
25. How high, usually, is the side-thrust pressure between piston and cylinder?
26. Enumerate the main advantages of bearings with rolling contact.
27. What types of bearings are designated as having *rolling contact?* How else are these bearings referred to?
28. What are the disadvantages of ball and roller bearings?
29. Enumerate the four component parts of a ball bearing; illustrate by a sketch.
30. What is the advantage of bearings with conical rollers?
31. What bearings are called *needle bearings?* Where are they used in diesel engines?
32. Of what magnitude is the apparent coefficient of friction of various bearings with rolling contact?

CHAPTER 6

FLYWHEELS

6-1. Flywheel Action. The purpose of a flywheel is to store up energy during the moments when the energy developed by the gases in the engine cylinder is greater than the engine load—namely, during the power stroke—and to return it to the crankshaft during the moments when the gases in the cylinders do not develop energy—namely, during the exhaust, suction, and compression strokes in four-stroke engines, or during the compression stroke in two-stroke engines. In general, when the speed of the crankshaft tends to increase, the flywheel absorbs energy; when it tends to decrease, the flywheel gives up energy to the crankshaft. Thus, the flywheel serves to keep the crankshaft rotating at a uniform speed.

Actually the flywheel serves several purposes:

1. It keeps the unavoidable variations in speed during each cycle within desired limits.

2. It limits the momentary rise or fall in engine speed during sudden changes of the load.

3. It carries the piston over the compression pressure when running at low or idling speed.

4. It helps to bring the engine up to speed when starting.

5. With alternators running in parallel, it keeps their angular advance or retardation, compared to a perfectly uniform angular speed, within desired limits.

Flywheel Effect. The amount of energy that a flywheel stores up depends upon the weight and diameter of the wheel and upon the engine speed. Thus a high-speed engine requires a lighter and smaller flywheel than a lower-speed engine of the same power. For a variable-speed engine the flywheel is designed to fulfill the minimum requirements at the lowest speed at which the engine may be operated. At higher speeds the flywheel will have greater energy than is actually required.

An important factor, called the *flywheel effect*, is expressed as WR^2,

FLYWHEELS

where W is the weight of the flywheel rim, lb, and R is the distance of the center of gravity of the rim section from the center of the flywheel, ft. The distance R is called the *radius of gyration* of the rim. Since the thickness of the rim of the flywheel is usually not very great compared with the wheel diameter, instead of the accurate radius of gyration it is permissible to take simply the outside radius of the flywheel. Such a simplification gives a slightly greater value of WR^2. On the other hand, the arms of the wheel also have a certain flywheel effect which is not taken into account, so that in the final count WR^2 with R taken as one-half the outside diameter D is close to an accurately computed value.

Flywheel-rim weight may be computed from the expression of its volume times the specific weight of the metal of the rim

$$W = a\pi(12D - h)w \tag{6-1}$$

where a = the cross area of the rim, sq in.
D = the outside wheel diameter, ft
h = the radial thickness of the rim, in.
w = the specific weight, 0.26 lb per cu in. for cast iron and 0.282 lb per cu in. for steel, cast or forged

6-2. Uniformity of Rotation. Since the energy developed by the combustion of the fuel and transmitted to the crankshaft acts intermittently but the useful work done by the crankshaft is more or less uniform, the instantaneous crankshaft speed will fluctuate between a highest value of n_1 rpm and a lowest value of n_2. The mean engine speed n may be considered as an average between these two values:

$$n = \frac{n_1 + n_2}{2} \tag{6-2}$$

The ratio of the speed fluctuation, the difference $n_1 - n_2$, over the mean speed n is called the *coefficient of uniformity* of rotation and designated as u

$$u = \frac{n_1 - n_2}{n} \tag{6-3}$$

The smaller the speed fluctuation $n_1 - n_2$ for a given n, the more uniform is the crankshaft rotation.

Coefficient of Steadiness. When investigating the action of a flywheel it is convenient to use the reciprocal of the coefficient of uniformity of rotation, called *coefficient of steadiness* and designated as m, namely,

$$m = \frac{1}{u} \tag{6-4}$$

or, with expression (6-3),

$$m = \frac{n}{n_1 - n_2} \tag{6-5}$$

Evidently, the smaller the speed fluctuation, the higher the uniformity of rotation of the engine.

For a given engine, whose flywheel effect WR^2, engine speed n, rpm, and horsepower N are known, the coefficient of steadiness may be calculated from the formula

$$m = \frac{WR^2(n/100)^2}{NC} \tag{6-6}$$

In this formula C is a coefficient which depends upon the engine type, the number of strokes per cycle, and the number of cylinders. Its approximate values are given in Table 6-1.

TABLE 6-1. COEFFICIENT C IN FLYWHEEL FORMULA (6-6)

Number of cylinders	Cylinder arrangement of single-acting engines	Two-stroke engines		Four-stroke engines	
		Angle between cranks	Coefficient C	Angle between cranks	Coefficient C
1	Vertical or horizontal	...	150	...	375
2	Vertical or horizontal	180	38	180	225
2	Vertical or horizontal	...	...	360	158
2	Opposed cylinders	180	128	180	158
3	In line	120	21	120	90
4	In line	90	13	180	26
6	In line	60	4	120	7.5
8	In line	45	1	90	6

Formula (6-6) and Table 6-1 show clearly that all other things being equal, a two-stroke engine has a higher uniformity of rotation than a four-stroke engine and the uniformity of rotation increases with an increase in the number of cylinders in an engine.

The influence of speed n is very great as m is proportional to the square of n. Therefore, when the speed of an engine is reduced, m drops very rapidly. Engines operating with variable speeds, such as automotive engines and boat and ship engines connected to a propeller, have a very low uniformity of rotation at idling speed and a very high uniformity at full speed.

Actually, the value of m depends not only on the weight W of the flywheel rim but also on the weight, or rather mass of all other rotating

FLYWHEELS

parts such as the cheeks of the crankshaft, large end of the connecting rods, and the rotating masses of the driven machines, generator rotors, propellers, etc. The rotating masses of a multicylinder engine may be so great that no flywheel is needed at all. Some diesel-electric units have no separate flywheel and the generator rotor is given the proper weight and acts as a flywheel. However, generally diesel engines have at least a light flywheel to turn the engine over and to carry valve- and injection-timing marks.

The coefficient of steadiness required for the satisfactory operation of a diesel engine driving various machinery depends upon the type of

Table 6-2. Minimum Values for Coefficient of Steadiness m

Machinery driven	Type of drive	m
Concrete mixers, crushers, hammers	Belt	5
Compressors, excavators	Belt	10
Reciprocating pumps, shears	Belt or flexible coupling	20
Metalworking and woodworking machines	Belt	30
Flour, paper, and textile mills	Belt	40
Compressors, pumps, and similar machines	Gear transmission	50
Spinning machinery—coarse to fine	Belt	50–65
D-C generator, single	Belt—direct-coupled	35–50
D-C generators, parallel	Belt—direct-coupled	50–70
A-C generator, single	Belt—direct-coupled	60–90
A-C generators, parallel	Direct-coupled	100
Boat propeller, up to about 100 hp	Direct-coupled—by gears	20–30
Boat propeller, 100–1,000 hp	Direct-coupled—by gears	30–50
Automobiles, engine idling	Clutch disengaged	5
Automobiles, normal speed	Gears	100

drive and the type of machinery driven. With a flexible drive, such as a belt or a flexible coupling, the cyclic fluctuations of the engine speed are to a certain extent absorbed by the flexibility of the drive, or partially damped out, and a lower coefficient of steadiness and hence a lighter flywheel are permissible. With a rigid drive, such as by gears or a rigid coupling, all the engine-speed fluctuations are transmitted to the driven machinery and a higher m and hence a heavier flywheel is required. Naturally, if the coefficient of steadiness happens to be higher than necessary, there is no harm in it, except that it means an excessively heavy flywheel and therefore a slightly greater friction loss in the main bearings.

Table 6-2 gives the minimum values of the coefficient of steadiness as established in actual practice for different engine applications. This means that for a given engine, knowing W, R, n, and N and having picked out the proper coefficient C from Table 1, one may compute the value of m from Eq. (6-6) and compare it with that recommended in Table 6-2. If the computed m is equal to or greater than given for the particular conditions in Table 6-2, the flywheel is adequate. If the computed m is found to be lower than the corresponding value given in Table 6-2, the flywheel is too light or too small and there is a danger of unsatisfactory operation of the engine-driven machinery.

6-3. Flywheel Rim Markings. The main markings indicate top dead center for all pistons, the marks being stamped accordingly, such as, 1 d.c., 2 d.c., etc. When the flywheel is turned so that one of these marks lines up with a registering mark or pointer on the engine crankcase, that piston is on top dead center. Most modern diesel engines have an integral forged camshaft and therefore no valve-timing marks are required on the flywheel, except possibly for the opening and closing of one valve, usually the intake valve of cylinder 1. If the timing of this valve is found to be correct, all other valves will also be correctly timed.

Establishing Dead Center. If the flywheel is not marked, the top dead center for each cylinder may be established by the following procedure: first the inlet valve of the cylinder whose dead center is to be found is removed; then the flywheel is turned until that particular piston is about one-fourth of its travel from top dead center and the distance from the piston top to the valve-port edge in the cylinder head is measured; for example, it is found to be 4 in. A mark is made on the flywheel rim at a certain distance from a point on the foundation as established by a steel rod used as a trammel, as shown in Fig. 6-1. The flywheel is then turned farther, past dead center, until the distance from the piston top to the valve-port edge is again exactly 4 in.; using the steel trammel rod, a second mark is made on the flywheel rim. The distance between the two marks is divided in two. The middle mark is then the point at which the piston is at top dead center when this mark coincides with the end of the trammel.

6-4. Rim Velocity. The rim velocity v, feet per minute, is usually referred to the outside diameter D, feet, of the wheel, and evidently

$$v = \pi D n \tag{6-7}$$

This product should not exceed 5,000 fpm for cast-iron wheels of engines under 100 hp and 6,000 feet per min for larger engines. With a special design of the rim, which eliminates the danger of blowholes in it, the rim velocity may be increased to 7,000 ft per min; with a special arm design, rim speeds as high as 10,000 ft per min are used. Cast-steel wheels can operate with speeds up to 12,000 ft per min.

Large flywheels with cast-steel arms and laminated steel arms and flywheels of all-welded construction with forged steel rims operate with rim speeds up to 15,000 ft per min. In automotive engines, rim speeds of cast-iron flywheels machined all over reach 10,000 ft per min, speeds of wheels made of semisteel may reach 15,000 ft per min and speeds of steel wheels even 20,000 ft per min.

If the above given limits are exceeded, the stresses created in the flywheels by the centrifugal force will be increased above their safe values and may cause bursting of the flywheel with all its catastrophic consequences, including loss of life and great property damage.

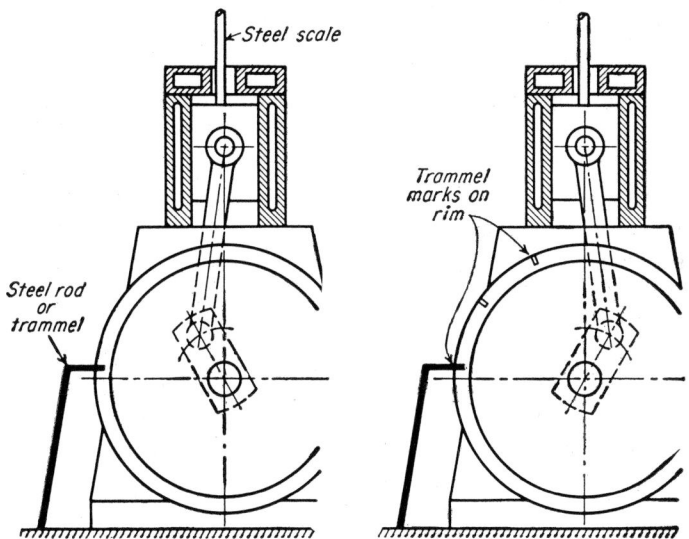

FIG. 6-1. Establishing top dead-center position.

6-5. Balancing. Flywheels of automotive and other high-speed engines are machined all over and additionally balanced on special machines to insure freedom from vibration. Flywheels of low-speed engines with cast arms should be balanced at least statically. This is done by drilling axial holes near the heavier point of the rim and filling these holes with wooden plugs. After the wheel is painted, the holes are hardly noticeable.

6-6. Flywheel Knock. Flywheels which are fastened to the crankshaft by a taper key have a tendency to pound the sides of the key and of the key-ways. When, after a certain time, as a result of pounding, a clearance is formed between the sides of the key and the keyway, the flywheel begins to produce an unpleasant knocking sound, called *flywheel knock*. This knock, if not stopped at once, will steadily increase in intensity, the damaging effect on the keyways will progress at an accelerated rate and at the same time the torsional stresses created in the crankshaft by the pounding may cause its failure. Sometimes the flywheel hub develops a crack in the corner of the keyway. Trying to stop the knock by driving the key deeper into the hub usually does not help. The knock will appear again after a few days of engine operation. The best remedy is to put in a new key and, if the sides of the keyways are noticeably damaged, to have them remachined.

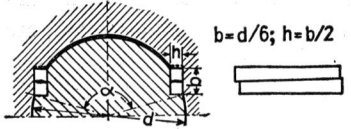

Fig. 6-2. Shaft keys for heavy duty.

When a new key is fitted, care must be exercised to have not only a good contact area at the tapered top of the key but also to have a very close fit on the sides. The key must be driven home with a heavy sledge hammer until it does not move any further after several full-force blows.

If the flywheel again develops a knock with the new key, the next step is to insert a second taper key, putting it at 90 deg or, with a large shaft, at 60 deg to the first key, fitting both keys still more carefully and driving them home as hard as possible.

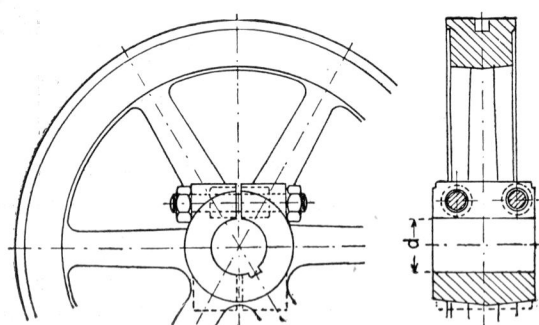

Fig. 6-3. Flywheel with a split hub.

With a large engine it is advisable to replace the taper key by two tangential keys (Fig. 6-2), each made of a pair of tapered keys with the tapers in opposite directions. To make the keyways by hand, with chisel and file, is a difficult job that requires a skillful mechanic; however, it is worth doing because a continuously knocking flywheel may result in a broken crankshaft.

6-7. Construction Features. Flywheels under 8 ft in diameter are cast solid; above this size they are made in halves. The length of the wheel hub should be at least 1.5 shaft diameters in order to give steady support and to prevent wobbling of the rim.

To prevent the key from working loose, the hub may have one split (Fig. 6-3) or may be split clear through, as shown by dotted lines. A split hub is bored slightly smaller (0.0005 in. per inch of shaft diameter) than the shaft, and a steel wedge or wedges are driven into the slots when the wheel is mounted on the shaft. Such a split hub clamps the shaft and helps prevent the key from working loose.

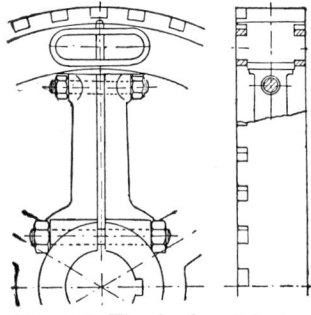

Fig. 6-4. Flywheel cast in two halves.

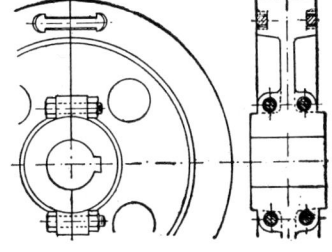

Fig. 6-5. Split-disk flywheel.

Flywheels made in two halves usually are parted on arms. The halves are connected by bolts through the hub and near the rim and mostly have additional shrink links (Fig. 6-4) or shrink anchors (Fig. 6-5, which also shows round holes cut in the disk to facilitate handling the heavy wheel). An anchor connection with wedge keys (Fig. 6-6) is used to avoid the difficulties encountered with shrink anchors, especially if the flywheel has to be taken off eventually.

In order to avoid the troubles connected with the working loose of the key, some small engines have the crankshaft end and the hub bore made on a slight taper and the hub tightened on the shaft by a nut (Fig. 6-7). For the same reason, marine-engine flywheels are made with a web fastened by fitted bolts to a separate steel-forged hub or to a flange forged in one piece with the crankshaft. Figure 6-8 shows such a flywheel of a large two-stroke opposed-piston eight-cylinder engine. Teeth cut in the outside surface of the rim are engaged with a worm for turning the engine over. This engine really does not need any flywheel at all, hence the small size and light weight of the flywheel.

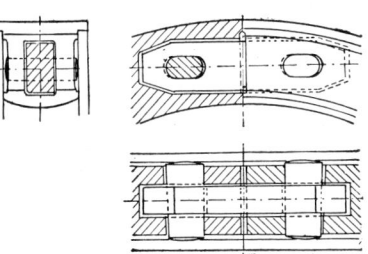

Fig. 6-6. Anchor with wedge keys for flywheel rim of large engine.

Figure 6-9 shows a section of a large flywheel for a diesel engine direct-connected to a generator. The web of the flywheel is clamped between the flanges of the crankshaft a and of the generator shaft b.

DIESEL-ENGINE COMPONENTS

Finally, Fig. 6-10 shows a flywheel of a truck diesel engine which is typical for all automotive cast-iron flywheels. The wheel is machined all over, has a register to assure its concentricity with the shaft, and is held by four or mostly six capscrews with spring washers under their heads. The starter teeth are cut in a steel ring shrunk onto the rim.

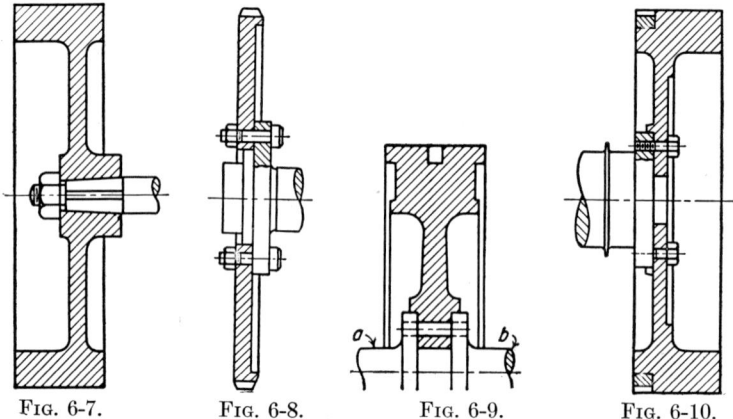

Fig. 6-7. Fig. 6-8. Fig. 6-9. Fig. 6-10.

Fig. 6-7. Flywheel for tapered shaft end.
Fig. 6-8. Flywheel bolted to a flanged shaft.
Fig. 6-9. Flywheel clamped between shaft flanges.
Fig. 6-10. Flywheel for a truck diesel engine.

QUESTIONS

1. Describe the action of a diesel-engine flywheel.
2. Enumerate the different purposes of a diesel flywheel.
3. State upon what factors depends the amount of energy that a flywheel stores up.
4. What is the radius of gyration of a flywheel?
5. What is the flywheel effect of a flywheel?
6. What is the coefficient of uniformity of rotation of a diesel engine?
7. What is the coefficient of steadiness of a flywheel?
8. What markings are usually put on the rim of a flywheel?
9. Describe the procedure of establishing the dead-center mark on a flywheel.
10. What are the rim velocities used with cast-iron flywheels (a) of smaller engines, under 100 hp, (b) of larger engines, over 100 hp, and (c) of special design which eliminates blowholes?
11. What are the rim velocities used with automotive-engine flywheels of cast-iron, semisteel, and steel?
12. How are flywheels balanced?
13. What is flywheel knock?
14. What are the results and dangers of a flywheel knock?
15. How is flywheel knock taken care of?
16. What type of flywheel hub is used to prevent working loose of the key?
17. Draw a sketch of a typical automotive-type flywheel and show the method of fastening it to the crankshaft.

CHAPTER 7

VALVE GEAR

7-1. Definitions. The term *valve gear* is used to designate the combination of all parts which *control* the admission of the air charge and the discharge of exhaust gases in four-stroke engines, the discharge of exhaust gases in some uniflow two-stroke engines, the admission of fuel in air-injection and some mechanical-injection engines, and the admission of compressed air for starting most of the larger engines. Another term, *valve-actuating gear*, refers to the combination of those parts which only *operate* or actuate the various intake, exhaust, fuel, and air-starter valves, open and close them at the proper moment in respect to the position of the piston and crankpin, and hold them open during the required time. The word *actuating* means producing action, or moving a certain part.

The valve gears of diesel engines vary considerably in their construction, depending on type, speed, and size of the engines. The action of the various parts of a valve gear may be best explained using Fig. 7-1: the rotating camshaft (14) with the cam (15) pushes the roller-cam follower and push rod (17) upward and thus transmits the cam action to

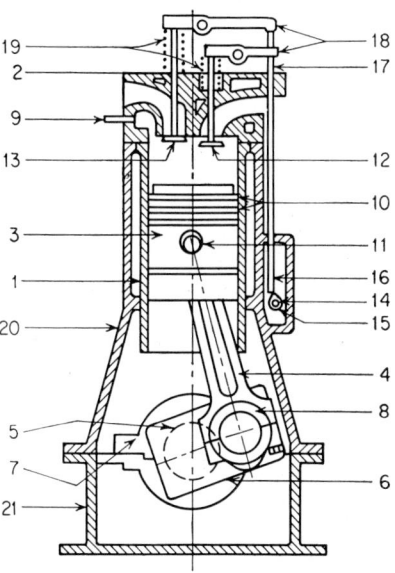

FIG. 7-1. Cross section of a typical diesel engine: 1, cylinder liner; 2, cylinder head; 3, piston; 4, connecting rod; 5, crankshaft; 6, crankpin and bearing; 7, main bearing cap; 8, connecting-rod bearing; 9, fuel nozzle; 10, piston rings; 11, piston pin and bearing; 12, intake valve; 13, exhaust valve; 14, camshaft; 15, cam; 16, cam follower; 17, push rod; 18, rocker arms; 19, valve springs; 20, cylinder block or crankcase; 21, bedplate.

the rocker arm (18); this latter changes the motion to a downward motion of the valve and the valve spring (19) returns the valve to its seat and closes it; the spring action is controlled by the closing side of the cam through the same intermediate members. In some engines the camshaft is located at the cylinder-head level; this arrangement eliminates the need for push rods and the roller-type cam follower is made part of the rocker arm.

7-2. Cams. A cam is an eccentric projection on a revolving disk used for the opening and closing of a valve through various intermediate parts, as described above. Originally cams were made as separate pieces and fastened to the camshaft. However, at present most diesel engines, even some in larger sizes, have cams forged or cast integral,

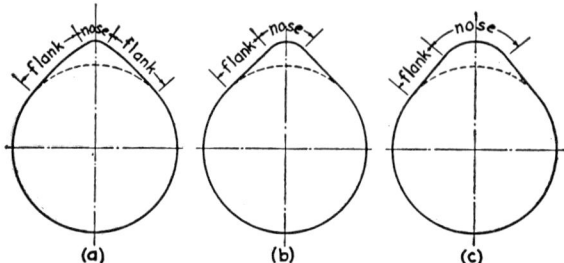

FIG. 7-2. Profiles of intake and exhaust cams.

meaning as one piece, with the camshaft and then machined, usually ground to the required exact shape. Thus diesel-engine camshafts are similar to automobile-engine camshafts. The advantage of such an integral camshaft is that if one valve of one cylinder is timed correctly, all the valves in all cylinders will be timed correctly. On the other hand, any change in timing will affect all valves and cylinders.

In operation, cams are subjected to impact and are hardened in order to reduce wear. The shape of the cam determines the points of opening and closing of the valve, the velocity of opening and closing, and the amount of the valve lift from its seat. The desired cam shape or profile is obtained by accurate grinding. The grinding stone repeats the shape of a master cam and thus insures accuracy of all cams.

The sides of a cam are called *flanks*, and the highest point is called the *nose*. Depending how fast a cam action is desired, the flanks are made either curved outward and called *convex* flanks (Fig. 7-2a) or straight, called *tangential* flanks (Figs. 7-2b and c).

Figures 7-2a and 7-2b show typical intake- and exhaust-cam profiles for four-stroke engines. Figure 7-2c shows an exhaust cam for a two-stroke engine. Figure 7-3a shows an adjustable fuel-injection cam for an engine with a common-rail or similar mechanical injection system. In this system the injection must be very accurately timed and the nose of the cam which is subject to wear must be exchanged when the wear exceeds a certain limit. The nose is made in the form of a hardened steel insert i held by set screws s and blocks or keys k which can be filed to obtain the exact location of the nose. Figure $3b$ shows a fuel-injection cam as used on two-stroke engines with a unit injector.

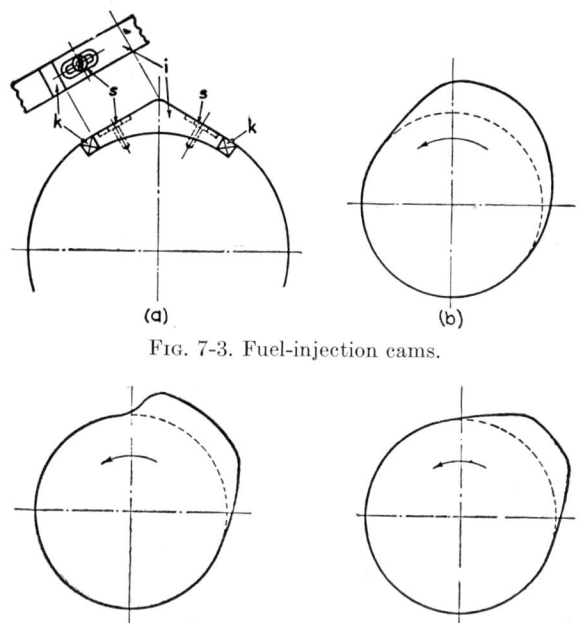

FIG. 7-3. Fuel-injection cams.

FIG. 7-4. Air-starting cams.

Figure 7-4a gives a typical shape of the cam for the air-starter valve of a four-stroke engine: it has a quick and abrupt lift which gives full valve opening almost immediately and prevents throttling of the compressed air through the starting valve. A gradual seating of the valve is obtained by the smoother shape on the closing side. Figure 7-4b shows an air-starting cam for a high-speed two-stroke diesel engine.

In a four-stroke engine there are at least two cams for each cylinder, one for the inlet and the other for the exhaust valve. Depending upon the fuel-injection system used, there may be another cam to operate the fuel injector, and if air-starting is used, still another cam for the air-starting valve. If the engine is reversible, there will be another set of cams for the other direction of rotation of the engine, resulting in as many as eight cams per cylinder. In two-stroke engines there is no cam for the intake, but if exhaust valves are used, at least two exhaust valves per cylinder are present which may be operated either by a common or two separate cams, so that the number of cams is about the same as in a four-stroke engine.

Camshaft. As already mentioned, in some engines the camshaft is a straight round shaft and the cams are separate pieces, machined and keyed to the shaft. However, in most modern diesel engines the cams and shaft are forged or cast in one piece. In some larger engines the camshafts are made up of two or more sections bolted together by

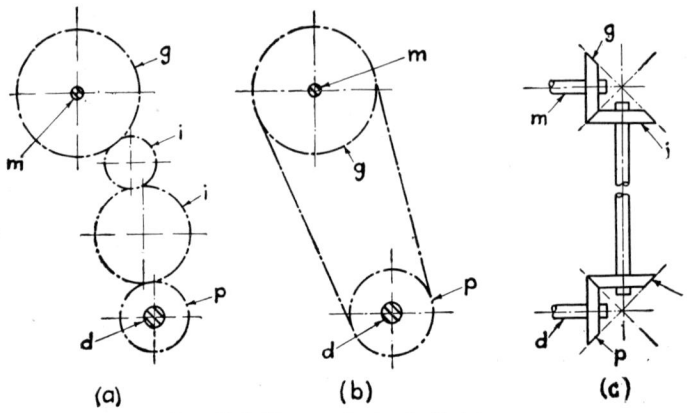

FIG. 7-5. Types of camshaft drives.

flanges with fitted reamed holes to assure accurate timing. Most camshafts are made of forged steel, usually of nickel-chromium alloy steel, and the larger camshafts are often bored hollow. They are heat-treated and the cams usually are surface-hardened. The camshafts are carried in plain bearings.

Camshaft Drives. Camshafts are driven from the engine crankshaft by various means. Figure 7-5a shows a drive by a train of straight spur or helical spur gears: d is the crankshaft, p the pinion keyed to it, i, i intermediate gears, g the camshaft gear, and m the camshaft. Figures 7-1 and 7-5b show a chain drive, and Fig. 7-5c a drive by two pairs of bevel gears and an intermediate vertical shaft.

In two-stroke engines the camshaft rotates at the same speed as the crankshaft, while in four-stroke engines it rotates at one-half the crankshaft speed.

Cam Followers. The cam follower is the engine part that is in contact with the cam and transmits the action of the cam to the push rod, as shown in Fig. 7-1.

Modern diesel engines use several types of cam followers:

1. The roller-type follower (Fig. 7-6a), used in medium-sized and large engines in combination with a tangential or a concave-flank cam.

2. The flat or mushroom follower (Fig. 7-6b), used in smaller and high-speed engines and operated by a convex cam.

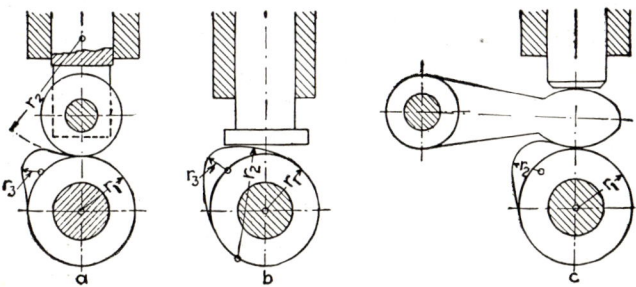

FIG. 7-6. Types of cams and cam followers.

3. The pivoted follower (Fig. 7-6c), which may be used with a cam of any shape.

4. The pivoted follower combined with a roller, shown in Fig. 7-7.

The combination of the convex-flank cam with the flat mushroom follower provides a considerably faster valve opening and closing and is used on smaller engines, running at higher speeds. The disadvantage of this type of follower is higher inertia forces. On the other hand, this combination gives a lower deceleration of the valve assembly as the latter approaches the top or nose of the cam and consequently requires somewhat smaller spring forces to prevent bouncing of the valve gear. Bouncing of the valve gear occurs after it has reached the maximum acceleration point during the lifting of the valve and from then on must be decelerated as the follower nears the top of the lift at the nose of the cam. The inertia of the valve gear at this point tends to lift the valve faster than the cam action. If the valve spring does not have enough force to decelerate the moving parts of the valve gear at the same rate as the contour of the cam, the follower will leave the surface of the cam, if only for an instant. When it makes contact again, it does so with an impact, causing pounding and excessive stresses and wear in various parts of the valve gear. This is called *bouncing* and is very undesirable. The sliding part t of a fol-

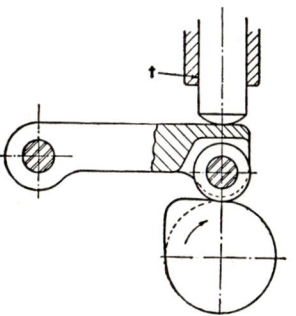

FIG. 7-7. Pivoted cam follower with roller.

lower (Figs. 7-6 and 7-7) which moves up and down in a bored hole above the cam is called a valve *lifter* or *tappet*.

The pivoted-type follower (Figs. 7-6c and 7-7) in its action resembles the roller follower. Its main advantage is that the side thrust from the cam is taken by the pivot of the lever arm leaving only a small thrust acting upon the sliding tappet due to the curved path of the end e.

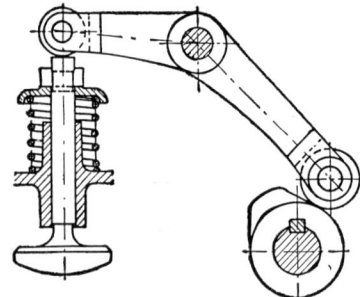

Fig. 7-8. Cam and rocker-arm mechanism.

The *rollers* of the followers are made of steel, are hardened and accurately ground to size and to a true circle. It is important that there be little or no play in the rollers and that they remain concentric, meaning that the centers of the outside circle and of the hole in the roller are at the same point.

7-3. Rocker Arms. *Push Rods.* The rocker arm has one end in contact with the top of the valve stem and the other end, through a hardened-steel roller, with the cam profile, if the camshaft is located

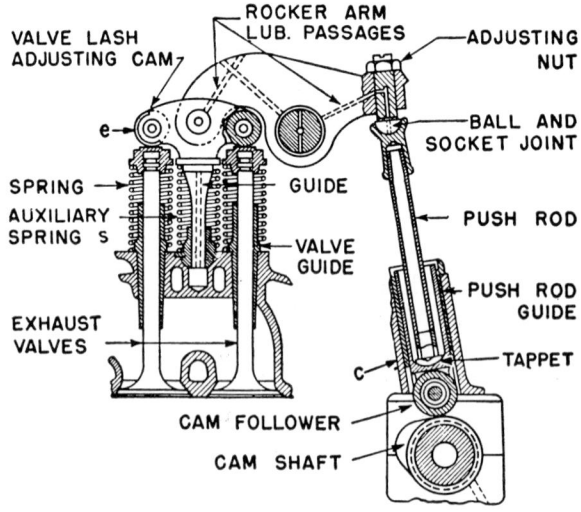

Fig. 7-9. GM valve-actuating gear.

near the cylinder head (Fig. 7-8). If the camshaft is located much lower, as in Fig. 7-1 or 7-9, the other end of the rocker arm is in contact with the upper end of the push rod; the lower end of the push rod is in contact with the rotating cam through a cam follower. The rocker arm is pivoted at or near its center and the pivot pin is held in brackets secured to the cylinder head.

When two or more valves of a cylinder must be opened and closed at the same time, as in the case of the exhaust valves of General Motors two-stroke engines, then one rocker arm may be used to operate two valves by acting upon a so-called *bridge* between the valve stems (Fig. 7-9). The bridge has special cams e, e in contact with the valve stems and a lower extension f acting as a guide. An auxiliary spring s is used under the bridge to offset its inertia and that of the rocker arm and assist the valve springs in maintaining the cam follower in contact with the cam at all times. Figure 7-9 shows also the arrangement of the push rod to take care of the slight swinging of its upper end. The lower end of the push rod has a hardened spherical insert which rests in the bored tappet which is accurately lapped into the bushings called *push rod guides* that are pressed into cast-iron supports c. The upper end of the push rod is connected to the rocker arm by a ball-and-socket joint. The entire assembly is lubricated from the force-feed lubrication system of the engine.

7-4. Valves. *Valve Requirements.* Getting the fresh air into the engine cylinder and the exhaust gases out of it requires power, referred to as the *pumping loss.* In order to reduce the back pressure during the exhaust process, the exhaust valve openings are made as large as practical. This is particularly important in the case of two-stroke engines since the entire exhaust process occurs in a small fraction of the piston stroke and scavenging must be accomplished entirely by the pressure of the fresh-air charge. For these reasons, two-stroke diesel engines generally use two to four exhaust valves per cylinder.

With four-stroke engines, the exhaust valve opening is not as critical because the exhaust gases are forced out under positive action of the piston during an entire stroke. The inlet-valve opening is more important since all the intake air enters the cylinder through it. Restrictions to the air flow not only increase the pumping loss but also reduce the density of the air charge. Reduction in the density of the air charge means that less weight of oxygen is available per intake stroke; consequently, less fuel can be burned and the maximum power which can be developed will therefore be smaller.

These conditions become more pronounced as engine speeds are increased; the pumping losses increase rapidly owing to the higher velocity of gas flow and the density of the air charge also decreases. For these reasons the power of every high-speed engine reaches the maximum or peak of its brake-horsepower curve at a certain speed beyond which an increase in speed results in a drop in power output.

Because of these conditions, intake valves are made as large as possible, and in some engines are made larger than the exhaust valves.

Poppet Valves. Intake and exhaust valves used in diesel engines are all of the so-called *poppet* type, consisting of a disk or *head* at the end of a rod or *stem*.

Poppet valves have heads with conically shaped, or bevelled, edges and bevel seats, which give them a self-centering action. Most poppet valves have bevel seats ground at an angle of 45 deg with the plane of the valve head. This has proved to give the best service, especially under severe operating conditions such as are encountered by exhaust valves. In order to obtain a slightly greater opening area for a limited valve lift, some inlet valves are made with a smaller angle, about 30 deg. For exhaust valves an angle smaller than 45 deg makes the edge of the heads too thin and more subject to the corrosive action of the hot exhaust gases.

The maximum diameter of a poppet valve is limited, chiefly by considerations of weight and cooling, to approximately two-fifths and not over one-half the diameter of the cylinder bore.

The service conditions of poppet valves, especially in high-speed engines, are most severe. The valves must be opened as quickly and as widely as possible; they must remain open as long as possible and then close as quickly as possible. In order to reduce the stresses in the valve gear due to inertia forces which appear when a valve is accelerated during opening and closing, the valves themselves should be as light in weight as possible. On the other hand, the valves must be made sufficiently sturdy to take the constant pounding due to continued opening and closing. They must be able to withstand the extreme temperature and pressure occurring in the engine cylinder and maintain a gas-tight seal.

The construction of diesel-engine valves, both intake and exhaust, follows the same general practice as that developed for heavy-duty automotive engines. Exhaust valves are usually made of silicon-chromium steel (silchrome) or steel alloys containing a high content of nickel and chromium to resist the corrosion by high-temperature gases. Sometimes a hard alloy such as *stellite* is welded to the seating surface of the valve head as well as to the tip of the valve stem to increase the hardness of the surfaces subjected to the constant pounding of closing and opening. To improve the wearing qualities of the valve stem, which usually operates in a bushing of softer metal, the stems of exhaust valves often are made of a different steel than the heads, to which they are welded to form so-called built-up valves.

Inlet valves are not subjected to the corrosive action of the hot exhaust gases, and are usually made of cheaper, low-alloy steels.

7-5. Replaceable Parts. *Valve-seat Inserts.* All poppet valves formerly were seated directly on the surface of the cast-iron cylinder head. This practice is satisfactory for most inlet valves but will not usually give good service for exhaust valves of high-output engines. At the high temperatures resulting from heat transfer from the exhaust gases and contact with the hot exhaust valves, cast-iron seating surfaces tend to soften and erode under the constant pounding of the valves. In order to increase the service life of exhaust-valve seats between reseatings, it is now common practice to employ valve-seat inserts made of alloys specially designed to resist these high-temperature operating conditions. In some high-output engines valve-seat inserts are used on both inlet and exhaust valves.

Valve-seat inserts consist of rings of a heat-resisting alloy which fit into counterbored recesses in the valve-port opening (Fig. 7-10) and are generally held in place by a shrink fit. To obtain a shrink fit the diameter of the insert is made slightly larger than the counterbore into which it goes. To install the insert, it is shrunk by cooling with dry ice, which has a temperature of -110 F, or the cylinder head is expanded by being heated in boiling water or oil, or both methods are used, after which the insert will go readily into the counterbore. As soon as the insert and cylinder head assume the same temperature, the expansion of the valve insert or contraction of the cylinder head will hold the insert firmly under slight compression. Valve-seat inserts are made of alloys with approximately the same rate of heat expansion as the cylinder head. The insert, therefore, will be held in place firmly at all operating temperatures so long as its temperature is not lower than the temperature of the cylinder-head.

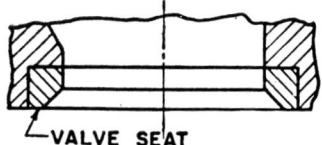
FIG. 7-10. Valve-seat insert.

Valve Guides. The holes in the guide for the valve stems wear from the continuous up-and-down motion of the stems, and in modern engines separate bushings are inserted in these holes to act as valve guides. These bushings not only can be renewed easily when worn, but they can be made of material which wears less than the cast iron of the cylinder head. Valve-guide bushings are commonly made of a cast iron which has good wearing qualities when used with steel valve stems. In order to obtain improved heat conductivity from the valve stem, some engines use bronze bushings for valve guides.

Valve guides for exhaust valves are sometimes made longer, *i.e.*, extend farther into the valve passage, than guides for inlet valves. The longer guide protects the valve stem better from the hot exhaust gases and also provides more area to assist in cooling the exhaust valve.

Valve Cages. To make for easier grinding and reconditioning of valves and their seats, some large low-speed engines have their entire valve assemblies mounted in cages. The valve cage consists of a hollow cylindrical casting containing the guide for the valve stem. The valve is seated on the lower edge of the cage, and the valve spring is mounted on the top. The valve cage fits into a hole bored through an inlet or exhaust passage in the cylinder head and is mounted with the head of the valve flush with the bottom surface of the head. Openings are provided in the walls of the cage which communicate with the corresponding passages in the cylinder head.

7-6. Valve Springs. A valve spring serves to close the valve. Valve springs used on diesel engines are made of round steel wire wound in a coil of cylindrical shape. Springs of this type have a force which is directly proportional to the amount they are compressed. Only a small portion of the maximum valve-spring force is necessary to keep the valve tight on its seat. The principal duty of the valve spring, as was mentioned before, is to provide sufficient force during the valve-lifting process to overcome the inertia of the valve gear and keep it in contact with the cam without bouncing and this is obtained by putting the spring under compression when the valve is installed. When the valve is opened, this force is increased by the additional compression of the spring.

The space available for the valve spring is very limited and it is therefore not easy to design and make a spring which will exert the necessary force and will not break under the constantly repeated change of stresses. A spring made of a small-diameter wire is subjected to a smaller stress than one made of a larger-diameter wire, for the same deflection, but it also has a smaller force. In some engines, therefore, two or three concentric valve springs of smaller wire diameter are used to operate the valve gear.

Spring Surge. Another factor which influences the operation of valve springs is the vibration of the spring. Whenever a sudden force is applied to one end of a coil spring, as by the action of the valve-operating cam, it will tend to start the coils to vibrate. The vibrations are transmitted through the coils of the spring at the rate of its natural frequency of vibration.[1] When repeated impulses are applied

---

[1] The terms natural frequency of *vibration* and *resonance* and the nature of vibration in general are explained in Sec. 18-1.

to the end of a valve spring at a rate which is equal to the natural frequency of vibration, or a multiple or a simple fraction of it, the amplitude of the vibration will be increased with each impulse. Such a condition is known as *resonance* and may result in spring surge.

Spring surge is the name given to an occurrence in which the center coils of a spring, being the farthest from the supports, move back and forth due to their own inertia; the spacing between the coils then ceases to be uniform and changes all the time. As a result, the effective force of the entire spring is reduced and may become insufficient to prevent the valve gear from bouncing. Another serious result of spring surge is breaking of the spring caused by an excessive deflection of a few coils at the end of the spring when they come together with a shock instead of having an equal compression of all coils as in normal operation.

Valve-spring surge will occur at certain engine speeds when the impact of the cams on the valve gear happens to be in resonance with some harmonic of the natural frequency of vibration of the spring. In order to eliminate surge, the natural frequency of vibration of valve springs is made as high as possible, usually several times as great as the maximum camshaft speed.

Where a single spring cannot be designed with a sufficiently high natural frequency, two or three smaller concentric springs are used, since smaller springs have a higher frequency of vibration. Multiple valve springs usually have different natural frequencies so that surge will not affect all of them at the same time and if one spring should break, the engine could still operate temporarily with the remaining spring or springs.

Spring Retainers. Valve springs are mounted between supports at their ends known as *spring seats*. The lower spring seat may be simply a recess in the top of the cylinder head or valve cage or it may be a steel washer which rests on top of the cylinder head and is shaped to fit the bottom coil of the spring. The upper spring seat, called a *spring retainer*, is a steel washer shaped to fit the top of the spring and attached to the top of the valve stem by a removable fastening.

The most widely used type of spring retainer is provided with a conical recess in the upper seat in which the valve stem is locked by means of a conical split collar, called a *lock* or *keeper*, which fits around the stem and into one (Fig. 7-11a) or several grooves (Fig. 7-11b) turned in the valve stem. The pressure of the spring on this retainer tends to hold the locks tightly in place, yet they may be easily removed by depressing the spring while holding the valve in its closed position. The advantages of this type of retainer and keeper are that they will not loosen in service and do not weaken the valve stem as do other types of fastenings.

Some large diesel engines use valve spring retainers which are held in place by a lock nut which screws on threads provided on the upper end of the valve stem.

7-7. Valve Lash and Adjustment. The expansion of the valve stem and of other parts of the valve gear when the engine heats up has a tendency to hold the valves off their seats and some provisions must be made in the valve gear to take care of this condition. The most common method used to permit this expansion is to provide a clearance, or *lash*, between the top of the valve stem and the valve-lifting mechanism. The proper lash is determined at the factory and is indicated in the engine instruction book. It is important that the valve

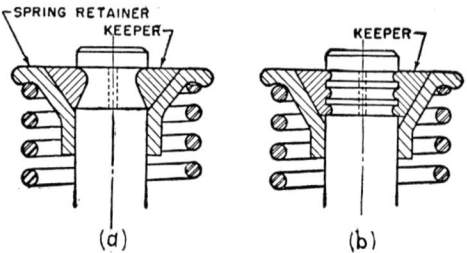

FIG. 7-11. Valve-spring retainers.

lash specified for the valve be maintained. Too much lash will cause noisy operation and excessive wear and will also result in improper valve timing, since the valve will open later and close earlier than it does with the proper lash. Too little lash is even more serious since it may prevent the valve from seating properly. This will result in valve leakage and burning of the valve seating surfaces and may even prevent combustion through loss of compression. All engines are provided with means for adjusting this lash in the valve gear somewhere between the cam follower and the valve stem. In most engines this adjustment consists of an adjustable screw and lock nut located at one end of the valve rocker arm. The clearance is measured directly by means of a feeler gauge inserted between the tip of the valve stem and the rocker-arm roller.

Automatic valve-lash adjusters are used to avoid the necessity of a clearance between the cam and the follower regardless of whether the engine is cold or warm and, by insuring a constant contact between cam and follower, to eliminate shock action at the beginning of the valve opening. They also eliminate the need for manual adjustment to take care of the wear at various points of the valve gear. At present two types of automatic adjusters are in use, mechanical and hydraulic ones.

Mechanical Adjuster. A mechanical valve-lash adjuster is incorporated in the GM bridge (Fig. 7-9): a spiral spring has a tendency to turn the cam, as shown by the arrow in Fig. 7-12, when the cylinder spring s (Fig. 7-9) pushes the bridge in its highest position. Spring s takes up any clearance between the various parts of the valve-actuating gear and the turning of the cams e takes up the clearance between the ends of the bridge and the upper spring seats, which are solidly con-

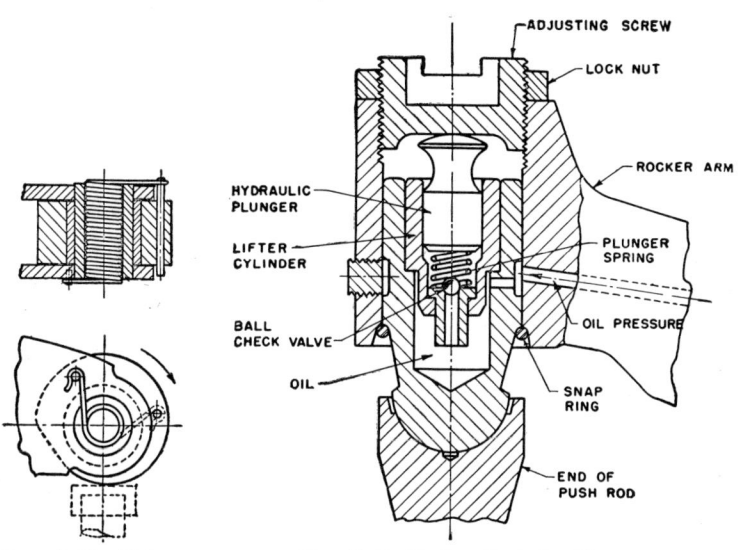

Fig. 7-12. Valve-lash take-up. *Courtesy of General Motors Corporation.*

Fig. 7-13. Hydraulic lash adjuster.

nected to the valve stem. This adjuster takes up the lash caused by wear but does not take care of the expansion in the valve mechanism.

Hydraulic Adjuster. Hydraulic lash adjusters may be built into the valve tappets. However, with valve-in-head engines they are generally built into the ends of the rocker arms or valve bridges which operate directly on the ends of the valve stems. Such a lash adjuster is shown in Fig. 7-13. Essentially it consists of a small cylinder, referred to as a *lifter cylinder*, containing a piston or plunger, a cylindrical spring, and a ball check valve, and located between the push rod and the rocker-arm end. In operation, oil under pressure from the lubricating oil system enters the lifter cylinder past the ball check valve and is trapped under the plunger. Any force exerted against the outer end of the plunger will be transmitted to the cylinder, mounted in the valve gear, by the entrapped oil. The valve is thus actuated as if the lash were taken up mechanically. Since the spring inside the cylinder acts to force the plunger outward, any clearance between the valve and its lifter will be taken up and the oil pressure will immediately fill up the lifter cylinder through the check valve. If the valve stem expands, there is sufficient leakage of oil past the plunger to permit it to move in slowly so there is no danger of holding the valve open.

QUESTIONS

1. What is the valve gear in four-stroke diesel engines? In two-stroke diesel engines?
2. What is the valve-actuating gear?
3. What is called a *cam* in references to the valve gear of a diesel engine?
4. Give a sketch of a diesel-engine cam and indicate the flanks and the nose.
5. Sketch a cam with convex flanks and one with tangential flanks.
6. Give a sketch of an adjustable fuel-injection cam for a common-rail injection system.
7. Give a sketch of an air-starting cam for a four-stroke diesel engine.
8. Enumerate the cams necessary for operation of each cylinder of a four-stroke engine.
9. What is the camshaft of an engine? What are the two types of camshafts now in use?
10. State the relation between camshaft speed and engine speed (*a*) in a two-stroke engine and (*b*) in a four-stroke engine.
11. Enumerate the different means of driving a camshaft. Illustrate the answer by schematic sketches.
12. Explain the action of a cam follower.
13. Enumerate the various types of cam followers used in modern diesel engines. Illustrate the answer by sketches.
14. With what cams can a flat follower be used?
15. With what cams can the roller-type follower be used?
16. With what cams can the pivoted follower be used?
17. When does a cam follower bounce, and why?
18. What is a valve lifter or tappet?
19. What is a rocker arm in a valve gear?
20. If one rocker arm is used to operate two valves simultaneously, what intermediate part is used? Give a sketch of such an arrangement.
21. What is a push rod in a valve gear? When are push rods necessary?
22. What are pumping losses in a four-stroke engine?
23. Which of the openings is more critical—the intake or the exhaust valve? Explain why.
24. Which of the valves should be and actually sometimes is made larger, the intake or exhaust valve?
25. Why do intake and exhaust valves have conical heads and seats?
26. What are the commonly used angles between the tangent to the valve seat and the plane of the valve head?

VALVE GEAR

27. What is the approximate limit for the maximum valve diameter as a function of the cylinder bore?

28. Of what material are the intake valves of a diesel engine usually made?

29. Of what material are the exhaust valves of a diesel engine usually made?

30. What material is often welded to the seating surface of an exhaust valve head? Explain why.

31. What is a valve-seat insert? Give a sketch and the advantages of using it.

32. What method is used to counteract wear of the valve-stem guides, particularly those of exhaust valves?

33. What is the purpose of using valve cages?

34. What is the main duty of a valve spring?

35. What is spring surge?

36. What are the purposes of providing engine valves with multiple springs?

37. What is a spring seat? Draw a sketch.

38. Draw a sketch of an upper spring seat with a conical split keeper.

39. What is a lash in the valve lifting mechanism?

40. State how the amount of valve lash is adjusted?

41. What is the object of automatic valve-lash adjustments?

42. What kind of automatic valve-lash arrangements are in use?

43. Draw a sketch of a mechanical valve-lash adjuster and explain its operation.

44. Draw a sketch of a hydraulic valve-lash adjuster and explain its operation.

CHAPTER 8

FUEL INJECTION

8-1. Methods. There are two distinct methods of fuel injection—*air injection* and *airless injection*. The latter method is known under different names, such as mechanical, solid, and hydraulic injection.

Requirements. The main requirements which a fuel injection system must fulfill are the following:
1. Accurate metering of the fuel oil.
2. Proper timing of the fuel injection.
3. Suitable rate of fuel injection.
4. Proper atomization of the fuel.
5. Good distribution of the fuel in the combustion space.

Metering. Accurate metering, or measuring, of the fuel means that the amount of fuel delivered for each cycle must be in accordance with the engine load and that exactly the same quantity of fuel must be delivered to each cylinder for each power stroke of the engine. Only in this manner will the engine operate at a uniform speed.

Timing. Proper timing means beginning the fuel injection at the required moment; it is essential in order to obtain the maximum power from the fuel and thus good fuel economy and clean burning. When the fuel is injected too early in the cycle, ignition may be delayed because the temperature of the air at this point is not high enough. Excessive delay gives rough and noisy operation of the engine and also permits some fuel to be lost owing to wetting of the cylinder walls and piston head. This in turn results in a poor fuel economy and smoke in the exhaust. When the fuel is injected too late in the cycle, part of the fuel will be burned with the piston well past top center. When this happens, the engine will not develop its maximum power, the exhaust will be smoky, and the fuel consumption high.

Rate of fuel injection means the amount of fuel that is injected into the combustion chamber in a unit of time or in one degree of crank travel. If the rate of injection is high, a given amount of fuel will be injected during a short time, or during a small number of degrees of crank travel. If it is desired to lower the injection rate, a nozzle tip with smaller holes must be used in order to increase the duration of fuel injection.

The rate of injection has a similar influence upon engine performance as timing. If the rate of injection is too high, the results will be similar to an excessively early injection; if the rate is too low, the results will be similar to an excessively late injection.

Atomization, or breaking up, of the fuel stream into mistlike sprays must conform to the type of the combustion chamber. Some chambers require a very fine atomization, others can operate with a coarser atomization. Proper atomization facilitates the starting of the burning and insures that each minute particle of fuel is surrounded by particles of oxygen with which it can combine.

Distribution. The distribution of the fuel must be such that the fuel will penetrate to all parts of the combustion chamber where oxygen is available for combustion. If the fuel is not properly distributed, some of the available oxygen will not be utilized and the power output of the engine will be low.

Additional Requirements. In order to be practical the fuel injection system, mainly the high-pressure pump, must have additional features. It must (1) keep its adjustment a reasonable time, not lose it from vibrations brought on by high engine speed or from excessive wear; (2) be economical of power; (3) be light and not too bulky, especially with small engines; (4) be quiet.

8-2. Air Injection. Air injection was used in early diesel engines. At present it is seldom used and only for large engines operating on heavy viscous fuels.

In air-injection engines the potential energy of compressed air is converted into kinetic energy. This kinetic energy of the expanding air is used to feed the fuel into the cylinder from the spray valve, to atomize the fuel, and to create turbulence in the combustion chamber for thoroughly mixing the fuel and air.

The air-injection system consists of four main components:
1. The fuel pump for metering the fuel.
2. The air compressor for supplying the injection air.
3. The spray valve.
4. The fuel cam and actuating gear.

Fuel Pump. In the type used most each cylinder of the engine has a plunger, and the quantity of fuel is controlled by varying the useful length of the plunger stroke. The only function of the pump is to meter the required quantity of fuel accurately and to deliver it to the spray valve.

Air Compressor. The compressor has either two, or more often three, stages and delivers air to an air receiver, called *air bottle*, from which a pipe leads the air to an air header connected to the fuel injectors. The same compressor serves to replenish the air bottles used for starting the engine.

The *spray valve* (Fig. 8-1) consists of a needle valve n with a conical end held on its seat by a heavy spring s, and several atomizer disks d with holes to break up the fuel and mix it with the injection air as both air and fuel flow through the valve and the flame plate p. This flame plate has an orifice through which the fuel-air mixture is admitted to the combustion chamber. The needle valve is lifted mechanically by a lever l actuated by a cam on the camshaft.

Fuel Cam. Only the timing of the injection is controlled by the fuel cam; the rate of injection, the atomization of the fuel, and the distribution in the combustion chamber are all controlled by the number and size of the orifices in the atomizer disks and in the flame plate and by the injection-air pressure. Injection-air pressures vary from 800 to 1,200 psi, depending upon speed and load. Therefore, the shape of the cam is not very important. The cam usually consists of a relatively small radial protrusion on the cylindrical surface of a disk.

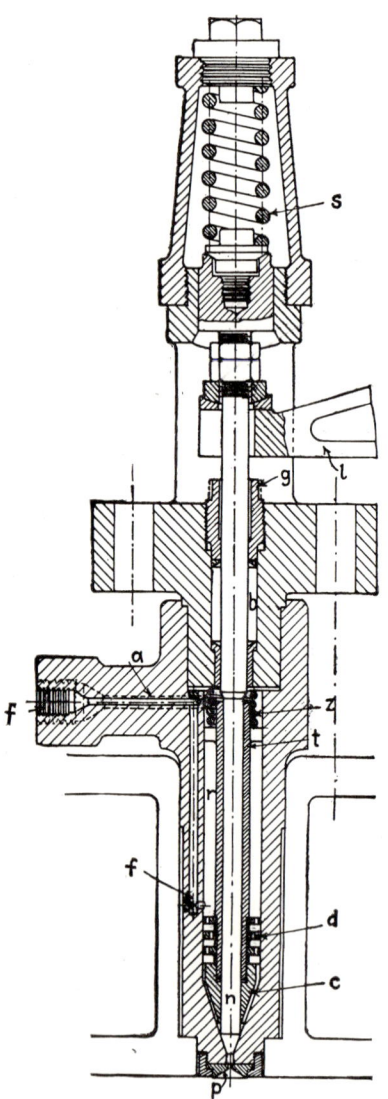

Fig. 8-1. Closed air-injection fuel nozzle.

Air injection fulfills the five main requirements rather well, but has the following practical drawbacks: (1) the compressor absorbs a comparatively large power and (2) is comparatively heavy and bulky, not suitable for high-speed, small-bore engines.

FUEL INJECTION

8-3. Mechanical Injection. *Classification.* All mechanical-injection methods may be subdivided into four main groups: (1) *constant-pressure*, or *common-rail*, system, (2) *jerk-pump* system, (3) *distributor* system, and (4) *precombustion-chamber* system. The differences in construction found in the engines of different manufacturers are such that it may help in their discussion to subdivide the main groups further. The common-rail system may be divided into the basic system and a modification used in Cooper-Bessemer engines. The jerk-pump system may be divided into the original system, with a separate pump and fuel injector for each cylinder, and a modification in which the pump and injector are combined in one unit.

Atomization. With mechanical injection, atomization is obtained as follows: the liquid fuel, subjected to a high pressure, passes through one or several small openings into the combustion space filled with air whose pressure is considerably lower; as a result, the fuel streams develop a high velocity and this creates great friction between the liquid streams and the air in the combustion space. Due to this friction minute particles of fuel are broken off the surface of the stream, then the freshly exposed particles are again broken off and so on until the whole liquid stream is broken up into very small particles, or globules. *Atomization* literally means breaking up into atoms. Actually each globule is not one atom, but consists of many atoms. Therefore *atomization* is an exaggeration but shows what the process aims at.

Penetration. In mechanical injection the distribution of the fuel in the combustion chamber is, generally speaking, obtained by two means, penetration and air turbulence. Penetration is the distance through which fuel particles are carried by the kinetic energy imparted to them when they leave the fuel nozzle. Friction between the fuel and the air in the combustion space gradually absorbs this energy. Penetration depends upon various characteristics of the fuel-injection system and is reduced by finer atomization. Thus the best conditions are found as a compromise between a minimum penetration and desirable atomization.

Air turbulence is practically independent of the fuel injection system and is an additional means in obtaining good combustion.

8-4. Common-rail System. This system consists of a high-pressure, constant-stroke and constant-delivery pump which discharges the fuel into a common rail, or header, to which each fuel injector is connected by tubing. A spring-loaded by-pass valve on the header maintains a constant pressure in the system, returning all excess oil to the fuel supply tank. The fuel injectors are operated mechanically and the amount of oil injected into the cylinder at each power stroke is controlled by the lift of the fuel-admission valve. The operation of the injection system is shown diagrammatically in Fig. 8-2, and a sectional view of the fuel nozzle is shown in Fig. 8-3: the fuel cam gives an upward motion to the push rod; through the rocker arm and intermediate lever l (Fig. 8-3) this motion is transmitted to the needle valve n; the space above the needle-valve seat is connected at all times with the fuel header through tubing and sealed from the top by a packing gland b. When the needle valve n is lifted from its seat, the

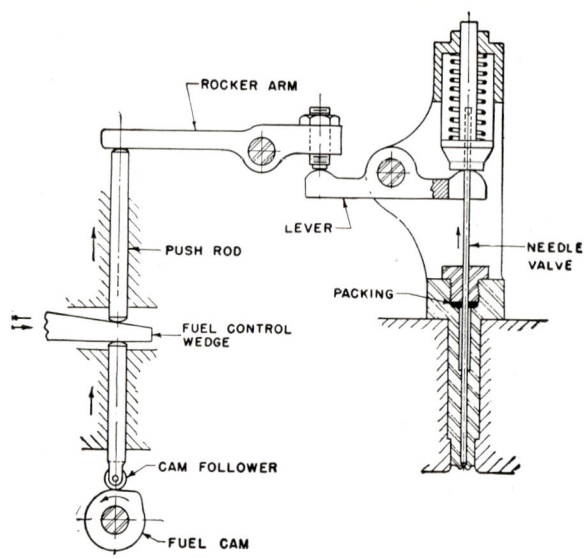

Fig. 8-2. Control of fuel injection in a common-rail injection system.

fuel is admitted to the combustion space through the small holes drilled in the injector tip p, below the valve seat q. Passing through these tiny holes the fuel is divided into small streams which are broken up, or atomized, as explained before. The amount of fuel injected is controlled in accordance with the power requirement by means of a wedge (Fig. 8-2) which changes the lash of the fuel valve. When the wedge is pushed to the right, the valve lash is decreased, the motion of the cam follower will be transmitted to the push rod earlier, the fuel needle will be opened earlier and closed later and its lift will be slightly greater, and more fuel will be admitted per cycle. When the wedge is pulled out, to the left, the valve lash is increased, the needle valve is lifted later and closed earlier, and therefore less fuel will be admitted. The position of the control wedge is changed either by the governor or, in variable-speed engines, by hand. The fuel-injection pressure is

FUEL INJECTION

adjusted to suit the operating conditions by changing the spring pressure in the by-pass valve, called fuel-pressure *regulator* or *unloader valve*.

Unloader Valve. The body b of the valve (Fig. 8-4) is connected at a to the high-pressure fuel line and at c to a discharge line leading

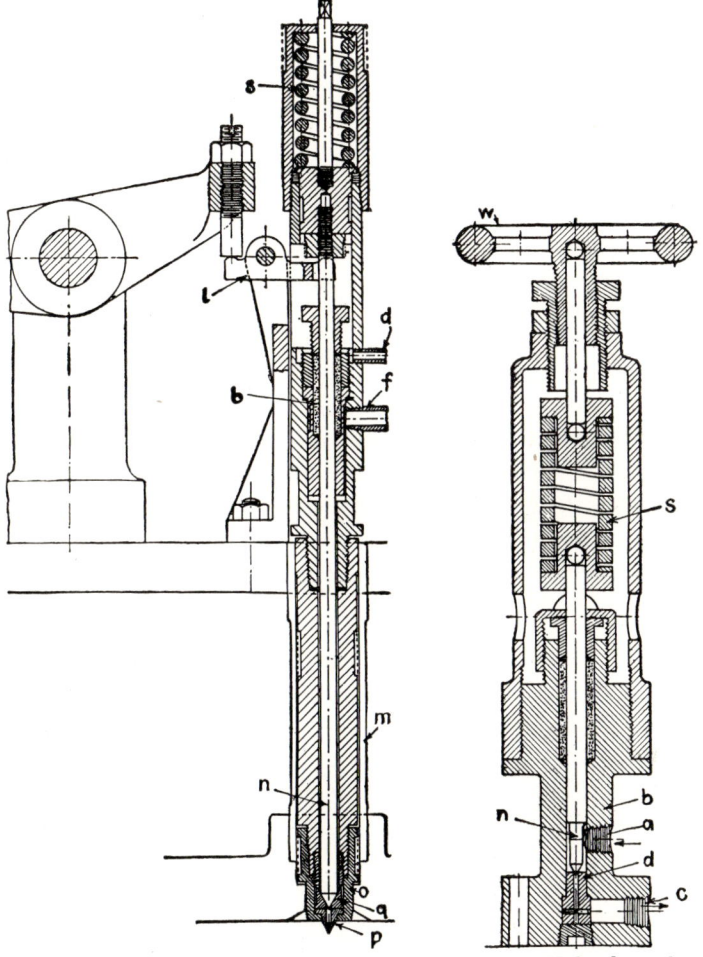

Fig. 8-3. Common-rail mechanical-injection fuel nozzle.

Fig. 8-4. Unloader valve.

to the fuel tank; the small opening in a hardened-steel plug d is closed by the needle valve n loaded by a stiff spring s. When the pressure in the fuel system increases above a certain value determined by the spring s, it lifts needle n and some oil is discharged through the line c until the pressure drops sufficiently and needle n is seated by the spring action. The handwheel *w* serves to change the compression of spring *s* according to the fuel pressure that is desired. Fuel-injection pressures of from 3,200 to about 5,000 psi are used at rated load and speed.

In order to reduce the pressure fluctuations in the system caused by the intermittent fuel discharge from the pumps and withdrawals by the fuel valves, the volume of the fuel in the system is increased by attaching to the fuel header an additional container that is called the *accumulator* and has a relatively large capacity.

The area past the needle-valve seat and through the passage between the valve seat and the valve tip containing the orifices is several times as large as the area through the orifices in the tip. The control of the fuel jets is largely by the orifices. The valve tip commonly used with this system has several holes or orifices.

The common-rail system is not suitable for high-speed small-bore engines because it is difficult to control accurately very small quantities of fuel injected into each cylinder at each power stroke.

8-5. Jerk-pump System. This system has two essential parts to each cylinder, the injection pump and the fuel nozzle. The requirements which a pump in this system must fulfill, in respect to both metering and timing, are such that only precision equipment can meet them.

Accuracy. In operation of a high-speed diesel engine, these injection pumps must measure and deliver, under high pressure and at exactly the required time, an exceedingly small quantity of fuel.

Metering. The volume of the fuel injected is very small compared with the size of the cylinder. At full load, the volume of the fuel injected is about 1/20,000 the piston displacement, and when the engine is idling the volume of fuel required is only about 1/100,000 the piston displacement.

Thus a $3\frac{1}{2} \times 4\frac{1}{2}$-in. engine has a piston displacement of 43.3 cu in. per cylinder. The amount of fuel required per injection at full load is about $43.3/20,000 = 0.00217$ cu in. and when idling only about one-fifth of that, or 0.00043 cu in. These volumes are equal to drops about $\frac{1}{8}$ in. and slightly over $\frac{1}{16}$ in. in diameter, respectively.

The difficulty of measuring very small quantities of oil accurately, so that all cylinders receive exactly equal amounts, is obvious. Only a precision-manufactured mechanism can fulfill these requirements.

Timing. An idea of the accuracy of timing required may be obtained from the following considerations: ordinarily the period during which the fuel is injected does not exceed 20 deg of crank travel; if the engine speed is 2,000 rpm, the whole injection period corresponds approximately to

$$\frac{60 \times (20/360)}{2,000} = \frac{1}{600} \text{ sec}$$

The start and end of injection cannot vary by more than approximately 1 deg of crankshaft rotation, or must occur within limits of about 1/12,000 sec.

Injection Pressures. Such minute quantities must be injected within such close time limits under pressures which may run as high as 6,000 psi, and in some injector types up to 30,000 psi, although most high-speed engines use injection pressures of 2,400 to 3,000 psi.

Precision Workmanship. All injection fuel pumps of the jerk-pump type have plungers fitted very closely to the pump barrels by lapping. Lapping is the term used to designate finishing metal surfaces by working them against so-called laps and using an exceedingly fine abrasive material. From a number of plungers and barrels which have been lapped truly round and cylindrical but differ slightly in diameter, the plungers are fitted to the barrels by selective assembly. In this way a fit can be obtained with a clearance of less than 0.0001 in. Such a small clearance gives very little leakage, even with the high pressures created, and no packing of any kind is necessary. Owing to the method of fitting a plunger to its barrel, these parts are not interchangeable. If a plunger or barrel is worn or damaged, both pieces must be replaced by new ones.

Pressure Waves. Fuel oil, as all other liquids, is compressible and also, as all matter, has inertia. Therefore, when the pump plunger at the beginning of the actual delivery strikes the oil in the barrel, all oil in the fuel line is not accelerated at once; the motion of the plunger increases the pressure first in the particles of the oil nearest to the plunger, this pressure increase is transmitted gradually through the line until it reaches the nozzle. On the other hand, since the pressure increase travels faster than the pump plunger, the liquid column in the line, due to its inertia, has a tendency to move away from the plunger. Thus the initial blow of the plunger sets up a compressive wave in the fuel line. When this wave reaches the end of the line at the nozzle, which presents a certain resistance, it is reflected and travels back to the plunger, increasing the pressure created by the plunger; after it reaches the plunger, it returns to the nozzle, and so on. The fluctuation of the pressure at the discharge end of the fuel line disturbs the fuel discharge through the nozzle. During the moments when the pressure at the nozzle is low, the fuel discharge is decreased and the atomization becomes poorer. The disturbances are particularly noticeable in engines operating at variable speed. In addition, the pressure waves may produce vibration of the tubing connecting the pump and the nozzle and cause its breakage. The building up of pressure waves is affected by many factors, the chief ones being the inside diameter and length of the fuel-line tubing. The proper length and diameter are found by the engine builders and should never be changed as a change may cause serious trouble.

8-6. Jerk Pumps. In order to obtain good atomization of the fuel spray, the injection pressure in the fuel line must be sufficiently high all the time, from the very start to the end of the injection. Since this pressure is proportional to the plunger speed, the latter must be also reasonably high during the whole injection period. This is obtained by using for the fuel delivery only part of the plunger stroke, after the plunger has acquired a certain speed, and discarding the initial and final parts of the stroke. This method produces a sudden

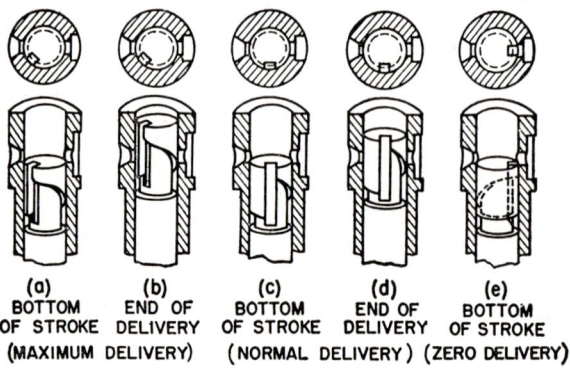

FIG. 8-5. Barrel with various plunger positions.

acceleration of the fuel in the line, acts with a *jerk*, hence the name for the pump.

Metering. In most jerk pumps the total plunger stroke is constant and the metering is controlled by varying the length of the useful part of the plunger stroke by one of the following two methods:

1. The fuel is admitted into the pump barrel through ports in the barrel controlled by a spiral groove or scroll on the plunger; the plunger can be turned in the barrel while moving back and forth and this changes the useful length of the plunger stroke during which the ports are covered and the fuel is delivered to the nozzle.

2. The fuel is admitted into the pump barrel through ports in the barrel which are controlled by a separate valve.

Plunger-controlled Pumps. The principle of this type of pump can be explained using Fig. 8-5: at the bottom of the plunger stroke (Fig. 8-5a) the suction and pressure-release ports are both in communication with the inner pump space. When the plunger has moved a certain distance and covered both ports, fuel delivery begins with a jerk and lasts until the lower edge of the spiral begins to uncover the release port

(Fig. 8-5b); at this moment the pressure drops and fuel delivery stops. The plunger continues to travel a short distance to the top of its stroke and then begins to move downward. If the plunger is turned about 45 deg (Fig. 8-5c), the distance between the top edge of the plunger and the edge of the spiral sliding over the release port is shorter and the fuel delivery stops earlier. Finally, if the plunger is turned 90 deg more (Fig. 8-5d), the release port stays uncovered all the time and no fuel is delivered.

Timing. In the scheme as shown in Fig. 8-5, the beginning of the useful stroke always occurs at the same moment but the end of injection changes, depending upon the engine load: it is later with a higher load, giving greater fuel delivery, and earlier with a lower load, giving less fuel delivery. The middle of the injection stroke is retarded with an increase of the load.

This condition can be reversed by making a spiral edge on the top of the plunger and a square groove at its bottom. In this case the beginning of the injection will be earlier with an increase of the load and vice versa, and the end will be always constant; the middle of the injection will be advanced with an increase of the load, which, however, is not desirable. By making both edges spiral, inclined in opposite directions (Fig. 8-7), the middle of the injection stroke can be kept constant for all loads.

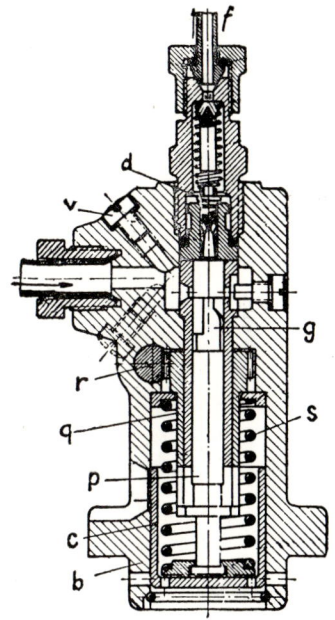

Fig. 8-6. Bosch individual injection pump

Naturally, the compressibility of the fuel and the mechanical flexibility in the pump mechanism retard the actual start of injection into the cylinder and produce a so-called *injection lag*.

Pump Drives. The plunger of plunger-controlled pumps is pushed by a cam with a cam follower during the delivery stroke and returned by a spring. The pumps are built either as separate units for each engine cylinder, or as a multiplunger in-line unit for all cylinders.

Figure 8-6 shows an American Bosch pump complete with body and governing mechanism. The actuating mechanism, not shown in Fig. 8-6, consists of a cam and cam follower that pushes the sleeve tappet *c* and with it the plunger p upward; the spring s returns the tappet c and plunger p to the bottom position. The fuel delivery is controlled by the governor which moves the rack r, that is engaged with gear teeth on the sleeve q. The sleeve q has a slot in which the cross piece on the plunger p slides. Thus turning of the sleeve q turns also the

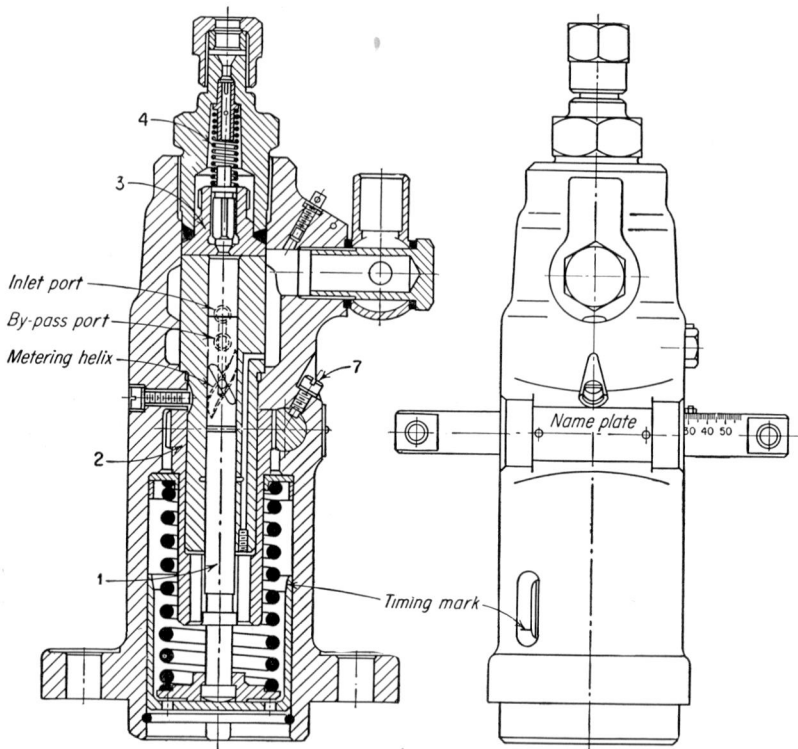

Fig. 8-7. Bendix-Scintilla injection pump: 1, plunger and barrel assembly; 2, control sleeve; 3, discharge valve assembly; 4, discharge-valve spring; 5, air-bleed screw; 6, inlet pipe connection; 7, control-rack screw; 8, plunger follower; 9, plunger return spring; 10, control rack.

plunger p with the spiral groove g and this, as explained before, changes the effective length of the pump stroke, and hence the amount of fuel delivery.

The Bendix-Scintilla pumps are very similar to the Bosch pumps and differ only in details, as may be seen from Fig. 8-7.

The Demco pump (Fig. 8-8) is based on the same principle, with a helix cut in the plunger. However, being designed for automotive engines, it differs in several respects: it is more compact, presents a

FUEL INJECTION 109

self-contained unit with a built-in camshaft and provisions for a primary pump, and is light, having an aluminum housing.

Pump with Control Valve. Figure 8-9 shows an injection pump particularly designed and built for small diesel engines. The pump is also of the jerk type and has a constant stroke and a turning fuel-control valve. When the upper edge of the pump plunger, on the upward stroke, covers port (4), the taken-in fuel is compressed and shortly

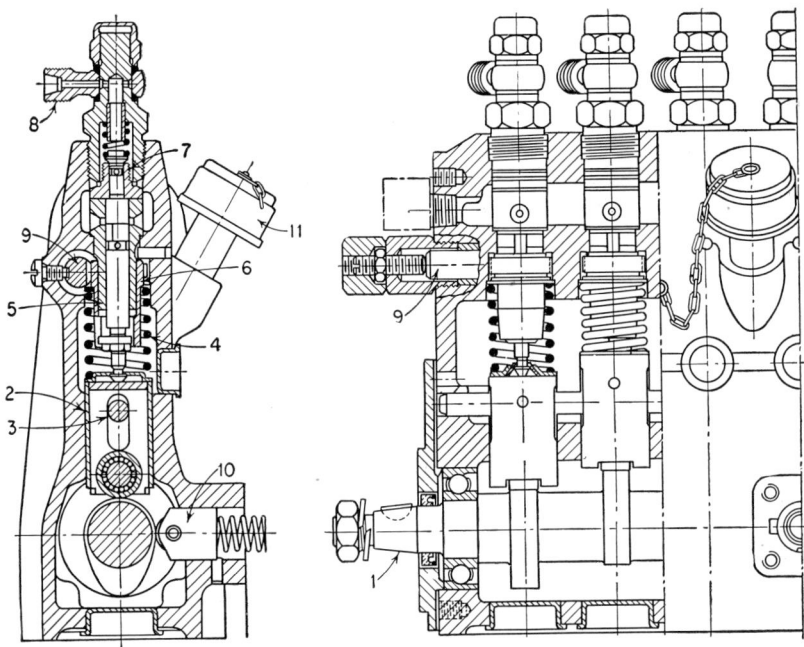

FIG. 8-8. Demco automotive injection pump: 1, camshaft; 2, tappet assembly; 3, tappet guide; 4, tappet spring; 5, plunger and barrel assembly; 6, control pinion; 7, delivery-valve assembly; 8, delivery-line connection; 9, control rack; 10, primary pump tappet; 11, breather assembly.

afterwards begins to be injected through the delivery valve (5) until the upper edge of the groove on the plunger uncovers the lower port (6). This groove communicates with the space above the plunger by holes drilled axially and crosswise and the delivery pressure drops at once and the rest of the fuel is spilled back into the header. On the return stroke the plunger covers port (6), and a vacuum is created above the plunger. When port (4) is uncovered, this vacuum sucks a charge of fuel into the space above the plunger. The fuel is metered by throttling (see Fig. 8-10, which shows a cross section through *a-a* of Fig. 8-9): from the filter the fuel is admitted to a ring-shaped header (11) on the outside of a stationary sleeve (8); from there, through a control port (9) and cross-and-lengthwise holes drilled in the control shaft (10), the fuel is admitted to the suction port (4). The shaft (10) is lapped in the sleeve (8) and has a flat spot (12) ground across the radial hole. When the fuel supply must be decreased, the control shaft is turned clockwise by hand or by a governor. This throttles the flow of fuel and thus decreases its amount. Because of the flat spot the cutting-off of the fuel proceeds gradually until it is cut off completely, as shown by the dotted lines; (13) is an air-bleed vent.

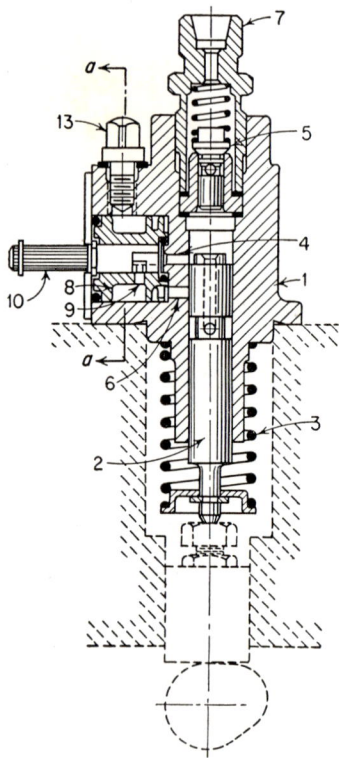

The advantages of this design, as compared with a helical groove on the plunger, are (1) simplicity and cheapness of the construction, (2) considerably less wear and longer life of the plunger and barrel, (3) ease of servicing and absence of adjustments to be made, and (4) a very small resistance to turning of the control valve, which permits the use of a small, inexpensive governor.

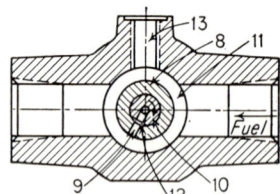

FIG. 8-9. Demco injection pump for a small diesel engine.

FIG. 8-10. Section through Demco small-engine pump.

8-7. Fuel Nozzles. The nozzle may be either of the open or of the closed type. The open type usually is a simple spray nozzle with a check valve which prevents the high-pressure gases in the engine cylinder from passing to the pump. It is simple but does not give a very good atomization and is not generally used. The closed-type nozzle is used more commonly. Basically it is a hydraulically operated, spring-loaded needle valve. Most closed nozzles open inward under the pressure acting on the differential area of the needle valve —which is a cylinder lapped in with the body, and seated by a spring—

when the pressure is cut off, Fig. 8-12. There are two main types of such nozzles—the *pintle-type* and the *hole-type* nozzle. The pintle diameter is only slightly smaller than the hole and the fuel delivered by such a nozzle must pass through a narrow ring-shaped orifice. The spray is in the form of a hollow cone the outside angle of which may be made, by selecting certain dimensions, as large as 60 deg. A valuable

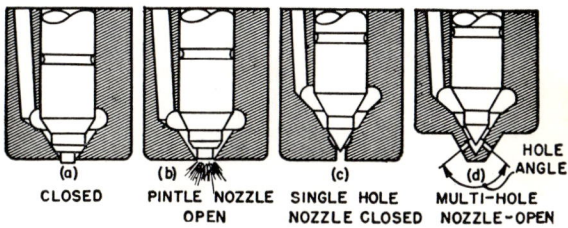

Fig. 8-11. Types of closed fuel nozzles.

feature of the pintle nozzle is its self-cleaning property, which prevents carbon deposits from building up in and around the orifice.

In the hole-type nozzle there are one (Fig. 8-11c) or several spray orifices (Fig. 8-11d), in the form of straight, round holes drilled through the tip of the nozzle body below the valve seat. The spray from the single-hole nozzle is relatively dense and has a greater penetration. The general spray pattern of a multihole nozzle, which may or may not be symmetrical, is determined by the number, size, and arrangement of the holes. Orifices are used from 0.006 to 0.0225 in. diameter, and their number may vary from three in small engines to eighteen in nozzles for large-bore engines. Multihole type nozzles are used mostly in engines with an undivided combustion chamber.

Figure 8-12 shows a pintle nozzle assembled in a nozzle holder with the spring and connections for the fuel-pressure line and leakage drain. The pressure necessary to open the needle valve varies from 1,500 to 3,000 psi.

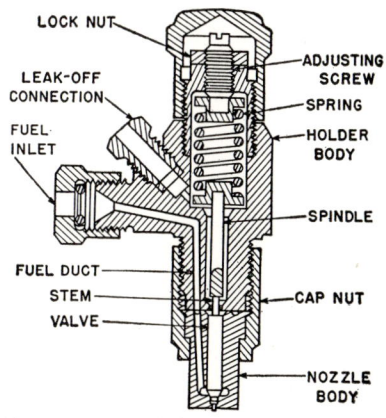

Fig. 8-12. Bosch fuel-injection nozzle.

8-8. Unit Injector. The unit injector combines a pump and a fuel-spray nozzle in one unit (Fig. 8-13). The pump is of the jerk type with ports controlled by helical-groove edges in the plunger, and the fuel amount is controlled by turning of the plunger. The nozzle l is of the

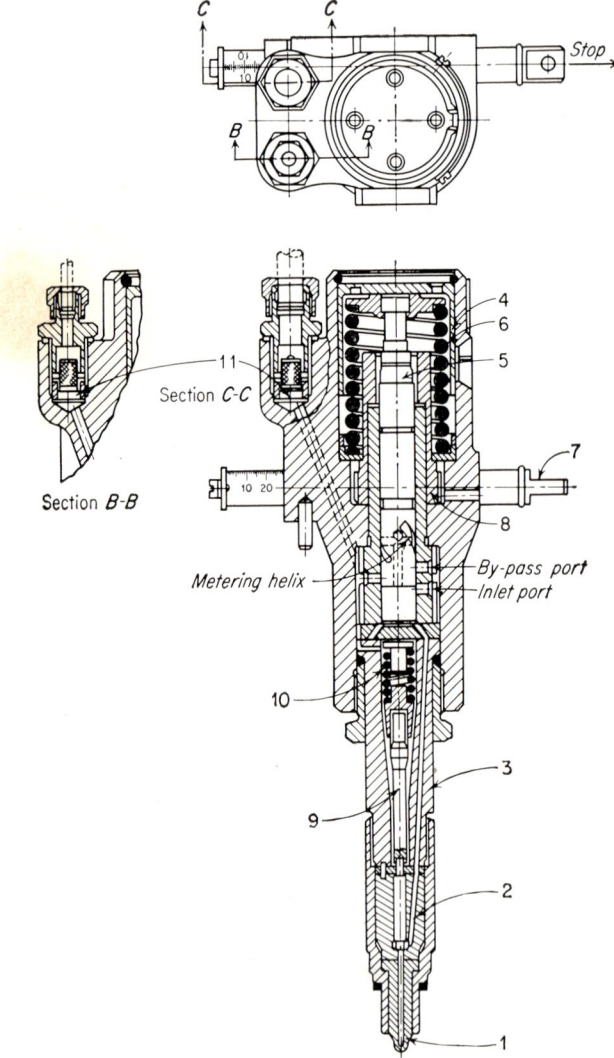

Fig. 8-13. Bendix-Scintilla unit injector: 1, spray tip; 2, nozzle-valve assembly; 3, nozzle-holder body; 4, plunger follower; 5, plunger and barrel assembly; 6, plunger return spring; 7, control rack; 8, control sleeve; 9, pressure rod; 10, pressure spring; 11, filter.

open type with a spherical check valve; the spray tip has several small holes. However, unit injectors are also in existence with closed-type nozzles with hydraulically operated differential needle valves and multihole nozzle tips. The pump plunger receives its downward motion, the delivery stroke, from a fuel cam through a rocker arm which acts off the plunger follower (4). The plunger (5) is returned by the plunger spring (6). The fuel, under a pressure of about 20 psi, is admitted through a small filter (11), and fills the ring-shaped supply chamber around the pump plunger barrel, called a *bushing*. As the plunger moves downward, fuel is displaced into the supply chamber, at first through the lower port and later, when the lower edge of the plunger closes this port, through a central hole in the plunger and a hole drilled crosswise in it and through the upper port. When the upper helical edge has covered the upper port, the fuel from the pump barrel is forced down into the nozzle body, opens the spherical check valve, and is forced past the flat check valve into the spray tip and through

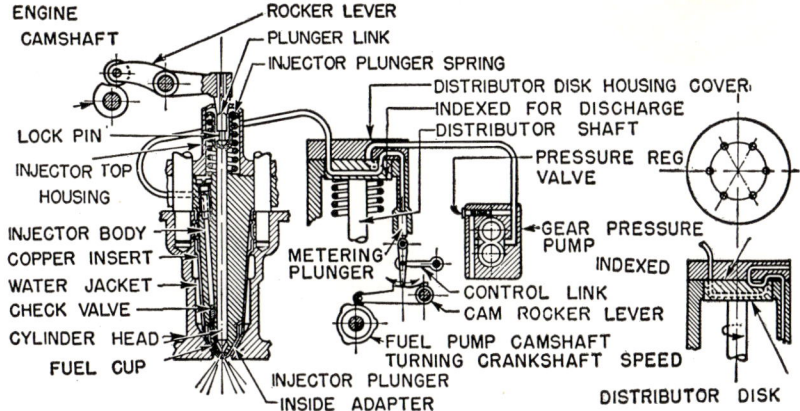

Fig. 8-14. Diagram of Cummins distributor-type low-pressure fuel-injection system.

the holes into the cylinder. The maximum fuel-injection pressure may reach 10,000 psi and higher. Injection continues until the lower helix on the plunger uncovers the lower port in the plunger barrel; from this moment the fuel begins to by-pass through the holes in the plunger and through the lower port back into the supply chamber. This releases the pressure on the fuel in the barrel, and the check valve spring causes the spherical check valve to seat. On the return stroke, the upward movement of the plunger fills the barrel with fuel oil, which flows from the supply chamber through the lower port. The function of the lower flat check valve is only to close the inside of the nozzle against gases from the cylinder.

In order to change the effective length of the stroke the plunger is turned by means of a gear and rack connected to the governor or hand throttle, along the same lines as in separate pumps described before. The plunger slides through the gear but has a feather key which forces it to turn with the gear. The effective stroke is determined by the relative positions of the helices and the upper and lower ports.

The advantage of the unit-injector construction is in the absence of long fuel lines, which cause pressure waves and sometimes mechanical troubles. However, its disadvantage is in the high injection pressures which cause faster wear of the spray orifices and in the necessity of dismantling the whole valve gear in order to take out one unit injector.

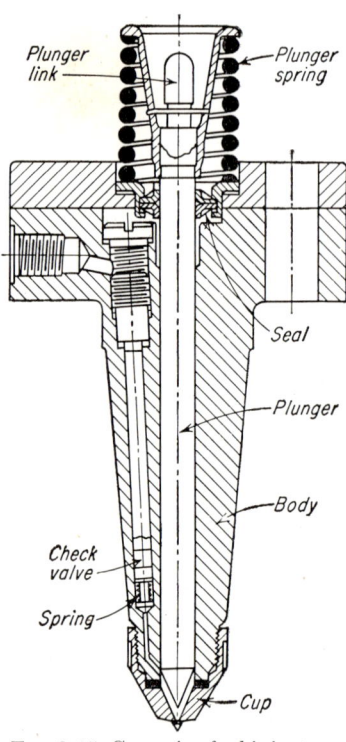

FIG. 8-15. Cummins fuel injector.

Another disadvantage is a greater chance of fuel leaks into the engine sump and a resulting dilution of the lubricating oil.

8-9. Distributor System. Of the several engine makes using distributor systems, the Cummins engines probably present the greatest number. However, the Cummins system has also the characteristic features of a unit injector and is sometimes classified as such. In this system, fuel under 50 to 60 psi pressure is supplied by a gear transfer pump (Fig. 8-14) through an indexed rotating distributor to a metering plunger during its downward stroke. This plunger has a variable stroke controlled by the governor and receives its upward motion from a multilobe cam and its downward motion from a spring. During the upward stroke the plunger sends the fuel through other passages in the same distributor to the individual injectors on each engine cylinder.

The fuel enters at the top of the injector (Fig. 8-15), flows through an inlet passage past a spring-loaded ball check valve, and fills the chamber under the injector plunger which, except at the end of injection, is off its seat. The injector plunger is operated by a cam through a rocker lever and link. While the fuel is delivered from the distributor to the injector during the suction stroke of the engine, the injector plunger is gradually lifted and the fuel fills the space in the cup under the plunger. At the time of injection the plunger is pushed downward and the fuel, prevented by the ball check valve from returning to the distributor, is injected into the combustion space finely atomized by being forced under a relatively high pressure through six small holes, 0.006 to 0.008 in. in diameter, depending upon the size of the engine.

The advantage of this system is in the absence of high-pressure lines and pressure waves. Its disadvantage is the relatively large inertia of moving parts, making it not suitable for very high speeds. Another disadvantage is in the often occurring dilution of the lubricating oil in the engine crankcase due to leakage of fuel oil past and up the injector plunger.

8-10. Precombustion Chambers. The compression space is subdivided into two parts, one formed by the space between the piston top and the cylinder head and the other formed by a special recess in the cylinder head and connected with the main chamber by a more or less

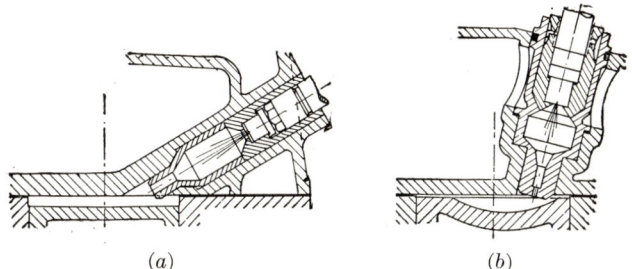

Fig. 8-16. Precombustion chambers of American engines.

restricted passage. The fuel is injected into the precombustion chamber sometime during the compression stroke, the earlier the smaller the passage to the precombustion chamber. Because of an insufficient amount of air in the chamber, only part of the fuel is burned when ignition occurs. However, this first combustion raises the pressure in the precombustion chamber and throws out the rest of the fuel into the main combustion chamber, where it is burned and raises the pressure in the cylinder.

The oil-injection pressure may be considerably lower, 1,000 to 1,500 psi. However, some engines use high-pressure injection.

The advantages of this system are that (1) the timing of injection does not need to be exact, (2) the fuel charge is given a longer time element to become ready for combustion, and (3) some turbulence is created by the precombustion discharge.

All these features are very helpful in the variable- and high-speed engines used in automobiles, trucks, and tractors. Figure 8-16 shows two typical precombustion chambers. The disadvantages of precombustion chambers are (1) less good control of the combustion and (2) comparatively low mean effective pressures.

QUESTIONS

1. What two distinct methods of fuel injection are used in diesel engines?
2. What are the different names under which airless fuel injection is also known?
3. Enumerate the five main requirements which a fuel-injection system should fulfill.
4. Why is accurate fuel measuring essential for good operation of a diesel engine?
5. What happens if the timing of fuel injection is not quite right?
6. What happens if the fuel is injected too early?
7. What happens if the fuel is injected too late?
8. What is meant by the term *rate of fuel injection*?
9. What is the influence of the rate of injection upon the performance of an engine?
10. How can the rate of fuel injection be changed?
11. What is meant by the term *atomization* of the fuel?
12. Do all engines require the same kind of atomization?
13. Why is proper distribution of the fuel essential?
14. Enumerate the additional features that a fuel-injection system must have in order to be practical.
15. What diesel engines use air-injection of the fuel?
16. Enumerate the four main components of air-injection systems.
17. How is the quantity of fuel controlled in an air-injection system?
18. What is the usual pressure of the injection air in air-injection fuel systems?
19. Give a sketch of a spray valve used in an air-injection system.
20. What are the functions of the fuel cam in an air-injection fuel system?
21. What are the advantages and disadvantages of the air-injection system?
22. Enumerate the four groups into which the airless fuel-injection methods at present in use may be divided.
23. Explain the process of atomization with mechanical injection.
24. By what means is the distribution of the fuel in the combustion chamber of an engine with mechanical injection obtained?
25. Give a diagrammatical sketch of a common-rail fuel-injection system and explain its operation.
26. Give a sketch and explain the operation of the unloader valve in the constant-pressure fuel-injection system.
27. Why is the common-rail system not suitable for high-speed small-bore engines?
28. What are the essential parts in a jerk-pump injection system?
29. What is the volume of fuel injected at each cycle as compared to the piston displacement, at full load and no load?
30. How accurate must the timing of fuel injection be in a high-speed diesel engine?
31. Compute the duration of fuel injection at full load for a diesel engine running at 1,800 rpm, if the total period of injection is 20 deg of crank travel.
32. What are the commonly used injection pressures in a jerk-pump fuel system? What are the limits of fuel pressure encountered in some engines?
33. By what method is it possible to obtain a fit of a plunger to a pump barrel with a clearance of less than 0.0001 in.?
34. Explain how pressure waves may be set up in high-pressure fuel systems.

FUEL INJECTION

35. Explain why high-pressure fuel-injection pumps are commonly called jerk pumps.

36. Indicate how the metering is controlled in jerk pumps using a constant plunger stroke.

37. Explain the operation of a pump with a spiral groove on the plunger.

38. What causes the injection lag in high-pressure fuel pumps?

39. Give a sketch of a valve-controlled fuel-injection pump and explain its operation.

40. What means are used to operate plungers of fuel pumps?

41. What are the two main types of fuel nozzles used in hydraulic-injection systems?

42. Draw a sketch of a closed fuel nozzle used in hydraulic-injection systems.

43. What are the two main types of a hydraulically operated fuel nozzle?

44. With how many orifices are hole-type nozzles made? Give sketches of these nozzles, explaining their operation.

45. Give a sketch of a pintle-type fuel nozzle.

46. What is a unit injector?

47. What is the maximum fuel-injection pressure in a unit injector of the GM diesel engines?

48. What are the advantages and disadvantages of high-pressure unit injectors?

49. Give a sketch of the Cummins injection system and explain its operation.

50. What are the advantages and drawbacks of the Cummins injection system?

51. Explain the operation of fuel-injection systems with precombustion chambers.

52. What is the fuel pressure range commonly used with precombustion chambers?

53. Give a sketch of a precombustion chamber and explain its operation.

54. What are the advantages and disadvantages of fuel injection into a precombustion chamber?

CHAPTER 9

GOVERNORS

9-1. Engine Loads. The purpose of a governor is to keep the engine running at a desired speed regardless of the changes in the load carried by the engine. Naturally, the load cannot exceed the maximum load that the engine can carry. The governor action is often referred to as engine control and depends upon two factors: (1) the performance characteristics of the engine and (2) the characteristics of the load which the engine drives.

In order to establish the performance characteristics of the engine, it must be brought out that the governor controls the total power developed by the engine, which is called the *indicated horsepower*. Part of the indicated horsepower is used up to overcome various friction losses. These consist of (1) friction between various moving parts; (2) windage losses, or resistance of air to the fast-moving pistons and connecting rods, and to the rotating flywheel; (3) pumping losses due to the resistance of the valve passages to the flow of intake air and exhaust gases, and (4) work to drive the camshaft, to operate the valves, and to drive pumps and other auxiliary mechanisms. The sum of all losses is called *friction load* of an engine and designated N_f.

The indicated horsepower N_i minus the friction load N_f, also expressed in horsepower units, gives the brake horsepower N_b or useful output of the engine:

$$N_b = N_i - N_f \tag{9-1}$$

N_b must at all times be equal to the external load connected to the crankshaft.

Thus the indicated horsepower which the governor must control is equal to the sum of the brake and friction horsepower:

$$N_i = N_b + N_f \tag{9-2}$$

The indicated horsepower depends on the pressures developed during combustion of the fuel and upon the resulting mean indicated pressure, often designated as *mip*. The combustion process and the mean indicated pressure depend primarily on the quantity of fuel injected at the beginning of the power stroke, all other conditions remaining the same. Diesel-engine governors are therefore direct-connected to the fuel-injection pumps.

Diesel fuel-injection pumps are so designed and built that the quantity of fuel delivered at each injection at a certain setting or position of the fuel control is approximately the same, regardless of the engine speed. Therefore, the total load can be balanced at any speed by obtaining the fuel-control setting corresponding to the required mean indicated pressure.

Thus the problem of controlling the speed of an engine whose external load varies but whose speed is to remain substantially constant is rather easily solved by changing the setting of the fuel control to correspond to the imposed load.

If the engine must operate at different speeds, the conditions become more involved, because the three terms, indicated horsepower, friction horsepower, and load, all increase with an increase of speed, but in different proportions. The indicated horsepower, for a given mip, increases in direct proportion to the speed, or as the first power of the speed. The friction load increases more rapidly, but no accurate expression can be given for it except that the increase is proportional to a speed increase in the power of from 1.1 to 1.5, depending on the engine type, speed range, and other conditions. The useful load increases with the speed usually still faster; in engines driving marine and air propellers, or centrifugal pumps or blowers, the load from the propeller, pump, or blower increases approximately as the cube of the speed. In vehicle engines the load increase in general is slightly less than the third power but at the same time is very uncertain as it depends upon the condition of the road, surface, grade, etc. In general greater mip is required to balance the total load at higher speeds. An example of a variable-speed external load is a direct-connected boat propeller. The power required to drive the propeller at various speeds, the external load, is represented by curve a (Fig. 9-1); the friction or internal load is represented by curve b; the sum of the ordinates of these two curves gives curve c which represents the total load or indicated power required. The indicated horsepowers available with full-load, $3/4$-load, and $1/2$-load fuel setting are given by the three dotted inclined lines. Their intersections with curve c shows that at 500 rpm the total load is about 75 ihp and will be balanced by the $1/2$-load setting; at 750 rpm the total load is about 185 ihp and will be balanced by the $3/4$-load setting; and, finally, running the propeller at 1,000 rpm requires 400 ihp, which will be balanced by the full-load setting of the fuel control. With this type of load the speed of the engine will increase only if more fuel is injected and will decrease if less fuel is injected. This type of load, which increases with speed faster than the engine output, is self-governing: any fuel setting will produce the combustion pressures that balance the load on the engine at only one speed. Under such conditions the speed will vary in accordance with the fuel setting and, if the load were perfectly constant, it could be regulated by manual control of the fuel. However, the propeller load on a ship is not constant, even at a constant engine speed, because of the pitching of the ship and uncovering of the propeller in a seaway. To prevent over-

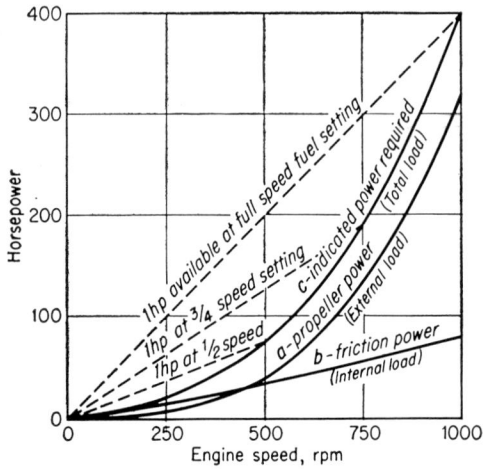

FIG. 9-1. Loads at different speeds on an engine driving a boat propeller.

speeding of the engine when the propeller load suddenly decreases, every marine diesel engine is equipped with a variable-speed governor.

When a diesel engine is operating at a constant speed, the friction losses remain practically the same. In this case, the combustion pressures necessary to balance the total load will vary directly with changes in the external load. An example of such a type of load is the direct-connected electric generator, which normally must operate at a constant speed. An increase or decrease in the electric load connected to the generator will require correspondingly higher or lower combustion pressures to maintain the engine speed constant. Therefore, the amount of fuel injected must be changed almost in direct proportion to the changes in the external load.

9-2. Functions of a Governor. When an engine is *idling*, meaning running without a load at a low speed, the friction power is relatively small and consequently the amount of fuel which is injected at each firing stroke is very small, especially in small engine cylinders. As such minute quantities of fuel cannot be very accurately metered and regularly injected, the engine speed usually varies. This fact is described by saying that the idle-speed stability of a diesel engine is not very good. At the same time, the slightest change of the fuel-control setting at idle position, when the effective stroke of the fuel pump is very small, causes a relatively large change in the amount of fuel injected. This, in turn, results in a correspondingly large change in engine speed, making it difficult to obtain close regulation of engine speed under idling conditions.

Also, when two or more engines are working in parallel, *i.e.*, are connected electrically or mechanically to the same load, their speeds cannot vary independently. Examples of this type of load are the diesel-electric a-c power plants with several units and the multiple-unit marine propulsion systems in which two or four engines drive a single propeller shaft. Under such conditions the fuel injected in each engine must be regulated very accurately so that the total load is distributed proportionately between the engines while their relative speeds are maintained.

Manual regulation of a diesel engine is, therefore, not satisfactory, and an automatic governor is necessary.

Governor Classification. According to their main functions, diesel-engine governors may be classified as follows:

1. *Constant-speed governors*—to maintain the same, or nearly the same, engine speed from no load to full load.

2. *Variable-speed governors*—to maintain any desired engine speed from idling to maximum speed, regardless of load changes; the speed itself is set manually.

3. *Speed-limiting governors*—to control the minimum engine speed and to limit its maximum speed or to limit the maximum speed only; the fuel setting for intermediate speeds is done manually.

4. *Load-limiting governors*—to limit the load which the engine can take at any speed.

Some governors are designed to perform two or more of the above functions by incorporating the required operating mechanisms in the same unit.

9-3. Governor Characteristics. Diesel-engine governors must have certain characteristics to fit the type of load which the engine drives. The main characteristics which determine the degree of governor control of the engine are the following:

1. *Speed droop*, or the decrease in speed of the engine from no load to full load, expressed in rpm or as a percentage of normal or average speed.

2. *Isochronous governing*, which means maintaining the speed of the engine truly constant at all loads—a perfect speed regulation with zero speed droop.

3. *Sensitivity*, or the change in speed necessary before the governor will make a corrective movement of the fuel control—generally expressed as a percentage of the normal or mean engine speed.

4. *Stability*, or the ability of the governor to maintain the required engine speed without fluctuations or hunting.

5. *Hunting*—the continual fluctuation of the engine from the required speed, even when the load does not change.

6. *Promptness*—the speed of action of the governor, usually expressed as the time in seconds required for the governor to move the fuel control from no-load to full-load position.

7. *Power* of the governor, or the force which the governor can develop to overcome the resistance in the fuel-control system.

The correct understanding of these characteristics and particularly of their influence one upon another is important for satisfactory diesel engine operation; they will therefore be discussed at a greater length.

Speed Droop. If n_0 designates the no-load speed, n_1 the full-load speed, often called *normal* or *rated* speed, the speed droop, rpm, is simply equal to $n_0 - n_1$.

As stated above, speed droop is usually expressed as a percentage of the normal speed. If the speed droop is designated by p, it can be computed from the expression

$$p = \frac{100(n_0 - n_1)}{n_1} \qquad (9\text{-}3)$$

Sometimes speed droop is referred to the mean engine speed n, which is evidently

$$n = \frac{n_0 + n_1}{2} \qquad (9\text{-}4)$$

Using n instead of n_1 in expression (9-3) and designating speed droop by p' in order to mark the difference, gives

$$p' = \frac{200(n_0 - n_1)}{n_0 + n_1} \qquad (9\text{-}5)$$

since $n_0 > n_1$, therefore $p' < p$. However, since $n_0 - n_1$ is small, the difference between p and p' is also very small and practically $p' = p$.

GOVERNORS

To avoid confusion the value of p' by expression (9-5) is also referred to as *closeness of regulation*.

In most cases it is considered desirable to have only a small speed droop. However, in some cases, as for engines driving electrical generators in parallel and equipped with mechanical governors, a greater droop, about 4 per cent, is desirable as it helps the governors to divide the load between the units. Where the load conditions require closer regulation, a speed droop of 2 per cent can be obtained without great complications. Engines driving alternators operating in parallel or when a standard frequency is required must have some means for speed adjustment, such as changing slightly the compression of the governor spring or the fulcrum of a floating lever.

Speed droop and closeness of regulation of an engine installed in a power plant are determined by running it from no load to full load, or a desired overload, as the case may be, and back to no load, and measuring the engine speed at various loads. Figure 9-2 shows the results of such a test of an engine whose rated speed was $n_1 = 505$ rpm; the no-load speed was $n_0 = 520$. By expression (9-3) the speed droop

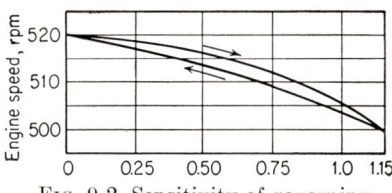

Fig. 9-2. Sensitivity of governing.

$$p = \frac{100(520 - 505)}{505} = 2.97\%$$

The closeness regulation, by expression (9-5),

$$p' = \frac{200(520 - 505)}{520 + 505} = 2.93\%$$

as may be seen, the difference between these two values is insignificant.

Isochronous governing can be obtained only with a relay-type governor equipped with special compensating devices and will be discussed separately.

Sensitivity. If an engine is running and its governor is in equilibrium, it requires a certain change of the engine speed, when the load changes, before the governor begins to act and to adjust the fuel delivery to correspond to the new load. This lag of governor action, caused by friction and lost motion in the governing mechanism, is called the *sensitivity* of the governor. Sensitivity is determined by testing an engine with an increasing and then with a decreasing load, as illustrated by Fig. 9-2. Sensitivity is designated by q and is expressed as the speed difference at a certain load over the average speed at this load:

$$q = \frac{2(n' - n'')}{n' + n''} \qquad (9\text{-}6)$$

For the governor represented by Fig. 9-2 the greatest speed difference is at 0.75 of the full load, with $n' = 509$ rpm when the load is decreasing and $n'' = 512$ when the load is increasing. Therefore by expression (9-6)

$$q = \frac{2(512 - 509)}{512 + 509} = 0.0059$$

or $0.0059 \times 100 = 0.59$ per cent.

It should be noted that while it is desirable to have a governor with a small internal friction, a low value of q, q should not be smaller than the coefficient of uniformity of rotation u, expression (6-3), which depends upon the mass and size of the flywheel. The value of u for a certain engine expresses speed fluctuations during each cycle, and if q is smaller than u, the governor will be continuously fluttering and will cause wear of various pivots.

Stability is the ability of the governor to assume and hold a definite position for any given speed and to return to this position of equilibrium after it has been displaced from it. Stability is obtained by using a spring whose force increases faster than the centrifugal force of the governor weights. A stable governor does not have any appreciable movement caused by reactions from the fuel-control mechanism.

If a governor lacks stability, it may change its position without any appreciable change of the engine load, will hunt, or race, and will cause the engine speed to fluctuate.

Hunting occurs because of the lag in action of the control mechanism. It may be caused by insufficient sensitivity or by lack of power. In either case the engine will slow down or speed up too much before a corrective regulation of the fuel is made. When the controls begin to move, an overcorrecting of the engine speed in the other direction will result. After that the governor will start to act in the opposite direction.

Power. A mechanical governor must furnish the energy necessary to do the work of regulating the admission of the fuel. All the requirements of good speed control depend upon this energy. The power of a mechanical governor can be increased by increasing the centrifugal force of its weights, as will be shown later. The power of a hydraulic governor is usually adequate and, if necessary, can be easily increased by raising the pressure of the oil which moves the piston of the servo-motor back and forth.

9-4. Governor Types. As already stated, governors used on diesel engines operate by controlling the amount of fuel admitted into the engine cylinders. The agent that actuates a governor is the centrifugal force of rotating weights. In most governors this centrifugal force is all the time balanced by the force of compression of a coil spring. Therefore they are commonly known as *spring-loaded centrifugal* governors. However, some governors, as the Massey hydraulic governor, balance the centrifugal force by a varying oil pressure.

Depending upon how the centrifugal force is put to work, governors may be divided into two groups: (1) direct-acting or mechanical governors and (2) relay-type or hydraulic governors acting through a servomotor.

In direct-acting governors, the control mechanism which regulates the fuel supply is operated by the flyballs themselves by means of a mechanical linkage.

In hydraulic governors the flyballs actuate a relay or a supply of external power. In most cases the flyballs operate a cylindrical slide valve which admits oil under pressure to the one or the other end of a power cylinder with a piston in it. The oil moves the piston which operates the fuel-control mechanism. The power cylinder and the piston in it is the servomotor mentioned before.

9-5. Mechanical Governors. A spring-loaded centrifugal mechanical governor is illustrated diagrammatically in Fig. 9-3, showing the low- and high-speed positions of the weights. The governor has two, sometimes four, rotating weights, called *flyweights* or *flyballs*. These are located at the upper ends of bell-crank levers mounted on pivots at their corners to the yoke. The yoke is usually connected by gears to rotate with the engine. The inner ends, or toes, of the flyweight levers bear against the thrust bearing of the control sleeve, which operates the fuel-regulating mechanism. The speeder spring, often referred to simply as the governor spring, bears against the upper end of the control sleeve and tends to move it, together with the fuel-regulating mechanism, downward in the direction of *greater* fuel supply. The centrifugal force acting outward on the flyweights has a tendency to move the control sleeve, together with the fuel-regulating mechanism, upward against the action of the spring, in the direction of *less* fuel supply.

When the centrifugal force of the rotating flyweights is exactly balanced by the force of the spring, the control sleeve assumes a certain position which determines a definite setting of the fuel-regulating mechanism, and the engine speed remains constant as long as the load does not change.

If, however, the load on the engine decreases, the engine will begin to speed up because the fuel-regulating setting supplies more fuel than is necessary for the reduced load. As the speed of the engine increases, the speed of the governor is also increased, and this increases the centrifugal force of the flyweights. This increase in centrifugal force moves the control sleeve in the direction of smaller fuel supply, compressing the spring further until the centrifugal force again is balanced

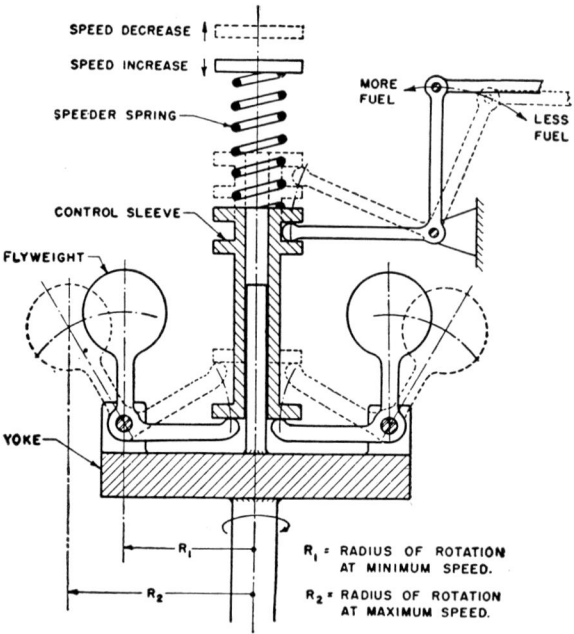

Fig. 9-3. Elementary spring-loaded centrifugal governor.

by the increase in spring force. The reduction in the fuel supplied prevents the engine from increasing its speed more than is necessary for the governor to actuate the fuel-control mechanism by the amount required to balance the reduced load.

If, on the other hand, the load on the engine increases, the engine will begin to slow down because the fuel supplied is not sufficient for the increased load. As the speed of the engine decreases, the centrifugal force of the governor flyweights also decreases. This unbalances the spring force which then moves the control sleeve, in the direction of greater fuel supply, until the decrease in centrifugal force is balanced by the decrease in spring force due to the increase in its length.

Constant-speed Governors. The acting force in a direct-acting governor is the change of centrifugal force of the flyballs. The speed at no load must be greater than at full load; so a speed droop is unavoidable. Therefore, a direct-action governor cannot be built as a constant-speed control mechanism in the strict sense of the word. However, the speed droop, as was already discussed, can be kept within a small percentage, and therefore in practice such governors which are built to maintain a certain rated speed are considered to be of the constant-speed type.

Variable-speed Governors. If the control speed of an engine must be changed while it is in operation, a variable-speed governor is required which may be adjusted to maintain any desired speed within the operating range of the engine.

The simplest method of obtaining variable-speed governing with a spring-loaded centrifugal governor is to change the initial compression of the spring as shown diagrammatically in Fig. 9-3. Thus, if the initial spring force is increased by compressing the spring to a shorter length, the engine speed must increase before the centrifugal force developed by the flyweights can balance the greater spring force. Or, if the initial compression of the spring is decreased by increasing the spring length, the engine speed required for the centrifugal force to balance the reduced spring force will be decreased.

Two-speed Governors. In some installations the engine has to run for prolonged periods without any load but cannot be stopped entirely. To reduce the wear on the engine and also the fuel consumption it is desirable in such cases to reduce the engine speed considerably while it is idling. With a direct-action governor this can be accomplished by having two different springs: a soft spring for idling, when the centrifugal force is small, and a stiffer spring for acting at the higher speeds when the engine operates under load. The springs may be arranged to act either separately or in combination. Figure 9-4 shows such a governor diagrammatically. For idling or, generally speaking, for low-speed operation, the governor hand control is put in the position shown in full lines, in which only the inner soft spring is acting. For high-speed operation the handle is pushed in the position indicated by the dotted lines, compressing both springs, which then begin to act together and require a higher centrifugal force and higher speed for balancing them.

Pivotless Governor. One of the troubles with conventional mechanical governors is that the fulcrums and fingers of the bell-crank levers acting on the spring-loaded sleeve are subject to heavy wear. The second trouble is that the power of the governor decreases very fast with a decrease of the speed of rotation. However, the lower speed corresponds to a greater load, when the power is particularly needed. A simple governor which avoids the above drawbacks is shown schematically in Fig. 9-5. The governor has six balls b, b which are forced to rotate around the axis of the stationary shaft a by the driver cage d, itself driven by a pinion p. At the same time the balls roll on the surface of the thrust cup c. The thrust plate t is free to rotate on its

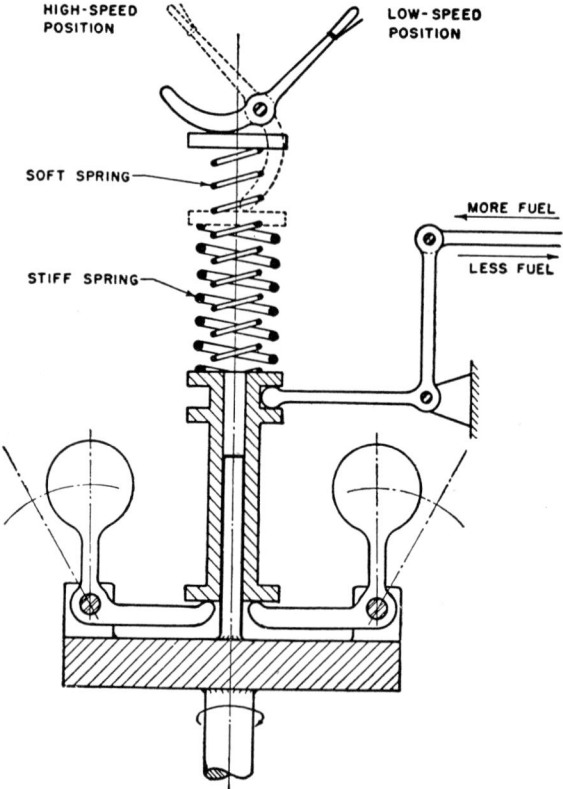

FIG. 9-4. Two-speed centrifugal governor.

bearing and is forced by the balls to rotate in the same direction as the ball driver d. When the engine speed increases, the balls, under the action of the centrifugal force, move farther from the axis of the shaft a and, owing to the shape of the cup c, push it against the spring s, and move the yoke pin y to the right. Once the governor has been adjusted there is only one position of the balls for each engine speed.

The shape of the thrust cup is made such that at low speed, when the balls are close to the axis, a relatively small centrifugal force produces a considerable thrust against the governor spring, as shown in Fig. 9-5b. As the balls fly farther apart with an increase of the engine and governor speed, as shown in dotted lines, they roll on that portion of the thrust cup which has a smaller radius of curvature. Therefore, the axial component of the centrifugal force and hence the governor power increases much less than in a governor with flyballs fastened to bell cranks as illustrated by Fig. 9-5c. Conversely, if at a high speed this governor has the same power as a conventional governor, at a low speed it will have a much greater power. The cup c can slide axially on the shaft but is prevented from rotating by a finger g.

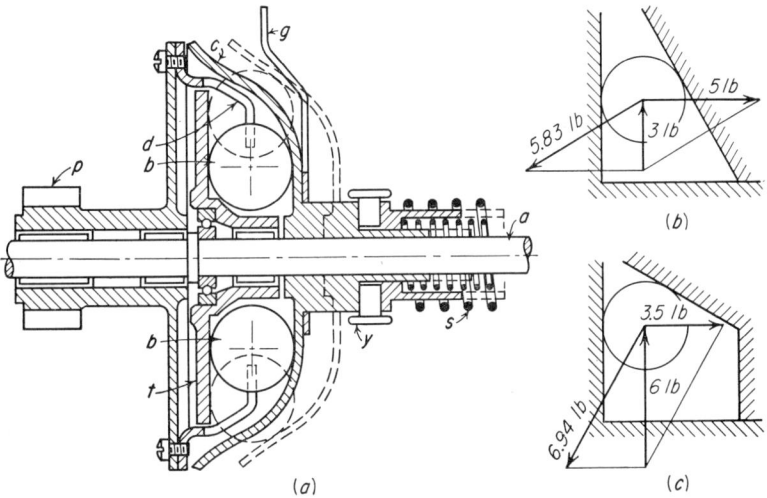

Fig. 9-5. Pivotless small-engine governor.

Figure 9-6 illustrates the action of the governor upon the fuel control and shows the speed control shaft d in the position corresponding to full-throttle operation. The speed control fork f is turned so that the pump rack is in full-load position with the yoke e against the adjustable stop screw h. The engine is running at its maximum full-load speed. If the load decreases, the engine speed will begin to increase. The higher speed moves the balls b, b out and this pushes the thrust cup c and with it the yoke stud y to the right. As a result the upper end of the yoke e moves to the right and pulls the pump rack in the same direction. The extreme right position of the yoke, when the engine is idling, but at full speed, is shown in dotted lines.

When the shaft d is turned clockwise, the engine speed will be decreased for all loads. If the engine load is increased above the full-load capacity, the engine speed will decrease. If, as in a truck engine on a steep hill, the engine slows down and it is necessary to increase the engine torque, the control shaft can be turned counterclockwise from its full-load position. This will increase the fuel delivery, the mean effective pressure will go up, and the engine will continue to pull—although it will be overloaded. The flyballs will be in their inner position. Under these conditions the engine is no longer controlled by the governor, until it picks up speed and the flyballs begin to act.

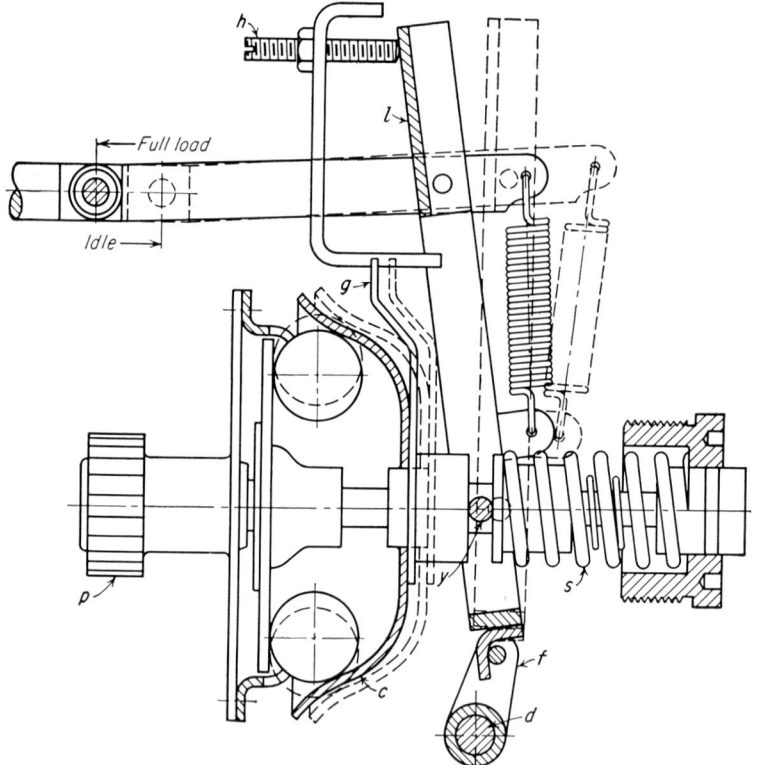

FIG. 9-6. Operating scheme of pivotless governor.

Advantages and Disadvantages. The advantages of a mechanical governor are simplicity, small size and weight, and low cost.

On the other hand, a mechanical governor has the following disadvantages: (1) inability to obtain a close speed regulation, particularly to maintain a constant speed, (2) limited power, and (3) poor sensitivity, since the speed-sensitive element, the spring, must also furnish the force to operate the engine control.

However, in most applications where the engine is used for driving machinery other than electric generators, the less expensive mechanical governor will usually suffice.

9-6. Relay-type Governors. To overcome the shortcomings of a direct-action centrifugal governor, the governor must have an independent source of power and use the centrifugal force only to put this power to work. Such an arrangement is naturally much more complicated than a mechanical governor. However, if the load conditions demand an exceptionally close speed control, the complication and expense of a relay-type governor are justified. With it almost any desired degree of speed control may be obtained by adding a number of compensating features to the control mechanism. A relay-type governor can be made to operate even on such a small speed change as 0.01 per cent. The relay-type governors are machines in themselves and, as such, are built by only a few special manufacturers. However, with the wide introduction of electric clocks, whose speed depends on the frequency of the current supplied, most of the utilities were forced to use this type of governor on their prime movers.

The method of obtaining truly constant speed or isochronous governing from a spring-loaded centrifugal governor basically consists in bringing the compression of the speeder spring back to a constant value after each speed change, regardless of the movement of the fuel-control mechanism and regardless of the engine load. This may be accomplished by using an indirect connection between the control sleeve and the fuel-control mechanism. Such an indirect connection may be provided either by an independent energy source or by the same source which is used to operate the fuel-control mechanism. The latter is conveniently furnished by oil pressure, produced by a special gear pump. Hence governors of this type are often called *hydraulic governors*.

Hydraulic Governor. A basic or simple hydraulic governor is illustrated by Fig. 9-7. The element which responds to a change in speed consists of a pair of flyballs b, b loaded by a spring s. The balls operate a pilot valve v which controls the flow of oil to and from a hydraulic power piston p. The piston p is connected to the fuel-control mechanism. When the governor is operating at control speed, the edges of the pilot-valve register with and just close the ports in the pilot-valve bushing and there is no flow of oil. When the engine and governor speed increase, the flyweights move out and raise the valve v. This connects the space under the power piston with the oil outlet to the sump. The spring q pushes the piston p to the left toward its no-load position and displaces some oil into the sump, through the line c past the needle valve t. The movement of the piston p reduces the fuel supply, the engine slows down and the speeder spring s, overcoming the reduced centrifugal force of the flyweights, moves the pilot valve v downward, covers the ports in the bushing, and stops the oil drain into the sump. When the engine speed decreases below the control speed, the flyweights move inward, lowering the pilot valve v and admitting oil under pressure from the oil supply into the space under the power piston. The latter moves to the right and pushes the fuel control toward the full-load position.

The length of the lower end of the pilot valve, called the *land*, is exactly equal to the width of the ports in the valve bushing. Since the diameter of the lower end of the pilot valve is exactly the same as that

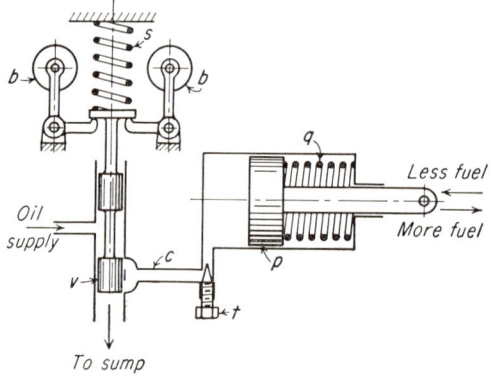

FIG. 9-7. Floating-action hydraulic governor.

of the upper end, the pilot valve is perfectly balanced axially and there is no axial thrust on the valve. The ports in the bushing are cut symmetrically so that there is no side thrust either. In order to increase further the sensitivity of the governor, the pilot valve is slowly rotated in the bushing to maintain an oil film between the valve and bushing and thus prevent sticking. The throttling action of valve t is necessary to assure stability under all loads, but results in a sluggish response to major load changes.

Compensation. If throttling is not used and the engine returns to control speed after a speed change and the fuel controls are set as required during the speed change, the governor will overcorrect the speed, will overshoot it, and then will begin to act in the opposite direction, and so on, or will *hunt*. The simple hydraulic governor without throttling has this fault. As long as the speed is below or above the control speed, the governor will continue to adjust the fuel controls. There is always a lag between the moment that a change in fuel setting is made and the time that the engine reaches a new equilibrium speed. Therefore, the engine will always return to control speed with the fuel delivery overcorrected and hunting due to overshooting will result. To avoid overshooting, a governor mechanism must anticipate the return to normal speed and must stop changing the fuel-control setting slightly before the new setting, required for sustaining the control speed, has been reached. A mechanism which enables a governor to anticipate the return to control is termed a *compensating device*.

The simplest method of compensating is to provide a speed droop with an increase in load. While this method does prevent truly isochronous governing with this type of hydraulic governor, the speed droop can be held to a minimum and the governor will still possess the advantages of fine sensitivity and of a large regulating force.

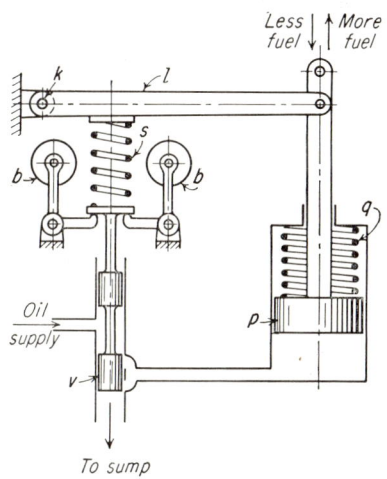

Fig. 9-8. Proportional-action hydraulic governor.

Figure 9-8 shows such a compensating device: when the speed decreases, the flyballs draw closer together and the spring s moves valve v downward, admitting oil under the piston p. This pushes it upward, compressing the spring q and moving the fuel control toward *more* fuel. The movement of the lever l, pivoted at k, slightly decreases the force of the spring s and returns v to its neutral position. This is called *primary* compensation.

Speed-droop governors are stable at any load change from small gradual ones to large sudden ones, but on variable-load service they require constant resetting to maintain a steady speed. They are therefore not suitable for certain applications such as a-c systems. Combining the floating control with the proportional action gives a stable isochronous governor. The isochronous or *secondary* compensation mechanism slowly resets the governor after each load change.

Figure 9-9 shows a small diesel-engine governor equipped with such a mechanism. Oil from the pump l, which can rotate in either direction, is stored in the accumulator a. The pressure in the accumulator is maintained constant by the spring-loaded piston c and relief opening d and is applied continuously through the line m to the left smaller surface of the differential piston p. When the speed decreases, the spring *s* overcomes the flyweights *b*, *b* and forces the pilot *v* downward. This admits oil past the edge *e* to the line *g* and the larger piston surface and moves *p* to the left, in the direction of greater fuel supply. At the same time the movement of *p* uncovers an opening *n* to the oil line *o* and the oil pressure acting on piston *k* compresses the spring *q* opposing the action of spring *s*, raising pilot *v* and cutting off the oil feed at *e*. At this moment the oil bleed through the needle valve *t* is insignificantly small. In time, oil bleed or feed through *t* under the push or pull of *q* relieves the force of *q* and causes pilot *v* to come to

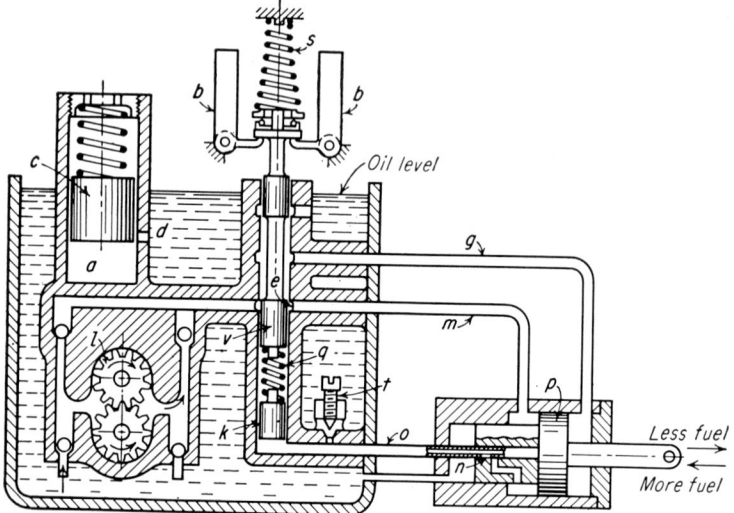

FIG. 9-9. Isochronous diesel-engine governor.

the set point, isochronous equilibrium, with corresponding further correction of the fuel delivery and speed.

Load Distribution. An engine equipped with an isochronous governor can carry any load between no load and the maximum load or overload that the governor and engine will permit. If two engines are coupled to a single load they cannot both be equipped with isochronous governors. Since isochronous governors will permit any fuel setting within the capacity of the system, providing the speed remains constant, they are incapable of distributing the load between two engines in a predictable manner. Engines which are to be operated in parallel must use governors with speed droop. Not more than one engine in such a system may be isochronously governed.

When engines equipped with isochronous governors must be used for parallel operation, some method for introducing speed droop must be employed. This is ordinarily done by a linkage between the fuel control mechanism and the speeder-spring compression regulator. The lever linkage must be made adjustable to obtain the speed droop required to distribute the load adequately.

9-7. Load-limit Governors. At any particular engine speed there is a maximum sustained load which an engine can carry without damage. An ordinary governor, whether isochronous or having a speed droop, is sensitive only to speed. If the engine slows down, the governor will increase the fuel supply even though this may overload the engine. To prevent the governor from increasing the fuel supply beyond the limit required for a safe load, the governor of a variable-speed engine such as that on a diesel locomotive must be equipped with a maximum-fuel stop. If a fixed stop were used, it would permit the engine to be overloaded at low speeds if the stop were set to allow the maximum safe load at full speed. To give full protection throughout the full engine speed range and at the same time to allow the engine to develop its maximum permissible power, the governor must be equipped with a variable maximum-fuel stop which will permit the delivery of the maximum safe fuel supply at any engine speed. Such a device is termed a *bmep limiter* or *torque limiter*.

The maximum fuel delivery permitted by a torque limiter may be set slightly below the smoke limit for the engine, so that the engine will automatically give practically smokeless operation.

If a very heavy load is put on an engine so that the maximum-fuel stop comes into operation, the engine speed will drop below the governor control speed and the engine will operate with a fixed rack setting as though it had no governor. Most types of load demand less driving torque as speed decreases, and in the usual case the engine will be able to carry the load, without stalling at some reduced speed. The engine will develop a somewhat reduced torque and a greatly reduced power as its speed decreases.

A typical load-limit hydraulic governor is illustrated in Fig. 9-10. It consists of a spring-loaded centrifugal governor which operates a hydraulic pilot valve which controls the load-limiting mechanism. This mechanism consists of a plunger, or piston p, to whose right end is attached a cam c which limits the movement of the main fuel controls. Spring q tends to pull piston p to the left. There is also a sloping cam surface cut into the lower side of the plunger p on which rides the roller attached to the top of the speeder spring s. In operation, when the speed increases, the centrifugal force of the flyweights b, b moves the pilot valve v up and uncovers the control port n, permitting oil trapped in the cylinder r to drain and thus allowing the plunger p to move to the left. This moves the cam c to the position permitting maximum opening of the fuel controls. When the engine slows down, pilot valve v moves down and opens the control port n so that it connects cylinder r with the supply of oil under pressure. The oil pressure acting on the plunger p will move it to the right until the control port n is again covered. As the plunger p moves to the right, the roller riding on the piston-cam surface will move up to reduce the compression of the spring s so that the centrifugal force of the flyweights b, b will be balanced at a lower speed. Thus, for every

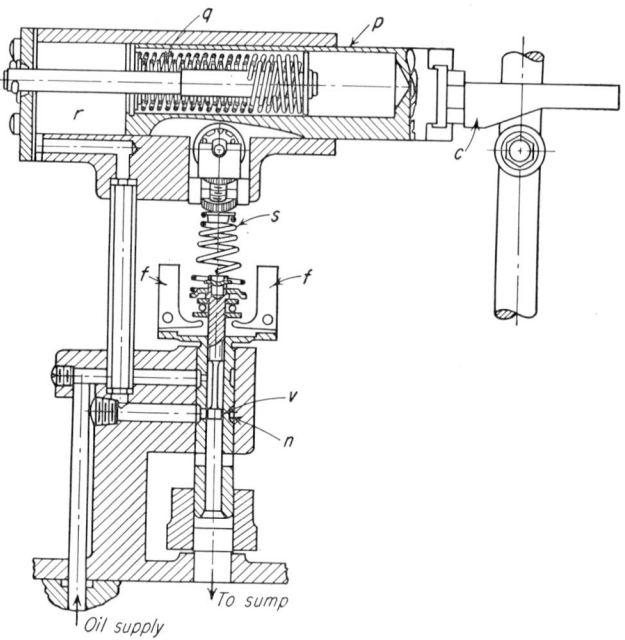

Fig. 9-10. Load-limit hydraulic governor.

engine speed there is a corresponding position of the power plunger which will provide the necessary spring compression to balance the centrifugal force of the flyweights and maintain the pilot valve in its control position covering the control port.

At very low speeds, plunger p will have to be moved all the way to the right before the compression of spring c can be reduced sufficiently to balance the low centrifugal force of the flyweights. This will cause cam c to stop the movement of the fuel control at its lowest limit, by means of the steep slope on the end of the cam. This will prevent overloading of the engine at low speeds, where the engine is more likely to smoke. By proper design of the slope of the cam c the load for each position corresponding to a certain engine speed will thus be limited to prevent overloading.

Another type of load-limit governor is sometimes used to prevent too much fuel injection during engine starting. This consists of a fuel control stop which is operated by compressed air and functions only when the air-starting system is operating. The air-starting system may also be used to permit the hydraulic governor to function immediately upon starting. A booster servomotor, meaning auxiliary power supply, is employed to furnish oil pressure to operate the controls until the regular oil pump supplying the governor can build up sufficient pressure. The booster servomotor is operated by compressed air from the air starting system during the starting period of the engine.

9-8. Overspeed Governors and Trips. Overspeed governors are employed as safety devices to protect engines from damage due to overspeeding from any cause. When an engine is equipped with a regular speed governor of any type, the overspeed governor will function in the event of failure of operation of the regular governor. If the engine speed is manually controlled, the overspeed governor will function if the speed should increase beyond a safe limit before the operator can control it.

Since overspeed governors are essentially emergency controls, they must operate either to stop combustion or to limit the combustion pressures in the engine cylinder in order to slow it down. This control of combustion may be obtained by regulation either of the fuel or of the air supply. Most overspeed governors function to cut off or limit the fuel supplied to the engine cylinders. In some two-stroke engines, however, it is possible for an engine to run away by burning lubricating oil which may happen to be taken in with the fresh air. Where this may occur, the governor is arranged to cut off the air supply to the cylinder and thus to stop the engine. Overspeed governors which bring an engine to a full stop by cutting off all the fuel or air supply are commonly termed *overspeed trips*. If the overspeed control merely slows the engine down but allows it to continue to run at safe operating speeds, it is more properly termed an *overspeed governor*. In marine service, particularly for main propulsion diesel engines, overspeed governors are preferred to overspeed trips since the latter type may leave a ship without power at a critical moment.

Overspeed governors and trips of all types depend upon a spring-loaded centrifugal-governor element for their action. The spring is preloaded to a force which will overbalance the centrifugal force of the flyweights until the engine speed rises above the prescribed maximum. When this speed is reached, the centrifugal force overcomes the spring force and puts to action the controls which cut off or limit the fuel or air supply.

The actual operation of the fuel or air controls may be accomplished directly by the centrifugal force of the overspeed governor, as in a mechanical governor, or it may be supplied by oil pressure, as in a hydraulic governor. In an overspeed trip, the shut-off control may be operated by the force of a power spring which is put under tension when the trip is manually reset and is held there by means of a latch. When the maximum speed limit is exceeded, a spring-loaded centrifugal flyweight will move out and trip the latch, allowing the power spring to operate the shut-off control.

QUESTIONS

1. What is the purpose of a diesel-engine governor?
2. On what two factors does the governor action depend?
3. Enumerate the losses that absorb part of the indicated power developed by combustion of the fuel.
4. What is the name given to the useful output of an engine after deduction of all losses?
5. By what means is the speed of a diesel engine kept practically constant in spite of a varying external load?
6. Enumerate the four main classifications of diesel-engine governors.
7. State the two different methods of action of governors upon the fuel-system of diesel engines.
8. Enumerate the main characteristics of a governor that determine the degree of control of the engine.
9. What is the speed droop of a governor?
10. What is isochronous governing?
11. What is the sensitivity of a governor?
12. What is the stability of a governor?
13. What is hunting?
14. What is the promptness of a governor?
15. What is the power of a governor?
16. Explain, giving a sketch, the action of a variable-speed spring-loaded governor.
17. Explain, giving a sketch, the action of a two-speed direct-action governor.
18. Enumerate the advantages and disadvantages of a mechanical governor.
19. What is the main feature and advantage of a relay-type hydraulic governor?
20. Explain, giving a sketch, the function of a simple hydraulic governor.
21. What is compensation in a hydraulic governor?
22. Explain, with a sketch, the action of a compensating device in a hydraulic governor.
23. What is the object of load-limit governors?
24. Give a sketch of a load-limit governor.
25. By what two methods do overspeed governors act?
26. What is an overspeed trip?
27. Explain the action of an overspeed trip on fuel or air supply.

PART 3

FUNDAMENTALS OF DIESEL ENGINE WORK

CHAPTER 10

FUELS AND COMBUSTION

10-1. Diesel Fuels. Diesel fuels used in this country are obtained by the distillation or the cracking of petroleum, or crude oil. Crude oil is a dark-brown liquid which is a mixture of a large number of various compounds. The main chemical elements which form all these compounds are hydrogen and carbon. For this reason the compounds are called *hydrocarbons*. The amount of hydrogen in the compounds varies from 11 to 15 per cent by weight, and the balance is carbon.

The crude oil found in a certain locality usually has some properties which distinguish it from crude oils found in other places. Thus some crude oils, such as those found in the Eastern states, contain much light oil or gasoline, considerable amounts of paraffin wax, and very little asphaltic material. After distillation of these crude oils, the residue consists mostly of paraffin wax, and for this reason they are called *paraffin-base* oils. Other crude oils, such as those found in California, contain smaller amounts of gasoline, large amounts of asphaltic materials, and often relatively high percentages of sulfur. After distillation these oils leave a residue that consists mostly of asphalt and are therefore called *asphalt-base* oils. Many crude oils, such as those found in the mid-continental states, have some features of one and some of the other type and are classified as *mixed-base* oils.

Distillation. The crude oil is separated into its products by a process called *fractional distillation*, which may be briefly described as follows: crude oil contained in a closed vessel is heated by a coil through which steam or hot gases are circulated. At first the compounds of low boiling points are driven off as vapors. These vapors are led away by a pipe connected to the top of the vessel, condensed by cooling with a coil through which cold water is circulated, and drained into a tank. The temperature of the crude oil is maintained constant. After all compounds which boil under and at this temperature are driven off, or distilled, the flow of hot gases through the heating coil is increased, the temperature of the crude oil is raised, and the vapors distilled are drained, after condensation, into another tank, and so on. The products obtained by distillation, in the order of increasing boiling points, are gasoline, distillate, kerosene, gas oil, and diesel fuel oil. Lubricating oils are distilled off later or left in the unvaporized residue. Additional gasoline and diesel fuel oil may be obtained by *cracking* the residue. *Cracking* is a process in which complex hydrocarbons composing the residue or crude oil are broken up by heat and pressure into lighter hydrocarbons of simpler molecular structure.

Diesel Fuels. Low-speed diesel engines can operate on almost any liquid fuel, from kerosene to Bunker C oil. Modern high-speed diesel engines, owing to the short time interval available for combustion at each cycle, require a more special, comparatively light fuel oil. A suitable diesel oil may be obtained by straight distillation, by cracking, or by the blending of several oils.

Specific gravity is the ratio of the weight of a certain volume of a substance to the weight of an equal volume of water. Another scale of density is the Baumé scale modified by the American Petroleum Institute. The relation between degrees API and specific gravity is

$$\text{Degrees API} = \frac{141.5}{\text{sp gr}} - 131.5 \qquad (10\text{-}1)$$

or

$$\text{Specific gravity} = \frac{141.5}{\text{deg API} + 131.5} \qquad (10\text{-}2)$$

Specific gravity has no direct bearing upon the burning quality of a fuel oil. However, other requirements confine the limits of specific gravity to about 0.82 to 0.89, or 41 to 27° API, for high-speed engines, to 0.91 or 24° API for other mechanical-injection engines. Air-injection engines can use fuels as heavy as 0.94 or 19° API and in some cases even heavier.

10-2. Properties of Fuel Oils. The following properties influence the performance and reliability of diesel engines: (1) volatility, (2) carbon residue, (3) viscosity, (4) sulfur content, (5) ash, (6) water and sediment, (7) flash point, (8) pour point, (9) corrosiveness and acidity, and (10) ignition quality. However, ignition quality is important only for high-speed engines and for this reason it is listed last in spite of its great importance for these engines.

Volatility of a diesel fuel is measured by the 90 per cent distillation temperature. This is the temperature at which 90 per cent of a sample of the oil has distilled off. The lower this temperature, the higher the volatility. For small engines a higher fuel volatility is necessary than for larger engines in order to obtain a low fuel consumption, low exhaust temperature, and a minimum of smoke. United States Navy specifications for high-output diesel engines give a maximum 90 per cent distillation temperature of 675 F.

Carbon residue is the carbon left after the evaporation and burning off of volatile matter from a sample of oil by heating; it indicates the tendency of the fuel to form carbon deposits on engine parts. A maximum carbon residue of 0.10 per cent is allowable.

Viscosity of a fluid is the measure of its resistance to flow or internal friction. Viscosity of an oil is expressed by the number of seconds which it takes a certain volume of oil to flow through a hole of a certain small diameter. The smaller this number of seconds, the lower the viscosity. The apparatus used in this country for determining the viscosity of oil is the *Saybolt viscosimeter* and the *universal* orifice and the data it gives are designated as so many SSU (seconds Saybolt Universal). Lubrication, friction between moving parts, their wear, and leakage—all these factors are influenced by viscosity. Lubrication of the parts of the fuel-injection system, chiefly of the plungers and barrels of the high-pressure pumps, depends entirely on the fuel oil, and its viscosity should therefore not be below a certain minimum value. Fuel-oil leakage past the packingless plunger of a high-pressure pump is in inverse proportion to the oil viscosity. Therefore a fuel oil with a very low viscosity is undesirable also because it is apt to give considerable pump leakage. The specifications usually prescribe a viscosity of 34 to 45 SSU at 100 F.

On the other hand, the viscosity should not be appreciably higher than the one specified because an increase of viscosity in a fuel oil means a greater resistance to breaking up during injection. This undesirable excess in viscosity may be overcome with relatively light fuels, such as those used in high-speed airless-injection engines, by raising the injection pressure until the desired atomization is obtained, and with very heavy and viscous oils, such as those sometimes used in air-injection engines, by heating the oil in special heaters.

Sulfur in the fuel burns with the oil and produces highly corrosive gases that are condensed by the cooled cylinder walls, especially when the engine operates under a light load and the cylinder temperature drops. Corrosion due to sulfur gases is frequently found in the exhaust system of diesel engines. Various specifications do not permit a sulfur content over 0.5 to 1.5 per cent.

Ash and *sediment* in a fuel are a source of abrasive material which will cause excessive engine wear. Sediment may also cause clogging of the fuel system. Wear may be increased due to corrosion if the fuel contains water, especially salt water. A maximum permissible ash content is 0.01 per cent and of water and sediment, together, 0.05 per cent.

Flash point is called the lowest temperature to which an oil must be heated in order to give off inflammable vapors in a sufficient amount to flash or momentarily ignite when brought in contact with a flame. A fuel oil having a low flash point is dangerous in storage and handling. The minimum flash point for diesel fuel is 150 F.

Pour point is the temperature at which an oil congeals, or ceases to flow. The pour point is important in respect to cold starting an engine and for handling the oil between the storage and engine. The maximum pour point for diesel fuel is 0 F.

Corrosiveness. A fuel oil must not be corrosive, must not contain free acids. Otherwise it may damage metal surfaces with which it comes in contact in storage or in the engine.

Ignition Quality. This name designates the ability of a fuel to ignite when it is injected into the compressed-air charge in the diesel-engine cylinder. A fuel with a good ignition quality ignites readily, with a small ignition delay; a fuel with a poorer ignition quality ignites with a greater delay.

Ignition quality is one of the most important properties of a diesel fuel for use in a high-speed engine. The ignition quality of a fuel determines not only the ease of ignition and of starting a cold engine but also the kind of combustion obtained from the fuel. A fuel with a better ignition quality gives smoother, less noisy engine operation—qualities particularly noticeable at light loads.

Cetane Number. Ignition quality is measured by an index called *cetane number*. Present high-speed diesel engines require a cetane number of about 50. The value of this number as a diesel-fuel characteristic is similar to that of the octane number for gasoline. The cetane number of a fuel is the percentage of cetane by volume in a mixture of *cetane* and *alpha-methyl-naphthalene* that has the same ignition quality as the fuel which is being tested. Both cetane and alpha-methyl-naphthalene are hydrocarbons, produced chemically from tar oil. Cetane has an excellent ignition quality and alpha-methyl-naphthalene has a very poor ignition quality. The scale runs from 0 to 100, pure alpha-methyl-naphthalene corresponding to 0 and pure cetane to 100. A cetane number of 48 would mean a fuel equivalent to a mixture composed of 48 per cent cetane and 52 per cent alpha-methyl-naphthalene.

The cetane number of a fuel sample is determined by testing it in a special single-cylinder test engine with a variable compression ratio. The test procedure is based on the fact that the ignition-delay period in a given engine at a fixed engine speed decreases with an increase of the compression ratio. The delay period is measured from the moment the fuel-injection valve leaves its seat until the ignition of the fuel produces a measurable pressure rise in the cylinder. An ignition-delay period of standard length, 13 crank-angle degrees, is used as reference; the test fuel is burned in the engine, and the compression ratio is raised in the engine until the 13-deg ignition-delay period, as indicated by special instruments, is reached and the required compression ratio is noted. Next the engine is run using two mixtures of cetane and alpha-methyl-naphthalene, one having a cetane number about five units higher and the other about five units lower than the expected cetane number of the fuel. The compression ratios of these mixtures to obtain a 13-deg ignition delay are found and by proportioning or interpolation the cetane number of the sample is computed. Fuels with good ignition qualities require lower compression ratios for the 13-deg ignition delay and have a higher cetane number. Fuels with poorer ignition qualities require higher compression ratios for the 13-deg ignition delay and have a lower cetane number.

Additives. The ignition quality of a diesel fuel can be improved, *i.e.*, its cetane number raised, by the addition of small percentages of more easily ignitable hydrocarbons. Fuel oils with such additives are available, as are additives which, used in small percentages, will improve a diesel fuel oil for use in high-speed engines.

Importance of Specifications. A fuel oil which does not comply with the specifications outlined above may have very harmful effects upon the engine. The main effects of unsatisfactory fuel-oil properties may be enumerated as follows:

1. Low *volatility* reduces maximum power output, increases fuel consumption, gives smoky exhaust—with the consequences explained in Sec. 20-13 and 20-18—and makes the starting of a cold engine more difficult.

2. High *carbon residue* produces deposits of carbon and of a gummy substance on pistons and cylinder liners and may cause sticking of piston rings and valve stems.

3. High *viscosity* may cause smoky exhaust; low viscosity gives excessive wear of injection-pump plungers and barrels and may cause pump leakage and contamination of crankcase oil by fuel oil.

4. Excessive *sulfur* content, *ash*, and *sediment* cause wear of pistons, piston rings, liners, and fuel-injection equipment.

5. High *pour point* may interfere with the starting of a cold engine.

6. *Corrosiveness* and *acidity* will cause rapid wear of various engine parts.

7. Low *cetane number*, in high-speed engines, causes difficult starting of a cold engine and rough and noisy operation.

Gasoline as Diesel Fuel. A good gasoline with a high octane number does not knock in a spark-ignition engine because it burns comparatively slowly, hence does not ignite as readily and has a low cetane number. In general the higher the octane number of a fuel, the lower its cetane number, and vice versa. Thus gasoline is not a desirable diesel fuel. However, tests have shown that diesel engines can run on gasoline, although the engine will be rather noisy and rough and the fuel consumption will be higher than with regular diesel fuel oil. By adding 10 to 15 per cent of regular diesel fuel the performance of the engine can be considerably improved. Another drawback of running a diesel engine on gasoline is the danger of wear and seizure of the fuel-injection pumps because of the low viscosity of gasoline; the addition of lubricating oil to gasoline will raise the viscosity and help the situation.

10-3. Ignition. In diesel engines, the fuel is ignited and combustion started by the heat of compression. The procedure is as follows: shortly before the end of the compression stroke fuel is admitted into the cylinder as a fine mist or fog; the cylinder at this moment is filled with practically pure air whose temperature was raised by the compression to about 900 to 950 F; the particles of fuel pick up heat from the air, begin to turn into vapor, and soon some vapor particles ignite; this develops additional heat and helps to ignite other vapor particles.

Ignition Delay. The time which it takes to heat the fuel particles, turn them into vapor, and start combustion is called *ignition lag* or *delay.* It depends upon the ignition quality of the fuel and certain conditions, such as engine speed and compression ratio.

In high-speed engines the ignition delay is of the order of 0.0012 to 0.0018 sec. It decreases with an increase of the engine speed owing to improvement of air movement, or turbulence in the cylinder, which results in better heating of the admitted fuel.

10-4. Combustion. Combustion is a chemical reaction in which certain elements of the fuel after it is ignited combine with oxygen, developing heat and thus increasing the temperature and pressure of the gases. The main combustible elements are *carbon* and *hydrogen;* another combustible element, undesirable and contained in small amounts, is *sulfur.* The oxygen necessary for combustion is obtained from air, which is a mixture of oxygen and nitrogen. Nitrogen is an inert gas and does not participate in the combustion process. During the process of combustion, the fuel oil particles are split into their component elements, hydrogen and carbon, and each element combines with the oxygen of the air separately. Hydrogen combines with oxygen to form water, and carbon and oxygen combine into carbon dioxide. If there is not enough oxygen present, part of the carbon, combining with oxygen, will form carbon monoxide. When carbon monoxide is formed, the amount of heat developed by each unit of carbon is only 30 per cent of the heat developed by the formation of carbon dioxide. Actual diesel engines always operate with excess air and produce only a very small amount of carbon monoxide.

Air-fuel Ratio. Theoretically, about 14.5 lb of air is required for the combustion of 1 lb of fuel oil. However, under such conditions, some particles of oxygen, diluted by nitrogen and products of combustion, will not be able to participate in the process of combustion, owing to the exceedingly short time during which combustion must take place; some carbon monoxide will then be formed or carbon particles will be left unburned. Therefore, to insure complete combustion of the fuel and avoid the heat loss due to formation of carbon monoxide and unburned carbon, an excess of air must be present in the cylinder. The ratio of the weight of air present to the weight of fuel injected during each power stroke is called the *air-fuel ratio*. The air-fuel ratio is a very important factor in the operation of a combustion engine. When a diesel engine operates at a light load, the actual air-fuel ratio is several times greater than the theoretical value of 14.5. As the load is increased, more fuel is injected; however, the amount of air in the cylinder remains practically constant and, as a result, the air-fuel ratio decreases. But even when the engine is fully loaded, the air-fuel ratio must be at least 25 to 30 per cent greater than 14.5. A considerable excess of air over the minimum required for complete combustion must thus be present in the cylinder.

Combustion Products. When fuel oil is burned in the diesel-engine cylinder, the gases present after combustion, called *products of combustion*, consist of water vapor, carbon dioxide, unused oxygen, and nitrogen. In addition, they may contain very small amounts of carbon monoxide, hydrogen, and traces of other gases formed at high temperatures. However, when diesel-engine exhaust gases are analyzed by means of an Orsat apparatus, usually only carbon dioxide and oxygen are measured and nitrogen is determined as balance.

Knowing the exhaust gas analysis, the amount of excess air may be computed and from it the air-fuel ratio.

Incomplete Combustion. Smoke. Even in the presence of excess air, there is a possibility that some particles of fuel will not come in contact with oxygen. However, the fuel particles are cracked into hydrogen and carbon molecules by the high temperatures that prevail during combustion in a diesel engine. The hydrogen molecules combine with oxygen more readily than do the carbon molecules, leaving unburned carbon molecules which appear as smoke in the exhaust or are deposited as greasy soot in the combustion space or the exhaust system. A certain amount of smoke is formed also by cracking and incomplete combustion of lubricating oil. However, smoke formed by lubricating oil has a bluish color, whereas smoke formed by fuel oil has a gray to black color, depending upon the air-fuel ratio and the thoroughness with which the fuel is mixed with the air.

10-5. Heat Values. The chemical reaction between the fuel and the oxygen of the air produces heat. The heat thus generated when one pound of fuel is completely burned is called the *heat value* of the fuel.

High Heat Value. Heat values are determined experimentally by calorimeters. In these instruments the products of combustion are cooled to practically room temperature, at which most of the water vapor formed by combustion of the hydrogen of the fuel is condensed and gives up its latent heat. The heat value thus determined is called *high heat value* and designated Q_h.

Low Heat Value. The heat value of the fuel without the latent heat of condensation of the vapor is called the low heat value and designated Q_l. In calculating the thermal efficiency of internal-combustion engines, it is logical to use the low heat value of the fuel because at the high temperature at which the exhaust gases leave the engine no condensation of water vapor can take place. However, the high heat value is often used because the corresponding figure is usually more readily available. Also the ASME code of testing prescribes the use of high heat values, whereas the SAE rules prescribe low heat values.

Diesel Fuels. The heat values of diesel oils vary slightly with their composition, mainly with the hydrogen content. For an oil with a known gravity in degrees API, the heat values Q_h and Q_l may be computed from the following formulas:

For heavy and medium oils, 19 to 30 °API,

$$Q_h = 17{,}680 + 60 \times °\text{API} \tag{10-3}$$
$$Q_l = 16{,}580 + 60 \times °\text{API} \tag{10-4}$$

For light oils, as used in high-speed engines, 31 to 41 °API,

$$Q_h = 18{,}030 + 40 \times °\text{API} \tag{10-5}$$
$$Q_l = 16{,}780 + 40 \times °\text{API} \tag{10-6}$$

Thus, the average value of the low heat value for the heavier fuel is 18,000 Btu per lb, and for the lighter fuels it is 18,200 Btu per lb.

The difference between the low and high heat values is about 5.5 per cent, as may be computed by Eqs. (10-3) to (10-6).

10-6. Combustion in Diesel Engines. *Air-injection Engines.* The breaking up, called *atomization*, and distribution of the fuel in air-injection engines is so efficient that the ignition lag is very small and no problems concerning combustion arose until airless injection became more widely accepted and the rotary speed began to exceed considerably the speed of air-injection engines.

Airless-injection Engines. When injection begins, finely atomized fuel particles come in contact with air preheated by the compression stroke. First, their temperature increases; then they begin to vaporize, and the temperature of the vapor particles increases. When the temperature reaches the ignition point, a quick chemical reaction starts, causing an increase of pressure and temperature and spreading to the rest of the fuel in the combustion space. Ignition does not

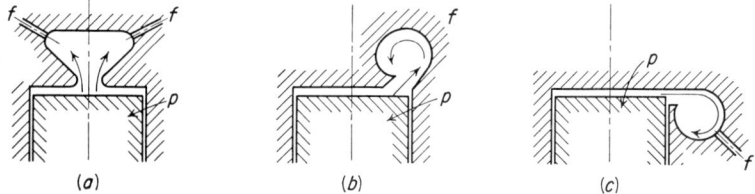

Fig. 10-1. Combustion spaces with turbulence.

always start at the same point, but at a place or places determined by conditions of temperature and fuel distribution; and it may start at several points simultaneously.

Turbulence. An essential condition for efficient combustion, particularly in high-speed engines, is a sufficient motion between the fuel droplets and the air. When the fuel is broken up in the form of a mist, the velocity of the spray and range of its penetration into remote points of the combustion chamber are reduced to rather low values. Thus distribution of the fuel and its mixing with air must depend upon motion of the air. This motion, called *turbulence*, is obtained by various means, such as by giving a particular shape to the combustion space or to the top of the piston or by directing the flow of the incoming air charge in a certain way, etc.

Figure 10-1a shows a Price head in which turbulence is produced by restriction through which the air is pushed when the piston p moves upward. The air velocity at the restriction is several times higher than before and after it, and the change of velocity results in a powerful general turbulence in the combustion chamber into which the fuel is injected from the fuel nozzles f, f. Figure 10-1b shows a Ricardo-

Comet head used in Waukesha engines. In this design the turbulence is produced both by a restriction and by forcing the air to travel in a circular path. Figure 10-1c shows a similar turbulence head used in Hercules engines which has an additional feature: when the piston approaches dead center, it begins to cover gradually the air passage between the cylinder and turbulence chamber. This increases the air velocity in the passage and counteracts the slowing down of the piston.

Figure 10-2 shows a combustion chamber with a dished piston. The rim of the piston, at dead cen-

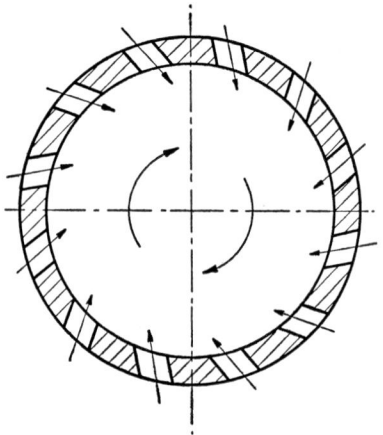

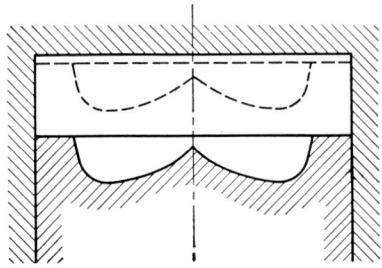

FIG. 10-2. Combustion space with a dished piston.

FIG. 10-3. Turbulence in a two-stroke diesel engine.

ter, comes close to the cylinder head and forces air to flow from the circumference toward the center and thus produces a certain turbulence.

In two-stroke engines a turbulence may be created by making the scavenge-air ports tangential[1] (Fig. 10-3). It should be noted that a circular movement of the air started during scavenging continues up to the time of fuel injection, regardless of the fact that the air has been moved from one end of the cylinder to the other and highly compressed—to a few per cent of its original volume.

High-speed Engines. A better understanding of what happens in a diesel-engine cylinder during the combustion period may be obtained by means of a graphic presentation. The changing pressures are plotted as ordinates against time as abscissas. Since the rotation of the crankshaft can be considered for most purposes to be uniform, the degrees of crank travel can be considered proportional to time and the abscissas can be conveniently expressed in angles of crank travel. A typical pressure diagram is shown in Fig. 10-4; it indicates pressure changes during 180 deg, from 90 deg before to 90 deg after top dead center. The first half of the diagram, the full line up to point 2 and

[1] A definition of this word is given in the Glossary.

the dotted part to point O, represents the change of pressure in the cylinder during the compression stroke. The symmetrical dotted curve on the right side shows the expansion of the air charge if no fuel is injected.

When fuel is injected and combustion takes place, the process in a high-speed diesel engine may be considered as divided into four separate stages or periods. The first period begins at point 1 when injection begins, the fuel starts to enter into the cylinder, and lasts to point 2. This is the *delay period;* it corresponds to a crank-travel

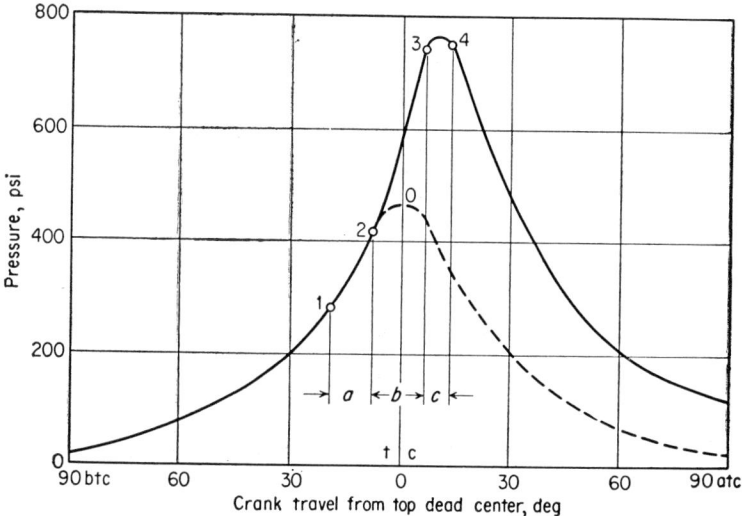

Fig. 10-4. Combustion stages in a high-speed diesel engine.

angle a. During this period there is no pressure rise above that produced by the compression of the air by the piston. Fuel is coming in through the nozzle all the time and at point 2 there is, in the combustion space, an appreciable amount of fuel, finely divided and partly evaporated, getting ready for combustion. When the fuel at last ignites (point 2), it burns rapidly causing a sudden pressure rise until point 3 is reached. This period of *rapid combustion,* corresponding to a crank angle b, forms the second stage. After point 3, fuel which has not yet burned and fuel which continues to be injected burn at a rate that depends upon the rate of injection and the amount and distribution of the oxygen left in the air charge. This period is the third stage of gradual or *controlled combustion;* it ends at point 4 with the end of injection. During this stage the pressure may increase, remain constant, or decrease. The angle c depends upon the load which the engine carries—the greater the load, the greater is c. In the fourth stage, called *after-burning*, fuel left in the cylinder when injection stops is finally burned. After-burning is not apparent on the diagram, since the receding piston causes the pressure to drop in spite of the heat developed by the burning of the last part of fuel.

Combustion Knock. If ignition is delayed so long that a large amount of the fuel is admitted, heated by the air, and vaporized, the primary ignition may be followed by such a rapid combustion that it approaches an explosion of the entire fuel mass. This explosion may set up a pressure wave, which, traveling at the speed of sound, strikes the metal walls and produces a sharp knock called *combustion knock*. This knock is sometimes called also *fuel knock* as it occurs when the fuel has a low ignition quality.

Rough Operation. Even when no real combustion knock is present, an engine may be noisy or rough if the rate of pressure increase during the second stage of combustion is too high. The allowable rate of pressure increase, in the presence of which the engine will still operate smoothly, depends upon the engine itself, its mass, rigidity, and other design features. For some engines a rate of 35 psi per degree of crank-travel is the limit, for others this rate may be as high as 55 psi per deg and even higher. An engine which begins to be rough with a 35 psi per deg pressure rise will require a fuel with a very good ignition quality, a high cetane number. An engine that can stand a high rate of pressure rise may operate smoothly on almost any fuel.

QUESTIONS

1. From what raw material are diesel fuels obtained in this country?
2. What are the main chemical elements that form petroleum?
3. What is a hydrocarbon?
4. What is the approximate percentage by weight of hydrogen in petroleum products?
5. What is the difference between paraffin-base and asphalt-base oils?
6. What diesel engines can operate on heavy fuel, such as Bunker C oil?
7. What fuel do modern high-speed diesel engines require for satisfactory operation?
8. What is the specific gravity of an oil?
9. What are *API degrees?*
10. What is the influence of specific gravity upon the burning quality of fuel oil in a diesel engine?
11. Enumerate the properties of fuel oil that influence the performance of a diesel engine.
12. How is volatility of a diesel fuel measured?
13. How is carbon residue determined? What is the permissible maximum carbon residue in fuel oil?
14. What is viscosity and how is it measured?

FUELS AND COMBUSTION

15. How is the viscosity of oils expressed numerically, and what instrument is used to determine it in the United States?
16. How does sulfur in the fuel affect certain parts of a diesel engine? What parts are affected most?
17. What is the maximum amount of sulfur in diesel fuel oil permitted by specifications?
18. What is the maximum permissible ash content in diesel fuels?
19. What is the maximum permissible content of sediment and water in diesel fuels?
20. What is the flash point of an oil?
21. What is the pour point? What is the maximum specified pour point for a diesel fuel?
22. What is ignition delay?
23. What is the approximate duration of the ignition delay in high-speed engines?
24. What is meant by ignition quality?
25. How is ignition quality measured?
26. What is the cetane number of a diesel fuel?
27. How is the cetane number of a diesel fuel determined?
28. How can the cetane number of a fuel be improved?
29. Enumerate the harmful effects that may result when the engine is operated on fuel oils that do not comply with the specifications as outlined in this chapter.
30. Can a diesel engine run on gasoline?
31. What are the undesirable features, besides higher cost, of operating a diesel engine on gasoline?
32. Describe the process of igniting the fuel oil injected into a diesel-engine cylinder.
33. What happens during combustion of fuel oil in a diesel-engine cylinder?
34. What becomes of the carbon in the fuel oil if there is not enough oxygen present in the combustion chamber of a diesel engine?
35. What is the air-fuel ratio? In what units is it expressed?
36. What is the theoretical air-fuel ratio of an average diesel fuel?
37. What actual air-fuel ratio is necessary to obtain complete combustion at maximum load?
38. What influence has a decrease of load in a diesel engine on the air-fuel ratio?
39. What are the products of combustion? Of what gases do they consist?
40. What causes incomplete combustion?
41. What may cause smoke in the exhaust of a diesel engine?
42. What is the heat value of a fuel?
43. What is high heat value?
44. What is low heat value?
45. What are the advantages of air injection in respect to ignition and combustion, compared with airless injection?
46. What is the mechanism of igniting the fuel in airless-injection engines?
47. Where does ignition start in a combustion chamber of an airless-injection engine?
48. What is turbulence in a combustion engine?
49. In what respect does turbulence help efficient combustion in a diesel engine?
50. Enumerate the various means of obtaining turbulence in the combustion chamber of a diesel engine.

51. Explain how turbulence is created in a two-stroke engine.

52. Into what stages may the combustion process in a high-speed airless-injection engine be divided?

53. What may cause noisy or rough operation of a diesel engine?

54. What is combustion knock, or fuel knock, in a diesel engine?

55. What causes combustion knock in a diesel engine?

56. What are the allowable rates of pressure rise during combustion that permit engines to operate smoothly?

57. On what engine features does the allowable rate of pressure rise in a certain engine depend?

58. How does the permissible rate of pressure rise in a certain engine influence the selection of fuel for this engine?

CHAPTER 11

FUEL SYSTEM

11-1. Fuel System. The fuel system of a diesel-engine installation is defined as the equipment which is necessary to handle fuel oil from the point where it is delivered to the plant and until it reaches the fuel-injection pumps. This equipment consists of strainers and filters, transfer pumps, storage and day tanks, fuel-tank level indicators, piping, and fuel meters.

Clean Fuel. Fuel oil, as produced at the refinery, is clean. However, during the transfer from the storage tank at the refinery into the tank car, barge, or tank truck, during transportation to the plant, and during transfer to the storage tank at the plant, it often becomes contaminated with dust, scale from the tanks, water, and products of oxidation.

A very important condition of successful operation of a diesel engine is the admission of absolutely clean fuel to the high-pressure precision pumps and injection nozzles. Dirt in the fuel acts as an abrasive; when dirt is present, a pump with its closely lapped-in plunger and barrel will begin to leak and will no longer be able to act as an accurate fuel-measuring device. In a multicylinder engine with a separate pump for each cylinder the even load distribution on each cylinder will be disrupted; at full load, as one or two of the pumps begin to leak and thus decrease the load on their respective cylinders, the other cylinders will have to carry more than their share. The cylinder whose pump is in the best condition will become considerably overloaded, with the dangerous attending consequences: high pressure and high exhaust temperature, insufficient cooling, excessive wear of the exhaust valve, and even danger of cracking of the cylinder head or of seizing of the piston. Thus, a relatively minor defect in a fuel pump, caused by dirt in the fuel, may have serious and expensive consequences.

Dirty fuel is also responsible to a great degree for excessive wear of cylinder liners, pistons, and piston rings.

156   FUNDAMENTALS OF DIESEL ENGINE WORK

11-2. Strainers and Filters. The first step in cleaning fuel oil is to install a fine strainer on the suction side of the injection pump. This strainer may be either of fine-mesh screen cloth or be of the metal-edge type and usually is in duplicate, with proper by-pass piping and valves to permit cleaning of each strainer without shutting down the engine. A 200-mesh monel-metal wire cloth should be used, as brass and copper screens will soon corrode in fuel oil and, if the damage to the screen is

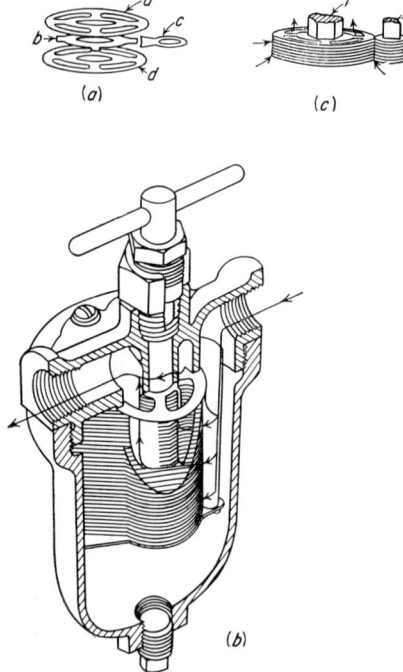

Fig. 11-1. Metal-edge strainer.   *Courtesy of the Cuno Engineering Corporation.*

not caught in time, will result in running the engine on unstrained oil. The 200-mesh screen will not let through particles larger than 0.0029 in. A 100-mesh wire cloth will catch all particles larger than 0.0035 in. and is sometimes preferred for its heavier, stronger wire.

Metal-edge strainer consists of a stack of round, thin perforated disks d (Fig. 11-1a), each separated from its neighbors by a thin metal spacer b, 0.003- or 0.0035-in. thick. Disks d and spacers b are assembled on a hexagonal rod t to form a cylindrical element. This element is inserted into a housing (Fig. 11-1b) with a baffle plate which supports the strainer element and closes it from the bottom end, with tapped inlet and outlet holes, and with a drain plug at the bottom.

Fuel oil enters the housing at the top and is forced by the baffle to pass from the outside of the element to its inside through the narrow slots between the disks; it then flows up through the inner space, and out to the outlet. Any particles of solid matter or sludge larger than the distances between the disks are held against the outside surface of the element and thus prevented from passing through it. In time a layer of sludge and dirt accumulates on the outer surface of the filter element. This layer acts as an additional filter bed and improves the effectiveness of the filter. This accumulation, however, increases the flow resistance and must be periodically removed. This is done by means of a built-in cleaning device which consists of a fixed square rod r (Fig. 11-1c) on which are assembled a series of cleaning blades c, one of which extends into the space between each pair of disks as far as the inner edge of the disk rim. When the hexagonal spindle t with the disks is turned, the ends of blades c scrape off the layer of sludge on the outside of the element and comb out any dirt that may have lodged between the disks. One turn of the element is sufficient for cleaning it. Sludge thus scraped off sinks to the sump below the element and is removed through the drain hole.

The strainer disks, spacers, and scraper blades are of monel metal or stainless steel to prevent corrosion.

Filters. The *bag-type filter* (Fig. 11-2) consists of a close-mesh cloth bag held in the shape of a cylinder with a helicoidal surface by two helical springs, an inside spring made of rather heavy wire and an outside one of lighter wire. The use of a helicoidal surface increases the filtering area, reduces the fluid velocity through the cloth, and thus gives a better filtering efficiency with a smaller pressure drop. The bag should be of woolen yarn because a cotton cloth may allow the fuel to pick up lint. The bag is cleaned by taking it out and washing it in kerosene.

Sintered Filter. Some engine builders use small filters made of sintered metal and placed directly in front of the nozzle. The filter element, made of finely powdered copper or cast iron, or of other metals, is formed under high pressure and has a porous structure which retains very small—under 0.001 in.—particles of solids from the fuel.

Fuller's-earth Filter. Filters resembling the bag-type filter (Fig. 11-2) by their general construction, but with a different filtering element, are used to remove very fine solids. Such an element (Fig. 11-3) consists of a cylindrical outer layer a of long-fibered cellulose which retains by absorption the coarse solid impurities down to about 0.001 in. The middle cylindrical layer b is made of molded fuller's earth and absorbs the soluble impurities, being of the same material as that used in the final step of refining oil in the refineries. Finally the fuel passes through the inner layer c of cellulose which removes by

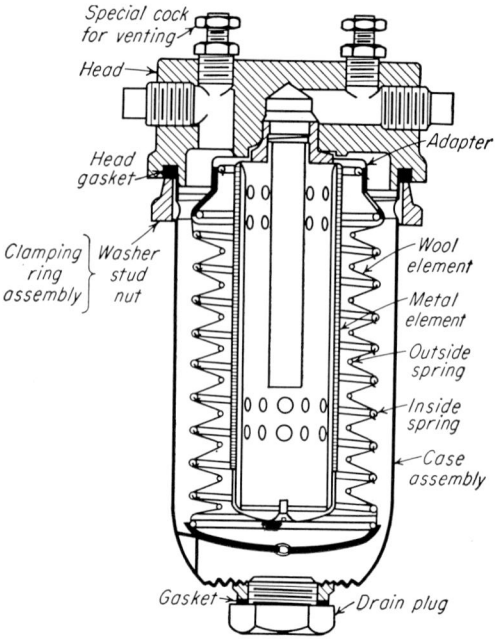

Fig. 11-2. Bag-type filter.

absorption solid impurities as small as 0.0003 in. The perforated metal tube d provides the inner support for the cellulose.

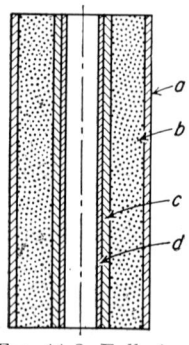

Fig. 11-3. Fuller's-earth cartridge.

When the element becomes clogged by absorbed dirt, as indicated by an increase of the pressure drop through it from 0 to about 5 psi, the element must be taken out and a new one inserted.

Such filter elements are built in three sizes, with filtering capacities of 5 and 10 gal per hr for small engines, and with a capacity of about 25 gal per hr for larger engines. Filters with the large elements are built to hold from one to 20 elements in one case, thus giving a range of a combined capacity of from 25 to 500 gal per hr, corresponding to the requirements of any medium and large diesel engine.

11-3. Centrifuging. One of the best methods of insuring the cleanness of fuel for large engines, especially those using heavy fuel oils, often contaminated with water, is to centrifuge it. Centrifuging is more widely used for lubricating oils and therefore is discussed in greater detail in Sec. 12-10. Centrifuging fuel oil reduces repair expenses up to one-half, and raises engine efficiency. The same centrifuge as used for lubricating oil may be used for fuel oil with a minor change in the adjustment to take care of the difference in gravity. How much centrifuging improves fuel oil may be seen from Table 11-1.

Table 11-1. Change in Fuel-oil Characteristics by Centrifuging

Fuel-oil characteristics	Before	After
Specific gravity, at 68 F	0.896	0.892
Flash point, closed-cup test, Gray, deg F	200	231
Ash, per cent	0.09	0.00
Sulfur, per cent	0.44	0.275
Water, per cent	1.3	0.0
Heat value, lower, Btu per lb	17,603	18,081

As with lubricating oil, the efficiency of cleaning the fuel by centrifuging is materially increased if the oil is heated to about 200 F in order to decrease its viscosity. Heating may be done by a hot-water or steam-pipe coil in the storage tank or by an electric immersion heater with a thermostatic control to avoid overheating when the oil begins to foam.

Combination Centrifuging. The expense of maintaining separate centrifuges for lubricating and fuel oils may be prohibitive in a small diesel plant. Figure 11-4 shows how a single centrifugal purifier may be connected to handle both lubricating and fuel oils.

First the centrifuge is set to purify fuel oil; pump 1 picks up the dirty oil from the storage tank and puts it through the centrifuge; pump 2 delivers the clean fuel oil to a daily service tank 1 and, when this is filled, to service tank 2. After that the centrifuge is readjusted to handle lubricating oil, which is picked up by pump 1 from the engine sump and returned by gravity either to the sump or to a separate clean-oil tank. Naturally, the change in the valve setting is made only after pump 1 has emptied the corresponding stretch of pipe from any fuel oil and put it through the centrifuge. A similar procedure must be observed before changing from handling lubricating oil to handling fuel oil. The purification of fuel oil and filling of both service tanks takes, with a small engine, only a comparatively short time since even the smallest centrifuge is built to handle 60 to 100 gal per hr. A 100-hp engine will normally use about 7–8 gal per hr of fuel oil and, if its service tanks each hold about 40 gal or a 10-hr supply, will require less than 1 hr for fuel cleaning. Thus about 9 hr are left for purification of the lubricating oil on the continuous sump scheme. The lubricating-oil circulation of a 100-hp engine will be about 300 to 600 gph, only three to ten times more, which is satisfactory.

In a system designed to handle two different kinds of oil particular attention must be given to preventing the two oils from mixing. This may be accomplished by taking a few simple precautions. First, a drain cock must be installed at the lowest point on the line through

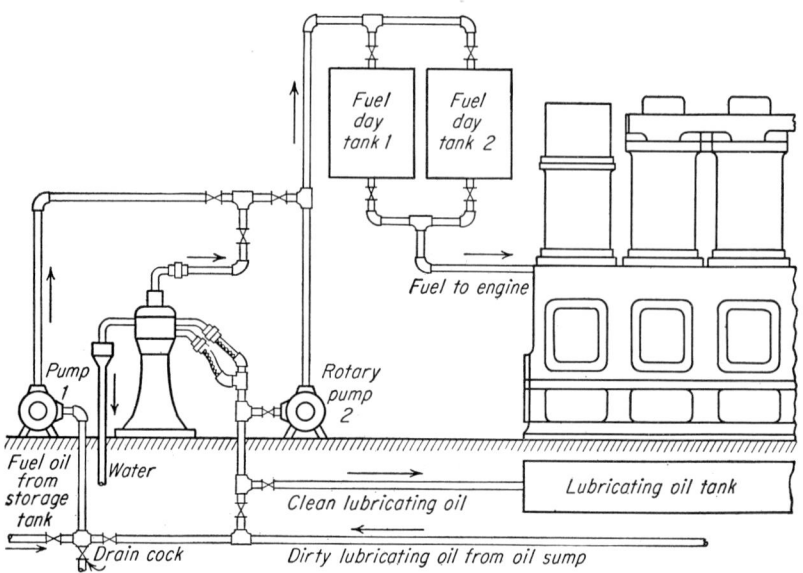

Fig. 11-4. Combination system of centrifuging fuel and lubricating oil.

which both fuel and lubricating oil are handled, as shown in Fig. 11-4. Draining the dead oil in the line will eliminate waste of expensive lubricating oil when the purifier is started on fuel oil and will prevent fuel oil from being put into the lubricating system when the procedure is reversed. Second, every valve must be marked either by a tag fastened to the valve or, better, by an inscription plate fastened to the pipe next to the valve telling in what position the valve must be when handling one or the other oil, as CLOSE FOR FUEL or OPEN FOR LUB OIL, or vice versa.

11-4. Fuel Transfer Pumps. Every diesel plant usually has several fuel-oil pumps, namely: one to unload a tank car or barge, if the unloading cannot be done by gravity, one to transfer the fuel from the main

FUEL SYSTEM

storage tank to the day tanks, and one to deliver fuel to the injection pump, if the day tank does not give a sufficient head.

The first two pumps are generally driven by electrical motors, but a hand-operated emergency unit should be installed for pumps of the second type so that oil can be supplied to the engines in case of trouble with the electric drive. However, small plants sometimes have only hand pumps. Pumps of the third type are generally driven by the engine itself.

Oil-unloading pumps must be of a sufficient capacity to unload a tank car or barge within 48 hr to obviate demurrage charges. However, it is generally advisable to provide enough capacity to empty a tank car in one 8-hr shift.

Plant and day-tank capacity determines the size of pumps which should be selected to transfer fuel from storage. Good practice with nonautomatic pumps is to provide sufficient capacity for the day tank to be filled in about 30 min pump operation. Thus, if a 1,000-hp engine has a 450-gal day tank, the transfer pump should be able to deliver $450/30 = 15$ gpm. When installing pumps in a plant with several engines, it is advisable to give the pumps a capacity that will enable them to fill more than one day tank at a time.

Pumps for unloading oil to above-ground storage tanks operate against a head that continually increases as the storage tank fills. Pumps for this service should have little variation in capacity for considerable change in the discharge head. The same characteristic is desirable for day-tank pumps, as their delivery head also varies considerably. When underground storage tanks are used, the transfer pump should be capable of lifting fuel oil out of the tank and should be self-priming.

Fuel-unloading pumps and transfer pumps are frequently placed in a pump house separate from the main plant building. It is good practice to locate the pump house as close to the storage tank as it is conveniently possible; this also ensures a flooded suction in the case of above-ground storage. When transfer pumps are located some distance from the storage tanks, the friction resistance of the piping and the suction conditions at the pumps should be investigated, as capacity of rotary pumps tends to drop off rapidly with high suction lifts. These pumps will not lift more than about 20 ft including friction loss, and less than this when the fuel oil contains entrained gas.

When transfer pumps are in a pump house or other remote location, switching equipment should be arranged so that the pump may be started and stopped from the plant. Sometimes starting equipment is placed at each day tank so the pump can be stopped as soon as the tank is filled, thereby minimizing the danger of overflow.

When separate pumps are provided for each day tank, starting switches, operated by the head of oil in the day tank, may be arranged to start and stop the pump automatically as the oil level in the tank varies.

Types of Transfer Pumps. Fuel-oil transfer pumps must be of the positive-displacement type; centrifugal pumps are not suitable. Some plants use reciprocating-piston pumps, but most employ various types of positive-displacement rotary pumps.

Some plants use spur or herringbone-gear pumps (Fig. 11-5). In these pumps the gears, shafts, and end plates are made of corrosion- and-wear-resisting nitrided alloy steel with hardened surfaces. The

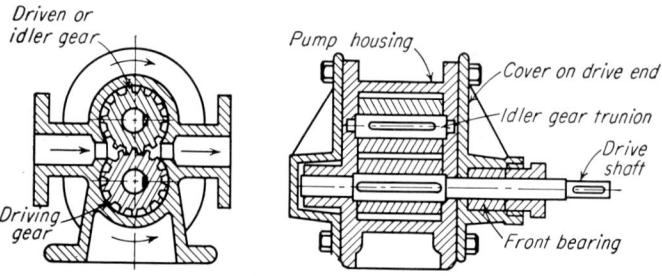

Fig. 11-5. Rotary pump with sliding vanes.

pump housing is made of nickel-chromium cast iron. The shafts run in bronze bushings pressed into the end plates. Another type of rotary pump used is shown in Fig. 11-6: the hollow rotor has an eccentric axis of rotation and two vanes sliding in it; the vanes are tightly pressed by coil springs against the housing.

Figure 11-7 shows a modification of the rotary-eccentric-type pump: the rotation of the eccentric moves the displacer cylinder so that at all times one of its sides is almost touching the round housing. Such an up-and-down and left-and-right motion acts as a piston, similar to that of Fig. 11-6. The motion of the displacer to the left and down is the suction stroke; the space between the pump housing and displacer is connected to the upper right part of the housing by an opening in the sliding valve; the check valve in the discharge line is closed. During the upward and to-the-right motion of the displacer, which constitutes the discharge stroke, the fuel oil is displaced through the opening in the slide valve, the pressure increase opens the check valve, and fuel is delivered into the line.

FUEL SYSTEM

Figure 11-8 shows a two-lobe pump identical in principle with the Root's blower or a gear pump. The shafts carrying the displacer lobes are connected by a pair of spur gears; one shaft is driven by an electric motor, and the other rotates in the opposite direction as a follower.

Finally, Fig. 11-9 shows a rotary pump of the internal-gear type. All the pumping is done by the smaller spur gear; the internal follower

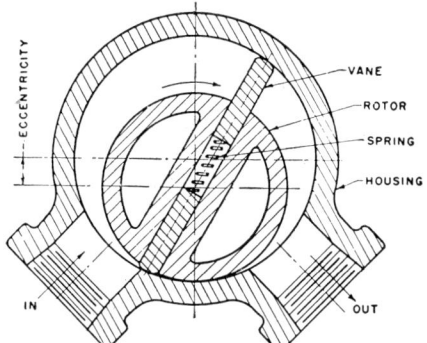

FIG. 11-6. Gear-type oil pump.

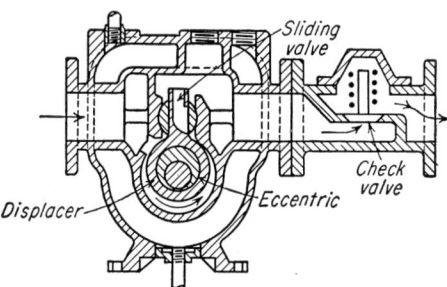

FIG. 11-7. Eccentric-and-piston-type pump.

gear provides only a seal between the suction and discharge spaces of the housing. The spacer with the moon-shaped cross section is stationary and serves also only as a seal between the two gears and the suction and discharge spaces.

All rotary pumps are built with very small clearances in order to insure a sufficient pump efficiency. Their internal surfaces are lubricated by the oil pumped. In medium- and high-speed diesel engines the fuel oil has a relatively low viscosity and has a rather indifferent lubricating quality. Therefore it is particularly important to prevent particles of dirt, sand, grit, etc., from getting into the pump and destroying the close clearances upon which the pump efficiency depends. A screen, filter, or strainer should be placed on the inlet side of each pump and periodically inspected and cleaned.

Booster Pumps. The transfer pump which delivers the fuel to the high-pressure injection pump is often called *booster pump* because its main function is to increase, or boost, the pressure in the suction line to the injection pump. Some engines have an overhead day tank and the fuel is fed to the injection pump by gravity. However, many modern diesel engines require a pressure from 10 to 30 psi at the injection-pump inlet. Booster pumps are usually part of the engine

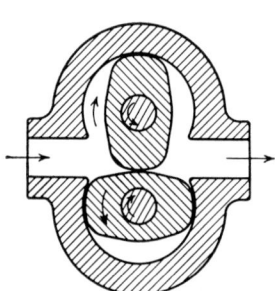 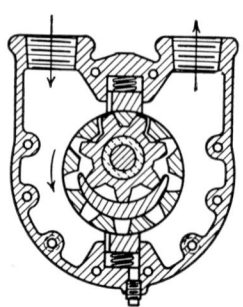

FIG. 11-8. Lobe-type pump. FIG. 11-9. Internal-gear-type pump.

equipment furnished by the engine builder. On larger engines the booster pump is usually of the gear type, along the same lines as the sort used for fuel-oil transfer, only of a smaller size. Some engines use rotary pumps with an eccentric rotor and sliding vanes, as illustrated in Fig. 11-6.

Direct Reversible Engines. With a direct reversible engine the direction of rotation of a direct-connected rotary pump, whether of the gear or vane type, is also reversed, and the suction end of the pump becomes the discharge end, and vice versa. However, the flow of oil in the system is maintained in the same direction by special check valves which automatically connect the suction and discharge sides of the pump with the corresponding oil lines.

Reciprocating Booster Pumps. Some smaller engines have a booster pump with a reciprocating plunger built as part of the injection pump. Other engines have the booster pump with a reciprocating plunger built as a separate unit bolted to the side of the injection-pump housing and operated by a cam on the camshaft of the injection pump. During the delivery stroke the plunger (Fig. 11-10) receives its motion from a cam through the roller tappet and rod connecting the tappet with the plunger. During the suction stroke the plunger, rod, and tappet are returned by the action of two helical springs. A hand lever is provided to enable the operator to prime the piping. When starting an engine after a long shutdown, it is always advisable to prime it by hand until fuel oil shows at each of the cylinder injection-nozzle bleed-off valves.

11-5. Tanks. *Storage.* The main storage tank at the plant may be located above or below the ground. An above-ground tank is usually

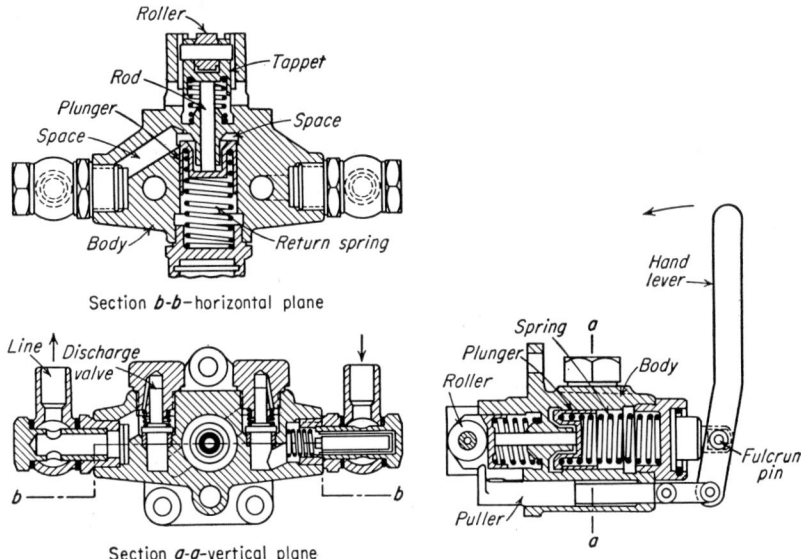

FIG. 11-10. Fuel transfer pump.

a cylindrical steel tank; an underground tank may be a steel or a reinforced-concrete tank. In either case provision must be made for draining off water and cleaning out sediment. The end of the pipe through which the oil is withdrawn must be above the point to which the sediment may rise—at least 2 to 3 in. above the bottom. The tank must have a vent with its top equipped with a standard rainproof cap.

Day Tank. The fuel-supply tank is often called *day tank*, as it should contain enough oil for operating the engine for one whole working day, or 8 to 9 hr. For a large engine the fuel-supply tank should contain as much fuel as Fire Underwriters Rules permit. Often the limit for storage within a building is 200 gal or less. In such a case, a larger tank must be added outside the building.

The day tank may be overhead or below the ground. Overhead day tanks are used mostly with gravity flow to the injection pump and are installed 10 to 15 ft above the engine injection pump. A tank located below ground must be set not more than $6\frac{1}{2}$ ft below the transfer pump, as shown in Fig. 11-11. With a greater suction lift, the vacuum created in the pipe near the entrance to the transfer pump may produce gas bubbles, vapor-locking the piping.

Fuel-level Indicators. For the successful uninterrupted operation of diesel engines and the elimination of unnecessary shutdowns due to

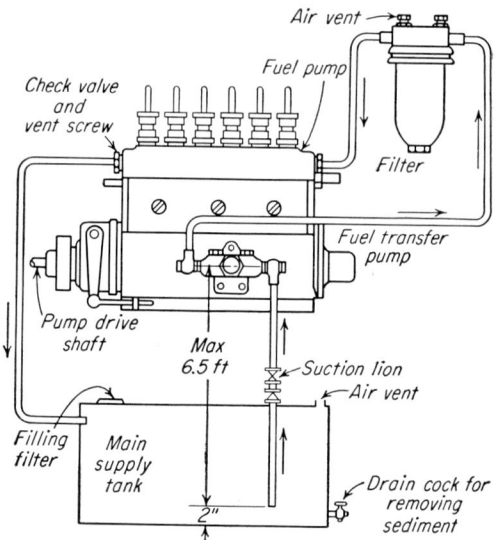

FIG. 11-11. Fuel system for a small engine.

lack of fuel, it is important that all tanks have level indicators. Since storage tanks are sometimes located 50 ft and farther from the engine room, instruments must be equipped with remote indicators to be really helpful. There are many such instruments on the market, but they all fall into one of two groups: pressure-type gauges and volume-type gauges with a float.

The principle of operation of the pressure-type indicator is shown in Fig. 11-12: variations in the tank level change the air pressure in the vertical drop pipe and cause a proportional rise or fall of the mercury in the U tube. The upper spaces a in the tank and the left column of the U tube are connected to the atmosphere by the vent v; neglecting the weight of air in the drop pipe d, the depth m of the liquid in the tank is balanced by the mercury column n. To insure a proper balance, the drop pipe d must not contain any liquid; the latter is expelled before each reading by means of the pump p. The air pressure in pipe d can never exceed that due to the depth of the liquid, because any excess air will escape through the liquid. Instead of the U tube a metal-spring pressure gauge may be used as indicator.

The principle of operation of a float-type indicator is shown in Fig. 11-13: the float k floats in the tank and when the level of the liquid falls, the float sinks and through the bell-crank level i, link l, and lever j expands the bellows a and compresses bellows c; this change of volume content of the bellows causes

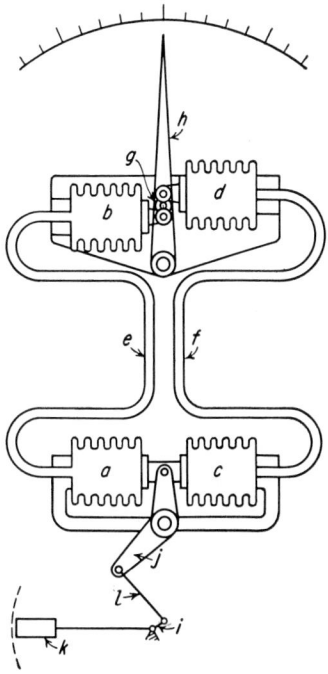

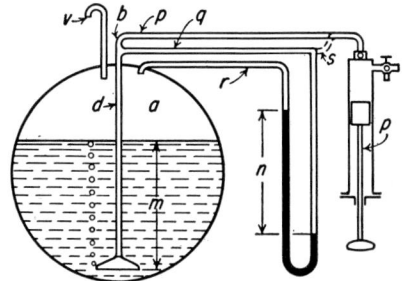

FIG. 11-12. Pressure-type tank gauge.

FIG. 11-13. Volume-type tank gauge.

the bellows b to contract and bellows d to expand; both these actions push the indicator pointer h to the left. When the float rises with the level, the pointer h is moved to the right. The liquid in the bellows and connecting tubes e and f is practically incompressible and the position of the float k is indicated accurately even with a considerable length of the connecting tubes e and f.

A simple day-tank level indicator, sometimes used where fire regulations permit it, consists of a glass tube connected to the bottom of the tank and open at the top. The clarity with which the level in the tube can be read is enhanced by placing a white-background plate behind the glass tube. Another type of indicator is shown in Fig. 11-14: the float f is connected to a weight w by a cord or thin pliable wire running over two grooved pulleys p, p; the weight has a pointer sliding along a scale with the 0, empty, at the top. The advantage of such a device is that by selecting the proper length l between the pulleys p, p and the distance s to the top of the scale, the latter can be placed where it is most easily seen by the operator.

Sometimes the level indicator is connected with an alarm, usually an electric bell, which begins to ring when the weight w (Fig. 11-14), approaching the top of the scale, closes a switch.

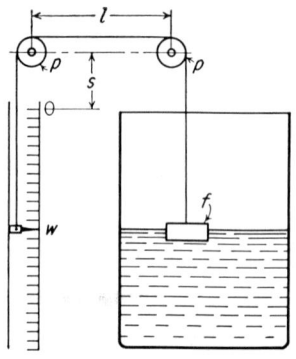

FIG. 11-14. Float-type level indicator.

11-6. Fuel Lines. Standard steel pipes with screw fittings are commonly used in fuel-handling systems, although in small plants steel tubing is sometimes used. Copper and brass tubing may be used for high-grade fuel oils but not for oils containing sulfur compounds which corrode copper. In making up the piping, especially where it is possible that mid-continental fuel oils will be used which precipitate wax at lower temperatures, it is advisable to use tees with a plug instead of elbows at all 90-deg turns as this will make it possible to clean out the pipe. Where unions are used they must have either leather gaskets or metal-to-metal ground joints, since rubber gaskets are attacked by fuel oil and become leaky.

Fuel Velocity. Fuel oil, particularly fuel oil of high viscosity, offers considerable resistance to flow in pipe lines. Hence the velocity should be kept low to prevent excessive power requirements. Good practice indicates velocities between 1.5 to 5 fpm (feet per minute) depending upon oil viscosity, length of line, and whether the oil is flowing by gravity or under pump pressure. To reduce the friction loss, pipe ends should be reamed to remove burrs, or if the pipes are cut with a hacksaw, the ends should be filed smooth. The following tabulation gives an idea of the pressure loss due to friction and the effect of increased viscosity and velocity.

TABLE 11-2. FLOW OF FUEL OILS

Flow, gpm	Pipe size, in.	Velocity, ft per min	Friction loss with fuel oils at different viscosity, psi per 100 ft		
			100 SSU	400 SSU	1,000 SSU
10	1	3.75	4.6	18	46
10	1½	1.58	1.2	5	12
15	1	5.6	6.6	26	66
15	1½	2.4	1.5	6	15
15	2	1.4	0.5	2	5

It is recommended that a riser, not less than five times the pipe diameter, or an air dome be installed at the pump discharge to prevent the forming of an air or gas pocket. A small vent valve at the top of the riser is also advisable. A check valve on the discharge side of transfer pumps will prevent backflow from the day tank through the pump when it is shut down and the inlet valve at the tank is not closed. Check valves should be placed on the pump side of the stop valve in the pump discharge so that it can be repaired without having to drain oil from the entire line.

A fuel system with overhead day tanks and gravity feed to the injection pumps, for a large municipal plant with air-injection engines using No. 5 heavy diesel fuel oil, is shown diagrammatically in Fig. 11-15. Fuel oil from three outside above-ground storage tanks is metered and

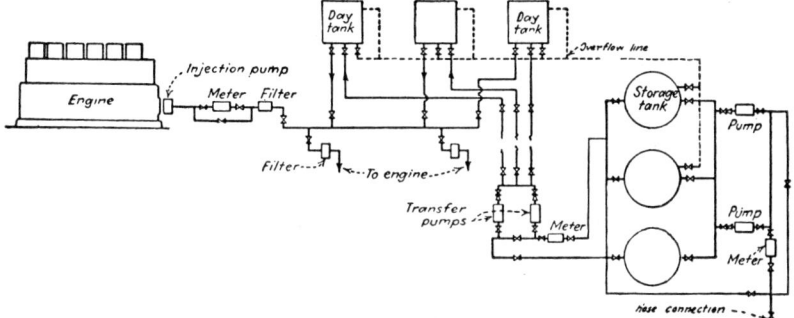

Fig. 11-15. Complete fuel system for large plant.

pumped through separate lines to three 500-gal. service tanks installed at a considerable elevation in the plant. Overflow and drains from these tanks are piped back to the storage tanks. The service, or day, tanks supply a plant header from which branches are taken to each engine. A typical branch connection is shown, with a filter and fuel-oil meter. As the fuel supply to the injection pumps is by gravity, no overflow return connection from these pumps is required, and the meter measures actual fuel consumption of the engine. All meters are provided with by-passes and the filters are duplex.

A system in which a day tank supplies an auxiliary transfer pump on the engine in another municipal plant is shown diagrammatically in Fig. 11-16. Fuel is delivered by a railroad tank car from which it is unloaded by motor-driven pumps piped so that any of several above-ground storage tanks may be filled. The same pump can transfer oil from one storage tank to another so that any tank may be emptied for cleaning. A smaller motor-driven transfer pump is used to deliver oil from the storage tank to the plant day tanks. This pump is in the pump house, together with the tank-car unloading pump; its operation, however, is controlled from the plant. Both pumps are provided with by-pass relief valves to protect the pump in case the discharge valve is accidentally closed. A hand pump is provided so that oil could be pumped to the plant if power to the motor-driven pump should fail. A check valve is placed in the pump discharge piping so that oil will not flow back from the plant. The transfer pump delivers the fuel to a plant header from which branches are taken to each engine day tank. A typical branch connection is shown; it will be seen that for each engine two connections are taken from the header, each having a screen filter

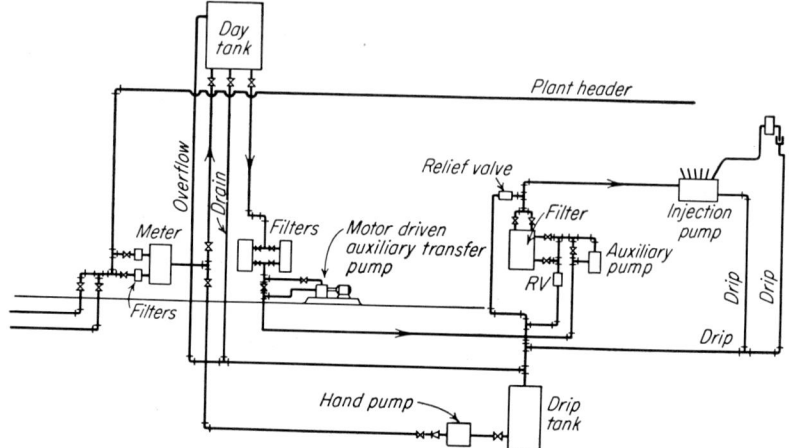

FIG. 11-16. Diagram of fuel system for a municipal plant.

and connecting to a meter that measures all oil delivered to the day tank. Fuel flows from the day tank by gravity through duplex strainers to the auxiliary transfer pump driven by the engine. The auxiliary pump delivers to the engine injection pump through additional filters. A relief valve is provided between the engine transfer pump and the filter to protect the pump in case the filter valves are closed. A second relief valve is provided after the filter to take care of pressure that may build up on the injection-pump suction if the transfer pump delivers more oil than is required by the engine. Both relief valves discharge to a drip tank in the basement of the plant. This same tank receives the overflow from the day tank and drips from the injection pump and nozzles. A hand pump is provided to transfer oil from the drip tank to the day tank. Thus this oil, which has been metered, is made available to the engine without again passing through the fuel-oil meter. Each engine has its separate drip tank. An emergency motor-driven auxiliary transfer pump has been provided for use in case the engine-driven pump gives trouble.

The piping connections between a below-ground day tank and a relatively small engine are shown in Fig. 11-11. As mentioned before, the suction lift should not be over $6\frac{1}{2}$ ft. The transfer pump, driven from the injection-pump camshaft, discharges through a filter to the injection pump. A relief valve on the suction side of the injection pump maintains a constant suction pressure and permits excess fuel to flow back to the supply tank. By filling this tank to a definite level, a meter in the tank supply line will give engine fuel consumption data for such periods as may be required. Hourly rates may be found by the use of a tank-measuring rule.

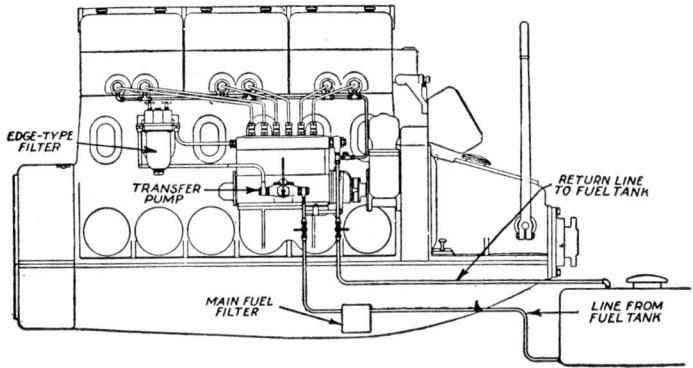

Fig. 11-17. Marine fuel-oil piping diagram.

A similar piping diagram for a small marine engine is shown in Fig. 11-17. In this case the main filter is placed between the fuel tank and transfer pump which acts also as a booster pump. The indicated location of the filter makes more convenient its inspection and cleaning or the exchange of the filtering element, depending upon the type of filter used.

11-7. Fuel Metering. The fuel oil supplied to each engine in a plant should be measured and a record kept of the fuel consumption in order to enable one to determine its rate from time to time and to compare it with the engine guarantee. The most satisfactory way is to install a meter for this purpose. With large engines, the meter may be placed in the supply line between the day tank and the engine, but with small engines, below about 150 hp, the flow of fuel is too small to permit accurate metering by a volumetric meter. Therefore, the meter is installed to measure the supply delivered to the day tank. In both locations care should be taken to account for the overflow and drip fuel metered but not consumed by the engine.

Fuel-oil meters are of the same type as those used for measuring water. Positive-displacement type meters for storage tanks are available with capacities from 10 to 500 gpm. Small special oil meters for use with engines are available with capacities from 10 to 150 gal per hr.

Rotameter. For a quick determination of the rate of fuel consumption it is very convenient to install an instantaneous-reading meter of the flow-rate type. Such a meter consists of a vertical, very slightly conical, or tapered, glass tube, increasing in diameter toward the top, and a metallic cylindrical float, tapered at the lower end. At the bottom of the tube its inside diameter is practically equal to the float diameter, and there is no flow of fluid. When the fluid begins to flow it raises the float and the higher the position of the float, the greater is the ring area through which the fluid passes and the greater is the rate of flow, meaning the amount of fluid flowing through this ring area and through the whole meter. The float has helical grooves on the outer surface which make it turn, or rotate, slowly; hence the float is called a *rotor* and the whole instrument a *rotameter*. The slow rotation of the float eliminates the danger of its adhesion to the tube and gives a more stable reading.

The flow resistance of a rotameter is very small and remains practically constant at all positions of the rotor. This results in a uniform scale. The instrument gives very accurate readings from the maximum flow, when the rotor is at the top, down to about $\frac{1}{10}$ of this reading, when the rotor is near the bottom. Rotameters are available for any amount of liquid and thus can be used for the smallest and up to the largest diesel engine.

QUESTIONS

1. Enumerate the equipment comprising the fuel system of a diesel-engine plant.
2. Explain the effects of dirty fuel on the operation and maintenance of a diesel engine.
3. Enumerate the various strainers that are used for fuel oil.
4. Explain the arrangement and operation of cleaning a metal-edge strainer.
5. Sketch and describe a bag-type filter.
6. What is a sintered filter?
7. Explain the action of the fuller's-earth filter.
8. What are the advantages of centrifuging diesel fuel oil?
9. Can the same centrifuge be used for alternate cleaning of fuel and lubricating oil?
10. If the same centrifuge is used for purifying both fuel and lubricating oil, approximately what part of the time must the centrifuge be operated for cleaning the fuel and what part for cleaning the lubricating oil?

FUEL SYSTEM

11. When the same centrifuge is used for both fuel and lubricating oil, what point requires particular attention in operation of the system?

12. What are the places around a diesel engine where fuel-transfer pumps are used?

13. What should be the capacity of the pump for unloading a tank car?

14. List the types of pumps generally used for fuel-oil transfer.

15. Give sketches of various fuel-transfer pumps.

16. What is a booster pump?

17. What provisions must be made in installing a storage tank in order to ensure clean fuel?

18. What is a day tank?

19. What should be the capacity of a day tank?

20. What types of fuel-level indicators are generally used in diesel plants?

21. Draw a sketch of a pressure-type fuel-level indicator and explain its action.

22. Draw a sketch of a float-type fuel-level indicator.

23. Why is it recommended to use tees with a plug instead of 90-deg elbows in a pipe line?

24. What are the recommended limits of flow-velocity in a fuel-line?

25. How is the forming of an air or gas pocket in the fuel line avoided?

26. Draw a diagram of a complete fuel system for a stationary diesel power plant.

27. Draw a diagram of a fuel system for a small marine engine.

28. Enumerate the types of fuel-oil meters used in diesel plants.

29. Draw a diagrammatic sketch of a rotameter and explain its action.

CHAPTER 12

LUBRICATION AND LUBRICANTS

12-1. Objects of Lubrication. The introduction of lubricating oil between two bearing surfaces, *i.e.*, surfaces that are in contact, under pressure, and in motion in respect to each other, is called *lubrication*. Lubrication may accomplish one or more of the following objects: (1) reduce the wear of bearing surfaces by lowering the friction between them; (2) cool the bearing surfaces by carrying away heat generated by friction; (3) clean the surfaces by washing away metal particles resulting from wear; and (4) assist in sealing the space adjoining the bearing surfaces, such as the engine cylinder with its piston or the crankcase

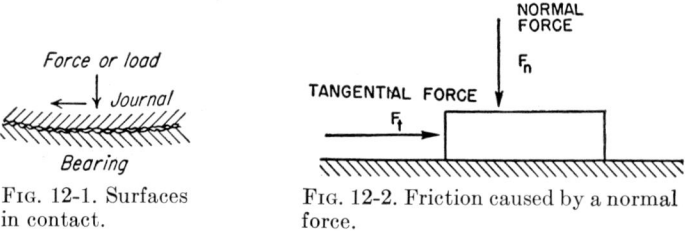

Fig. 12-1. Surfaces in contact.

Fig. 12-2. Friction caused by a normal force.

space with the rotating crankshaft. Of all these objects the most important is the first.

12-2. Principles of Lubrication. *Friction.* Regardless of how smooth and true a metal surface may look and feel, it is actually not even but consists of high and low spots; when seen under great magnification, the cross section looks as shown in Fig. 12-1. When one surface slides over the other and a force presses it against the other surface, the high spots on both surfaces interlock and resist the relative motion. In sliding and overcoming this resistance, the harder surface will remove some high spots of the softer surface but at the same time may lose some of its own high spots. This resistance to sliding is called *friction;* the removal of the high spots causes *wear*. When a load or force F_n (Fig. 12-2) normal to the direction of motion presses a moving object to another object, a force F_t must be applied in the direction of motion in order to overcome the friction that results from the action of the force F_n. Numerically, friction is measured by the so-called *coefficient of friction*, designated by f and expressed as the ratio of the tangential force F_t to the normal force F_n:

$$f = \frac{F_t}{F_n} \qquad (12\text{-}1)$$

If the surfaces are in direct contact without any lubrication, it is called *dry friction*.

Oil Film. All oils have a tendency to spread on a metal surface and to adhere to it. Thus a very thin film can be established between moving surfaces, such as between a journal and its bearing or between a piston and a cylinder. Depending upon the thickness of the oil film, the contact between the metal surfaces may be either reduced or eliminated entirely. However, even when metal-to-metal contact is eliminated, some resistance to sliding of one surface over the other one will remain, but it will be of a different nature. Under these conditions it will be the resistance to sliding of particles of oil adhering to one metal surface with its high and low spots in relation to other particles of oil adhering to the other metal surface. This is called *fluid friction*. The force necessary to overcome fluid friction, all other conditions being equal, is only a small fraction of the force necessary to overcome dry friction.

The thickness of the oil film which will exist between two metal surfaces in relative motion depends upon several conditions, of which the main ones are (1) the finish of the metal surfaces; (2) the bearing pressure; (3) the viscosity of the oil; and (4) the relative speed of the moving surfaces.

In the case of a rotating journal and its bearing, the oil-film thickness depends also on (5) the clearance or difference between the diameters of the bearing and journal, and (6) the shape of the bearing surface, or the ratio of its length to the diameter and the arc of the actual bearing surface.

An oil film which completely separates two metal surfaces is called a *thick film*. In engines oil-film thickness may vary from 0.0001 to about 0.0007 in. Under these conditions there is no wear of the metal surfaces. If the oil viscosity decreases or if the bearing pressure is increased, some of the oil will be squeezed out and the film thickness will become smaller until some of the high spots on both surfaces will come in contact, but the main load will continue to be supported by the oil film. Such a condition is called *imperfect* or *thin-film lubrication*.

With thin-film lubrication some wear takes place, depending upon how much the oil-film thickness is reduced compared with that of a thick, or perfect, oil film.

12-3. Wedge Action Lubrication. The crankshaft is the backbone of a diesel engine. A clear understanding of the way it must be protected by proper lubrication is important and will therefore be discussed in more detail. According to the so-called *hydrodynamic theory* of lubrication of plain bearings, the rotating journal acts as a pump, owing to adhesion of the oil. Figure 12-3 shows diagrammatically what occurs in a bearing at various speeds of the journal. The horizontally shaded areas represent the clearance space filled with oil. In Fig. 12-3a the journal is shown at rest and the oil film has been squeezed out by the load. When the journal begins to rotate (Fig. 12-3b), friction causes the journal to climb up the wall of the bearing in a direction

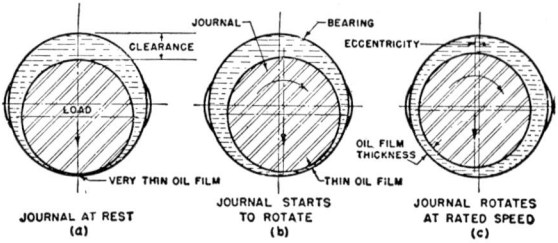

FIG. 12-3. Oil film at various journal speeds.

opposite to the direction of rotation. However, one layer of oil will adhere to the journal and another layer to the bearing surface, and oil will begin to be drawn into the wedge-shaped clearance between the rubbing surfaces, at first as a thin film, producing thin-film lubrication. As the journal speed increases, more oil will be drawn between the surfaces, the journal acting as a rotary pump and thus creating a certain pressure in the oil wedge. If the journal speed is sufficiently high, the pressure in the oil between the surfaces will become such that it will be able to support the full load acting on the journal. The journal will begin to float and the oil pressure will push it toward the other side (Fig. 12-3c). Figure 12-4 shows a polar diagram of the pressures in the oil film produced by the pumping action of the rotating journal. It should be noted, however, that the pressure vectors are drawn from the circumference of the bearing. The pressure is zero at point m; it gradually increases to a maximum at point p, which lies somewhere between point l, the point of load action, and point f, the point of minimum oil-film thickness; from point p to point g the oil pressure gradually decreases to zero, owing to leakage.

The oil-film pressure must not be confused with the oil pressure in the lubricating system. There is no direct relation between them. The lubricating-system pressure is usually maintained from about 30 to 70 psi, whereas the oil-film pressure depends upon several factors, such as oil viscosity, oil temperature, journal speed, bearing clearance, and load. The oil-film pressure, at the point of maximum pressure, may reach several hundred pounds per square inch

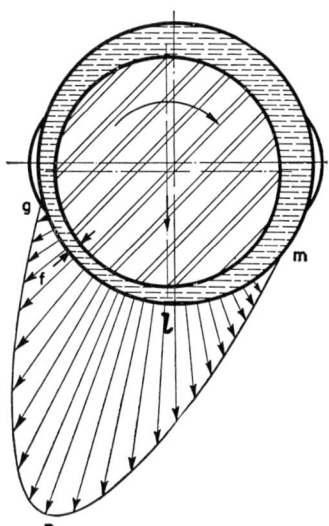

Fig. 12-4. Oil-film pressure.

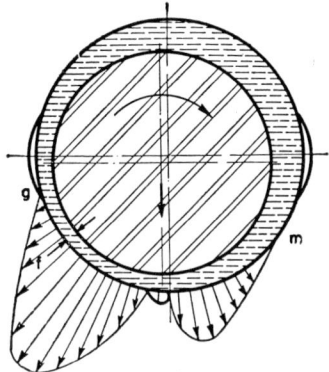

Fig. 12-5. Oil-film pressure with an oil groove.

and in heavy loaded bearings running at a high speed may easily exceed 1,000 psi.

Oil Grooves. If the bearing has an oil groove in the region of the oil-film pressure, the groove will interfere with the pumping action of the journal and the pressure at the groove will drop to zero, as shown by the curves of Fig. 12-5; and while it will be built up again after the oil groove, it will not reach the value it would have reached without the groove. For this reason, although oil grooves may help to distribute oil over the bearing length, they should never be located in the region of load application. In main bearings of diesel engines the direction of the load changes during the cycle and also from bearing to bearing, and therefore it is best not to use any oil grooves at all. However, for lengthwise oil distribution it is necessary to have side reliefs where the two shells meet, as shown in Fig. 5-9. Such a side relief not only serves as an oil reservoir and distributor but helps to wedge the oil between the surfaces and thus to maintain the necessary oil film. Neither lengthwise nor circumferential, or spiral, grooves should extend to the pressure region of the bearing surface. To avoid a scraping off of the oil, the edges of oil grooves must be well-rounded.

Clearance. The above discussion has brought out the fact that proper lubrication by the formation of a wedge-shaped space which permits to obtain the required oil-film pressure is possible only with a certain clearance between the journal and bearing shell. For best results the clearance must be of a more or less definite size, about 0.001 in. per inch of journal diameter. If the clearance is greater, too much oil will leak out at the bearing ends, the adhesion of the oil to the metal surfaces will not extend through the whole clearance, the pumping action will be reduced, the oil-film pressure will be lower, and, as a result, the load capacity of the bearing will be smaller. The

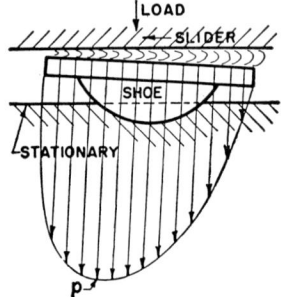

Fig. 12-6. Element of Kingsbury thrust bearing.

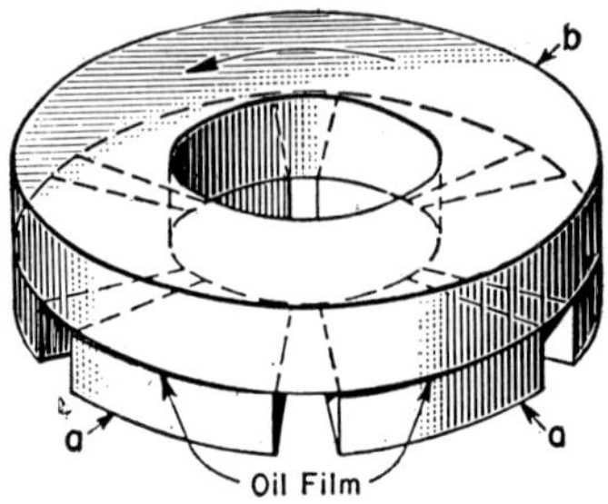

Fig. 12-7. Kingsbury thrust bearing.

clearance most suitable for each bearing and the limit to which it may increase by wear without becoming dangerous are given by the engine manufacturers in their instruction books and should be carefully watched and strictly observed in engine operation.

Kingsbury Thrust Bearing. The taking up of an axial load such as that produced by a propeller direct-coupled to a large marine diesel engine presents some problems in respect to lubrication and wear. The problems are materially simplified by the use of the Kingsbury thrust bearing, so named for its inventor. The lubrication of this bearing is again based on the principle of oil adhesion to metal surfaces and on the tendency of pressure to build up in the oil film separating two metal surfaces if the oil is drawn into a wedge-shaped space.

Figure 12-6 shows an element of this bearing: the stationary part has shoes which are supported on a ball-and-socket-joint and therefore can tilt; the center of the joint is slightly nearer the discharge edge. Oil drawn in by adhesion to the slides and moving in the direction indicated by the arrow, owing to the difference in the supporting areas, tilts the shoe, and thus produces a wedge-shaped space into which it is drawn and raises the pressure as indicated by the pressure curve. The oil-film pressure drops, due to leakage, after reaching a high at point p. Figure 12-7 shows the shoes and the slides in the form of a collar fastened to the shaft.

12-4. Ball-bearing Lubrication. Ball and roller bearings are lubricated with grease or oil. However, although the lubricant acts as a cushion and reduces somewhat the pounding action of the balls or rollers on the races, the main object of lubrication of this type of bearings is not reduction of friction but protection from corrosion. When lubricated with grease, they must not be filled more than about $2/3$ of the space available, otherwise the grease may develop internal friction and overheat the bearing, causing its failure. Both ball and roller bearings are rather sensitive to particles of foreign matter, such as dust, minute metal chips, carbon, etc. Such particles can easily damage the working surfaces by acting as wedges between the balls or rollers and their races. In order to prevent such conditions, some bearings are made with the inner space effectively sealed from the outside. Due to absence of leakage, sometimes they may not require any refilling with grease during their entire life.

12-5. Lubricants. All lubricants used in diesel engines may be divided into two groups: the important group of oils and a small group of greases. Oils can perform all four objects listed in Sec. 12-1; on the other hand, greases can be used for only the first object—separation of metal surfaces in motion—and to a limited extent for the fourth object—sealing one space from another, such as the inside of a water pump from the outside atmosphere. Therefore, oils will be discussed in greater detail than greases.

12-6. Lubricating-oil Characteristics. The working conditions, such as bearing pressure, temperature, and rubbing speed of different engine parts, vary considerably and, theoretically, these different parts should be lubricated with different oils. However, the use of several or even two different kinds of oil in the same engine would complicate its servicing. The oil refineries have developed a great number of different lubricating oils with different viscosities, resistances to heat, and other characteristics. From these different oils a compromise oil can be selected which is satisfactory for all parts of a certain engine.

Good Oil. It is impossible to give a specification of a good oil. An oil giving satisfactory operation in one engine type operating under a certain set of conditions might be unsatisfactory for an engine of a different design or for a similar engine but running under different operating conditions.

Lubricating oils used in diesel engines are obtained by distillation of the stock left after the lighter fractions, gasoline, kerosene, and gas oils, which include diesel fuel oils, are driven off the crude petroleum. Such oils are called *mineral oils* to differentiate them from *vegetable* and *animal oils* used with some other machinery. Lubricating oils are hydrocarbons, as is diesel fuel oil, but they differ from it by the internal structure of their particles, as shown chiefly in their greater viscosity and greater specific weight, or gravity. The desired properties of lubricating oils are obtained by mixing, or more accurately, blending, oils distilled from different stocks and at different temperatures and also by admixing so-called *additives*.

Properties of lubricating oils, both physical and chemical, are determined by tests similar to those used for testing fuel oils. They will be discussed approximately in the order of their importance.

1. *Viscosity.* This, the most important property, indicates the relative fluidity of a given oil. Therefore it is a measure of fluid friction, or of the resistance, which the molecules or particles of an oil will offer to one another when the main body of oil is in motion, as in a circulating system. Heavier, or more sluggish, oils have a higher viscosity. As the relative fluidity decreases, the molecular, or internal, friction will normally increase.

Viscosity decreases with an increase of temperature and is determined with the Saybolt viscosimeter with the Universal orifice. Viscosity of diesel oils for various engines varies from 100 to about 500 sec, designated as 100 to 500 SSU at 130 F.

Friction, engine wear, and oil consumption depend principally on the viscosity of the oil.

For some types of diesel service, oils with a higher viscosity are necessary, as, for example, when cylinders are lubricated separately from the bearings by means of a mechanical forced-feed lubricator. Oils such as bright stock have a viscosity of about 150 SSU at 210 F and as high as 3,000 SSU at 100 F and are suitable for cylinder lubrication of such engines. The relatively high operating temperature of the cylinder walls causes such oils to become amply fluid, with the result that there is no unnecessary power consumption due to excessive fluid friction. A similar oil placed in the crankcase, however, will cause a considerable increase in power consumption and possibly higher bearing temperatures, because of its resistance to flow through the bearing clearances and the load it may impose upon the lubricating-oil pump. For this and other reasons, a crankcase oil should have a viscosity of from 200 to 300 SSU at 130 F.

A diesel operator should, therefore, understand the meaning of viscosity, its value as an indication of the ability of an oil to function properly in his engine, and the means by which this property is customarily determined. He can then check any particular oil against the specific requirements of his engine established by the engine builder.

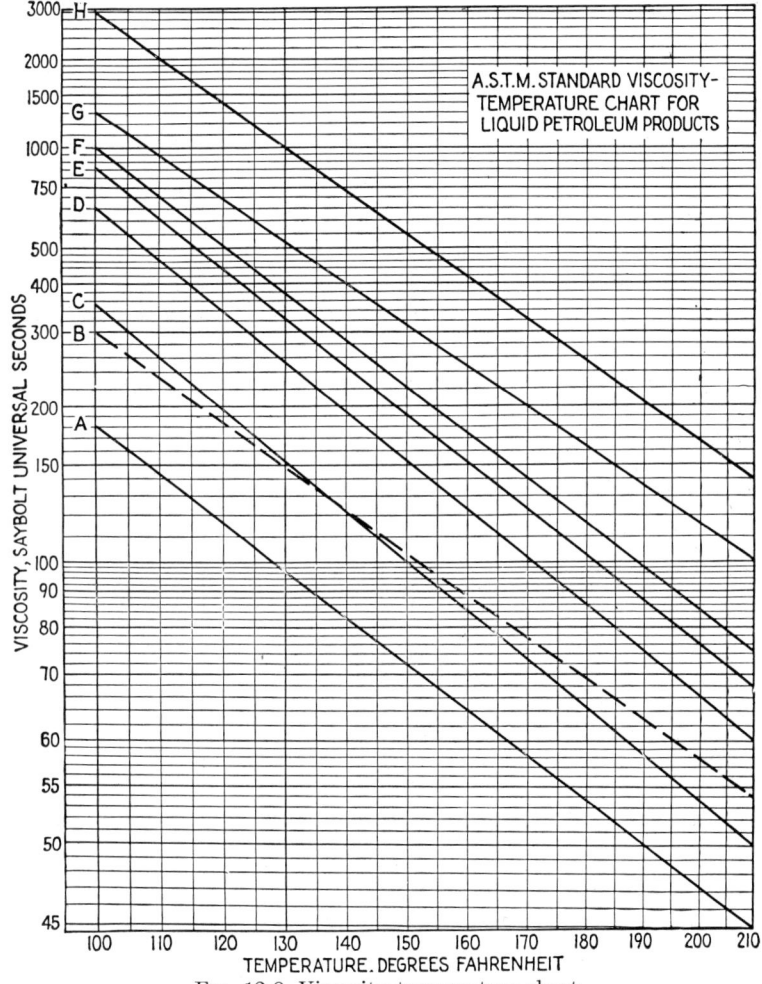

Fig. 12-8. Viscosity-temperature chart.

The Saybolt Universal viscosimeter is the standard apparatus in this country for such tests. The viscosity of diesel crankcase lubricating oils is normally determined at 130 F. Oils for cylinder lubrication, however, are frequently run at both 130 and 210 F, to facilitate drawing a viscosity-temperature curve with which to estimate the probable operating viscosity at crankcase or cylinder-wall temperatures. Viscosity thus determined is spoken of in terms of time or, in other words, of the number of seconds required for 60 cu cm of the oil to flow through the viscosimeter orifice at a designated temperature and is designated by SSU.

Figure 12-8 shows the relation between viscosity and temperature for several lubricating oils as used in diesel and other combustion engines. Table 12-1 gives additional data about these eight oils: their SAE numbers and their specific gravity at 60 F.

Table 12-1. Characteristics of Lubricating Oils, Fig. 12-8

Designation	Oil specifications	SAE No.	Specific gravity at 60 F
A	Automobile oil, light.................	10	0.8894
B	Automobile oil, all-year.............	20	0.9036
C	Automobile oil, medium.............	20	0.9254
D	Diesel oil, medium...................	30	0.9250
E	Automobile oil, heavy...............	40	0.9275
F	Diesel oil, heavy.....................	40	0.9285
G	Airplane oil 100.....................	60	0.8927
H	Transmission oil, bright stock, diesel cylinder oil...............................	110	0.9328

SAE numbers constitute a classification of lubricating oils in terms of viscosity only. Other factors of oil quality, or character, are not considered. The numbers of crankcase-type oils used in diesel engines start with SAE 10 for light oil and increase gradually to SAE 70, heavy aircraft-engine oil.

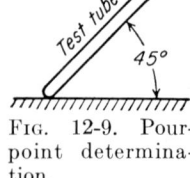

Fig. 12-9. Pour-point determination.

2. *Pour point* is the temperature at which the oil refuses to flow when the test tube is placed at 45 deg from the horizontal (Fig. 12-9). A relatively high pour point affects the ability to pump the oil through the engine lubricating system with the numerous small-sized tubings and orifices. A relatively high pour point may also cause starting difficulties in cold weather.

The pour test of a diesel lubricating oil is important whenever the engine must be operated at low temperatures and run intermittently. This test is also of importance where oil must be handled from storage or external lubricating devices in cold weather and will indicate the lowest temperature at which the oil will pour or flow to the transfer pump. The pour point may also be important if the oil is to be used to lubricate engine cylinders by means of a mechanical forced-feed lubricator. Cylinder oil or bright-stock grade oil may have a pour point of from 30 to 55 F.

It is obvious that an oil should have no tendency to congeal either in the lubricator or in the crankcase during stand-by periods; otherwise, the pumping mechanism may fail to handle the oil, and a lack of lubrication, a scoring of cylinders, and bearing troubles will result.

3. *Carbon Residue.* Carbon residue is the amount of carbon left after the volatile matter has been evaporated and burned off by heating the oil. It gives an indication of the amount of carbon that may be deposited in an engine and thus lead to operating troubles.

An excess of carbon residue, as determined by testing an oil, will usually indicate a tendency to deposit carbon around rings, on valve seats, piston heads, and in the combustion chamber.

Carbon formation in the diesel engine, air compressor, or any other type of high-temperature service is the result of a chemical change. The average petroleum lubricant contains from 83 to 87 per cent of carbon by weight, from 11 to 15 per cent of hydrogen, and very small quantities of other elements, such as sulfur, oxygen, or nitrogen.

In lubricating oils the carbon and hydrogen are bound together in a wide variety of hydrocarbon combinations of varying volatility and stability. Refining removes the lighter and more volatile fractions as well as the unstable components which would impair the lubricating ability of the oil when used in an engine.

The object is to produce stable lubricants over a comparatively wide range of viscosity, which will be highly resistant to chemical change and to the deposition of free carbon when subjected to high temperatures.

Obviously, it is impossible to eliminate the formation of carbon residues entirely, for all petroleum oils will develop volatile products and carbon deposits when exposed for any length of time to temperatures considerably above the flash point. Furthermore, where oils have been subjected to the same degree of refinement, more carbon will result from those oils which remain longer in the engine or compressor cylinder or on the discharge valves. For this reason, a more volatile oil of low carbon content may prove more economical, even though consumption may be somewhat high.

Incompletely refined oils of comparatively low volatility may tend, through slow distillation, to form gummy deposits which will gather abrasive dust or metallic particles. This may lead to hard accumulations which may seriously affect engine performance. The ideal diesel lubricant should, therefore, show a narrow range of distillation, low carbon residue by the Conradson test, and not too high a viscosity. The amount of residual matter which may ultimately develop will vary directly with the volume of the oil which is exposed to engine temperatures and with its stability.

Carbonaceous deposits are poor conductors of heat. As a result, when they accumulate in cylinders or around piston rings, they may become heated considerably above the temperature of the cylinder walls. If the deposits become large, they may cause structural failure of certain parts of the engine through uneven heat transfer. Long before this, however, residual accumulations on the valves, valve seats, around piston rings, or in the ends of the cylinders, may cause gas leaks, scoring of cylinders, wear of rings, and loss of power from improper ring and valve action.

4. *Flash point* is that temperature at which the oil vapor above the oil will flash when a small flame is applied. The flash point of lubricating oil is determined by the same method as that used for fuel oil. The flash point of various diesel lubricating oils varies from 340 to 430 F.

The extent to which the flash point is an indication of lubricating ability of a diesel engine oil is frequently misunderstood. Normally it is of little value as a guide to the nature of the lubricating film which will be formed.

The flash point is an indication of relative initial volatility. It is much lower than the boiling point, or the temperature at which a petroleum oil will pass from the liquid to the vapor stage. It is arbitrary and dependent upon the type of instrument used for testing. Where an open cup is used, the flash-point temperature observed will be higher than when the same oil is heated in a closed cup. In the United States the open cup is normally used in testing diesel-engine oils.

In engine service it is desirable to use oils which will burn cleanly and leave as little carbon residue as possible. A certain amount of lubricating oil on the upper portions of the cylinder walls must burn when exposed to the hot combustion gases during the power stroke. At this time the temperature to which the lubricating film is exposed will be far above the flash point of the oil. Obviously, there is little or no advantage in insisting upon a high flash point as a specification, for some oil must be burned periodically from the upper parts of the cylinder walls anyway. It is far better that this occur at a lower temperature but with as little resultant carbon residue as possible.

5. *Water and Sediment.* The oil is tested by centrifuging and should be free of water and sediment. There should, of course, be no dirt in

the supply of lubricating oil. In spite of this obvious point, in a considerable percentage of existing diesel plants open oil barrels are left open in many diesel plants. In such a case, dirt is bound to get into the oil and is likely to settle in oil lines, stopping the flow to a vital bearing; it may also act as an abrasive.

6. *Acidity.* Lubricating oils should show a neutral reaction when tested with litmus paper. An acid oil tends to corrode or pit the engine parts and to form emulsions with water and sludge with carbon. In service, oils have a tendency to become acid through oxidation.

7. *Emulsion.* A mixture of oil and water which does not separate into its components, oil and water, is called an *emulsion*. Lubricating oil should not form an emulsion with water. If shaken with water it should readily separate from it. This ability to separate is particularly important after the oil has been used for some time. When exhaust gases, which always contain water vapor, enter into the crankcase owing to blow-by, water vapor is condensed and mixed with oil in the crankcase.

8. *Oxidation.* The oil should not have an excessive tendency to oxidize, since oxidation causes formation of sludge. Oxidation and sludge formation in the crankcase or elsewhere in the lubricating system of a diesel engine is objectionable in view of the possibility of interference with oil flow and impairment of lubrication in those parts where sludge accumulation may occur. All petroleum oils have a certain tendency to become oxidized in the presence of oxygen, which makes up over 20 per cent of the atmospheric air. Certain types, according to the degree of refinement, have a greater amount of oxidizable components and have this tendency more than others, especially when they are exposed to oxidizing conditions under higher temperatures. As these components become oxidized, they develop a gummy or resinous type of material which is relatively insoluble in oil and which may accumulate in certain parts of the engine lubricating system, thus preventing free circulation of the oil.

Whenever blow-by of gases or residual matter may occur, gummy sludge is considered to originate from the fuel rather than from the lubricating oil. If the condition or fit of the piston rings to the surface of the liners allows blow-by, unburned fuel and soot will work down into the crankcase of a trunk-type engine. This will occur whenever the engine is smoking. Under such conditions soot from the smoke may be thrown out and will work down past the rings even if there is no direct blow-by. Under high pressures, it is difficult and almost impossible to overcome this tendency. As a result, one must not conclude that, because there is evidence of sludge in the lubricating oil, the latter has been oxidized and is breaking down; more frequently sludge will be coming from the fuel as a product of incomplete combustion.

To prevent undue sludging of lubricating oils under normal operating conditions, every effort is made today to remove those materials which have the greatest tendency toward oxidation. The extent to which sludge formation may occur cannot be predicted by any of the usual physical tests which comprise the average oil specifications. All mineral oils are susceptible to oxidation when diluted with water under high temperatures and in the presence of air and metallic particles. Obviously, any or all of these conditions may be quite normal.

There is evidence, furthermore, that these conditions may be so contingent upon each other that it is difficult to isolate any one as being the most objectionable. In average engine service they will all be present, and must, therefore, be considered jointly. Where a particular factor, such as the entry or condensation of water or the development of higher operating temperatures can be corrected, it is reasonable to expect that the tendency toward oxidation of the oil will be reduced.

The earlier stages of oxidation in a lubricating oil will normally be indicated by emulsification, if water is present. When it is possible to keep such emulsions in temporary form, sludge development can be materially reduced. Unfortunately, under continued exposure to oxidizing conditions permanent emulsions and insoluble sludges may result.

Means of oil purification are available which will effectively remove a certain amount of sludge, especially where it is more or less bonded together with metallic particles. As a result, where a refined diesel lubricant is used in an engine which is so designed as to enable proper circulation of the oil without undue overheating and where the oil can be readily treated or reconditioned, sludge formation need not be regarded as alarming.

9. *Ash.* Ash in oil is a measure of the foreign matter which may cause abrasion or scoring of the moving parts in contact.

10. *Sulfur.* Free sulfur or corrosive compounds of sulfur are not allowed in lubricating oils, because they have a tendency to form acids with water vapor. Noncorrosive compounds of sulfur are permitted to a limited extent.

11. *Color.* The color of a lubricating oil has no relation to its lubricating quality.

12. *Gravity.* While in general gravity is higher in oils with a higher viscosity, there is no definite relation between these two oil characteristics, as may be seen from Table 12-1. Gravity of an oil bears no relation to its lubricating quality.

12-7. Additive Oils. Higher cylinder pressures and temperatures, higher loads and speeds, used in present-day high-output diesel engines, have increased the requirements as to the qualities of the lubricating oils. It is difficult to meet these requirements with the use of straight mineral oils. One of the main troubles encountered with high-output engines is the sticking of piston rings, which in turn causes a number of other troubles, such as reduction of power, smoky exhaust, and contamination of lubricating oil. In trying to overcome such troubles, the oil refineries found that certain chemicals added to the oil, called *additives*, increase the resistance to oxidation and that other additives help materially in keeping the piston rings from sticking, serving to cleanse, or to deterge, them. This explains why heavy-duty additive diesel engine oils are often called *detergent* oils. They have taken the place of the straight mineral oils in the operation of naval diesel engines and are gradually finding their way also into industrial installations.

The chemicals in the additive oils do their work by combining with the undesirable contaminations. Therefore, the amount of free additives in the oil gradually decreases, the oil "wears out," and its ability to resist oxidation and deterge the piston rings gradually diminishes. Worn-out oil must be removed and replaced by fresh oil.

In connection with this it should be noted that detergent oils should not be used in *sight-feed oilers* having a water-glycerine mixture for feeding the oil. The water-glycerine mixture reacts with many detergents and may cloud the glass and gum up the piping. The Dow Corning Fluid 200 is recommended in this case instead of the water-glycerine mixture.

United States Navy symbols for identifying lubricating oils consist of four digits, of which the first classifies the oil according to its use and the last three indicate the viscosity in seconds Saybolt Universal. Forced-feed oils with the viscosity referred to 130 F have the first digit 2. Thus, a Navy symbol 2190 oil is a forced-feed oil with an approximate viscosity of 190 SSU at 130 F. The digit symbol for detergent diesel oils is 9 and their viscosity is referred to 130 F. Thus Navy symbol 9370 oil is a detergent diesel oil with a viscosity of 370 SSU at 130 F. The Navy uses three grades of additive diesel lubricating oils, designated by the symbols 9170, 9250, and 9370. Their viscosity corresponds approximately to SAE viscosity numbers 20, 30, and 40.

12-8. Greases. Lubricating greases are emulsions or intimate mixtures of a lubricating oil with soap. With diesel engines greases are generally used only in two places—for the shafts of some centrifugal water pumps and on the inside of some reverse gears in marine engines.

The pump grease must contain soap which will not dissolve in water. Reverse-gear greases must resist the high temperatures created by friction in the gears. These greases may contain graphite, which has the ability of lowering the friction coefficient.

12-9. Filtration. The two most important conditions for proper lubrication of a diesel engine are the delivery of oil in a sufficient amount and the use of clean oil. Since the lubricating oil is being contaminated continuously while the engine is running, means must be provided to clean the oil to prevent it from becoming contaminated beyond a certain limit, as explained in the following chapter.

The most widely used apparatus for cleaning are called *filters* and *strainers*. In general strainers are understood to keep from the oil such objects as rags and nuts and to remove from it coarse particles of carbon or scale. However, with diesel engines there are several strainers in use with a very fine mesh or with slots which catch particles as small as 0.005 or 0.006 in. Therefore, in practice, some strainers are often referred to as *filters*.

Oil Filters. There are a great number of different oil filters in use and only a few typical ones will be briefly discussed.

Filter with Element. A filter with a replaceable filter element is similar to filters used with automobile engines. The oil is admitted to the filter shell either near the top (Fig. 12-10) or, sometimes, nearer the bottom and enters the filtering element through a large number of small holes on its cylindrical surface. The ends of the element are held by metal flanges. From the periphery the oil works to the center, enters through a small hole $\frac{1}{16}$ in. in diameter into the hollow center stem and is discharged from its lower end. The element itself consists of some porous, tightly packed material, such as cotton waste or cellulose, which catches and retains all impurities above a certain very small size, about 0.001 in. This filter is usually installed in a by-pass line, hence the necessity of the small orifice in the center and, working continuously, removes a comparatively large portion of sludge and dirt. When the inside of the filter element becomes loaded with removed sludge and dirt, the oil begins to get dirtier and darker; this indicates that it is time to install a new element and to change the oil. The element is taken out by unscrewing the hold-down screw at the top and taking off the lid. Care must be exercised not to damage the cork gasket under the lid. This small filter is made 8 in. high and $4\frac{1}{2}$ in. in diameter.

Pressure Filter. Figure 12-11 shows a pressure filter using cloth or fabric as filtering medium. The filtering element consists of four fabric bags, with flexible-link metal separators between them, and the whole

LUBRICATION AND LUBRICANTS

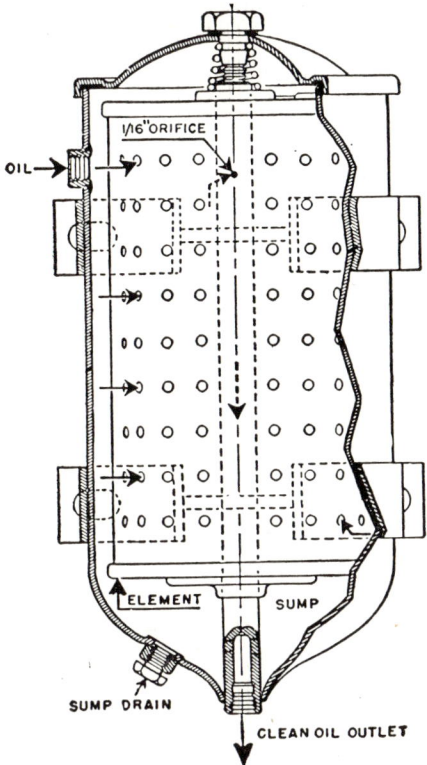

Fig. 12-10. Small lubricating-oil filter.

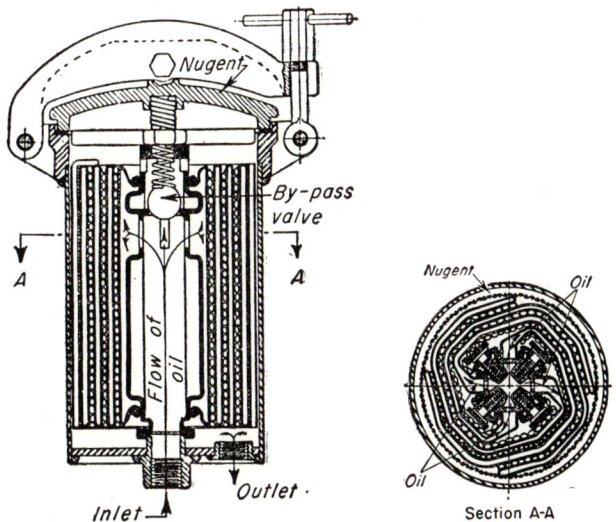

Fig. 12-11. Fabric-type filter.

is rolled up and inserted into a cylindrical container. Oil is admitted at the bottom in the center under pump pressure, flows along the path indicated by the arrows, and leaves again at the bottom. A spring-loaded relief valve by-passes part of the oil in order to prevent an excessive pressure drop when the filtering element becomes clogged with sludge. The filtering bags can easily be taken out, washed in kerosene, and put back in operation. Filtering bags are of wool, linen, or cotton, linen and cotton giving a higher degree of filtration with viscous oils. The filter is built for a working pressure up to 100 psi.

The filtering-element area varies from 2 to 21 sq ft, giving a capacity of 1.1 to 12 gpm, based on oil of 182 SSU viscosity.

Metal-edge Filter. Such a filter, or rather strainer, is similar to the metal-edge strainer used for fuel oil and described in section 11-2, Fig. 11-1. The main differences are in the greater thickness of the spacers, which for lubricating oils are made 0.006 in. thick, and in the over-all size. The amount of lubricating oil circulated in a diesel engine is from 80 to 1,000 times greater than the amount of fuel burned, and this proportion must exist in the relative capacities of the respective strainers if all lubricating oil is strained. To obtain such large capacities the strainer elements are made larger in diameter and length and several—one to four—elements are built into one filter housing. This gives capacities from 5 to 900 gpm, based on a viscosity of 300 SSU, a spacing between the disks of 0.006 in., and a pressure drop across the filter of 5 to 10 psi. The internal parts are made of various metals, in order to meet specific operating conditions: mild steel, stainless steel, bronze, and monel metal.

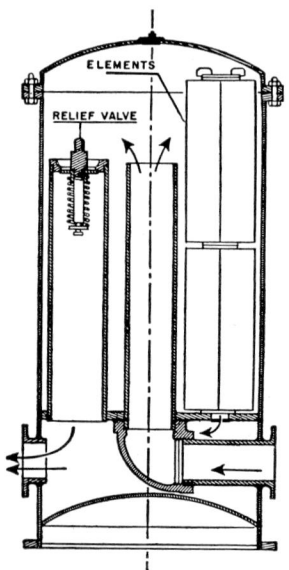

FIG. 12-12. Metal-edge strainer.

A filter with metal-edge filtering elements is used in United States naval small- and medium-sized engines. It is built to Navy standard dimensions of 18 in. over-all height and $7\tfrac{5}{16}$ in. over-all diameter (Fig. 12-12). The Navy type has no device for cleaning the elements but has a relief valve through which the oil will by-pass if the resistance to oil flow of the elements begins to exceed the allowable pressure drop because of clogging of the slots.

Paper-edge Filters. Filters based on the same principle as the metal-edge filter are made also with paper disks and spacers. They are known under the trade name *clarifiers* and give good performance. However, they are bulkier in size. When the element becomes clogged, it is removed and a new stack of paper disks and spacers is inserted.

Gravity Filtration. If given sufficient time, the dirt in lubricating oil will settle out, owing to the difference in dirt and oil densities. Keeping the oil warm speeds up the settling out by decreasing its viscosity. Usually the oil is first passed through cloth bags to remove the coarser particles. In other designs settling is carried out first and the oil is put through cloth bags later. The cloth bags must be cleaned at regular intervals in order to function properly.

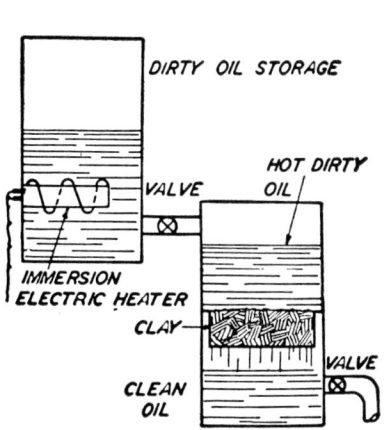
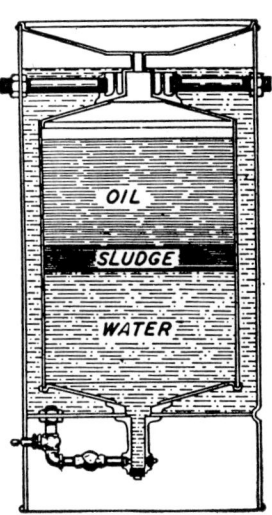

Fig. 12-13. Activated-clay oil filtration.   Fig. 12-14. Chemical filter.

Filter Bed. Excellent cleaning is possible by passing the dirty oil through a bed of activated clay. This clay removes all the dirt and carboniferous matter and even brings back the original color of the oil. The system, shown in Fig. 12-13, consists of a filter tank, an electric oil heater, and a clean-oil sump. However, the process is extremely slow and rather costly because the clay must be frequently renewed. About 20 per cent of the oil will not pass through the clay bed and is lost in the system. The cost of cleaning is about 15 cents per gallon treated.

Chemical Treatment. If dirty oil is warmed and then treated with a chemical that will cause the dirt particles to coagulate, or mingle, with the chemical, a high purification effect may be obtained. The De LaVergne chemical filter is shown in Fig. 12-14. Potassium hydroxide is used by some; others employ trade-mark preparations such as Oakite. The only objection to the system is that a supply of hot water must be available and the cleaning of the settling tank is a somewhat dirty job, not liked by the engine operator.

12-10. Centrifuging. The oil-purification centrifuge is built on the same principle as the familiar cream separator. The separation is effected by making use of the difference in specific gravity between solid and water particles and pure oil. When the dirty oil in the centrifuge bowl is rotated at a high speed, the centrifugal force created by the rotation is proportional to the specific gravity; it is greatest for the suspended particles of metal, which therefore are thrown out to the periphery of the bowl; smaller for carbon and other solid particles which are pressed against the metal particles; still smaller for water, droplets of which are thrown out next; then may come other impurities, leaving pure oil nearest to the center of the bowl. Solids lighter than oil cannot be separated centrifugally, but they do not get into the lubricating oil of diesel plants except in rare instances.

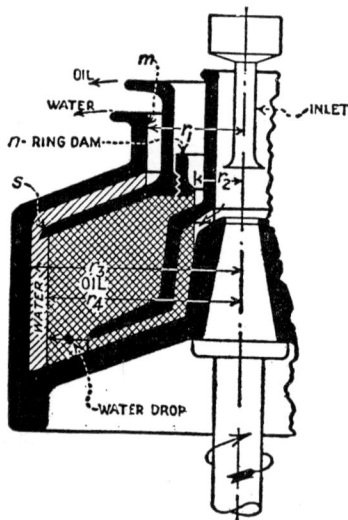

FIG. 12-15. Action in a centrifugal oil purifier.

Continuous separation is carried out by adding to the oil appreciable quantities of hot water in order to wash out the solids. The removal of the water and of the oil, after they are separated, is effected by narrowing the top of the bowl to two concentric necks, as shown schematically on Fig. 12-15: the water head $(r_3 - r_1)$, because of its greater specific gravity, balances the combined water and oil head $(r_3 - r_2)$. As long as no liquid is added, the inner diameter of the centrifugal ring of water will stand edge-to-edge with the deflector m while the inner diameter of the rotating oil will similarly register with the edge of the removable ring dam n for oil. The radius r_2 must be chosen according to the specific gravity of the oil and the temperature at which it is centrifuged. Let it be assumed that only water is now admitted to the machine: it will immediately be thrown into the outer ring of water, from which it will displace an equivalent quantity past the edge m while the diameter of the ring of oil will remain undisturbed and in registry with r_2. On the other hand, if oil only is added, the additional amount will be unable to disturb the depth of the water ring because of the greater specific gravity of the latter; hence the entering oil merely causes an equal quantity to be displaced past the edge r_2. Mixtures of oil and water fed to the machine will be similarly separated and discharged through the corresponding stationary outlets suitably located in relation to the rotating element. A machine set up and adjusted in this manner by means of a ring dam with a correct radius r_2 will not require special removal of water from the bowl, although solid matter will pile up on its outer wall up to the point where the machine should be stopped for cleaning. Normally the machine is stopped and the bowl cleaned at about 8-hr intervals.

Centrifuging is essentially identical with letting the dirt settle out by gravity. However, if gravity alone were to be depended upon for settling, the process would be too slow, particularly in view of the small difference in specific gravity existing between some of the solids and the oil and because of the considerable opposing effect of viscosity. The large force available in the centrifugal purifier, the magnitude of which, in the commercial centrifugal purifiers, is 7,000 to 13,000 times greater than gravity, reduces the time of separation practically in direct ratio to the gravity force, while it probably also precipitates a considerable class of solids so fine as to remain suspended if subjected to gravity settling during the time available in the average plant. It is considered also that those particles which are too fine to be taken out by centrifugal force are also of such minute dimensions as to come well within the thickness of ordinary lubricating oil films. Impurities of the latter class are therefore regarded as having no practical effect on the normal process of lubrication unless they are present in large quantities and then tend to clog the piping and small passages.

Centrifugal purification of lubricating oil is considered by many engineers to be the best method. It is used in most larger diesel plants in conjunction with either the continuous or the sump scheme of filtration, as explained later. Its main drawbacks are the first cost of the equipment and the cost of labor to operate it.

QUESTIONS

1. What is lubrication in reference to machinery?
2. Enumerate the four objects that may be attained by means of lubrication.
3. What causes friction between two metal surfaces?
4. What is the coefficient of friction? How is it expressed numerically?
5. What is dry friction?
6. What is fluid friction?
7. Enumerate the main conditions that influence the oil-film thickness between two metal surfaces in relative motion.
8. What is clearance in respect to a bearing?
9. What is thick-film lubrication?
10. Between what limits may the oil-film thickness in engine bearings vary?

194   FUNDAMENTALS OF DIESEL ENGINE WORK

11. What is thin-film, or imperfect, lubrication?

12. Explain the lubrication action in a plain bearing according to the hydrodynamic theory, and illustrate by sketches.

13. Explain the action of an oil groove according to the hydrodynamic theory.

14. Upon what factor does the oil-film pressure depend in a plain bearing?

15. What, if any, is the relation between the oil-film pressure in a bearing and the pressure in the lubricating system feeding oil to the bearing?

16. What is the required clearance generally required between the journal and the bearing?

17. Give a sketch explaining the principle of operation of a Kingsbury thrust bearing.

18. How are ball bearings usually lubricated?

19. State the two objects of lubricating ball and roller bearings.

20. How are ball and roller bearings protected against dust and other foreign matter?

21. Into what two different groups may all lubricants be divided?

22. How is lubricating oil of the necessary characteristics or properties obtained?

23. What is the viscosity of an oil?

24. What is the instrument used to determine the viscosity of lubricating oil in the United States?

25. How is viscosity measured by the Saybolt viscosimeter?

26. How is viscosity determined by a Saybolt viscosimeter designated?

27. At what temperature is the viscosity of crankcase lubricating oils determined?

28. How does temperature influence oil viscosity?

29. What is the pour point of an oil?

30. What is the pour point of diesel cylinder oil?

31. Why and when is the pour point important for diesel-engine operation?

32. What is carbon residue?

33. Why is it desirable to have a small carbon residue in a diesel lubricating oil?

34. What is the flash point of a lubricating oil?

35. What are the limits of flash point of lubricating oils used for diesel engines?

36. How is water and sediment content of an oil determined?

37. How is acidity of a lubricating oil tested?

38. What is an emulsion?

39. Why should a lubricating oil not have an excessive tendency to oxidize?

40. What does the ash content of an oil indicate?

41. What is the relation between the color of a lubricating oil and its properties?

42. What is the gravity of an oil?

43. Does gravity have any relation to viscosity and other properties of a lubricating oil?

44. What are additive, or detergent, oils?

45. How does an additive, or detergent, oil prevent piston rings from sticking?

46. What are the United States Navy symbols for straight mineral lubricating oils?

47. What are the Navy symbols for detergent diesel oils?

48. State how lubricating greases differ from lubricating oils in composition and application.

49. Explain the general difference between an oil strainer and an oil filter.

50. Give a sketch and explain the operation of an oil filter with a replaceable element.

51. Give a sketch and explain the operation of a so-called *pressure oil filter*.

52. Give necessary sketches to explain the operation of a metal-edge filter.

53. Explain the difference between a metal-edge and a paper-edge filter in respect to their size and to the method of handling them.

54. What type of oil filter is known under the trade name of Clarifier?

55. Draw a sketch of a filter that uses a bed of activated clay.

56. Draw a sketch of a De LaVergne chemical filter and explain its action.

57. Draw a sketch of a centrifugal oil purifier and explain its action.

CHAPTER 13

ENGINE LUBRICATION

13-1. Importance of Lubrication. Regardless of how well an engine is designed from the standpoint of thermal efficiency and strength, and how well it is built from the standpoint of materials and workmanship, if the lubrication of all moving parts is not taken care of properly, the engine will either not run at all or show heavy wear and have a short life.

The subject of lubrication is probably the most vital of all the details of operating a diesel power plant. Poor lubricating oils and incorrect application of good oils are the causes of much of the trouble experienced in the operation of an engine. No engineer can afford to be indifferent to the problems of lubrication. Among these problems those pertaining to the lubrication of rotating journals are the most important and therefore will be discussed at greater length.

13-2. Bearing Lubrication. Bearing lubrication methods may be listed as (1) intermittent lubrication, (2) continuous lubrication with a limited supply of lubricant, and (3) continuous lubrication with an abundant supply of lubricant. *Intermittent lubrication* may be obtained by (1) dropping oil from an oil can into an oil hole or into a plug of waste or (2) forcing grease from a grease cup or a pressure gun into a hollow space.

Both methods provide only imperfect lubrication and are used for places where the loads are small and velocities very low.

Limited continuous lubrication may be obtained by (1) grease cups with spring action, (2) oil reservoirs with a wick which carries the oil by capillarity and gravity, and (3) sight-feed drop oil cups.

All three methods are suited to light duty only.

Abundant lubrication alone insures perfect lubrication and may be obtained by (1) ring, chain, and collar oiling, (2) splash and bath lubrication, (3) forced-feed lubrication, and (4) pressure or flooded lubrication.

Ring, chain, and *collar lubrication* give satisfactory results only at low and medium speeds. At higher speeds the oil is thrown off by the centrifugal force.

Splash and *bath lubrication* are satisfactory for light and medium duty but have no provision for the control of oil temperature.

Forced-feed lubrication delivers oil to desired points under a very small pressure but in definite amounts, which can be regulated by means of sight-feeds. In diesel engines this method is used in the lubrication of the cylinders of low-speed engines and of the cylinder and piston rods of double-acting engines.

Pressure lubrication, is, in diesel engines, the standard method of lubrication of all important parts which do not get oil by splash from the crankcase. The oil is drawn from the crankcase bottom, called the *oil sump*, by a gear-type pump; it is then pumped through a filter and oil cooler to the header and through separate lines to the main bearings. The header is sometimes called the *engine-oil gallery.* From the main bearings most of the oil passes through drilled passages in the shaft and crank webs to the crankpin bearings. From them the oil usually passes through the rifle-drilled connecting rod to the piston-pin bearings. In some engines oil from the piston-pin bearings is used for cooling the piston. Separate lines from the header carry oil to the camshaft bearings, gears, and other parts requiring lubrication.

The oil circulated through the main and crankpin bearings serves not only to lubricate them by ensuring the formation of a wedge-shaped oil film, but also to cool these bearings, to carry away the heat into which the work of friction is transformed. The oil pressure in the header is maintained at from 30 to 60 psi, depending upon the engine type. The correct pressure for a particular engine is indicated in the instruction book. The oil-circulating pump has a larger delivery than is actually required and the excess oil is by-passed from the high-pressure side back into the sump through a spring-loaded adjustable pressure regulator.

13-3. Diesel-engine Lubrication. The satisfactory lubrication of modern high-speed diesel engines has become a difficult problem. The following are some of the causes for the encountered difficulties.

1. Higher speeds, which have resulted in smaller engines with smaller oil reservoirs and higher engine and oil temperatures.

2. Higher piston-ring temperatures, which have a tendency to cause ring sticking and piston seizures.

3. Higher engine temperatures, which have made more difficult proper cylinder and piston lubrication.

4. Increased use of two-stroke engines, with their increased power

output, and supercharging of four-stroke engines, which has increased the cylinder pressures and temperatures.

5. Higher cylinder pressures, which have resulted in greater bearing loads.

6. Higher engine temperatures, loads, and speeds, which have put new strain on bearing materials and increased their tendency to corrosion in spite of the use of special bearing metals.

7. Use of bearings with precision shells, which are more sensitive to solid particles in the lubricating oil and require improved oil purification.

8. Increased use of copper in bearings, which contributes to oil oxidation.

9. The use of an all-enclosed construction in present high-speed diesel engines, which has made more difficult the prevention of fuel-oil leaks and increased the rate of crankcase oil dilution with fuel oil.

10. The increased clearance necessary with aluminum-alloy pistons, which has resulted in higher oil consumption, particularly in larger engines operating at partial loads.

11. The increased use of mobile diesel engines, which has brought up the problem of the viscosity of lubricating oil when engines are started in subfreezing weather.

The best test of a lubricating oil is the way it behaves in the engine. The oil must have a viscosity which remains within suitable limits throughout the operating-temperature range of the engine. In addition to maintaining a sufficient oil film between the moving parts, the oil—which unavoidably burns in the cylinder—must leave a minimum of carbon residue. The oil must be stable and resist oxidation, acidification, and emulsification.

An ideal diesel engine lubricant must fulfill the following requirements:

1. Maintain a good oil film on cylinder walls and thus prevent excessive wear of cylinder liners, pistons and rings.

2. Prevent sticking of piston rings.

3. Seal compression in the cylinders.

4. Leave no carbon deposits on the crown and upper part of the piston and in exhaust and scavenge ports.

5. Not lacquer piston or cylinder surfaces.

6. Prevent wear of bearings.

7. Cleanse the interior of the engine.

8. Not form sludge, clog oil lines, strainers and filters, or leave deposits in the oil cooler.

9. Be usable with any kind of filter.

10. Have a low consumption rate.
11. Permit long intervals between renewals.
12. Have good cold-starting properties.

Evidently no single oil can meet all these requirements, and the oil that comes the nearest to meeting the most important requirements for a given engine and operating conditions must be selected. Usually the oils to be used in the various engines are specified by the engine manufacturer after exhaustive tests. Since the oil refineries are continuing to improve their products, new oils should be tested in an effort to find a better oil, especially by the management of larger diesel plants.

13-4. Oil Contamination and Dilution. The main source of contamination of the oil in the crankcase of a diesel engine comes from the lubricating oil itself. Most of the oil admitted to the engine cylinder, whether by special cylinder lubricators, as in large low-speed engines, or by splash and mist formation in the crankcase, is burned with formation of coke, or carbon. This carbon is scraped down into the crankcase, where it is joined by carbon from incomplete combustion of fuel oil.

A second important cause of contamination is in the particles of dust introduced with air where no air cleaners are used. Part of this dust adheres to the oil film which covers the cylinder walls and is scraped down with the oil into the crankcase.

Another cause of contamination is water formed by condensation of the water vapor formed by the combustion of the hydrogen of the fuel with the oxygen of the air charge. This water forms an emulsion with the less stable parts of the oil in the crankcase. Aided by the oxidation of other parts of the oil, this emulsion forms sludge, the danger of which has already been explained.

Finally, the oil is contaminated by metal particles loosened by wear—cast iron from piston rings, cylinder walls, and gears, steel from other gears, cams, and journals, and brass and babbitt from various bushings and bearings. The most dangerous are the cast-iron and steel particles which, when squeezed between moving surfaces, increase their wear. However, even the soft-metal particles are undesirable as they tend to increase sludge formation.

In order to operate an engine satisfactorily and prolong the life of its parts, contamination of the oil must not exceed a certain maximum and must be reduced, periodically or continuously, by means of filters of different design, centrifuges, gravity settling, or chemical methods.

Generally, fresh oil is used for the lubrication of the cylinders and the air compressors of large engines. This oil becomes contaminated in a very short time, usually right after entering the cylinders, due partially to the dust and dirt carried in with the intake air, and partially to the carbon formed.

In some crosshead-type engines, the oil that drains down the cylinder walls is caught by a stuffing box at the base of the cylinder and drained to waste. This keeps the dirtiest oil from returning to the system, and a smaller reclaiming outfit than that usually used will work satisfactorily. While the oil so gathered is thrown away, actually there is very little waste, as only a small amount of the oil could be reclaimed anyway.

Steel and cast-iron particles may be extracted by means of small permanent magnets incorporated in the oil drain plugs.

Dilution of lubricating oil by fuel oil may be caused by leakage from the high-pressure fuel injection system when the leakage is permitted to enter, through some not sufficiently tight joint, into the crankcase. In most diesel engines such dilution either does not exist or can be eliminated by careful watching of the tightness of all joints. In the few constructions where dilution is unavoidable, the crankcase oil should be changed after the dilution reaches a certain maximum, usually 5 per cent, or as stated in the engine builders instruction book. The rate of dilution may be determined by comparing a sample of the oil from the crankcase with a sample of clean lubricating oil mixed with 5 per cent of fuel oil. Naturally, both samples must have the same temperature. On the other hand, the operator's attention is called to an excessive dilution by a pressure drop in the lubricating system, for the viscosity of fuel oil is very much lower than that of lubricating oil. When the oil pressure at the rated engine speed drops to 20 psi or slightly below, which is considered the lower safe limit, and cannot be raised by screwing down the pressure-regulator valve, and when at the same time the whole lubricating system is in proper condition, with no excessive leaks, then this is an indication that the dilution is excessive and the oil in the whole system must be changed—replaced by fresh oil.

13-5. Used-oil Analysis. In large diesel plants it is recommended to have samples of lubricating oil, taken at the time when the oil is changed, analyzed in a properly equipped laboratory. Such an analysis can reveal conditions to which the oil has been exposed that may have escaped the attention of the operator. It tells, for instance, whether proper water and oil temperatures were maintained, how much dust entered the engine, whether there are water leaks, how effectively the oil filtering system is functioning, whether blow-by is considerable, and whether the oil is being changed as often as necessary.

The analysis should be conducted to determine the following characteristics:

Viscosity. Normally the viscosity of used oil is higher than that of fresh oil due to evaporation of the lighter portions and oxidation. However, dilution has a tendency to decrease the viscosity.

Oxidation products formed in the oil during a certain service period indicate how stable the oil is and whether it was kept too long between changes.

Flash Point. A noticeable decrease in the flash point of the oil together with a decrease of viscosity indicates dilution by fuel oil. If dilution is excessive, it may be due to fouled injection nozzles, poor or incorrect fuel, delayed injection, too low a compression ratio and compression temperature, or leakage from the injection system.

Water. Usually water in crankcase oil is due to condensation of combustion water vapor and a certain amount of blow-by. Excessive water content may be due to leakage of cooling water.

The *neutralization number* is the number of milligrams of potassium hydroxide necessary to neutralize one gram of the oil. However, the potassium hydroxide measures only the amount and not the kind of acid present. Some acids are more corrosive at elevated temperatures than others so that the neutralization number in itself, without other data, is no indication that the oil is no longer suitable for use. Laboratory studies show that a sudden rise in this number is a better indicator of possible corrosion trouble than is its absolute value, within certain limits of course. In most cases a neutralization number of 0.5 is considered as indication that the oil should be changed, and oil with a neutralization number of 0.9 is positively dangerous.

Precipitation Number. This test is made to determine the amount of sediment in used oil. It represents the number of cc (cubic centimeters) of precipitate formed when 10 cu cm of lubricating oil are mixed with 90 cu cm of petroleum naphtha (ASTM precipitation naphtha) and centrifuged under prescribed conditions. Fresh oil is completely soluble in this naphtha. After use in the engine it contains various contaminants which are insoluble in naphtha.

When judged in conjunction with other factors, the precipitation number is very helpful in deciding when an oil is no longer fit for use. Engines with a dry-sump system can stand an oil with a higher precipitation number, up to 0.5, than can engines with a wet-sump system. The reason is that a dry-sump system contains a larger volume of oil and generally uses a more elaborate filtering system.

Ash Content. This test gives the amount of mineral and metallic solids in used oil that do not burn. Since some materials are more dangerous than others, expert judgment is necessary to interpret the data obtained from this test.

13-6. Oil Circulation. All present diesel engines have enclosed crankcases and use pressure lubrication. The oil is circulated in rather large quantities, from 0.01 to 0.05 gpm or more per rated horsepower. The reason for such a large circulation is that the lubricating oil is at the same time used as a coolant for the bearings and, in some engines, for the pistons. The oil is delivered by a gear pump under a pressure of 20 to 60 psi into a manifold, or header, also called, particularly in marine service, an *oil gallery*, from which it is distributed through pipes to the bearings and other points which require lubrication.

In some two-stroke cycle engines, the oil circulation is two to three times greater, up to 0.15 gpm per rated horsepower.

13-7. Oil Consumption. The amount of oil actually used up varies between 0.00025 and 0.0005 gal per hr per rated horsepower in four-stroke engines with a properly designed and operated lubricating system. In two-stroke engines the oil consumption is about twice as great.

In practice it is customary to express the lubricating-oil consumption by the number of rated horsepower-hours per gallon of oil. The above given figures correspond to about 2,000 to 4,000 hp-hr per gal for four-stroke engines and 1,000 to 2,000 hp-hr per gal for two-stroke engines. For air-cooled aircraft engines the figures are 150 to 400 hp-hr per gal.

13-8. Friction Surfaces. The principal friction surfaces requiring lubrication in a combustion engine are (1) pistons and cylinders, (2) crankshaft and main bearings, (3) crankpins and their bearings, (4) crossheads and guides, (5) wristpins and their bearings, and (6) valve gear.

13-9. Pistons and Cylinders. *Splash Lubrication.* Trunk pistons of single-acting high-speed engines usually do not need special lubricating provisions. The oil, which flows from the main and crank bearings and is splashed by the cranks and connecting rods, and the oil mist, which is formed in the enclosed crankcase by the fast-moving parts, furnish sufficient lubrication for the pistons. These conditions exist not only when the oil in the crankcase is kept at a certain level to allow it to be picked up by the connecting rods but even with a so-called dry crankcase.

Positive-feed Oilers. Pistons of low- and medium-speed engines are lubricated by positive-feed mechanical oilers. Figure 13-1 gives a schematic section of such an oiler: The oil from a supply reservoir *a* drips into the chamber c, the number of drops per minute being regulated by a screw b; the plunger d on its downward stroke, having covered the connection to c, pushes the oil into the copper tube e which is discharging the oil to the cylinder surface; the check valve f prevents the oil from being sucked back by the upward stroke of d and also separates the space c from the engine cylinder.

In cylinders cast in one piece with the water jacket, special bosses b are cast (Fig. 13-2) to take the oil nipple e. In cylinders with inserted liners the oil tube is fastened to a nipple g (Fig. 13-3) with a conical

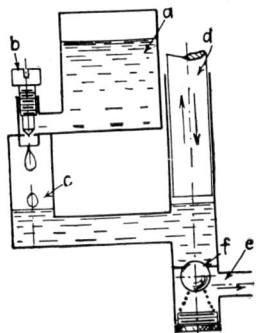

Fig. 13-1. Mechanical oiler.

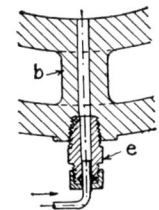

Fig. 13-2. Cylinder lubrication, integral water jacket.

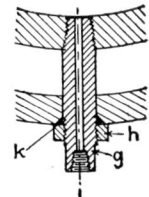

Fig. 13-3. Cylinder lubrication, with wet liner.

pipe thread screwed into the liner and a lock nut h and rubber ring k to seal the water from the outside. To avoid corrosion, the nipple g is made of brass or monel metal.

Location. In single-acting engines the hole discharging oil should come between the first and second piston rings when the piston is at the lower or outer dead center, so that the oil hole will never be exposed. In double-acting engines the oil holes are located halfway between the cylinder ends and become exposed at certain piston positions. To prevent the gases from getting into the oil tube a check valve is arranged either in the oil delivery nipple itself or close to it.

To avoid the possibility of discharging the oil during the expansion stroke, when the cylinder is filled with very hot gases, in some engines in which the oil openings are exposed the oil delivery is timed and occurs during the very beginning of the compression stroke, when the oil holes are covered by the piston.

In horizontal cylinders the oil is usually delivered at three points of the cylinder circumference—at the top and 90 deg from it to each side. In small vertical engines the oil is delivered at two points of the circumference, 180 deg apart, in a plane normal to the crankshaft center line. In larger vertical engines the oil is delivered in four to eight equally spaced points, all connected to one circuit (Fig. 13-4) to ensure uniform oil pressure and distribution.

Amount of Oil. The best results are obtained with a minimum amount of oil, just enough to maintain a film of oil on the cylinder surface. Sticking piston rings, much carbon in the circulating oil, and even excessive wear of the liner and piston rings can be traced to excess lubrication of the cylinders. Consumption of cylinder oil varies between 0.0001 and 0.0005 gal per hp-hr, depending on the engine type and the attention given it by the operators.

Even cylinders of medium-speed engines receive a considerable amount of oil from the crankcase splash and mist, so that it may be necessary to use scraper rings to remove the superfluous oil.

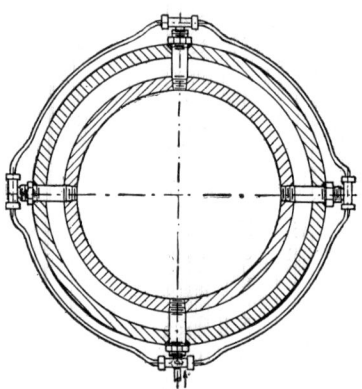

FIG. 13-4. Cylinder lubrication of large vertical engines.

13-10. Rotating Parts. *Main Bearings.* A film of oil can sustain a heavy pressure for a short time. However, when the film has broken down after the engine has been standing still for a certain time—several hours—a comparatively small pressure is sufficient to cause abrasion. The surfaces cannot receive a new film of oil until the pressure has been removed or until a sufficient peripheral speed of the journal has been developed. Therefore the lubrication of single-acting four-stroke engine main bearings presents less difficulty than the lubrication of bearings of two-stroke engines, because the pressure on the journals of a four-stroke engine is continually reversed. In a single-acting two-stroke engine the direction of pressure in most cases is not reversed and proper lubrication is more difficult. In a properly designed two-stroke engine the maximum bearing pressure is about one-third lower than that in a similar four-stroke engine.

In engines with an open crankcase or at least not very tight crankcases, as most horizontal engines, the main bearings are lubricated by positive-feed lubricators such as those used for cylinder lubrication. Engines with enclosed crankcases use pressure lubrication.

The oil is admitted either from underneath, as shown in Fig. 5-6, or through the cap (Fig. 5-7). Theoretically, the latter construction is better, as the oil is delivered at a point where the bearing pressure is at a minimum. However, its disadvantage is in the more complicated oil piping and in the necessity of breaking the pipe connections every time the bearing must be taken out for replacement or even inspection.

Crankpin Bearings. Small horizontal engines and crankcase-scavenging two-stroke engines use the centrifugal or so-called *banjo oiler* (Fig. 13-5). The oil hole leading to the surface of the crankpin is often drilled at an angle of about 30 deg in advance of the dead

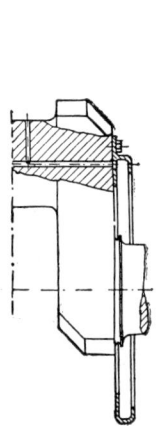

Fig. 13-5. Banjo oiler.

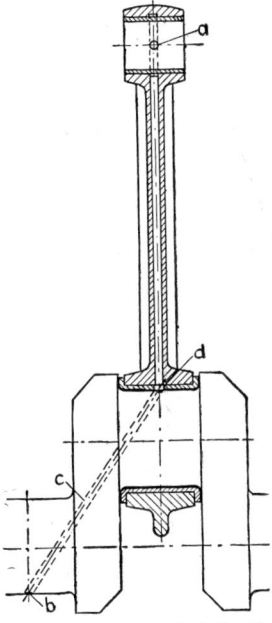

Fig. 13-6. Pressure lubrication of crankpin and piston pin.

center, so that the upper shell receives oil before the ignition stroke and at a point of relatively low pressure.

With pressure-feed lubrication either the shaft is drilled to conduct oil from the adjoining journal to the crankpin (Fig. 13-6) or the whole shaft is drilled and the crankpins receive oil from the inside of the shaft, which is fed by all main bearings.

Camshaft. Depending on the type of engine and other considerations, the camshaft bearings are lubricated by various methods: sight-feed oilers, ring oilers, automatic grease cups, and, when located inside an enclosed crankcase, by splash or feed lines from the main oil-pressure system. In large engines the long camshafts sometimes run in ball bearings that are grease packed and require the renewal of grease only once in several months.

Cams. On account of the blows that a cam, especially the exhaust cam, receives at each cycle, good lubrication is necessary to prevent rapid wear of the cam and its pin. In low-speed engines the cams often simply dip in an oil bath, the level in which is kept constant by a branch of the pressure oil pipe and an overflow. In engines with an enclosed crankcase, the oil splash and mist keep the cams and their bushings well lubricated.

Camshaft Drive. To decrease noise and wear, the drive gears must either run in an oil bath or receive a small stream of oil from the pres-

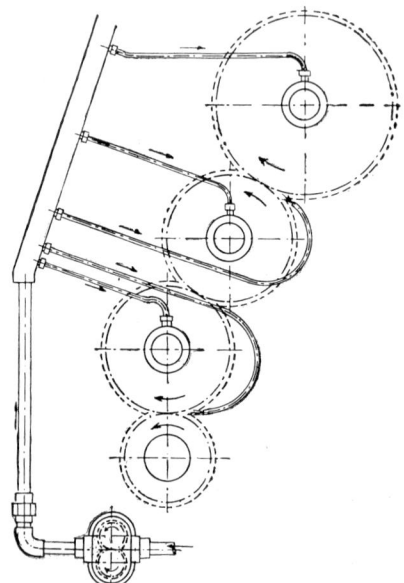

Fig. 13-7. Lubrication of a gear drive.

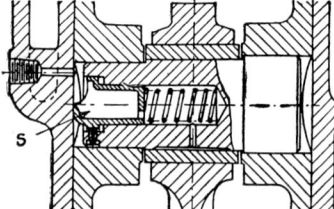

Fig. 13-8. Piston-pin lubrication by a scraper device.

sure system. In Fig. 13-7, all bearings have pressure lubrication, and small streams of oil taken from the main pressure system are directed on the teeth of mating gears. In enclosed crankcases this stream may be obtained from splash oil guided by special catchers. The same schemes are used for silent chain drives and governor drives.

13-11. Nonrotating Parts. *Wrist Pins.* In small vertical engines the wristpin may be lubricated by oil scraped from the cylinder walls by a special scraper scoop s (Fig. 13-8).

In vertical engines, with pressure lubrication delivered by a pump or with banjo-lubricated crankpins, the wrist pin is lubricated by excess oil from the crankpin bearing. The oil is delivered either by a hole drilled through the connecting rod, as shown in Fig. 13-6, or if the connecting rod has an I section that is too thin to be rifle-drilled, by a copper tube fastened with clamps to the inner web of the rod.

Valves. The guides and stems of valves must be lubricated but very sparingly, especially those of the exhaust valve. A drop per minute or a grease cup with spring pressure is all that is necessary.

Crossheads. Crosshead guides require only a small amount of oil as the specific pressures are usually very low. A single line from the force-feed lubricator is sufficient even in a large engine.

13-12. Air Compressor. The lubrication of the air compressor is subject to special requirements arising out of the thermal and chemical conditions set up by the relatively high air temperature, the moisture in the air after cooling in an intercooler, and the explosion danger if excess amounts of oil are fed to the cylinders. Splash from the force-feed oiling system in the crankcase should be carefully excluded, because it nearly always causes dangerously large quantities of oil to be pumped into the intermediate-pressure cylinder by piston ring action. For this reason the wristpin bearing is often mounted not in the intermediate-pressure trunk, but in a cylindrical guide situated immediately below it. The latter is isolated, so far as lubrication is concerned, from the body of the compressor and incidentally also serves as a barrier against oil thrown from the compressor crank and main bearings. When no separate guide is provided, it is advantageous to fit sheet-metal baffles directly underneath the compressor cylinder in order to reduce as much as possible the amount of oil thrown against the cylinder walls. However, the most effective remedy against oil splashing is to fit one or two scraper rings near the bottom end of the compressor piston. Rings of this kind act on the same principle as the scraper rings of the engine pistons—their upper beveled or rounded edges ride over the oil film on the upstroke, while their sharpened lower edges scrape the oil downward on the return stroke. Either the rings or the ring grooves are provided with recesses communicating through drilled holes in the piston wall with the interior of the latter.

The rate of oil feeding for air compressors should be reduced to a minimum, admitting oil only in such an extent as to prevent excessive wear of the pistons, rings, and cylinder walls. In compressors of moderate size, the minimum oil quantity is determined not so much by the occurrence of mechanical wear as by the minimum rate for which mechanical oilers can be adjusted. This lies somewhere between one and two drops per feed per minute, and unless the oiler is new and in perfect condition, attempts to reduce this further will incur the danger of having the oiler function irregularly or stop altogether.

Approximately 1½ drops of oil per feed per minute appears to be the minimum that can be relied upon, although as far as wear of the compressor is concerned this rate could be still further reduced. Most three-stage compressors have a total of five feeds, two on the low-, two on the intermediate-, and one on the high-pressure stage, so that the minimum rate of feed for the entire machine is about 7½ drops per minute.

13-13. Oil Filtering. The importance of using only clean lubricating oil and methods of purifying it after it has become contami-

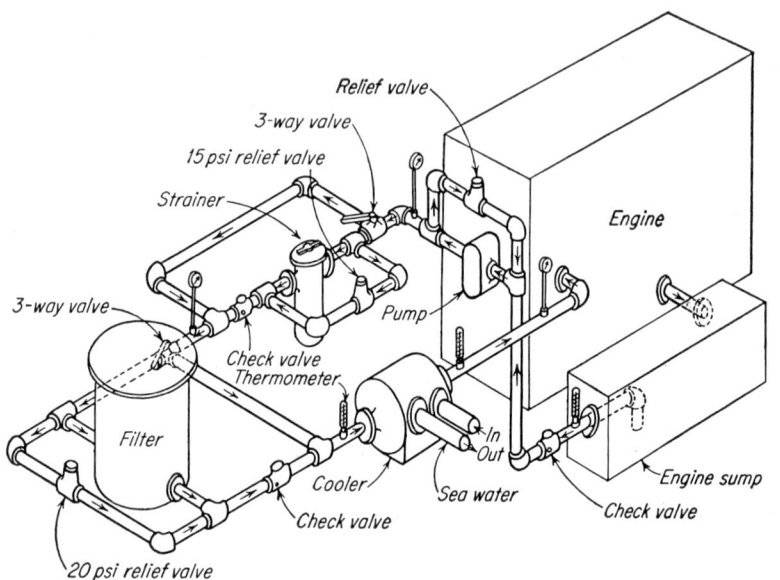

FIG. 13-9. Shunt system of oil filtering.

nated during engine operation are discussed in Chap. 12. The subject of this section is to discuss the different schemes of filter installation used with various types of engines and to describe other equipment necessary in a complete lubricating system.

Filtering Schemes. There are several filtering schemes in use: (1) continuous, (2) shunt, (3) by-pass, (4) sump, and (5) batch scheme.

The *continuous scheme* is the best—all the oil coming from the engine into the sump is pumped first through a strainer to catch the coarser impurities, then through a filter, and from there through an oil cooler to a clean-oil tank, usually located overhead. From this tank a second pump delivers the oil under the desired pressure to the lubricating-oil header, from which the oil is distributed by different pipes to all places requiring lubrication. The main disadvantage of this scheme is that it requires a large filter to handle all the oil for even a medium-sized engine and results in a comparatively large power loss to drive the pumps.

The *shunt scheme* is also essentially a continuous scheme but using one oil pump and with the clean-oil tank left out. This scheme, commonly used for large marine engines, is shown diagrammatically in Fig. 13-9. The oil pump delivers a constant amount of oil per revolution, but the resistance in the strainer and filter varies with the temperature of the oil and also gradually increases as they pick up the

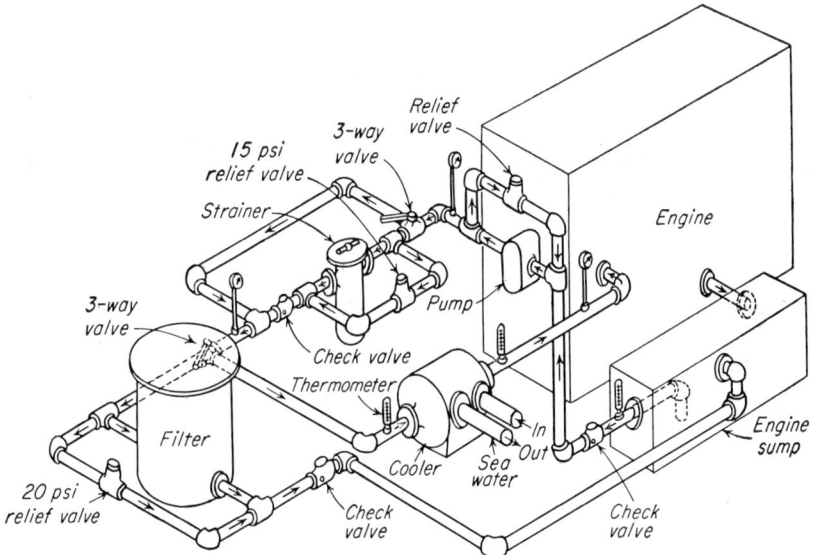

Fig. 13-10. By-pass system of oil filtering.

contaminants from the oil. The desired pressure in the oil header, also called *engine oil gallery*, is maintained automatically by by-passing, if necessary, part of the oil around the strainer and filter by means of pressure-relief valves built into the strainer and filter but shown outside of them in the diagram of Fig. 13-9. Three-way valves are used to by-pass these pieces of equipment when the strainer is cleaned or the filter element exchanged.

The *by-pass filter scheme* is normally used in small power-boat or vehicle installations and in moderate-sized stationary plants; it is shown diagrammatically in Fig. 13-10. The arrangement is the same as with the shunt scheme, except that the oil from the filter is discharged back into the engine sump. For sufficient oil to be supplied to the engine, the amount passed through the filter must be limited.

This is done in small filters by putting a 1/16-in. flow-restricting orifice in the standpipe through which the oil is discharged from the filter element, as shown in Fig. 12-10, or a relief valve, as shown in Figs. 13-9 and 13-10.

The *sump scheme* of filtration is shown in Fig. 13-11. The lubricating-oil pressure pump takes the oil from the engine sump and passes it through a metal-edge strainer and oil cooler into the oil gallery.

The filter is placed in a separate recirculating pipe line with a separate pump driven by an electric motor. This permits the oil to be

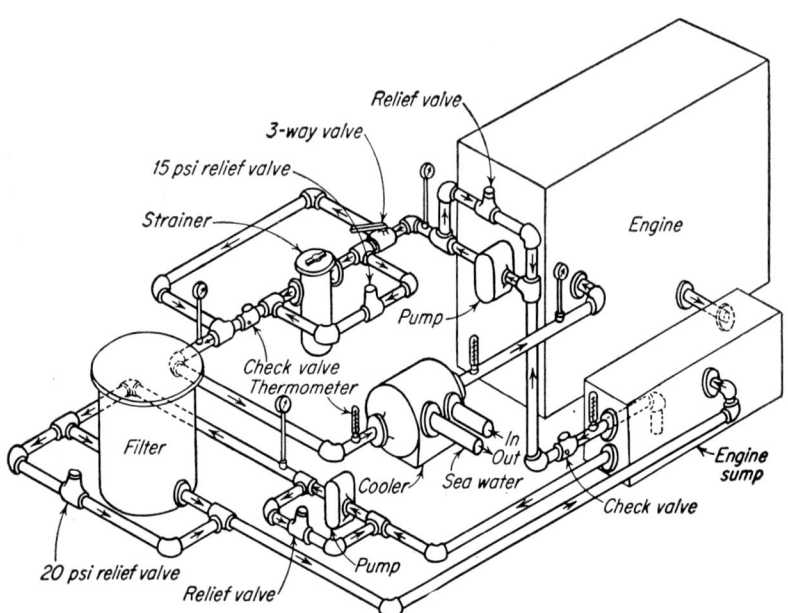

FIG. 13-11. Sump system of oil filtering.

filtered even when the engine is not running. The oil should be always filtered when hot, 140 to 180 F, when its viscosity is lower.

When using the sump scheme in connection with a centrifugal purifier, it is necessary to heat the oil, as shown in Fig. 13-12. However, the temperature of the oil entering the centrifuge should not be above 180 F in order to avoid fire risk. The advantages of the sump scheme are a smaller filter or centrifuge and lower power consumption for circulating the oil. The amount of oil circulated through the filter should not be less than 20 per cent of the oil circulated through the engine. While the quality of the oil in the system is not so high as with the continuous scheme of filtration, the oil never reaches a point near the danger line. The cost is so much less that this system seems to give the best results when all factors are taken into consideration. This explains its rather wide use in large power plants.

The *batch scheme* of filtering, as the name implies, works periodically: when the oil has become too contaminated, it is drained and the system filled with fresh oil. The drained oil is permitted to settle and later is purified by some method of filtering, chemical treatment, or centrifuging. The main advantage of this scheme is the low cost of the filtering equipment—but on the other hand, except for a short time after fresh oil is put into the system, the engine is running on contaminated oil which gradually gets worse and worse. This scheme is used mostly in

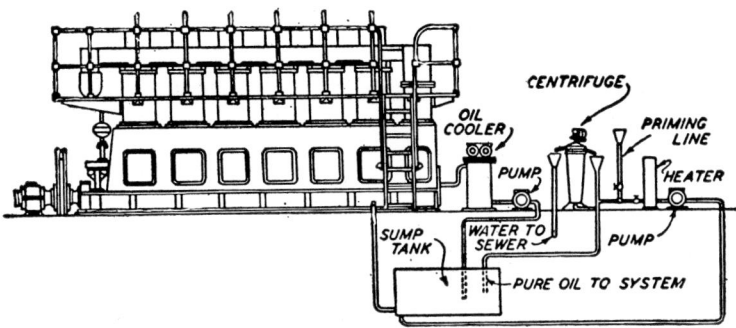

Fig. 13-12. Sump scheme of oil cleaning by centrifuging.

small power plants or plants operating very intermittently or during a certain season only.

13-14. Oil Coolers. As stated in Sec. 12-1, one of the important functions of lubricating-oil circulation is to cool the bearing surfaces by carrying away heat generated by friction. In addition, the oil in the engine sump is heated by the heat coming from the combustion both through blow-by gases and through heat conduction from the metal parts.

The temperature of the oil entering the engine header should not exceed 120 F. The oil leaving the crankcase should not be above 160 F under any conditions. Thus the heat picked up by the use in the engine corresponds to a temperature increase of about 40 F; although it is partly dissipated to the outside air, the major part of it must be removed by an oil cooler. However, many small engines do not have oil coolers.

Construction. Three basic types of cooler construction are in use: plate coolers, radiator coolers, and shell-and-tube coolers.

Plate Cooler. A plate-type cooler consists of hollow plates, 3 by 4 in., or 3 by 6 in. (Fig. 13-13), with the flat sides about $\frac{3}{16}$ in. apart and with inside baffles that reinforce the flat sides and, by keeping the oil in a turbulent flow, increase the heat transfer. The oil is admitted through the large hole at one end of the plate and discharged through the hole at the other end; the cooling water passes along the flat sides. Several plates placed one on top of the other in a housing (Fig. 13-14) provide a parallel passage for the oil. The size and number of plates used depend upon the amount of oil that must be circulated. The zinc pencil shown in Fig. 13-14 prevents corrosion of the thin-walled

Fig. 13-13. Plate of lubricating-oil cooler.

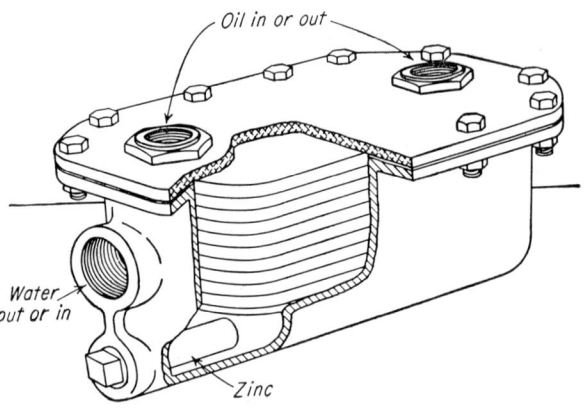

Fig. 13-14. Plate-type lubricating oil cooler.

plates, as explained in the discussion of heat exchangers for jacket-water cooling.

The plate cooler is rather efficient but on account of its comparatively small size is used only on smaller engines. Owing to its light weight it is preferred for mobile equipment.

Radiator Cooler. This type of cooler is made up of one or several—two on Fig. 13-15—rows of tubes, called the *core*, and an enclosing case divided into three parts, a centerpiece and two end covers. The tubes have oblong sections, about $1\frac{1}{16}$ by $\frac{3}{16}$ in.; their ends are brazed into tube sheets that are parts of the admission and discharge headers. Oil enters one header, passes through the tubes to the other header, which is connected to the pipe leading to the engine. The cooling water enters the housing on one side, flows across the tubes and leaves at the opposite side. The flat tubes are reinforced by rows of round cross tubes brazed into the flat sides. These cross tubes, or *struts*, also increase considerably both the inside and outside contact areas with the fluids. Figure 13-15 shows an oil cooler with two rows of 9 tubes in each row, and Fig. 13-16 a bigger cooler with four rows of

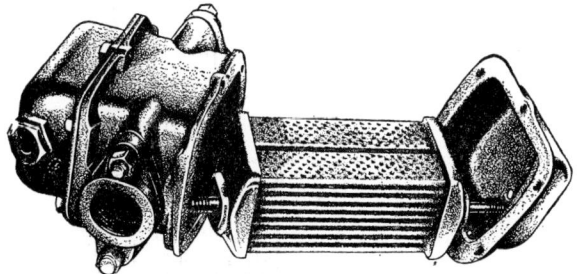

FIG. 13-15. Harrison radiator-cooler disassembled.

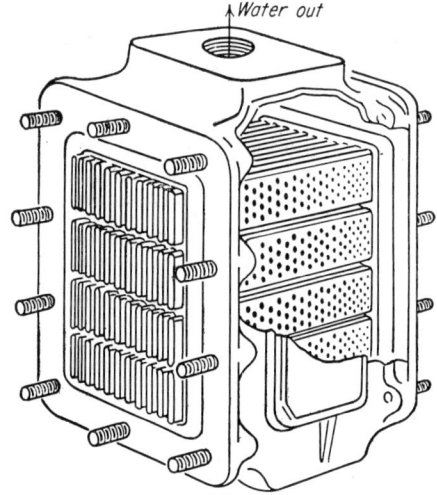

FIG. 13-16. Strut-tube heat exchanger.

11 tubes each. The larger cooler has the tube sheets integral with the center case, the oil headers (not shown on the illustration) being simple dished covers.

The name *radiator* is applied to the cooler because it is usually made up of tubes brazed to the headers (Fig. 13-15) and thus resembles an automotive radiator.

This type of cooler is extensively used for medium-sized and large marine engines.

Shell-and-tube Cooler. This cooler (Fig. 13-17) is built along the same lines as heat exchangers for cooling jacket water, only is smaller in size. The baffle plates force the cooling water to take a zigzag course, thus creating a turbulent flow with a high heat transfer.

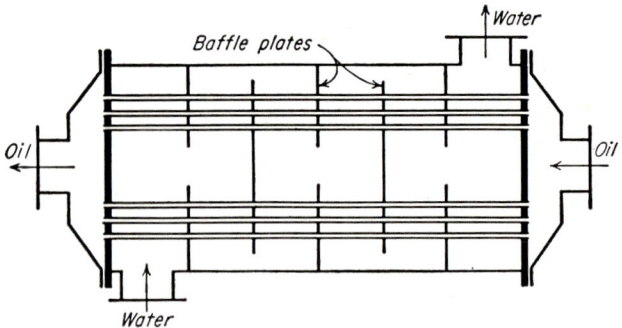

Fig. 13-17. Shell-and-tube oil cooler.

13-15. Pressure Regulators. Figure 13-18 shows a pressure-regulating valve b whose spring can be screwed down to obtain the required pressure. When the main bearings of an engine are in good shape, the pressure pump delivers more oil to the header than can flow through the small clearances and the excess oil is discharged through a spring-loaded piston valve from the pressure line a into the suction or by-pass line c. As the bearings wear down, their clearances increase and more oil begins to ooze out through them and less into line c. When the bearing clearances increase so much that no oil flows to c, the pressure gauge will begin to show a smaller oil pressure, which indicates that it is time to overhaul the bearings.

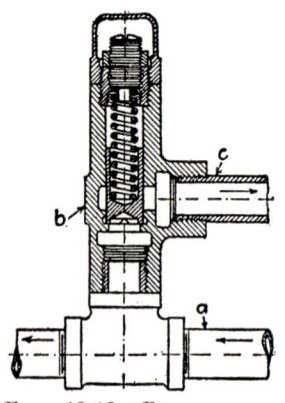

Fig. 13-18. Pressure-regulating valve.

A spring-loaded valve acts also as a safety valve protecting the whole system from an excessive pressure rise in case one or several discharge points should be plugged up. Spring-loaded valves are also used to by-pass oil around strainers and filters as they gradually become clogged by sediment.

Some pressure-relief valves have a spring-loaded disk with a conical seat instead of the piston-type valve shown in Fig. 13-18.

13-16. Pumps. Most diesel engines use gear pumps for delivering the lubricating-oil under pressure and also for scavenging the oil from the sump with dry-sump operation. The pumps are built with either spur or helical teeth and in general are similar to the gear pumps used for fuel-oil transfer (Fig. 11-5).

Some small and medium-sized low-speed engines use reciprocating plunger-type pumps.

13-17. Temperature Measurement. It is essential to keep a steady observation of the oil temperature, both in and out of the engine. As already mentioned, the temperature of the oil as it enters the engine should not be over 120 F, and the discharge temperature should not exceed 160 F. If the oil temperature begins to climb when the load does not exceed the normal rating of the engine, either the oil supply has decreased or excessive friction has developed at some bearing or in the cylinders. The cause should be investigated and, if it cannot be eliminated otherwise, the engine should be stopped for repairs.

The oil temperatures are indicated either by glass-and-mercury thermometers inserted in special wells in the pipe lines or by thermocouples with the indicator mounted on one panel with other instruments. While the thermocouples are very handy, damage to the long lead wires may yield incorrect indications.

13-18. Safety Devices. Two important items must be observed for satisfactory lubrication: the temperature of the oil leaving the engine and the oil pressure in the oil gallery, or header.

The devices protecting the engine in respect to oil temperature and pressure may be divided into two groups: (1) alarms—warning the operator that trouble is in the offing—and (2) devices which automatically shut down the engine.

Alarms. The simplest form of alarm operates by electric controls.

Temperature. When the oil temperature reaches the set limit of 160 F, the mercury column in the thermometer closes a switch which puts on a red light. If the temperature is permitted to go still higher, to 162 to 165 F, a second contact closes a switch to an electric bell or horn.

Pressure. If the oil pressure goes down below a predetermined minimum, the oil pressure-gauge handle closes a switch that turns on a red light or sounds a horn (the same horn as above).

Oil Level. When a separate clean-oil tank is used, as with the continuous-filtering system, it is good practice to provide the tank with visible and audible alarms, warning the engine operator in case of high or low levels of the oil supply. A green light shows when a high level is reached and a red light in case of a low level. An electric horn also sounds to warn the operator in case the lights are unnoticed.

Shut-down devices are often used in connection with a drop in pressure, indicating insufficient flow of lubricating oil. The simplest device works through a hydraulically operated valve installed in the low-pressure fuel line through which the fuel oil is delivered to the high-pressure injection pump. As shown in Fig. 13-19, the lubricating-oil pressure keeps the piston p and the valve v, connected to it, down and the spring s compressed; the fuel oil flows freely past the valve v. When the oil pressure drops beyond a certain amount, say 20 psi, the spring s pushes the piston p up and thus closes the valve v and cuts off

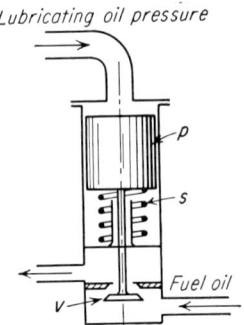

Fig. 13-19. Fuel cut-out device.

the fuel supply to the engine. Other shut-down devices are connected to the control shafts of the fuel-injection pumps.

QUESTIONS

1. What are the two most important conditions for proper lubrication of a diesel engine?
2. Explain what will happen to a diesel engine if its lubrication problems are not properly solved.
3. What are the three different methods of bearing lubrication in use?
4. How is intermittent lubrication of engine parts obtained?
5. Under what conditions can intermittent lubrication be used?
6. How is continuous limited lubrication of engine parts obtained? For what service are these methods suitable?
7. Enumerate the methods of obtaining abundant lubrication.
8. At what operating conditions do ring, chain, and collar oiling give satisfactory results? Why?
9. For what diesel-engine parts is forced-feed lubrication chiefly used?
10. What are the advantages of pressure lubrication as used in most diesel engines for all important parts?
11. Explain how pressure-feed lubrication operates in a diesel engine.
12. What is an engine-oil gallery?
13. What are the limits of oil pressure commonly used in diesel-engine lubrication?
14. What is the purpose of the adjustable pressure regulator and how does it operate?

ENGINE LUBRICATION

15. Enumerate the main causes that make the satisfactory lubrication of modern high-speed diesel engines difficult.

16. Enumerate the requirements that an ideal diesel-engine lubricant must fulfill.

17. What factors cause contamination of diesel lubricating oil?

18. How, during engine operation, can the lubricating oil be diluted by fuel oil?

19. What characteristics should be determined by an analysis of lubricating oil after it has been used for a certain time in a diesel engine?

20. How much oil is normally circulated in diesel engines?

21. How is lubricating oil consumption expressed?

22. Enumerate the principal friction surfaces requiring lubrication in a diesel engine.

23. Draw a sketch of cylinder-lubrication arrangement of a low-speed diesel engine.

24. Draw sketches of pressure lubrication of main and crankpin bearings.

25. Explain with sketches the lubrication of piston-pin bearings of small and large diesel engines.

26. Explain the lubrication of crankshafts in various types of engines.

27. What difference exists between the lubrication difficulties of a four- and a two-stroke single-acting engine?

28. How are the cams lubricated in various engines to prevent rapid wear?

29. Show by a sketch methods of lubrication of gear and silent-chain camshaft drives.

30. How should valve-stem guides be lubricated?

31. What is the main concern of piston lubrication of air compressors?

32. What is the minimum rate of oil feeding to air compressors? How many drops should be fed per minute to a three-stage compressor?

33. Enumerate the various filtering schemes used with diesel engines.

34. Draw a diagrammatical sketch of a continuous filtering scheme.

35. Draw a diagram of a shunt filtering scheme.

36. Draw a diagram of a by-pass filtering scheme.

37. Draw a diagram of a sump scheme of filtration.

38. Draw a diagram of a batch scheme of filtration.

39. Enumerate the three types of oil coolers in use.

40. Draw a sketch of a plate oil cooler and describe its action.

41. Draw a sketch of a radiator oil cooler.

42. Draw a sketch of a shell-and-tube oil cooler.

43. Draw a sketch of an oil-pressure regulating valve and explain its action.

44. Draw a sketch of an oil-circulating pump for use in diesel engines.

45. What devices are used to protect diesel engines in respect to lubricating-oil temperature and pressure?

CHAPTER 14

ENGINE COOLING

14-1. Necessity of Cooling. Part of the heat developed during combustion flows from the gases to the cylinder walls, raising their temperature. If, with an uncooled piston, the wall temperature is allowed to rise above a certain limit, about 300 F, the oil that lubricates the piston begins to evaporate rapidly, and both piston and cylinder may be injured. At the same time high local temperatures in certain parts of the engine, such as the cylinder head and piston, may cause excessive stresses and cracking of these parts. Additional heat is developed through friction between various rubbing surfaces, chiefly between the piston and piston rings and the cylinder walls. With oil-cooled pistons the limit for a safe cylinder-wall temperature is considerably higher.

The heat generated in an engine cylinder by the combustion of the fuel varies from about 6,000 to 10,000 Btu per hp-hr. Tests show that from 25 to 35 per cent of this heat in water-cooled and about 15 to 25 per cent in air-cooled engines finds its way into the cylinder walls and must be carried away. If some means were not provided for removal of this heat, the temperature of the metal would begin to approach that of the combustion gases as they leave the engine cylinder, or about 800 to 1200 F. Therefore, this heat removal, or cooling, problem is so vital that, if not taken care of properly, it can cause more engine trouble than any other phase of engine operation. For this reason cooling will be discussed in detail.

Methods. All heat carried away from an engine in the final count is conveyed to the atmosphere, even if at first it is given up to water in rivers, lakes, or ocean. However, the methods of cooling may be divided into two main groups: direct, or air cooling, and indirect, or liquid cooling. These two methods differ in construction details and in operating conditions, chiefly in the temperatures of the cylinder walls, and will therefore be discussed separately.

14-2. Heat Transfer. The three means of heat transfer—conduction, convection, and radiation—are used in cooling engine cylinders. Conduction plays an important part in carrying the heat through the metal walls and the thin layers of stagnant gas and water in contact with the walls; the rest of the heat is exchanged partly by radiation but chiefly by convection.

The heat flow between two fluids separated by a metal wall can be best explained using the diagram in Fig. 14-1. The temperature t_a of the gas at point a in the interior of the cylinder gradually falls to the value t_b at the surface of the inert gas film. The thermal resistance of this film is very great and a great temperature head, $t_b - t_c$, is required for conduction. The temperature head required to cause heat flow through the metal wall is $t_c - t_d$. The temperature head required for conduction through the outside film is $t_d - t_e$. Its value is comparatively small if the cooling fluid is water and large if it is air. Finally, the temperature of the cooling liquid drops to t_f at some distance away from the wall.

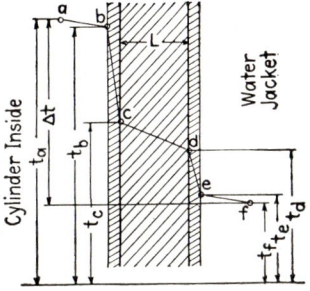

FIG. 14-1. Heat flow from hot gases to water.

The heat flow per unit area of surface in contact with hot gases on one side and the cooling medium on the other side thus depends upon the *inside film coefficient* h_1, the *conductivity* k of the metal, thickness L of the cylinder wall, on the *outside film coefficient* h_2 between the metal and cooling medium, and the difference between the gas temperature and the cooling-medium temperature.

Figure 14-2 shows as an example the values of the gas film coefficient h_1 for the whole cycle of an airless-injection diesel engine operating at 1,600 rpm at full load. The pressure and temperature curves complete the picture. T is the absolute temperature of the gas in the engine cylinder in degrees Rankine.

The value of the outside-film coefficient depends in the first place on whether the outside surface of the cylinder is cooled directly by air or by a liquid. The value of h_2 is rather small if the cylinder is cooled by air and considerably higher when it is cooled by water.

For reference the thermal conductivity k of metals is given in Table 14-1. However, the coefficient of conductivity k affects the heat flow very little, much less than the film coefficients h_1 and h_2.

Table 14-1 gives also the specific heats and coefficients of linear expansion of metals used for pistons, cylinders, and other engine parts.

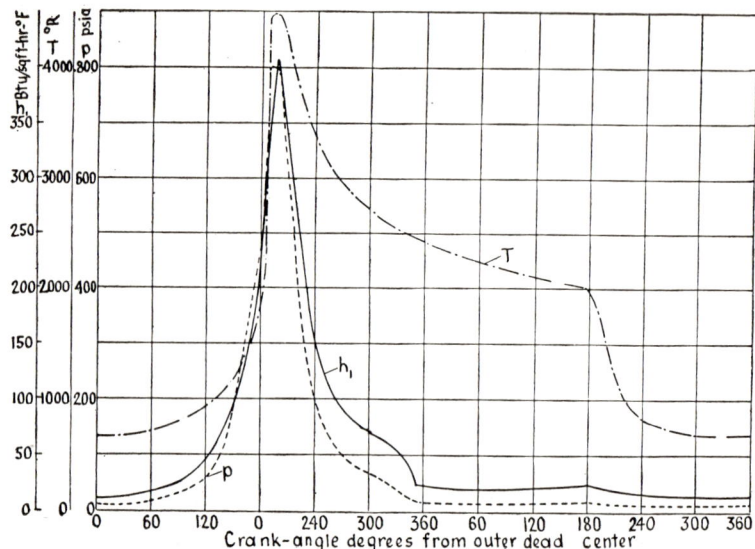

Fig. 14-2. Temperatures and gas-film coefficients in a diesel-engine cylinder at 1,000 rpm.

The temperature t_2 of the inside surface of the cylinder wall is not constant during a cycle but fluctuates following the variation of the gas temperature. Figure 14-3 gives a diagram of the temperature distribution through the wall. The temperature of the inner surface goes

TABLE 14-1. HEAT PROPERTIES OF METALS

Alloy	Specific heat, c, Btu / lb-deg F	Average conductivity, k, Btu-in. / hr-sq ft-deg F	Coefficient of linear expansion, $a \times 10^6$
Aluminum (Y alloy)	0.207	1,200	14.1
Brass, yellow	0.090	740	10.4
Bronze	0.102	710	9.8
Cast iron, gray	0.130	320	6.5
Monel metal	0.127	180	8.3
Steel, mild	0.110	310	6.5

up to t_{max} during combustion, and drops to t_{min} toward the end of the suction stroke. However, these fluctuations are not great: for a two-stroke oil engine at full load the fluctuation above the average value, $t_{max} - t_2$, is about 25 F and that below it, $t_2 - t_{min}$, is about 15 F. In a four-stroke engine the downward fluctuation will be about the same as the upward one owing to cooling during the suction stroke. This gives a temperature range of about 50 F.

The temperature fluctuation does not penetrate deeply; $3/8$ in. from the surface the range of fluctuation is less than 1 F.

Heat Flow. When the rotary speed of an engine increases, the duration in seconds of all events of each cycle decreases. However, the increased piston speed creates a greater turbulence, slightly increasing the heat flow, and as a result the percentage of heat of the fuel rejected to the jacket increases slightly with the engine speed.

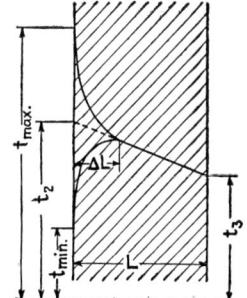

Fig. 14-3. Temperature fluctuation of the inner cylinder surface.

Tests have shown that the percentage of jacket loss is nearly independent of the engine load and decreases slightly with an increase in the cylinder diameter.

14-3. Liquid Cooling. Since water-cooled engines constitute the majority, they will be discussed first. Figure 14-4 shows the distribution of heat flow in four-stroke engines of different stroke-bore ratio l/d. An average figure of 30 per cent of the total heat developed is assumed as the basis for the distribution.

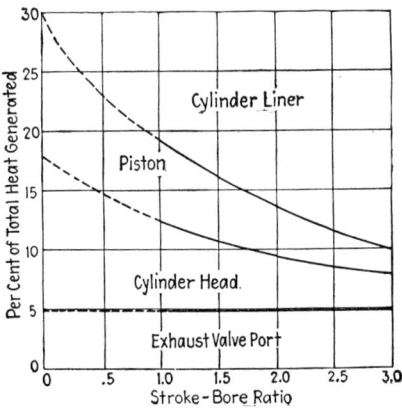
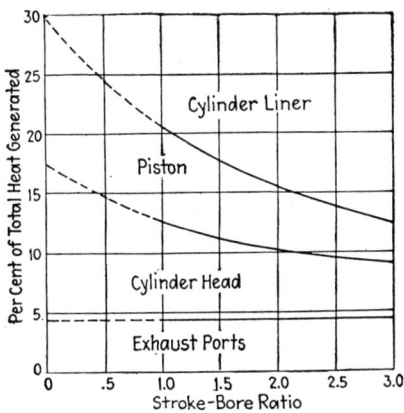

Fig. 14-4. Heat loss distribution in four-stroke engines.

Fig. 14-5. Heat-loss distribution in two-stroke engines.

Figure 14-5 shows a similar picture for two-stroke engines.

Temperatures. Formerly it was considered good practice to operate all engines so as to maintain a moderate outlet-water temperature of some 120 to 140 F and not over 160 or at the most 180 F, when using an enclosed cooling system. The object in using low water temperatures was mainly to reduce the formation of scale in the cylinder jackets. Scale is particularly dangerous in horizontal engines.

The dew point of the water vapor in the exhaust gases depends upon the pressure and hydrogen content of the fuel. Figure 14-6 shows that with the cooling temperatures given above a considerable condensation of water is bound to occur on the cylinder walls. The water causes corrosion, which seems to be one of the main causes of cylinder wear.

Numerous tests conducted since 1937 and careful observation of a number of installations in this country and similar tests in Great Britain have shown that permitting the water temperature to rise above the boiling temperature, to about 220 to 250 F, gives very important and far-reaching advantages:

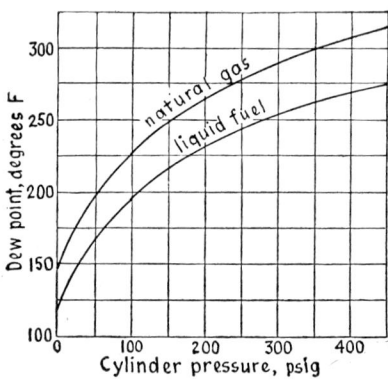

Fig. 14-6. Condensation of water vapor at different cylinder pressures.

1. It eliminates the condensation of the water vapor contained in the products of combustion, thereby (a) preventing, or at least reducing materially, the washing off of the lubricating-oil film from the cylinder and piston-ring surfaces and prevents the formation of sulfuric acid from sulfur dioxide, often contained in the products of combustion; these two factors reduce the wear of the cylinder, piston rings, and valves considerably—under certain conditions down to one-eighth of the usual amount; (b) eliminating crankcase condensation and sludging of the lubricating oil.

2. It lowers the viscosity of the cylinder-lubricating oil; this reduces the mechanical losses and raises the mechanical efficiency of the engine, thereby permitting a lower fuel consumption per horsepower-hour.

3. It reduces the amount of water which must be circulated, because part of the water is evaporated in the jackets and the cooling effect of each pound of evaporated water is about 970 Btu per lb, instead of the 10 to 20 Btu per lb absorbed by the water due to temperature difference; this fact reduces the fuel consumption still further.

4. It increases the temperature difference between the cooling water and the air to which the heat is rejected and, if a radiator is used, a

considerably smaller radiator surface and a smaller fan will do, and fuel will thus be saved.

5. At full load, the total fuel saving may reach 10 per cent.

Quite naturally, with an increase of the jacket temperature, the heat absorbed by the jacket from the gases in the cylinder decreases because of a smaller temperature difference, as illustrated by Fig. 14-7 for a typical engine. The heat not transferred to the jacket water increases the heat carried away by the exhaust gases and raises their temperature.

The use of higher jacket temperatures, up to 250 F, does not require a change in the construction of the engine or in the lubricating-oil specifications. However, it is desirable to have wide water passages and to eliminate possible vapor pockets.

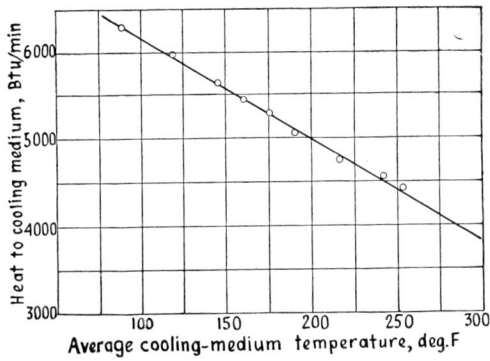

FIG. 14-7. Effect of cooling temperature on heat rejected to cooling medium in liquid-cooled engines.

Other Liquids. The use of ethylene glycol, or Prestone, which at atmospheric pressure has a boiling temperature of 387 F, instead of water, gives the same advantages, except No. 3, if the jacket temperature is maintained at the same level. The specific heat of Prestone is 0.675 Btu per lb at 212 F and about 0.775 Btu per lb at the boiling temperature.

The pistons of big double-acting engines are sometimes cooled by circulating oil instead of water through them. Pistons of high-speed single-acting engines are sometimes cooled by a jet of oil directed toward the underside of the piston top.

14-4. Construction Features. *Cylinders.* In some small- and medium-sized engines the water jacket is cast together with the cylinder. In many automotive and in all larger engines the cylinder is formed by a cast-iron liner inserted into a cast-iron jacket.

The water space between the cylinder proper or liner and the water jacket is made such as to obtain a fair velocity of water circulation, at least 5 ft per min in stationary engines and up to 60 ft per min in automotive engines.

Cylinder Heads. In a four-stroke engine the heat carried away by the water that cools the head comes from two places: from the bottom plate, which forms the upper wall of the combustion space, and from the exhaust passage and exhaust valve, if the latter is not water-cooled. Figure 14-4 shows the average relative amounts.

In cylinder heads good cooling is obtained by (1) eliminating air and steam pockets, (2) maintaining, as far as possible, uniform water velocities in all parts of the water space, and (3) avoiding narrow water passages that are apt to become closed by scale and thus to disturb proper circulation.

Exhaust valves need cooling only in large engines. With the use of heat-resisting steels or special cast iron for valve heads, even large engines are built with uncooled exhaust valves but then have water-cooled valve cages or valve seats.

Trunk pistons dissipate heat to the cylinder walls and to the lubricating oil quite satisfactorily and some engine builders therefore dispense with special cooling with pistons up to 22 in. in diameter. However, most engines from 6 in. up have oil-cooled pistons.

Pistons of many small and medium-sized diesel engines at present are cooled by lubricating oil delivered in comparatively large quantities through the rifle-bored connecting rods. In some engines the oil is admitted to one side of an enclosed space under the piston crown and discharged on the opposite side and allowed to flow down into the oil sump. In other engines the oil is discharged in the form of a jet from the top of the connecting rod and impinges on cooling ribs on the inside of the piston crown.

Barrel Pistons. With the improved design of the water-circulating system some large engines now use water for piston cooling. However, the majority use oil.

Piston rods in crosshead engines are cooled by water or oil that is admitted through the crosshead to the pistons.

14-5. Water Circulation. *Quantity.* The quantity of water that must be circulated depends upon the initial temperature and the desired temperature rise of the water. The initial temperature depends upon the atmospheric conditions, either directly, as in marine engines, or indirectly, if a recooling system is used and the water is recirculated over and over. In order to avoid excessive heat stresses, the temperature difference between the incoming and outgoing water should be about 20 F in small- and medium-sized engines and slightly less in large engines. The temperature of the outgoing water was usually not allowed to go above 140 F. For engines with a closed system a maximum temperature of 160 to 180 F was allowed. In automotive engines the cooling water often reaches the boiling point, about 212 F, without damage to the engine, but thermostats are usually set for 180 F. The results of investigations of cooling by evaporation discussed above, with jacket water temperatures from 215 to 250 F, will probably change the above limits.

If an engine is cooled by untreated water, which always contains dissolved salts and other foreign matter, the temperature should be kept low enough to prevent the precipitation of impurities and the formation of scale. If an engine uses salt water in the cylinder jackets, the temperature of the outgoing water should not exceed 110 to 115 F.

The water is usually circulated through the lubricating-oil cooler, through the cylinder jackets, then to the cylinder heads; after this, in large engines, a branch line leads water to the exhaust-valve cages. Pistons are usually cooled from a separate pipe line.

The quantity of water G that must be circulated, gallons per hour, is

$$G = \frac{Q}{8.32(t_2 - t_1)} \qquad (14\text{-}1)$$

where Q = the amount of heat rejected to the cooling water, Btu per hr
t_1 = the temperature of the incoming cooling water, degrees F
t_2 = the temperature of the outgoing water

For average conditions, the heat flow to the water jacket, in unsupercharged engines, is about 2600 Btu per hp-hr for large engines, increasing to about 3000 and to 3500 Btu per hp-hr for small and less efficient engines. In a supercharged engine the total heat flow, Btu per hr, is about the same as in an engine with natural aspiration of the same dimensions and speed. However, since a supercharged engine develops from 35 to 50 per cent more power, the specific heat flow, or heat flow referred to 1 hp-hr, is correspondingly smaller, about 1850 to 2300 Btu per hp-hr. The heat flow to lubricating-oil coolers, where these are used, is about 100 to 200 Btu per hp-hr, depending upon the amount of oil circulated and friction losses in the bearings.

Excessive water circulation resulting in low final water temperature is not desirable, since it will increase the fuel consumption and decrease the useful power.

A low cooling-water temperature increases the viscosity of the lubricating oil and, consequently, the piston friction. The difference between friction loss at high and low jacket temperature may amount to as much as 8 per cent of the power, if the piston is large and heavy, and drops to about 4 per cent, if the piston has small bearing area and weight.

Circulation. Two methods of water circulation are in use—gravity circulation and forced circulation. Gravity circulation, also called *thermosiphon* circulation, is based on the fact that when water is heated its density decreases and it tends to rise, the colder particles sinking to take the place of the rising, warmer ones. Circulation is obtained if the water is heated at one point and cooled at another. Gravity circulation is used in small engines only—seldom in those of more than 30 hp.

Figure 14-8 shows the gravity-circulation arrangement for a small horizontal engine. Water heated in the cylinder jacket j flows to a tank t, where it is cooled by radiation and convection, gradually descends to the bottom, and flows back to the engine. Figure 14-9 shows gravity circulation as applied to an automotive engine. To obtain proper water circulation the connections between the engine jacket j and the radiator t must present small resistance to the water flow and be wide, short, and have as few bends as possible. Even under favorable conditions circulation is slow, especially when the temperature difference is small, as at light loads. At heavy loads the jacket heat may exceed the heat dissipated by the radiator and the water in the jacket is apt to boil. This system is used only in smaller engines where simplicity is of importance.

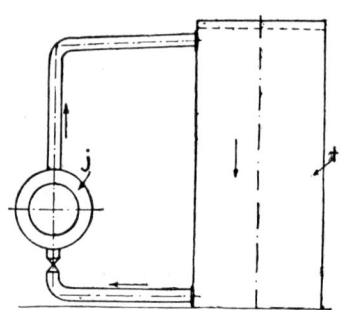

FIG. 14-8. Gravity-circulation cooling.

Most engines have forced circulation by pumps, of either the centrifugal or the plunger type. The advantage of the forced circulation is the ease of controlling the jacket-water temperature. This may be accomplished either by regulating the opening of the valve between the pump and the engine or by regulating the water-discharge valve of individual cylinders. Figure 14-10 shows a pump circulation system in an automotive engine with automatic temperature control by a by-pass b with valve v operated by a thermostatic element e.

Evaporative Cooling. If the water in the cylinder jacket is allowed to boil, 1 lb of evaporated water will absorb heat equal to the latent heat of vaporization, or about 970 Btu. This is from 24 to 48 times more than the heat carried away by 1 lb of circulating water with a temperature rise of 20 to 40 F. Neither pump nor radiator being required, this system has the advantage of simplicity and is used for small stationary and tractor engines. The water jacket is made large at the top, forming a so-called *hopper*. The quantity of water in the hopper must be sufficient to run the engine for several hours without the addition of water. The evaporative system is not advisable if the water contains impurities which form scale on the cylinder walls.

14-6. Recooling of Water. Only seldom, as in small marine engines, is the cooling water used once and discarded. The recooling of water for continuous use can be effected by one of the following means: (1) direct evaporation; (2) heat exchangers with a secondary water circulation; (3) radiators with atmospheric air as a coolant.

By the first method, called an *open cooling system*, the water from the jacket is discharged either into a cooling pond or to the top of a

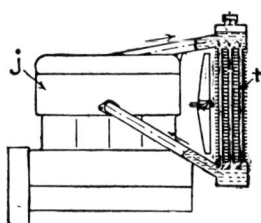

Fig. 14-9. Thermosiphon water circulation.

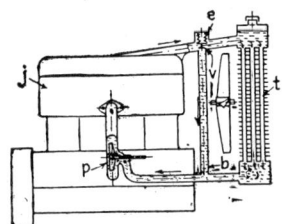

Fig. 14-10. Forced-water circulation with by-pass.

cooling tower and is cooled by the latent heat of evaporation of the part carried away by the air. The advantage of this system is its simplicity and the small expenditure of power needed for circulation of the water. Its big drawback is a gradual contamination of the water by salts. As pure water evaporates, leaving salts behind, and make-up water is added, with salts of its own, the salt concentration gradually increases. When it reaches a certain limit, all water must be drained and fresh water put into the system. However, even if this is done regularly, a certain amount of sediment is deposited in the engine jackets and forms scale, which eventually may cause cracks, usually in the cylinder head. At the same time, this system requires low jacket temperatures, with the ensuing drawbacks mentioned before.

A *closed system* normally uses distilled or treated soft water. However, raw water is also occasionally used because the original small mineral content in the raw water is not increased and therefore little scale is deposited. The cooling water from the engine is passed through a heat exchanger, where it is cooled and then led back to the jacket. The heat exchanger may be either simply a coil in the basin of a cooling tower or a shell-and-tube exchanger. In the latter the jacket water passes through the tubes and the cooling medium through the shell. In oil-pipe-line pumping stations, the pumped oil is used as a coolant. A closed system permits the use of any jacket temperature up to the highest desirable; if the amount circulated is large enough, the temperature difference between the incoming and outgoing water can be kept low, 10 to 20 F.

The drawbacks of the closed system are a slightly greater power requirement for the two pumps and a higher initial cost. However, the elimination of scale and the advantages of higher jacket temper-

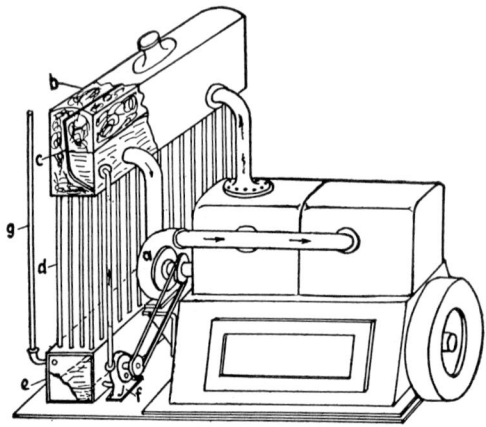

Fig. 14-11. Vapor-phase cooling with air-cooled condenser.

atures are so important that the use of the closed system has become almost universal.

Radiators. In a radiator the heat is rejected to the air which passes through it. In automotive engines the air is sucked through by a fan, assisted by the movement of the car. From the previous discussion of heat transfer between water and another liquid divided by a metal wall it is clear that the performance of a radiator depends upon the velocities of air and water. The increase of air velocity, in the first place, increases the quantity of air passing the radiator fins and, in the second place, by removing the inert air film sticking to the metal surface, it increases the outside-surface heat-transfer coefficient h_2. The water velocity acts in the same way and increases the inner surface coefficient. Since the specific heat of air is less than one-fourth that of water and the surface coefficient between metal and air is many times lower than between metal and water, the air-cooled surface should be considerably greater than the inner surface in contact with the water. This difference is obtained by adding thin metal fins to the water tubes t (Fig. 14-9 or 14-10) which form the water passages between the upper and lower radiator tanks. Radiators are used generally with mobile or portable engines and in temporary installations.

Vapor-phase Cooling. The advantages of high-temperature jackets, as explained before, apply particularly to a system developed in this country, called *vapor-phase* cooling (Fig. 14-11). The water is circulated by the pump a; when it is delivered to the overhead tank b, part of it boils out. The vapor rises over the partition c and, because of the condensing action of the radiator tubes d, flows down into the lower tank e, from which the condensate is picked up and returned to tank b by the small pump f. The vertical pipe g is a communication with the outside atmosphere to prevent the collapse of tanks b and e when the pressure inside them, owing to condensation, falls below atmospheric pressure.

For larger engines the condensation of vapor formed in the overhead tank b (Fig. 14-12) occurs in the heat exchanger c, cooled by a secondary water circuit, and the water returns to b by gravity.

14-7. Direct Air Cooling. Because of the low value of the heat-transfer coefficient h between metal and air, the wall temperature of air-cooled cylinders is considerably higher than that of the water-cooled type. In order to lower the cylinder-wall temperature, the outside surface must be increased by fins. Experiments have shown that for a satisfactory operation the cylinder-head temperature of most engines should not exceed 570 to 600 F.

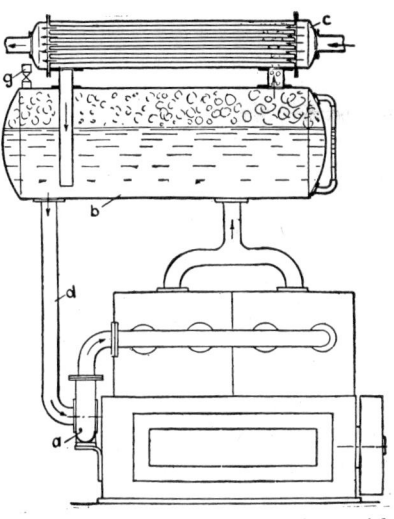

Fig. 14-12. Vapor-phase cooling with water-cooled condenser.

Air-cooled cylinders are used chiefly in aircraft engines and in some automobile and small stationary diesel engines. In automobile and stationary engines the cooling air is furnished by a blower. The blower is either of the centrifugal type of conventional design driven by a belt from a pulley keyed to the engine shaft or is of the axial type with blades formed by the spokes of the engine flywheel.

Heat Dissipation. The amount of heat that must be dissipated by an air-cooled cylinder at full load is from 1500 to 2300 Btu per hp-hr;

in addition, the lubricating oil gives up in the oil cooler 200 to 1200 Btu per hp-hr, the amount decreasing with an increase of the heat dissipated by the fins. Thus the total amount of heat that is extracted is about 2500 to 2700 Btu per hp-hr.

14-8. Cooling Equipment. The term *cooling equipment* covers accessories required for an effective cooling of a diesel engine. As explained in Sec. 14-6, the majority of diesel engines use a closed cooling system. Therefore the accessories comprising a closed water system will be taken as a basis for the discussion.

A complete system consists of
1. A soft-water circulating pump.
2. Pipe lines for soft-water circulation.
3. An expansion tank for soft water.
4. A soft-water cooler.
5. Thermometers for inlet and outlet water.
6. A temperature regulator for maintaining a desired outlet-water temperature.
7. Safety devices for protecting the engine against excessive jacket-water temperature or stoppage of water circulation.
8. A raw-water softener.
9. A raw-water circulating pump.
10. Pipe lines for raw-water circulation.
11. A raw-water cooler.

In some cases the last three items may be absent. Thus in pipe-line plants pumping oil or in water-works engines, no raw water is circulated. The soft water is recooled by putting it through a heat exchanger through which the pumped oil or water passes and thus carries away the jacket-water heat. The same is true when a radiator is used, similar to the installation shown in Fig. 14-10. However, in this last case the place of the raw-water pump is taken by a fan. In other cases the raw-water cooler may be omitted, when the supply of water is abundant, such as when the engine is located on the bank of a river or lake or on board a boat.

An open cooling system uses raw water in the jackets and the necessary equipment is reduced to
1. A water-circulating pump
2. Pipe lines for water circulation
3. A water cooler
4. Thermometers for inlet and outlet temperatures
5. A thermostat for maintaining a desired outlet-water temperature
6. Safety devices, the same as in a closed system

Figure 14-13 shows schematically a closed system as used in a stationary diesel plant with items (1) to (4) and (8) to (11) clearly indicated. The water from the cooling tower flows over an open coil-type heat exchanger and thus cools the jacket water in the closed system.

Figure 14-14 shows schematically a closed system used with a marine engine, with items (1) to (7). In order not to obscure the illustration, the sea-water circulation piping and pump, items (9) and (10), are shown separately in Fig. 14-15. The marine installation differs from a complete stationary installation in the absence of a cooling tower for recooling of the raw water.

Water Pumps. Some marine diesel engines use reciprocating plunger pumps for water circulation and drive them by gears from the crank-

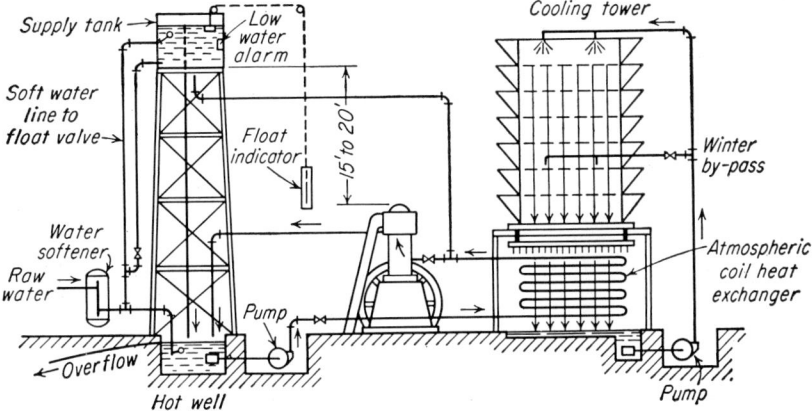

FIG. 14-13. Double-circuit closed cooling system.

shaft or camshaft. Some small diesel engines have water-circulating pumps of the gear type, similar to that shown in Fig. 11-6. However, the majority of diesel engines use centrifugal pumps for circulating both jacket water and the secondary cooling water furnished to the heat exchanger.

The conventional centrifugal pump consists of an *impeller*, with *vanes* curved in the direction opposite to the direction of rotation, and a spiral housing or *scroll*, with the cross section increasing toward the outlet (Fig. 14-16). The water inlet is at the center—axial—the outlet is tangential. The pressure necessary to push the water through the engine jackets and the heat exchanger is produced by the centrifugal force which, during the rotation of the impeller, throws the water toward the tip ends of the vanes at high velocity. When the water passes through the expanding spiral housing, its velocity is

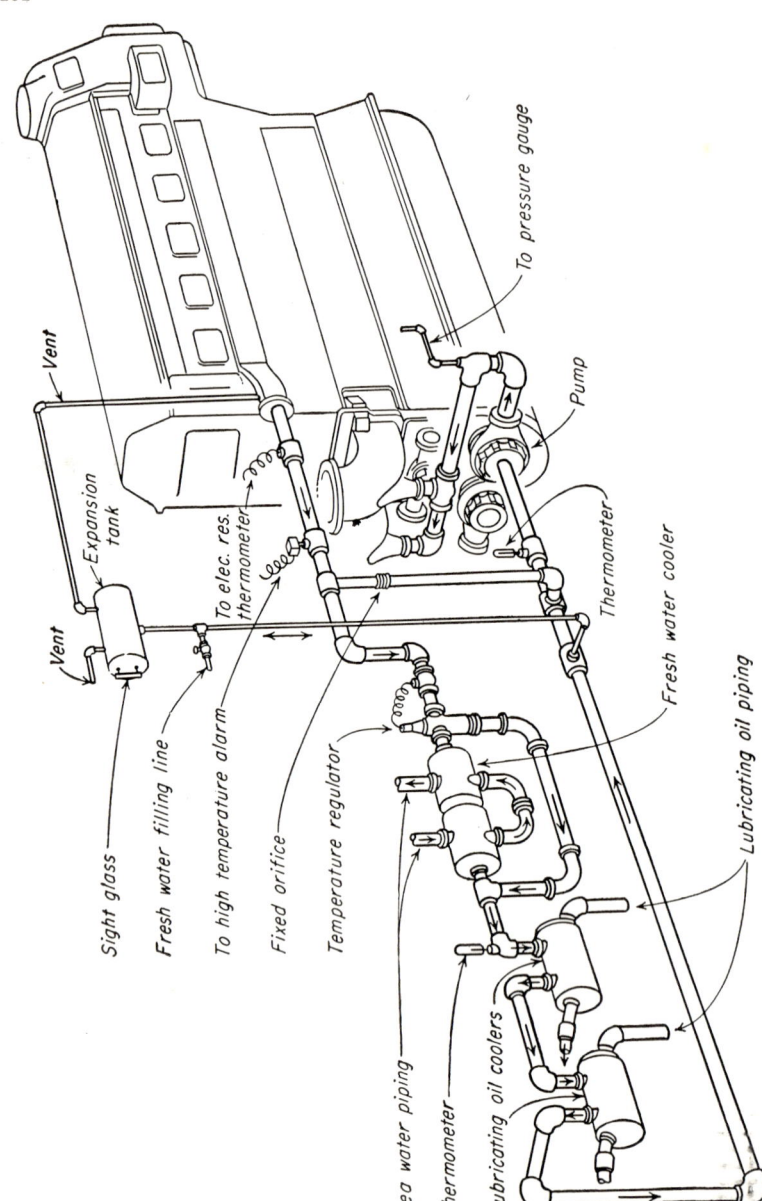

FIG. 14-14. Closed cooling system, marine engine, fresh-water circulation.

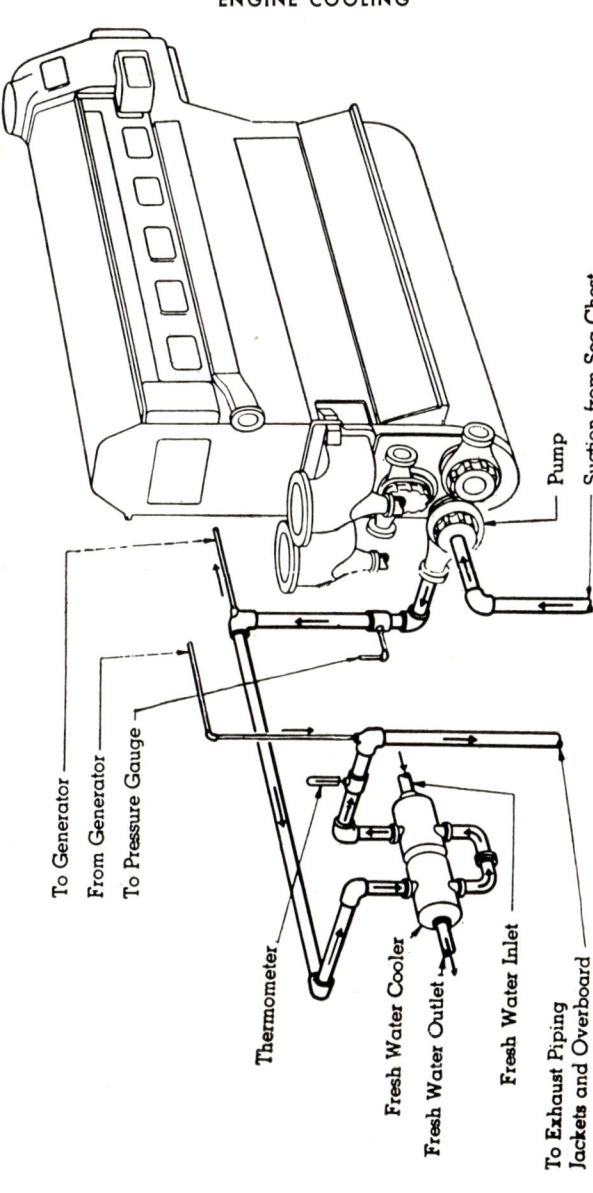

Fig. 14-15. Closed cooling system, marine engine, sea-water circulation.

reduced and the corresponding kinetic energy is transformed into pressure. The pumps operate at speeds from 1,200 to 3,500 rpm, depending on the size and design.

Centrifugal pumps driven by direct-reversible engines, such as those used for ship propulsion, usually have straight radial vanes and a concentric housing (Fig. 14-17). The efficiency of such a pump is lower

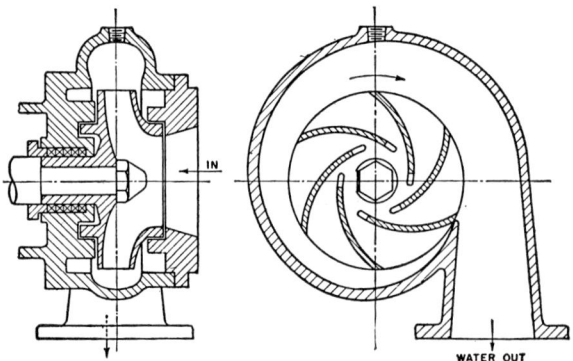

Fig. 14-16. Centrifugal water pump.

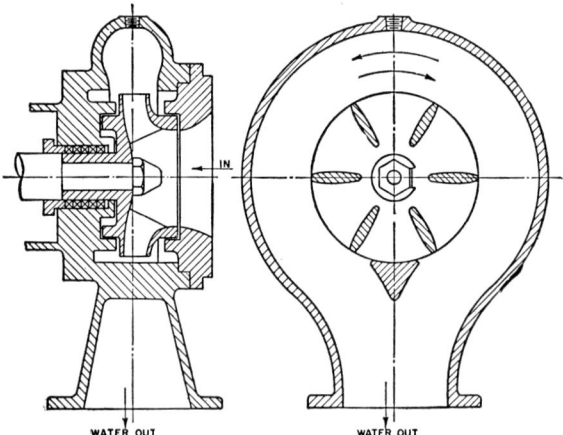

Fig. 14-17. Centrifugal pump for rotation in either direction.

than of a pump designed to rotate in one direction only, chiefly owing to the shape of the housing. In order to obtain the same pressure and capacity, the impeller speed must be increased, compared with that of a conventional pump. Some direct-reversible marine engines have conventional centrifugal pumps running always in the same direction. Such a pump is connected to the engine by a special reverse gear whose driving shaft turns in the same direction, regardless of the direction of rotation of the driven shaft.

It should be remembered that, when a centrifugal pump is not running, water will leak back through it, sometimes even if a check valve is put in the suction line. Centrifugal pumps are not self-priming—the water level in the sump or other source of supply must therefore be higher than the top of the pump, and water should flow into the suction end of the pump by gravity or under a pressure.

Centrifugal water pumps are driven from the engine crankshaft by means of gears or chains. In power plants with large diesel engines, the pumps are driven by electric motors.

Piping. Flow Resistance. Every pipe presents a certain resistance to the flow of the fluid, in this case water, which it conducts. The flow resistance in a pipe increases in direct proportion to the length and approximately as the square of the velocity of the fluid. Since the water velocity is inversely proportional to the cross section of the pipe, a reduction in the cross area or size of the pipe will increase the flow resistance and, with a given pressure head created by the pump, will decrease the flow rate. The resistance also increases with every elbow and valve through which the water must pass. The resistance of a valve depends upon its construction; thus the resistance of a globe valve is higher than that of a gate valve.

In renewing a pipe line, one must be careful not to increase the flow resistance by making changes in the original installation.

The following data may serve as a guide for piping layout and installation: suitable water velocities, on the suction side 60 to 200 ft per min and on the discharge side 120 to 240 ft per min; the smaller the pipe diameter, the lower should be the velocity. Velocities higher than those indicated give excessive resistance; lower velocities require excessively large pipe diameters and mean an unnecessarily high first cost. The resistance of fittings is usually considered to be equal to a certain additional length of the pipe: an elbow is equivalent to three pipe diameters, a gate valve to about five diameters and a globe valve to ten diameters, at most.

Figure 14-18 shows a diagram which permits one to pick out the size of a water pipe for a given flow in gallons per minute and a certain friction head or loss in feet of water. Good practice permits a friction head of 1.0 to 2.5 ft per 100 feet of effective pipe length. The values of Fig. 14-18 are for smooth, clean steel pipes. For cast-iron pipes the values of friction head from Fig. 14-18 should be increased by about 30 per cent.

Expansion Tank. The water in a cooling system expands as the water temperature goes up and the excess water goes into a so-called *expansion tank.* This tank is located at the highest point of the pipe line, maintains a constant pressure in the system, prevents formation of air or steam pockets in it, and serves to add make-up water to take care of unavoidable leaks in the system.

The size of the expansion tank depends upon the water capacity of the whole system, including the water space in the engine jackets. The volume of the tank should be not less than 5 per cent of the total water capacity in order to allow for the expansion from room temperature to the temperature of the water leaving the engine. A greater volume,

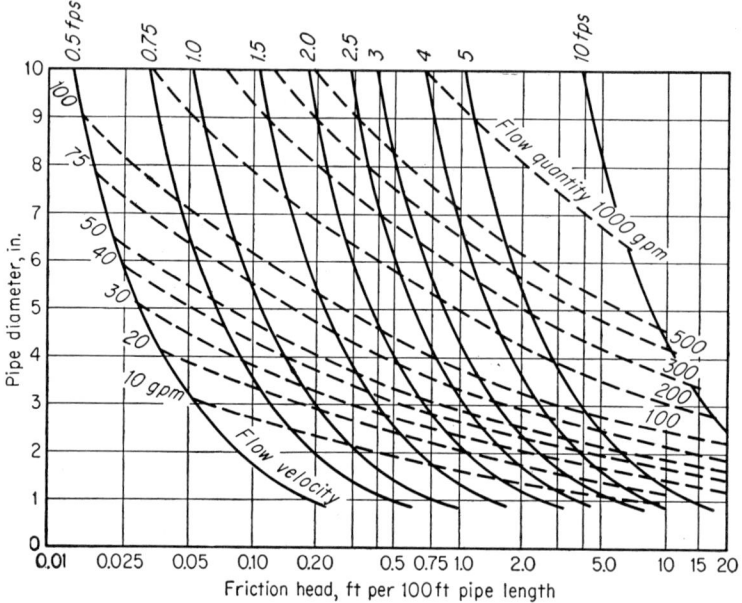

FIG. 14-18. Friction chart for water flow in steel pipes.

up to 10 per cent, is advisable in order to take care of the unavoidable losses through leaks, such as through pump glands, and evaporation.

The tank must be of well-galvanized steel in order to prevent rusting occasioned by the fluctuating water level. Sometimes the expansion tank serves also as a soft-water supply tank; it is then made considerably larger.

Soft-water Cooler. In stationary installations, the cooler is usually a pipe coil placed either flat in the sump of the cooling tower which is used for recooling the raw water or vertically and the raw water from the cooling tower runs over it.

The advantages of these coolers, which are of the *open* or *atmospheric* type, are the following: (1) the evaporation of the cooling water running over the coils helps heat dissipation; (2) there is no danger of raw-water leakage into the closed soft-water circuit, because the pressure inside the coils is higher; (3) good accessibility for cleaning scale and mud deposits off the coils; (4) low first cost.

Heat Exchangers. Sometimes the soft water is run through some kind of heat exchanger, usually of the *shell-and-tube* type, shown diagrammatically in Fig. 14-19. The soft water flows inside the tubes, the raw water from the outside of the tubes, directed in its flow by baffles. The baffles give better contact with all parts of the tube and increase the water velocity; these two conditions increase the heat transfer. Sometimes raw water is passed inside the tubes in order to make their cleaning easier. In heat exchangers used in pipe-line plants the cooling oil is passed through the tubes, in order to reduce the flow

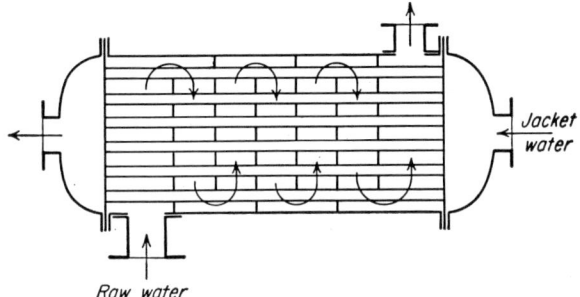

Fig. 14-19. Single-pass shell-and-tube heat exchanger.

resistance, and the tensile stress in the tubes produced by the big temperature difference is taken up by special tie rods and braces.

Shell-and-tube heat exchangers are used sometimes in other than marine and pumping plants because of the following advantages: (1) good heat transfer due to the use of small diameter thin-walled tubes and relatively high water velocities; (2) compactness—the exchanger takes up little space and may be placed in any position; (3) ease of cleaning the tubes from the inside and, in exchangers with an expansion joint or floating head, also from the outside.

Pressure. In order to prevent leakage of raw water into a closed system if the tube ends eventually become loose in the tube plates, the pressure of the raw water should be always less than that of the soft water; this precaution necessitates larger raw-water piping and smaller water velocities.

Zinc Electrodes. Cooling systems using sea water must have zinc electrodes inserted in the sea-water inlet line. This is necessary in order to control electrolysis which takes place in the sea-water lines of the cooling system from stray electric currents. The zinc provides a terminal which attracts the stray current and thus restricts the electrolytic action to corroding the zinc and leaves the other parts of the system intact. The zinc electrodes are corroded rather fast; they must therefore be inspected at regular intervals and replaced before they become too small, in about three to six months. In shell-and-tube heat exchangers the zinc electrodes are made of plates fastened inside the shell if sea water is circulated through the shell, or inside the headers if sea water is circulated through the tubes.

Radiator Units. Sometimes soft water is cooled by circulating it through a radiator unit, which consists of a radiator—similar to one used in automobiles, tractors, and trucks—mounted on a common base with the water-circulating pump and fan, both driven from an electric motor or from an extension shaft of the diesel engine. Such units are light and compact and are suitable for temporary or portable installations. However, they are not economical, because of the relatively large amount of power required to drive the fan. They are therefore seldom used in stationary power plants, except where their compactness is of particular importance, such as in a diesel-power plant installed in the basement of an office building.

Water Softeners. Except where distilled water is available, the treating of cooling water, *i.e.*, elimination of its temporary and permanent hardness, is done in so-called *water softeners*. A typical water softener consists of a metal shell or tank containing zeolite material which abstracts the hardness from the water as it flows through the tank. The zeolite exchanges its sodium for the calcium and magnesium in the water, leaving only soluble sodium salts in the water, which do not form scale. After a certain amount of hard water is run through the softener, its charge must be regenerated with common salt.

The proper size and type of water softener depends upon the raw-water analysis; this analysis should be made by the concern furnishing the softener.

The softener works practically automatically and all necessary instructions in its proper operation are furnished by the softener manufacturer.

Raw-water Cooling. Raw water is cooled by evaporation in cooling towers. A good estimate of the amount of water that must go over the tower is twice as much as is circulated through the engine jacket.

Atmospheric Cooling Towers. A typical tower is shown in Fig. 14-20a. It consists essentially of a system of distributing gutters, or troughs, which allow the water to trickle down through the successive decks of the tower, eventually collecting in a basin or sump at the bottom after having been exposed to the cooling effect of the air; this

ENGINE COOLING

effect is due chiefly to evaporation of a certain part of the water. The various decks of the tower are protected by louvers to prevent the wind from carrying away the falling streams of water. The water is admitted through sprinklers at the top and flows by gravity into the sump.

For service in cold weather a secondary distributing system is sometimes located nearer to the bottom: the sprinklers on the top are shut off and the water is passed only over a small portion of the tower when the atmospheric temperature makes it unnecessary to use all decks.

Some towers, particularly in smaller sizes, are made without troughs inside (Fig. 14-20b), and the water is broken up by spray nozzles to which the water is delivered under a pressure of 3 to 5 psi.

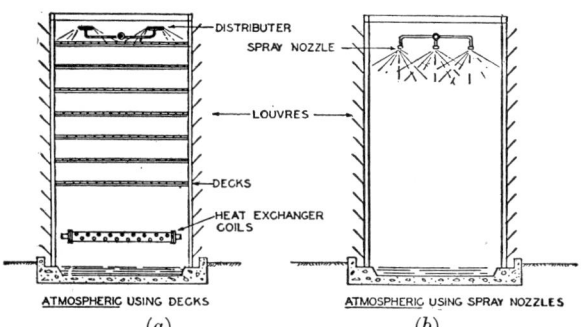

FIG. 14-20. Atmospheric cooling towers.

The above-described cooling towers are called *atmospheric towers*, because evaporation is assisted by the natural movement of atmospheric air, or *natural draft*.

Mechanical-draft cooling towers are made in the form of a tall steel box with spray nozzles near the top to break up the water and S-shaped steel baffles above the nozzles to prevent water drops from being carried away. Either the air is *forced* through the tower by a fan located near the bottom, as shown in Fig. 14-21a, or the draft is *induced* by a fan on top of the tower, as shown in Fig. 14-21b. Mechanical-draft cooling towers made of steel are more expensive than wooden towers. However, they are much lighter and therefore are used for temporary and semiportable installations and on the roofs of buildings.

Open Cooling System. The danger of scale formation and, therefore, of impaired cooling in an open cooling system can be materially reduced if soft water is used for jacket cooling and the make-up for evaporation is also treated in a water softener.

14-9. Cooling Controls. *Temperature Measurement.* Water temperatures are conveniently measured with ordinary glass mercury thermometers, usually of the industrial type, in a metal case protecting the glass from easy breakage. Dial thermometers are also used. In some installations, in order to centralize the observation of

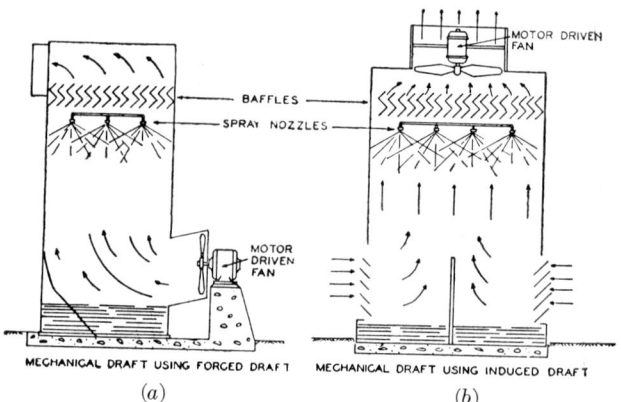

Fig. 14-21. Mechanical-draft cooling towers.

various temperatures by using thermocouples, the latter are used also for water-temperature reading.

Temperature regulators are automatic valves operated by thermostatic elements which are set to open or close at a certain temperature. The several types of thermostatic elements in use consist of the following elements: (1) a corrugated metal pipe, called a *bellows;* (2) a bimetallic coil; and (3) a cylinder with a readily evaporating substance and a piston.

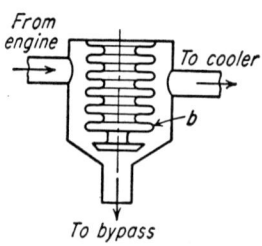

Fig. 14-22. Bellows control of by-pass.

Bellows. Such a thermostat consists of a brass or monel-metal thinwall pipe b with deep corrugations (Fig. 14-22). The inside of the bellows is filled with a volatile liquid, such as alcohol or ether, which evaporates readily with a rise of the temperature of the water in which the element is immersed. The resulting vapor pressure pushes the lower, free end of the bellows in respect to the fixed one. This motion is increased by the large number of corrugations, which make the element more sensitive to a change of pressure inside it, meaning to a change of temperature outside it. In small engines the bellows are fastened directly to the regulating valve; as the water temperature goes up, the by-pass is gradually closed, and more water is forced to the cooler or heat exchanger, until at maximum load the by-pass is closed entirely.

In large engines, the valve regulating the rate of flow to the cooler and operated by the bellows is placed in the soft-water piping where it is most convenient and the action of the bellows is transmitted by remote control. A temperature regulator with remote control consists of a steel bulb filled with a volatile liquid, inserted in the water outlet from the engine and connected to the bellows by a fine tube which transmits the pressure change from the bulb to the bellows.

Bimetallic Coil. This element consists of a strip made up of two metals, which have different coefficients of heat expansion, formed into a flat spiral coil. The outer end of the coil is fastened solidly to the perforated housing, and the inner end is fastened to a shaft. When the temperature of the bimetallic coil changes, its free end moves, rotating the shaft and thus opening or closing a flat hinge-type valve to the by-pass pipe.

Cylinder and Piston. This thermostatic element consists of a cylinder with a well-fitting piston and some readily evaporating substance between them. When, with an increase in temperature, the substance melts and begins to evaporate, it pushes the piston, which in turn operates a disk valve and at the same time compresses a spring. When the temperature decreases, the pressure under the piston begins to drop and the spring begins to return the piston and the disk valve toward its seat.

Safety Devices. These devices may be divided into two groups: (1) instruments which sound an alarm when the water temperature reaches a certain height and (2) instruments which stop the engine automatically when the water temperature reaches a predetermined value.

Alarm. One of the alarm devices consists of an electric bell or siren whose circuit is closed by a mercury-glass thermometer with a contact embedded in the glass tube at a certain point. Another scheme uses a switch in the circuit which is operated by vapor pressure through a capillary tube from a bulb immersed in the water-discharge pipe. Still another scheme uses a switch operated by a shaft fastened to a bimetallic coil similar to the one used as a thermostat element and also located in the discharge pipe.

Water-level alarm is sometimes installed in addition to the float indicator to call the attention of the operator when the water level in the soft-water supply tank becomes too low as shown in Fig. 14-13.

Automatic Stop Device. This device consists of a bulb with a volatile liquid, which is inserted in the water discharge pipe and connected by a capillary tube to a special control box; the box contains a valve which normally closes tightly a by-pass in the fuel line leading from the service pump to the injection pump. If the water temperature exceeds a certain maximum, the pressure in the bulb acts on the control box and opens the by-pass valve, thus shutting off the fuel supply to the engine and allowing any oil left in the fuel line between the control box and the injection pump to drain back into the fuel supply tank. An engine-stopping device usually is combined with an audible alarm signal, which begins to sound a warning slightly before the engine is stopped.

QUESTIONS

1. Explain why diesel-engine cylinders and heads must be cooled.
2. What two methods of cooling are used in diesel engines?
3. What are the three means of heat transfer that participate in the cooling of engines?
4. Draw a diagrammatic sketch of heat flow from one fluid to another separated by a metal wall.
5. What are the inside, or gas-film, coefficient and the outside film coefficient in heat transfer?
6. In what units are the heat-transfer coefficients expressed?
7. What is the heat conductivity of a metal? In what units is it expressed?
8. How do engine speed, load, and cylinder diameter each affect heat flow to the cooling fluid?
9. How does the stroke-bore ratio affect the distribution of heat flow among exhaust ports, piston, cylinder, and cylinder head? Show by a diagrammatic sketch.
10. What temperatures of the cooling water in cylinder jackets are recommended? What is the reason for preferring lower temperatures?
11. What are the five reasons that lead to the recommendation of high water temperatures in the cooling system?
12. What liquids, besides water, are used in cooling engine cylinders and pistons?
13. What velocities of water circulation are commonly used in diesel engines?
14. What conditions, besides velocity of water circulation, are essential to satisfactory cooling of cylinder heads of vertical engines?
15. Explain how the exhaust valves of large diesel engines are cooled.
16. What methods of cooling large-bore trunk pistons are in use? Illustrate by sketches.
17. What is used as the basis for computing the amount of water to be circulated through the engine cooling system?
18. What methods of water circulation are in use?
19. What is thermosiphon circulation?
20. What is the advantage of forced water circulation?
21. What is evaporative cooling and how does it function?
22. What is the hopper in evaporative-cooled engines?
23. What methods of recooling water used for engine cooling are in existence?
24. What is an open cooling system? What are its advantages and drawbacks?

ENGINE COOLING

25. What is a closed cooling system? Illustrate by a sketch.
26. What are the advantages and drawbacks of a closed cooling system?
27. What are the advantages and drawbacks of radiators?
28. What is vapor-phase cooling? Illustrate by a sketch.
29. What is the cylinder-wall temperature of air-cooled cylinders compared with water-cooled ones?
30. What are the limits of cylinder-head temperatures in air-cooled engines for satisfactory engine operation?
31. How much heat is dissipated by the fins of an air-cooled cylinder at full load? How much heat is carried away by the lubricating oil? What is the total amount of heat extracted from an air-cooled cylinder?
32. Enumerate the accessories that are used in a closed cooling system.
33. Enumerate the accessories that are used in an open cooling system.
34. What pump types are used for circulating cooling water?
35. Draw a diagrammatic cross-section of a centrifugal water pump and label the main parts.
36. What precaution must be taken to avoid running out of water from a centrifugal pump when the pump is standing still?
37. What factors influence the resistance of water flow through a pipe line?
38. What are suitable water velocities on the suction and discharge sides?
39. What is the purpose of an expansion tank?
40. What should be the size of the expansion tank?
41. Draw sketches of soft-water cooler arrangements.
42. What are the advantages of a so-called *open*, or *atmospheric*, *cooler?*
43. When are heat exchangers used for recooling soft water or cooling oil?
44. Draw a sketch of a typical shell-and-tube heat exchanger.
45. What are the advantages of shell-and-tube heat exchangers?
46. Which pressure must be higher in a heat exchanger, that of the raw water or that of the soft water? Give the reason.
47. Why and where are zinc electrodes inserted in a cooling system?
48. Where are zinc electrodes located in a shell-and-tube heat exchanger if sea water is circulated through it?
49. What are the advantages and drawbacks of radiator cooling units?
50. What is a water softener?
51. How is raw water cooled?
52. Give diagrammatic sketches of mechanical-draft cooling towers with different locations of the fan.
53. In an open cooling system, how can the danger of scale deposits in engine cylinders and heads be materially reduced?
54. Enumerate the types of instruments used to measure water temperatures.
55. Enumerate the various types of temperature regulators.
56. What are the two different types of safety devices used in diesel plants?
57. Explain the action of an automatic stop device.

CHAPTER 15

AIR-INTAKE SYSTEM

15-1. Air-intake System. The whole combination of devices which serve to supply air to a diesel-engine cylinder is called the *air-intake system*.

The purpose of the intake system is to supply the air needed for combustion of the fuel. However, in addition, the air-intake system of a diesel engine may have to (1) clean the intake air, (2) silence the intake noise, (3) furnish air for supercharging, and (4) supply scavenge air in two-stroke engines.

A complete air-intake system is made up of the following parts: (1) an air cleaner or filter, (2) an air-intake silencer, (3) a blower for supercharging or scavenging, (4) an air cooler for supercharging air, (5) piping connecting the air filter, blower, and air cooler with the air-intake header, and (6) an air-intake header.

However, not all these parts are present in every installation.

15-2. Air Cleaners. Formerly air from the atmosphere was drawn into the engine cylinder without any attempt being made to clean it of dust and grit. Dust, grit, sand, and other foreign matter carried with air into a diesel-engine cylinder during the suction period is one of the main causes of fouled valves and of worn piston rings and cylinder liners. Air filters are particularly necessary in power plants where dust and grit are present in the atmosphere, as in stone-working and mining localities or near highways with heavy traffic. In such places from 1.0 to 5.0 grains and more of dust per 1,000 cu ft of air can be found. This amount of air is going every minute through a 500-hp oil engine; on this basis the amount of abrasive matter which will enter the engine is from 30 to 150 lb per year. Even in localities where the air seems to be free of dust, an air filter will show a surprisingly high dust content. It is impossible to overemphasize the need of air filtration for any oil-engine installation.

Filter Requirements. The requirements which an air filter must fulfill may be listed as follows:

AIR-INTAKE SYSTEM

1. Small resistance to the passage of air in order not to reduce the volumetric efficiency of the engine.
2. High efficiency, that is, dust-retaining capacity, with a design which insures that particles of the filtering material cannot get loose and be sucked into the engine.
3. Ease of cleaning.
4. Ability to operate without constant attention or excessively frequent cleaning intervals.
5. Small space, compact shape.
6. Moderate first cost.
7. Low operating expense, if any.

Commercial air filters for diesel engines can be divided into three classes: dry, impingement, and wet. Dry filters depend for their cleaning effect upon actual filtration through some screening material—wool felt, steel wool, or cloth. Devices for small engines depend also upon centrifugal force created by a rotary motion imparted to the air stream.

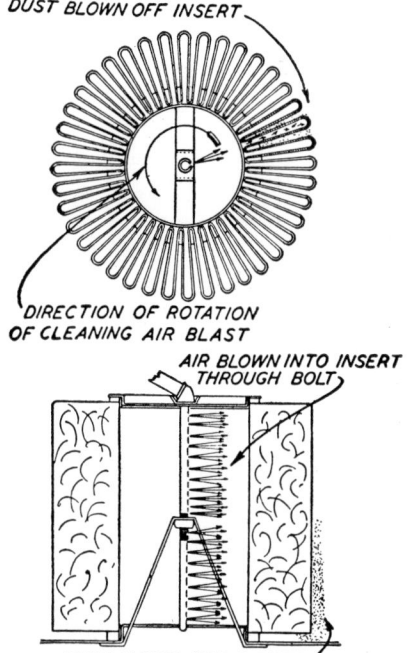

Dry Filters. Figure 15-1 shows the main part of a dry filter, which has a star-shaped wire frame covered with felt. It has a cylindrical hood with louvers for air admission, the latter being fastened to the suction pipe by a bolt with a thumb nut. Such units are built for capacities from 15 to 250 cu ft per min. They can be assembled in special boxes holding from two to twenty units, thus giving a capacity up to 5,000 cu ft per min, sufficient for a 2,500-hp diesel engine. A unit of 250 cu ft capacity has a filter area of 20 sq ft, is 17 in. high and has a 14-in. diameter felt insert.

FIG. 15-1. Cloth arrangement in air filter and cleaning it with air blast.

The method of cleaning the filter is also shown in Fig. 15-1: an air hose is connected to the top of the hollow central bolt that has a number of holes which admit air into the inside of the filter star. By slowly turning the bolt, with the air turned on, as shown by the arrow, the dust collected on the outer surface is blown off.

The smaller sizes of this filter do not have the central air pipe. For cleaning such a filter its insert must be removed from the case. The air blast should be directed along and parallel to the surface of the felt. This loosens the dust without forcing it into the pores of the felt. A final blast of air should be directed into each pocket from the inside, along the entire length of the fins. Gasoline should be used only in more severe cases. Care should be exercised not to damage the Feltex filter material. When placing the insert back into the hood, one must watch that the gasket under the insert rests firmly on the supporting base, and that the adjustment nut is in the correct position before tightening the locknuts. This is done by pressing the retaining cup

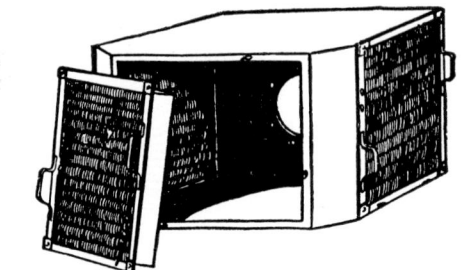

Fig. 15-2. Impingement air filter.

by hand until it touches the nut. The cup must be firmly locked between the two nuts before the upper locknut is tightened down.

With proper care the insert gives satisfactory performance for from two to three years. After that the insert frame may be recovered with new cloth or felt. These filters are installed outside the engine building, preferably where they are less exposed to wind, and are connected to the engine by a pipe or duct.

Impingement Filters. For large diesel engines better results are obtained with filters of the impingement type. Generally they consist of a frame fastened to a box (Fig. 15-2) through which air is drawn into the engine. One or more of the walls of the box are replaced by such frames, containing steel wool or equivalent material coated with a substantially nonevaporative dust-catching oil. Such a frame, or unit, occupies a rectangular space about 22 × 22 × 14 in. and has a capacity of around 150 cu ft per min, when used with a single-cylinder 60-hp four-stroke engine. With a six-cylinder four-stroke engine, the same unit can pass up to 800 cu ft per min owing to the greater uniformity of the air flow.

In these filter units, steel wool is packed in several layers and the incoming air first comes in contact with loose layers and then passes through layers progressively increasing in density. The air is broken up into fine streams which change their direction from twelve to eighteen times, each change being caused by an impingement of the air upon the steel wool coated with an adhesive oil.

When the filtering material becomes dirty, the complete filter cell is removed from the frame and dipped into a hot solution of washing soda in water. This solution cuts the coating oil together with the dust loose from the filter. The clean cell is then charged by dipping it into the adhesive oil and put back. The filter needs cleaning every four to twelve weeks, depending upon the amount of dust in the air.

In order to protect both the steel wool and the case from corrosion, the entire cell, after it is completely assembled, is immersed in a bath of molten lead alloy, which also binds the pieces of steel wire together in one coherent mass. The box frame receives a heavy coat of enamel and is oven-baked.

Another filter of the impingement type is shown in Fig. 15-6. It is a combination of an air cleaner with an intake silencer and is used mostly on marine engines. The filter part consists of a pad of metallic gauze wetted with an adhesive oil. The air drawn into the engine passes through this pad and particles of dust adhere to the oil gauze. The filter element must be periodically removed and washed either in a solution of washing soda in water or in light diesel fuel oil. The clean gauze, after drying, is dipped in adhesive oil and put back into the housing.

Oil-bath Filters. The oil-bath filter (Fig. 15-3) has also received wide application, especially for high-speed engines. The cleaner, of which several designs are available, consists of a tank containing oil and an upper compartment containing steel wool. When the cleaner is in operation, the atmospheric air enters through the center inlet whose lower end is submerged below the oil level. As the air enters, the oil is forced upward into the screen element which separates the oil from the air and returns it to the oil reservoir. The sudden reversal of air-flow direction removes a large portion of the dust, which is thrown to the bottom of the cup. The air is continually washed as it passes through the oil in the cup.

Only a portion of the oil is carried up into the screen element, while the balance is held in the outer cup. This oil is practically at rest, permitting the dust to settle to the bottom, leaving the oil surface sufficiently clean. All the cleaning is done in the oil cup and the lower screens, with the result that the upper screens remain clean. The air, upon entering the screens, travels toward the outside of the cleaner, automatically washing the screens.

The air cleaner should be serviced daily by removing the oil cup, emptying the oil, scraping out the settled dirt, refilling the cup with fresh oil of the same grade as that used in the engine lubricating system, replacing the oil cup, and fastening it securely. The oil should be filled to the level indicated by the bead.

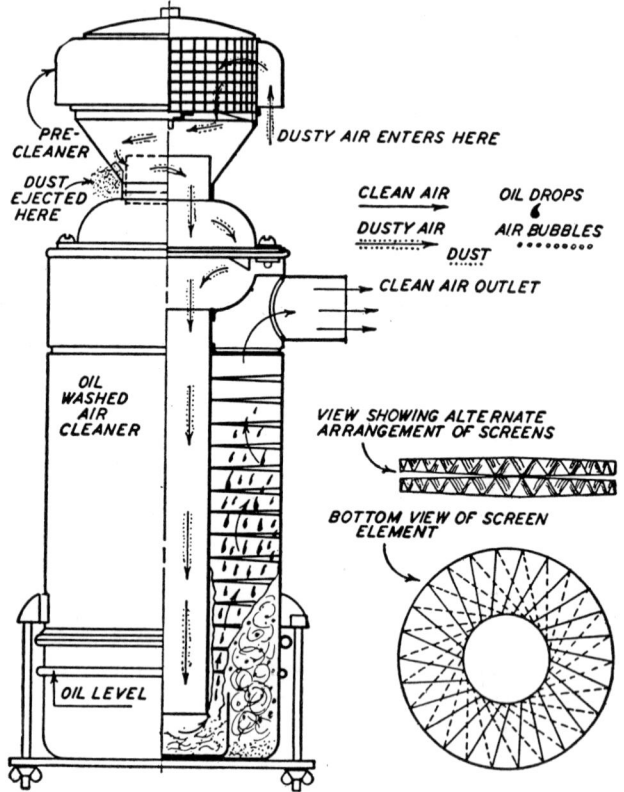

Fig. 15-3. Oil-bath air cleaner.

If the air is not very dusty or the engine is used only intermittently, the servicing may be done not every day, but only when it is actually needed. This is indicated whenever the level of the dirt accumulated in the bottom of the cup reaches the half-way mark or the oil becomes too thick or heavy to be sprayed properly by the air flow. The depth of dirt at the bottom may be measured with a stick or steel rod. However, a daily inspection is necessary to enable the engine operator to notice when any of the undesirable conditions have been reached.

Thus a slight disadvantage of this air cleaner is in that it requires regular and constant attention. This is outweighed by the fact that it has a high efficiency and low resistance to air flow and uses ordinary lubricating oil.

15-3. Intake Silencers. The air rushing in through the intake valve with a high velocity produces a disagreeable hissing noise, which in large engines can be heard over a great distance. A simple silencer that at the same time protects the intake valve from foreign matter, such as rags or loose nuts, may be made of a piece of thin-wall brass or aluminum pipe b (Fig. 15-4) with milled, very narrow slots c, not

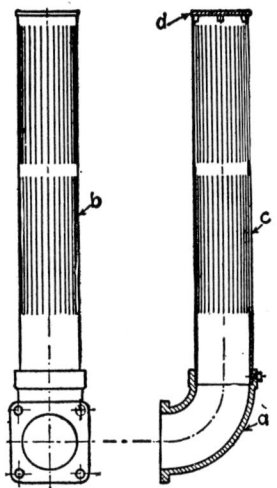

Fig. 15-4. Intake silencer.

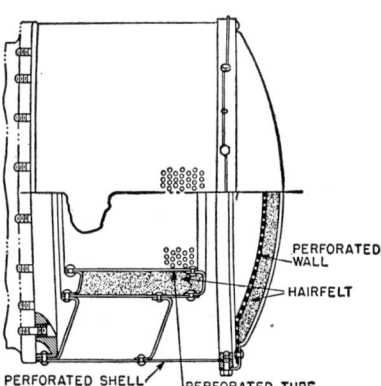

Fig. 15-5. Air-intake silencer for large engine.

over $\frac{1}{32}$ in. wide, and covered with a removable cap d; the cast-iron elbow a is bolted to the cylinder head. To be effective as a silencer, the slots should stop within 12 to 14 in. from the point where the elbow joins the cylinder head. To avoid excessive throttling, the total area through the slots should be 1.5 to 2 times the area of the pipe, which in turn must be slightly larger than the intake port area.

When the air is taken in through an air filter or cleaner, the latter is already acting as a silencer, at least to a certain extent. If better silencing is desired, an acoustic silencer may be used. These silencers are of two general types. One type uses a sound-absorbing material, such as fireproofed hair felt placed around air passages having perforated walls (Fig. 15-5). Sound passes through the perforations but is partly absorbed by the felt. The other type has two or more chambers connected to the air passage by openings. These chambers act as acoustic filters. Intake silencers are sometimes combined with an air cleaner, as shown in Fig. 15-6: the air first passes through an impingement air cleaner into a chamber, from which it flows through a large pipe to the engine intake. This pipe is connected by suitable holes to two acoustic filter chambers for high and low frequency waves; the top of the filter chamber is protected by sound-absorbing material.

15-4. Intake Headers and Piping. In order not to reduce the volumetric efficiency, the intake headers and piping should offer the least possible resistance to air flow. With a permissible pressure drop of 1 in. water or about 0.04 psi, the air velocity should not exceed 3,000 to 3,500 ft per min.

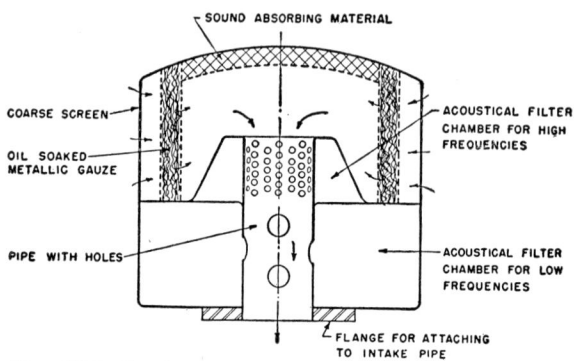

FIG. 15-6. Combination air cleaner and intake silencer.

When in an engine developing comparatively large power and equipped with an air filter the latter is located outside the engine building, the pipe connecting the filter to the engine becomes rather large. This pipe should be made of a sheet steel sufficiently heavy to prevent its collapse under the difference of pressures outside and inside. It should be remembered that a round welded pipe is stronger than a flat-walled duct of a square cross section of the same material and of the same sheet thickness.

15-5. Supercharging. As already brought out in Sec. 2-7, supercharging has for its object the increase of the power output of an engine; it is therefore also called *boosting*. Supercharging is used in the following cases: (1) to overcome the effect of high altitude, such as in aircraft engines or in stationary installations in the mountains; (2) to reduce the weight of the engine per horsepower developed; (3) to reduce the bulk of the engine to fit it into a limited space, such as in locomotive or marine engines; and (4) to increase the power of an existing engine when a greater power demand occurs, such as in pipeline pumping stations.

Thus it can be seen that supercharging has a wide and varied field of applications. Its use is steadily increasing.

Methods. The increased air pressure is obtained by using blowers, either of the positive-displacement type, such as Roots blowers (Fig. 15-7a) and vane blowers (Fig. 15-7b), or of the centrifugal, or turbo-type (Fig. 15-7c). Reciprocating-piston blowers are seldom used,

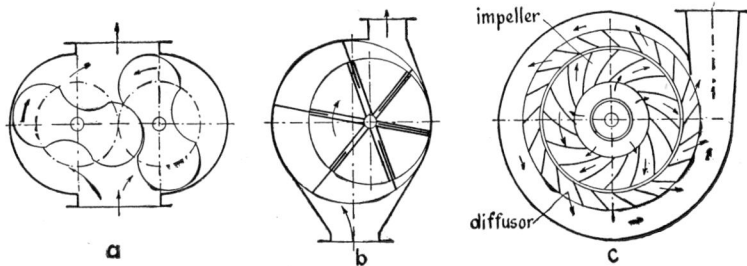

Fig. 15-7. Types of supercharger blowers.

because they are bulkier, more expensive, and less dependable than blowers of the rotary type.

Roots blowers are used with two lobes or with three and with cylindrical or helical surfaces. The three lobes and the helical surfaces are used to obtain a more uniform nonpulsating flow of air. The blowers are usually driven by the engine itself by means of spur, helical, or

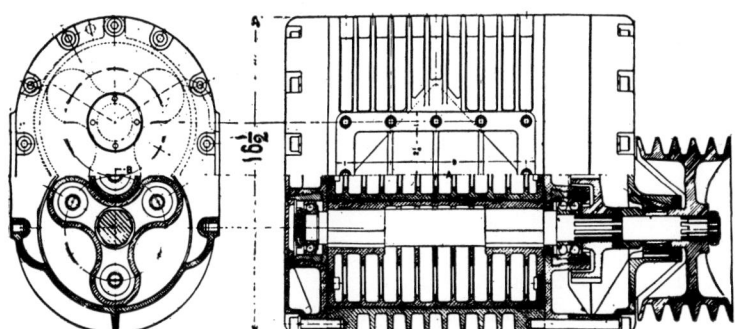

Fig. 15-8. MacCulloch positive-action supercharger.

herringbone gears, silent chains, or V belts at a speed two to three times that of the engine. Sometimes, particularly in case (4), the blower is driven by an electric motor.

Figure 15-8 shows a three-lobe cylindrical Roots-type supercharger for a 290-hp rail-car engine. The blower has skew-shaped inlet and outlet ports, which give a gradual change of velocity and thus eliminate the noise present in most cylindrical blowers. It is driven by five V belts.

Centrifugal blowers are used on both low-speed heavy-duty and high-speed, lightweight engines. They are driven either by gears from the engine or directly by an impulse exhaust-gas turbine. The speed of the centrifugal blowers must be comparatively high, around 10,000

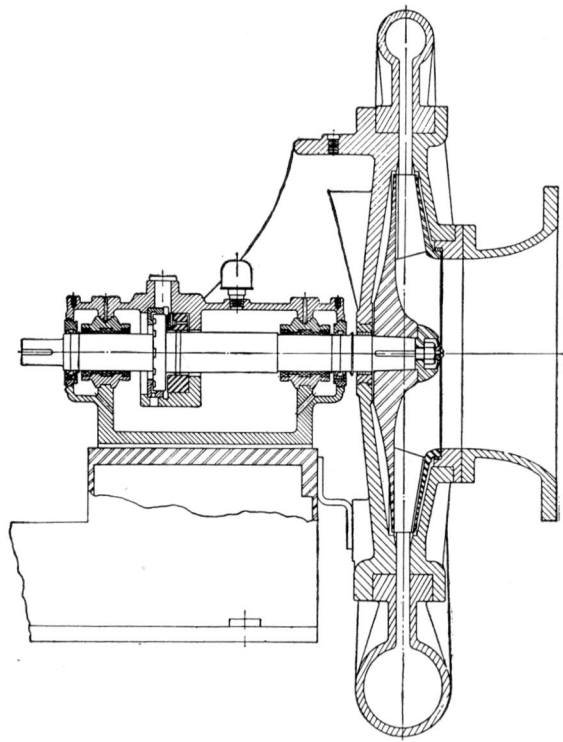

Fig. 15-9. Centrifugal supercharger.

to 15,000 rpm for low-speed engines and 15,000 to 30,000 rpm for high-speed engines. If they are driven by the engine, the speed step-up must be considerable, at least 5:1 and up to 11:1.

Figure 15-9 shows a section through a horizontal centrifugal blower with a spiral discharge header, designed for a mechanical drive. A section through a vertical Buchi turboblower driven by an exhaust-gas turbine is shown in Fig. 15-10. The kinetic energy of the air, when it leaves the blower wheel, is transformed into pressure by the vanes of a diffuser (Fig. 15-7c or Fig. 15-10).

The Buchi turbochargers are built for supercharging four-stroke engines of 250 hp and upward.

AIR-INTAKE SYSTEM 253

Field of Application. The positive-displacement blowers are used for low- and medium-speed engines, such as those found in stationary and marine installations. Their output per revolution is very little affected by speed; however, if the engine is reversed, the blower must be driven from an outside source, usually an electric motor. Positive-displacement blowers operate at moderate speeds—not over 4,000 rpm —and therefore are somewhat bulky and heavy.

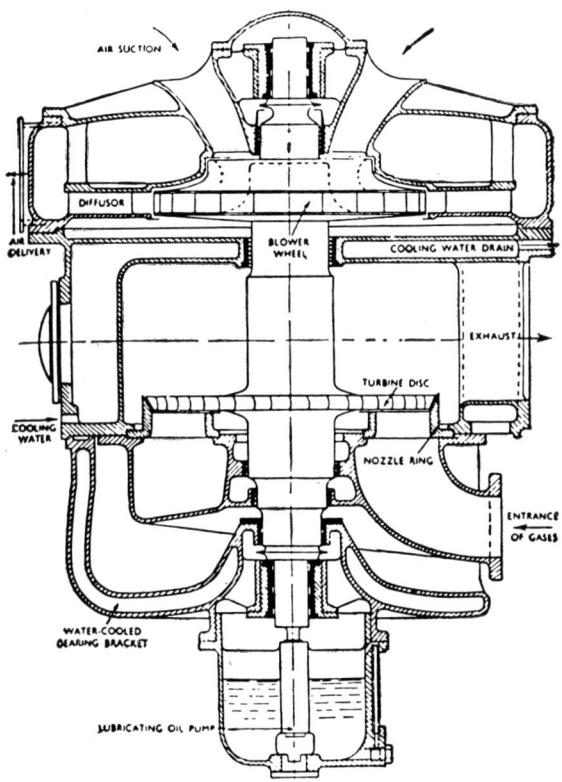

FIG. 15-10. Centrifugal supercharger driven by an exhaust-gas turbine.

Centrifugal blowers driven by exhaust-gas turbines are small and light and are used for stationary, locomotive, marine, and aircraft engines. With this supercharger an engine can be reversed without affecting the performance of the supercharger.

At reduced, down to half speed or less, the torque developed by the engine can be maintained constant without an excessive temperature increase.

Power Increase. An increase of the mean effective pressure by 30 to

50 per cent is easily obtainable. A number of tests are known which show that mean effective pressures of 350 psi have been obtained continuously, and pressure of about 440 psi for a short time.

When supercharging is used to counteract the effect of a suction vacuum due either to high altitude or to a high engine speed, both of which produce a low volumetric efficiency, the power increase through supercharging is about 1.3 times greater than the increase of the volumetric efficiency. For example, an increase of the volumetric efficiency from 70 to 90 per cent, or by 28.5 per cent, gives a 37 per cent increase of power.

The greater power increase is due partly to better air turbulence, and hence more complete combustion, and partly to an increased mechanical efficiency, as explained below.

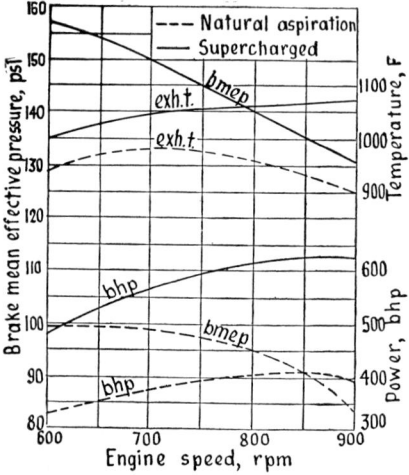

FIG. 15-11. Effect of supercharging by a Buchi-Elliot turboblower.

Figure 15-11 shows the power gained by supercharging an 8- by 10½-in. six-cylinder compression-ignition oil engine equipped with a Buchi-Elliot turboblower run at different speeds. In this engine the brake mean effective pressure when supercharged was limited not by the available air but by the fuel-pump delivery. As may be seen from Fig. 15-11, the increase of the exhaust-gas temperature is not very great, only about 150 F above the maximum temperature obtained with natural aspiration.

The exhaust gases proper are cooled by mixing with the relatively cold surplus air that goes through the engine cylinder during the time when both the exhaust and intake valves are open—a condition called *overlap* of timing, explained below.

Mechanical Efficiency. The increase of friction losses with a supercharger driven by the engine itself is considerably smaller than the power gained through supercharging. As a result the mechanical efficiency, referred to the maximum load, increases with supercharging. Figure 15-12 shows the mechanical efficiencies of a 7- by 10-in. six-cylinder compression-ignition Cummins engine at different speeds and mean effective pressures when operating supercharged with a Roots blower. The mechanical efficiency of the same engine with natural aspiration at 1,000 rpm when developing a brake mean effective pressure of 84 psi was found to be 73 per cent.

Fuel Economy. Owing to better combustion because of increased turbulence, better mixing of the fuel and air, and increased mechanical efficiency, the specific fuel consumption in most cases, though not in all, is lowered by supercharging.

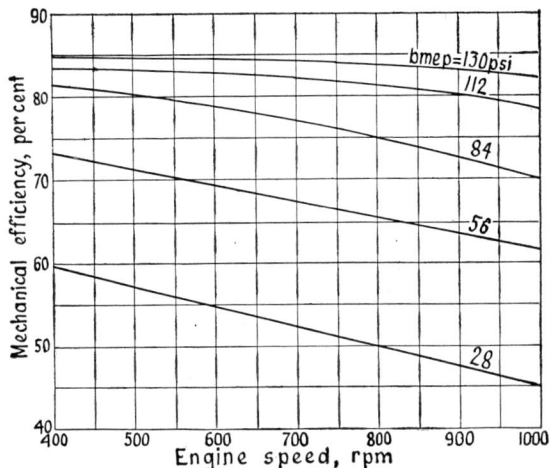

Fig. 15-12. Mechanical efficiencies of a Cummins supercharged diesel engine.

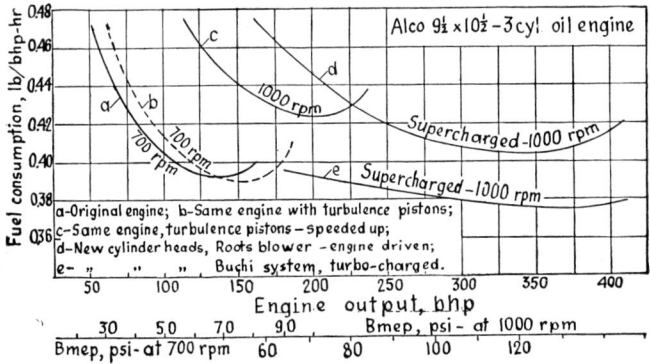

Fig. 15-13. Performance development of Alco 9½ × 10½-in. diesel engine.

Figure 15-13 gives an interesting comparison of the performance of a three-cylinder 9½- by 10½-in. Alco oil engine at 1,000 rpm: curve c, unsupercharged; curve d, supercharged with a Roots blower, power increase about 75 per cent, fuel consumption 5 per cent lower; curve e, supercharged by a Buchi turboblower, power increase about 87 per cent, fuel consumption 13 per cent lower than in the unsupercharged engine.

Curves *a* and *b* show how the performance of this engine was improved by putting in turbulence-creating pistons.

Compression Ratio. In compression-ignition engines, increasing the inlet pressure decreases the ignition lag and consequently the rate of pressure rise in the cylinder, thereby producing smoother operation. However, the maximum cylinder pressure also goes up and in order to prevent excessively high pressures—evidenced by sounds like sharp knocks—and overstressing of certain engine parts, the compression ratio must be lowered till the compression pressure is reduced to its value before supercharging. This will slightly lower the thermal efficiency, but the power output will be increased, because a greater amount of fuel will be burned. An engine operating with natural aspiration with a compression ratio of 17:1 should have, when supercharged, a compression ratio about 14:1.

Overlap. One of the main differences between supercharged engines and engines with natural aspiration is the greater overlap of the exhaust- and intake-valve timing used in supercharged engines. In compression-ignition oil engines, the overlap is made from 80 to 160 deg; excess air must be delivered from the supercharger to compensate for the air escaping through the exhaust valve. This excess may vary from about 40 to 80 per cent of the piston displacement, depending upon the amount of overlap, the shape of the cams, supercharger pressure, and engine speed. The advantages of a large overlap are a thorough scavenging of the combustion space and good cooling of the exhaust valves.

15-6. Scavenging of Two-stroke Engines. Filtering and silencing of the air intake is done by the same devices as in four-stroke engines.

Scavenge Pressure. The required scavenge-air pressure is obtained in some diesel engines by using the crankcase and the other end of the cylinder, including the back side of the piston, as a pump as shown in Fig. 1-9. Other engines use either reciprocating-piston or Roots-type rotary positive-displacement blowers, as shown in Figs. 1-6 and 1-11.

The advantages of the crankcase-scavenging scheme are that it is simple and comparatively foolproof. Its maintenance is very simple—the operator has only to inspect the suction valves from time to time, if the engine has any, to repair them if they are leaking, and also to see that the main bearings and the special seals next to them do not start to leak scavenge air from the crankcase.

However, since in engines of this type the pistons are lubricated by special feeds, it is rather difficult to have the exact necessary amount of oil admitted. If not quite enough oil is admitted, there is the danger that the piston may seize when the load exceeds a certain maximum. Therefore, most engine operators prefer to overlubricate the pistons slightly, with the usual results that the excess lubricating oil forms coke and that the latter, wetted with oil, is deposited in the exhaust ports and gradually decreases their openings. The pistons should be pulled at regular intervals and cleaned and the exhaust ports should be cleaned, and if too much coke deposit is found, the amount of lubricating oil fed to the cylinder should be reduced.

The scavenge-air pressure in crankcase-scavenging engines is rather low. However, this is not objectionable if the volumetric efficiency of the pump is sufficiently high. This efficiency depends upon the design of the ports and suction valves, and on the shape of the crankcase space and cannot be changed in a given engine. Since the amount of air taken in these engines is less than the piston displacement, its further decrease by the unavoidable resistance presented by air filters is undesirable. Besides, the construction of most engines of this type is such that it would be difficult to connect an air filter to the crankcase.

The reciprocating-piston blowers do not present any special problems. They normally have a high volumetric efficiency and furnish a sufficient excess of scavenge air, thereby permitting the use of air filters. There is, however, a certain problem with their lubrication: the lubricating oil is carried with the air into the engine cylinders. Therefore, the nearer to the safe minimum the amount of oil is kept, the less coke will tend to deposit in the exhaust ports. The temperature of the scavenge air in the blower is not very high, and a very small amount of lubrication will suffice. Also, there are known cases where the oil fed to lubricate the blower piston caused the engine to continue running after the fuel-oil supply was cut off. Such a condition not only interferes with the control of the engine by the governor, but may be dangerous in service when it is necessary to stop the engine as quickly as possible.

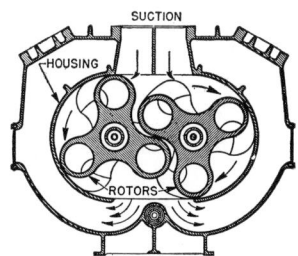

FIG. 15-14. Roots-type blower.

Roots Blowers. Figure 1-11 shows an air-intake system with a Roots blower. The air from the blower is delivered to a large receiver surrounding the cylinder and from there is admitted into the cylinder as scavenge air through round ports uncovered by the piston.

Figure 15-14 shows a cross section of a Roots blower from a larger multicylinder engine with a separate receiver for the scavenge air.

The air pressure created by a blower is about 2 psi in small engines, and may be as high as 4 to 6 psi in heavy-duty high-speed engines.

In the latter engines a high scavenge pressure is used in order to increase the amount of excess air and to provide a certain amount of supercharging to raise the power output.

The rotor shafts are connected by two helical gears, which space the rotor lobes with a slight clearance, 0.002 to 0.006 in., on the trailing side in small engines and correspondingly more in larger engines. The gears are therefore called *timing* gears. Since the rotor surfaces do not touch one another, they require no lubrication. Very effective seals prevent air leakage at the ends of the lobes and also keep the oil used for lubricating the timing gears and rotor-shaft bearings from entering the inside of the blower. One of the rotors is driven from the camshaft gear and the other through the timing gears. The rotor shafts are supported in the doweled end plates of the blower housing by ball bearings at each end.

The timing gears are splined and pressed on splines of the shafts and retained by nuts locked with a tongued lock washer. The gears can be assembled only in one position and have timing marks which insure their correct location with respect to rotor lobes. The position of the rotor lobes is also fixed by a pin through the rotor and shaft.

Servicing. The blower should be inspected at regular engine-overhaul intervals. Such an inspection will permit one to detect minor irregularities and to correct them before more serious damage occurs. A blower may fail to function properly as the result of one or several of the following conditions:

1. The clearance between the mating rotor lobes is not correct due to wear of the timing gears.

2. The rotor lobes or the housing are scored by dirt or foreign matter drawn through the blower.

3. Loose rotor shafts or worn bearings are causing contact between the rotor lobes, rotors, and end plates, or between rotors and housing.

4. Worn oil seals permit lubricating oil to enter into the rotor compartment.

To inspect the blower for any of these conditions, the blower must be removed from the engine and dismantled.

Clearance. The correct clearance is very important to ensure the necessary amount and pressure of the scavenge air. Normal gear wear causes an increase of rotor-to-rotor clearance between the leading flanks of the upper-rotor lobes and the trailing flanks of the lower-rotor lobes, as shown in Fig. 15-15. An increase of the clearance results in an increased leakage and reduced pressure of the scavenge air.

The clearance can be brought to normal by slightly moving one of the helical timing gears axially. However, when the backlash in the gears exceeds 0.004 in. at any point, or the teeth surfaces become uneven or pitted, the gears should be replaced.

Scored rotor or housing, if everything else is in order, indicates that an air filter should be installed or that the existing air filter should be made to operate properly.

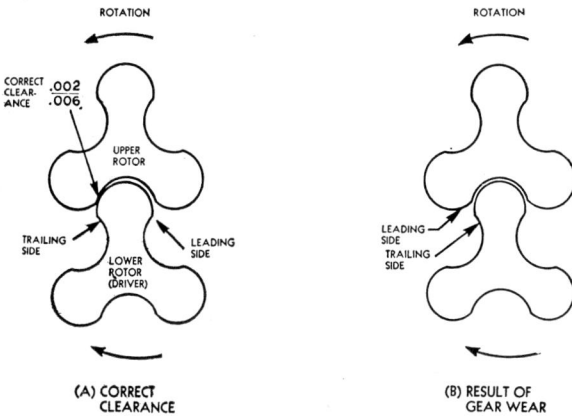

Fig. 15-15. Blower-rotor lobe clearance.

The other defects due to wear of the blower mechanism call for repair or replacement of the parts.

QUESTIONS

1. Enumerate the functions which an air-intake system may have to perform.
2. Enumerate the parts that comprise a complete air-intake system.
3. How much dust may be found per 1,000 cu ft of air in mining localities and near highways with heavy traffic?
4. On the basis of the dust content in the air, how much abrasive matter may go through a 500-hp engine per year if the intake air is not cleaned?
5. List the requirements which an air filter should fulfill.
6. In what three main classes can commercial diesel air filters be divided?
7. Explain the principle of operation of a dry air filter.
8. Explain the principles of operation of an impingement filter.
9. Draw a sketch of an oil-bath filter and explain its operation.
10. Explain the principles of operation of air-intake silencers.
11. What is the permissible pressure drop in an air-intake system, including the header and piping?
12. State in what cases and for what purposes superchargers are used.
13. Enumerate, showing by sketches, the various methods used for supercharging four-stroke diesel engines.
14. State in what cases particular types of superchargers are most suitable.
15. State how much the power output of a diesel engine may be increased by supercharging.
16. How does supercharging affect the mechanical efficiency of a diesel engine? Explain, using as an example Fig. 15-12.

17. How does supercharging affect the fuel economy of a diesel engine?

18. What must be done about the compression ratio of a diesel engine when changing it from natural aspiration to supercharging?

19. What is overlap in valve timing, and between what limits should it be in supercharged diesel engines?

20. What are the objects of a large overlap in supercharged engines?

21. Explain the advantages and drawbacks of the crankcase scavenging scheme in two-stroke engines.

22. Explain the problem of lubrication in a reciprocating-piston scavenge pump.

23. Explain the operation of a scavenge blower of the Roots type.

24. What might cause the failure of a Roots blower to function properly?

25. Explain how the clearance between the mating rotor lobes may become excessive and how it can be restored to the proper amount.

26. What does a worn or scored rotor or housing of a Roots-type scavenge blower usually indicate?

CHAPTER 16

EXHAUST SYSTEM

16-1. Exhaust in Four-stroke Engines. At the end of the expansion stroke the gases inside the engine cylinder are still under a comparatively high pressure—30 to 50 psig. When suddenly released into a pipe filled with gases at atmospheric pressure, the exhaust gases produce a pressure rise in the piping and impart a velocity to the gases in the pipe. Their flow and inertia produce a pressure drop in the cylinder and a pressure rise in the exhaust piping. This pressure rise, owing to the inertia of the gases, is followed by a pressure drop, then another pressure rise, and so forth. Such pressure fluctuations or waves occur not only in the exhaust pipe but may extend back into the engine cylinder.

FIG. 16-1. Weak-spring diagram.

Figure 16-1 shows an indicator diagram taken with a weak spring from a diesel engine in which two complete pressure waves are recorded. Theoretically, a sudden impulse from the released exhaust gases is always followed by a pressure fluctuation of an oscillatory, or wave, character in the exhaust system. However, with a short and wide exhaust pipe, the difference between the maximum and minimum pressure, called the *amplitude* of pressure fluctuation, may be so small that it does not appear on the indicator diagram, particularly on a regular full diagram taken with a hard spring.

Such pressure fluctuations, generally speaking, are undesirable and can be avoided by using short exhaust lines with as few elbows as possible and by having a small resistance in the whole exhaust system.

Exhaust Conditions. Back Pressure. An increase of the back pressure by 1 psi reduces the mean effective pressure approximately by one per cent owing to the extra work that the piston has to do during the exhaust stroke. The additional exhaust gases left in the compression space of the engine cylinder cause an increase in dilution which lowers the charge efficiency by another 0.5 per cent. Thus for a 1-psi increase in the back pressure, the power output decreases by about 1.5 per cent.

Thus, it pays to reduce the resistance in every part of the exhaust system by eliminating sharp turns and abrupt changes of gas velocity, by reducing as much as possible the length of the exhaust pipes, and by keeping a moderate exhaust velocity, not over 4,000 to 5,000 ft per min.

Exhaust Overlap. The average duration of the exhaust valve opening is about 220 deg of crank travel. This means that with a four-cylinder engine the valve opening of two cylinders will overlap on each end by

$$a_4 = \frac{220 \times 4 - 720}{4 \times 2} = 20 \text{ deg}$$

Since the valve lift is small at the beginning of opening and before closing, there will be no appreciable interference. With a six-cylinder engine the overlap on each end will be

$$a_6 = \frac{220 \times 6 - 720}{6 \times 2} = 50 \text{ deg}$$

This must increase the resistance created by the simultaneous discharge of gases from two cylinders, although the valves are not opened very much. With eight cylinders in line, the overlap becomes

$$a_8 = \frac{220 \times 8 - 720}{8 \times 2} = 65 \text{ deg}$$

and to prevent an increase of the back pressure, either two separate manifolds are used or, if a single manifold is used, its diameter is increased by some 10 to 20 per cent, depending upon the actual timing and valve lifts.

16-2. Exhaust System. The combination of devices through which the exhaust gases leave a diesel engine is called the *exhaust system*. The main purpose of the exhaust system is to conduct the exhaust gases from the engine cylinders to the atmosphere and to do this with a minimum of flow resistance. In addition the exhaust system may also perform one or more of the following functions: (1) silence the noise made by the escaping exhaust gases; (2) protect the neighborhood from exhaust gases and occasional smoke; (3) quench occasional sparks and remove them from the exhaust gases; (4) furnish energy to the exhaust-gas turbine driving the supercharger; (5) furnish heat for heating purposes, generating steam, or distilling water.

The exhaust system usually consists of the following parts arranged in the order given: (1) exhaust valves and ports in the cylinder head; (2) the exhaust manifold; (3) an exhaust pipe; (4) a muffler, also called *exhaust silencer;* (5) a tail pipe. The exhaust system may have also one of the following additional parts: (6) an exhaust-gas turbine to drive the supercharger; (7) exhaust-heat exchangers, boilers, or evaporators; (8) spark arresters.

The various component parts will now be discussed in the order given.

Exhaust Valves. The satisfactory operation of exhaust valves depends on two conditions: correct timing and proper seating. The timing may get out of order from excessive wear of the cam and more often from an increase of the clearance between the cam and cam follower or cam follower and pushrod. It should be checked from time to time against the specifications given in the instruction book furnished by the engine builder.

The seating surface of the exhaust valve and the seat in the exhaust port are swept by very hot gases which may carry incandescent particles of carbon. The seat in the port usually is less subject to damage because it is cooled by the circulating water. The temperature of the exhaust valve is higher than that of the seat and it is therefore the conical seating surface of the valve that is apt to be damaged and to start to leak. When even a very small leak develops, the hot gases begin to pass through it during the combustion stroke, erode the valve edges further, and do it so fast that sometimes a few hours after the first leak has developed the cylinder begins to loose compression and may even stop firing. Therefore, it is very important to inspect the valves periodically and to regrind them as soon as the least damage appears on the seat surface. During engine operation a leaky exhaust valve may be discovered by watching the exhaust temperature of the individual cylinders and noticing that the exhaust gas temperature of one of the cylinders begins to rise even though the fuel supply remains the same as to the other cylinders. Unfortunately, the temperature rise usually becomes noticeable only after the valve is burned beyond repair.

Exhaust Elbow. The exhaust port usually is connected to the exhaust manifold by an elbow to permit the removal of one cylinder head without disturbing the other heads or the exhaust manifold. Figure 16-2 shows a section of a water-jacketed exhaust elbow a, with the cross section of a cast-iron exhaust manifold b; c is a screw to regulate the water flow from the cylinder head, and d is a hole with a pipe tap for a $3/4$- or 1-in. test cock for checking the color of the exhaust gases. Sometimes the opening of a test cock will emit a short flame;

this would indicate a very protracted afterburning or a leaky exhaust valve; in either case, the cause must be investigated and the condition remedied.

16-3. Exhaust Manifold. Exhaust manifolds are used on multicylinder engines to connect the exhaust ports of each cylinder to the common exhaust pipe. In smaller engines exhaust manifolds are made of cast iron or cast steel and usually have a water jacket, as shown in Fig. 16-2. In larger engines the manifolds are mostly of welded construction and have water jackets all around for cooling them by water which comes from the cylinder heads. The water-cooling of the exhaust manifolds has several objects: (1) to make them less dangerous for persons working around the engine; (2) to reduce the fire hazard to the building; (3) to prevent excessive heating of the air in the engine room; (4) to reduce the back pressure in the cylinder by lowering the temperature of the exhaust gases; (5) to lower heat stresses in the manifold caused by temperature variations; and (6) to permit recovery of heat, if desired.

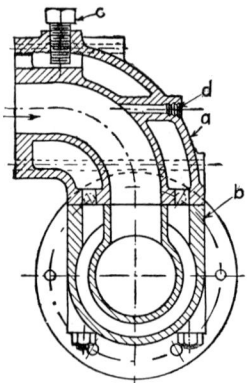

Fig. 16-2. Water-cooled exhaust elbow.

The temperature of an uncooled exhaust manifold varies considerably depending upon the engine load. In order to permit free expansion of the manifold when its temperature goes up and contraction when it goes down, the manifold should be fastened to the cylinder heads by flexible connections. Insulating the exhaust manifold by covering it with special heat-insulating slabs or by a paste of asbestos fibers with some binder, takes care of conditions (1) to (3) but does not help in respect to conditions (4) and (5).

A water-jacketed exhaust manifold should have openings for cleaning the water space, as scale deposits decrease the heat transfer, raise the temperature of the inside wall, and thus may cause dangerous heat stresses. The water jackets should be inspected at regular intervals and the scale removed.

16-4. Exhaust Pipe. The exhaust line connecting the exhaust manifold to the muffler should be made as short and with as few turns or bends as possible. Only long-sweep or 45-deg elbows should be used for bends. The best solution is to have the muffler connected by a horizontal pipe, as shown on Fig. 16-3, where a is the exhaust manifold, b the exhaust pipe, and c the muffler.

In order to reduce the transmission of engine vibrations, to avoid stresses set up by expansion of the hot pipe, and to simplify the installation of the muffler, the exhaust pipe, or at least part of it, can be made of flexible piping. Suitable flexible piping is obtainable in any length and with any inside diameter.

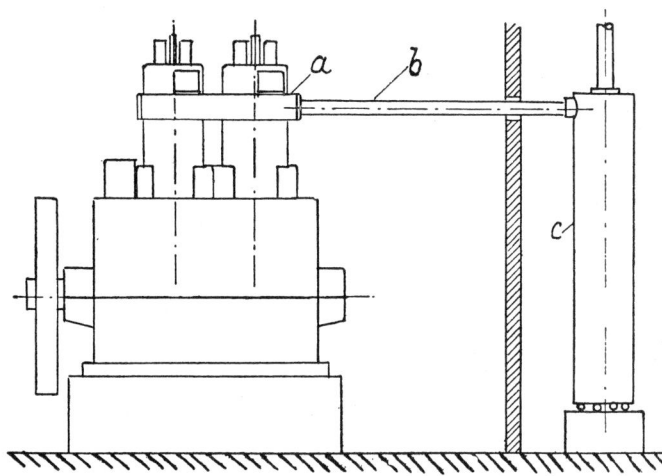

Fig. 16-3. Desirable muffler installation.

Often the exhaust pipe goes first down, then horizontally under the floor to the muffler. Provisions should be made to allow the pipe line to expand when becoming hot. Only in a very short horizontal line with elbows can one depend upon the bends to absorb the heat expansion. Sometimes a coupling is inserted, as shown in Fig. 16-4: the

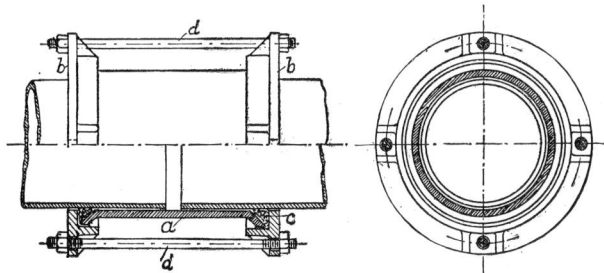

Fig. 16-4. Expansion coupling.

ends of the pipes to be connected are slipped into a cast-iron or steel sleeve a flared out at the ends; b, b are flanges acting as stuffing boxes when they are drawn together by four bolts d; c is an asbestos packing. Another type of expansion coupling made of corrugated steel pipe, with the ends welded on to flanges or expanded in them, is shown in Fig. 16-5.

With a very long horizontal exhaust pipe of a considerable diameter, it is advisable, in addition to the expansion coupling, to support the pipe on short pieces of gas pipe laid crosswise on the concrete bottom of the pipe ditch to act as rollers when the pipe is moved by expansion and contraction.

If two engines are connected to a common exhaust line, there should be gate valves in the connection from each engine, to prevent exhaust gas backflow from the running engine into the idle one.

Insulation. It is advisable to have the exhaust pipes inside the building insulated, whether they are exposed or in a ditch covered by floor grating. Insulation protects the engine room from excessive heat dissipated by the pipe and, when the exhaust line passes near wooden parts, protects them from becoming charred. The most common heat insulation is made of 85 per cent magnesia powder with 15 per cent asbestos fiber and comes in block or shell form, 3 ft. long and curved to go on pipes of various diameters. The pipe covering comes in different thicknesses, from 1 to 6 in.; however, 1 or $1\frac{1}{2}$ in. is sufficient for exhaust piping. The cracks between the shells are filled with a cement which consists of magnesia, asbestos, and a binder, and the whole pipe is wrapped in canvas, held by wire or hooplike strips of steel, and painted.

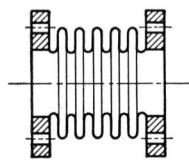

Fig. 16-5. Expansion coupling.

For temperatures above 600 F, which are common in four-stroke engines, it is better to use magnesia over a layer of blocks of diatomaceous earth for which a better name is *diatomite*. Mineral wool is also a good insulation for high temperatures, up to 1200 F.

Instead of blocks, especially for bends, insulation may be applied in the form of a paste made of insulating cement and water. The paste is applied in a layer about $\frac{1}{4}$ in. thick; when the first layer is dry, a second layer is applied, and so on, until the desired thickness is obtained. After this the pipe is wrapped in canvas in the same way as with blocks. Applying the paste in thin layers prevents the insulation from cracking.

Vertical pipes may be insulated by an air jacket by surrounding the hot exhaust pipe by a concentric galvanized sheet-iron pipe about 3 to 4 in. larger than the outside diameter of the exhaust pipe and providing sufficient openings at the bottom and top for air circulation in the jacket.

If the exhaust pipe is not insulated, it must not come closer than two diameters to any wooden part of the building. Where a bare exhaust pipe, horizontal or vertical, is less than 12 in. from a wall, sheet-iron shields should be placed so as to prevent combustible matter, such as rags, to be lodged between the pipe and the wall.

16-5. Mufflers. Mufflers, or exhaust silencers, are devices used for quieting the barking sound produced by the escaping exhaust gases. There are a great many different mufflers in use, but they can all be divided into two groups—dry-type mufflers and wet-type mufflers.

Dry-type Mufflers. These mufflers may be divided into two groups—one group based on the theory that exhaust noise is the result of the high velocity of the escaping gases, and the other based on the newer theory that sound is noticeable and irritating to the ear primarily when it produces vibrations of a high frequency and that the velocity of the gases has little to do with the noise. Mufflers of the first group reduce the noise by reducing the gas velocity by allowing the gases to expand or by changing the direction of their flow; mufflers of the second group reduce the noise by suppressing the high-frequency vibration.

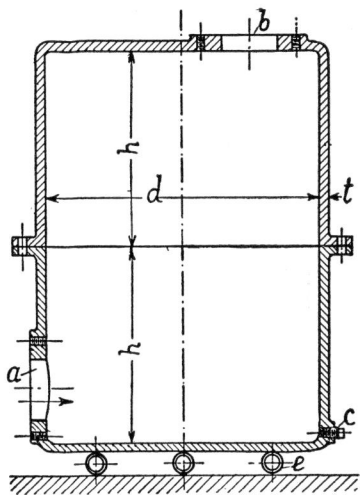

Fig. 16-6. Cast-iron muffler for a small engine.

In small diesel engines, a fairly satisfactory noise reduction is obtained by using a cast-iron pot, if its volume is at least six, and better yet ten, times greater than the piston displacement of one cylinder. Figure 16-6 shows such a muffler whose action is based on expansion of the exhaust gases, admitted at a, and their rather steady discharge at b. The opening c for draining condensed water may be provided either with a cock or simply with a pipe plug; in cold weather it is often left open without noticeable noise effect. The short pieces of $1\frac{1}{2}$- to 2-in. gas pipe e act as rollers to allow expansion of the exhaust pipe. This muffler is used mostly for low-speed engines rated up to 75 hp per cylinder.

The muffling effect of this simple pot can be improved by putting between the flanges of the two halves a cast-iron baffle plate with a great number of small holes, $\frac{1}{2}$- to $\frac{3}{4}$-in. in diameter, so that the sum of their areas is only slightly larger than the cross area of the exhaust pipe. The adding of a perforated baffle plate is an attempt to change the frequency of the exhaust waves to a lower one and one therefore less objectionable to the human ear. With a perforated baffle plate the volume of each half of the muffler must be at least five times the piston displacement and even then a slight increase of the back pressure may occur.

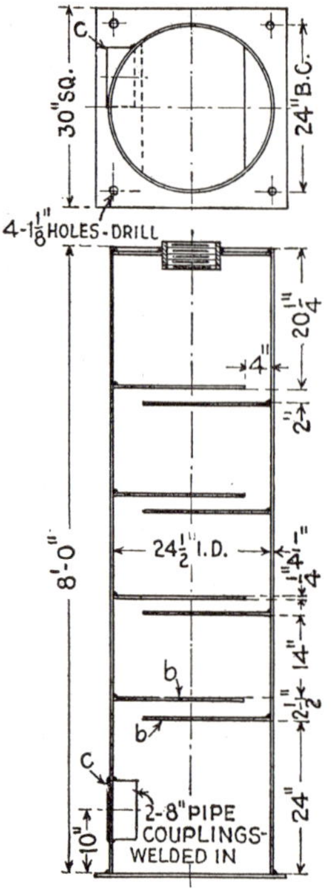

Cast iron is suitable for mufflers as it has the characteristic of damping out vibration. However, for larger engines cast-iron mufflers become too heavy and expensive and steel mufflers are used instead.

Figure 16-7 shows an efficient muffler welded of sheet iron. Baffle plates b, b leave openings that force the gases to change their direction of flow many times and allow the gases to expand gradually. The tangential location of the inlet pipe c prevents the impingement of gases on the opposite wall, which would give a ringing metallic sound. Covering the muffler with insulation also reduces the ringing sound.

Horizontal Mufflers. Horizontal mufflers are used chiefly where it is desired to conceal a muffler by putting it below ground. The simplest horizontal muffler is an old drum or tank buried in soft ground with the gas inlet at one end and the outlet at the other. The soft dirt surrounding the muffler absorbs the noise if the tank is sufficiently large, fifteen to twenty times the piston displacement of each cylinder.

FIG. 16-7. Welded muffler for a 200-hp engine.

Figure 16-8 shows the welded steel muffler of a four-cylinder 450-hp oil engine. The pipe couplings for gas inlet and outlet are welded in on a tangent; the central baffle plate may be tack-welded and must

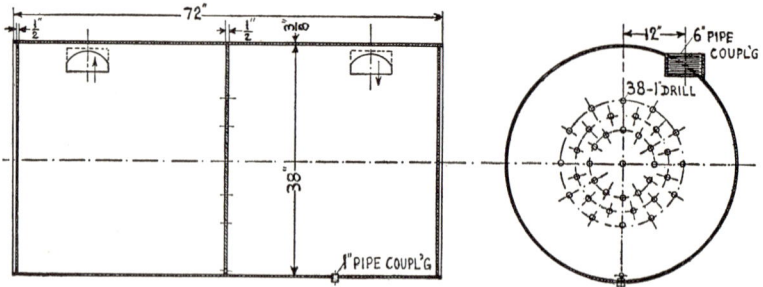

FIG. 16-8. Welded horizontal underground muffler.

have a hole near the bottom for the discharge of condensate through the 1-in. pipe coupling for a drain pipe, particularly in a cold climate. This muffler, with a volume ten times the piston displacement, gives as good a silencing effect, when buried below the ground, as a considerably larger drum without the central partition with holes.

Exhaust Snubber. A muffler based on reduction of noise by the reduction of the frequency of gas vibration is shown in Fig. 16-9: the high-pressure, high-velocity gases first expand in the chamber a, part

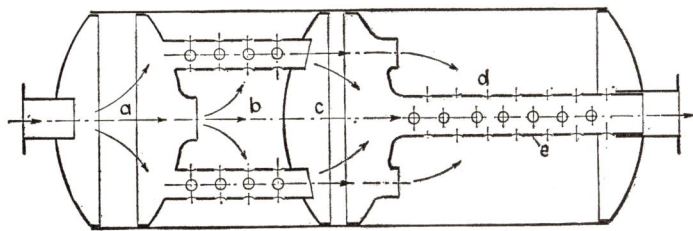

Fig. 16-9. Muffler with frequency reduction.

flows into chamber c directly, and part flows through chamber b and holes in the connecting tubes, thus interfering with the vibration waves; the process is repeated in the discharge pipe e and with the gases from the chamber d. The advantage of such mufflers is their small volume and light weight, hence their use in trucks and marine installations. These mufflers are known under the trade name *Burgess exhaust snubbers* and are produced commercially in different sizes for a wide range of engines.

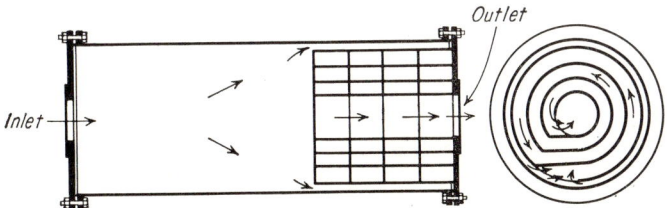

Fig. 16-10. Maxim silencer.

Maxim Silencer. Another commercial silencer, based on the same principle, is sold under the trade name of *Maxim silencer.* It consists essentially of a cylindrical tank (Fig. 16-10) with baffles in the far end, which permit the gases to expand, destroy the peak pressures, and eliminate the return waves. Noise will disappear if the dimensions of the silencer are correct.

The silencer is built in many sizes to suit any type and size of engine.

Absorbent Silencer. The absorbent type consists of a cylindrical tank lined with a layer of rock wool held in place by an inner cylindrical shell perforated with many holes. The pressure wave strikes the silencer but as the rock wool absorbs a part of the wave, the noise is reduced or entirely eliminated. In this design, also, dimensions must be correct and the silencer placed in the correct position in the exhaust line. Otherwise the silencer might be at one of the nodes or places where the pressure was at the minimum, with the result that no wave destruction will occur. Carbon deposits reduce, in time, the effectiveness of absorbent silencers. Absorbent-type silencers are built commercially and can be purchased to desired specifications.

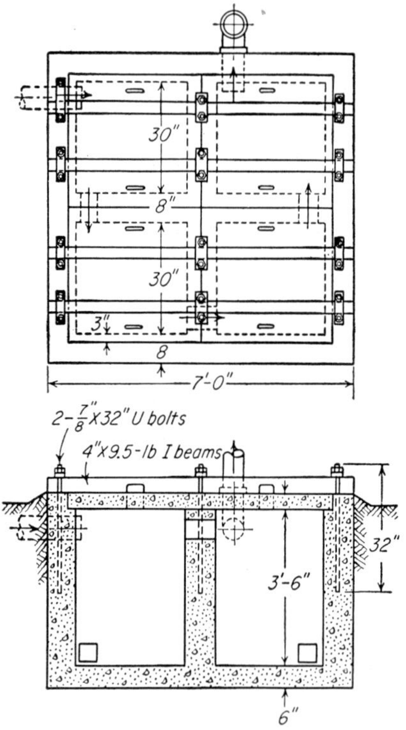

Concrete Mufflers. For stationary medium- and large-size engines it is common practice to use concrete pits with partitions as underground mufflers. The cost of a concrete exhaust pit of the same or even larger volume is considerably smaller than that of a cast-iron or welded-steel muffler. Therefore concrete exhaust pits are usually made considerably larger than metal ones. A good rule of thumb is to make the inside space equal to at least 1 cu ft per horsepower of one cylinder. The walls are made of plain or reinforced concrete 6 to 12 in. thick, depending upon the size of the pit, the top slab of reinforced concrete designed to withstand an inside pressure of 60 to 90 psi. The underwriter's specifications require that the pit be provided with a manhole and cover and with an exhaust stack one size larger than the exhaust pipe from the engine and at least 10 ft high.

FIG. 16-11. Concrete muffler for a 250-hp engine.

Figure 16-11 shows a concrete muffler with baffle partitions and concrete covers for a four-cylinder 250-hp engine which has a very good muffling effect. The four removable concrete covers are reinforced by I beams held by U bolts.

Wet Mufflers. An efficient means for reducing the noise of escaping exhaust gases is to admit a spray of water into the muffler chamber.

The water mixes with the hot gases, is evaporated, and the latent heat of evaporation cools the gases, reducing their volume and pressure and, hence, their kinetic energy and noise. Figure 16-12 shows such a horizontal wet-type muffler for a marine engine: water is admitted through a great number of small holes drilled in the tube s; baffle plates b, b are equipped with holes to cut down the frequency of vibration of the gas column; v is a safety valve or vent; and j is a water jacket whose main object is to protect the operating personnel from contact with the hot muffler walls.

In some wet mufflers an additional small stream of water is admitted into the tail pipe. Some wet-type exhaust systems used in small boats consist simply of an exhaust pipe into which water is admitted under a light pressure.

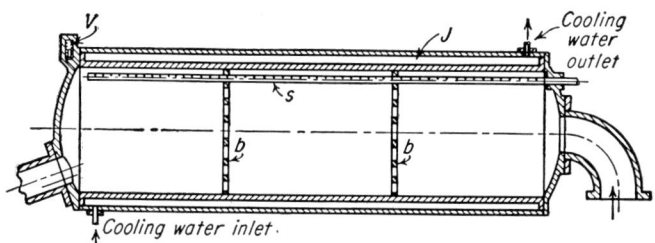

Fig. 16-12. Wet-type muffler for a marine engine.

Wet mufflers are sometimes used also in stationary installations: Fig. 16-13 shows a muffler for a 200- to 250-hp engine with a water reservoir in the lower part of the shell; water is gradually admitted through a ½-in. valve at a with an overflow at b; c is a coupling for a drain pipe for eventual condensate; the two square baffle plates d and e serve to break up the flow of gases. This muffler is exceptionally quiet.

Another muffler with water cooling is shown on Fig. 16-14. The muffler is open at the bottom and its lower part acts as a very effective expansion chamber; the water level inside the muffler goes down at every gas discharge and increase of pressure, and comes up again when the pressure decreases. This breathing action of the chamber equalizes the exhaust pressure and results in a more or less uniform flow of the gases. The water necessary to maintain a constant level is admitted to a, and b is the overflow. A grate c with a 6- to 8-in. layer of pebbles about 2 to 3 in. large divides the upper part and acts as an additional equalizer of gas pressure; d is the tail pipe cut on the bias to obtain a gradual release of the gases to the atmosphere. With a volume six times the piston displacement of each cylinder this muffler gives excellent results and a back pressure even smaller than with a dry muffler.

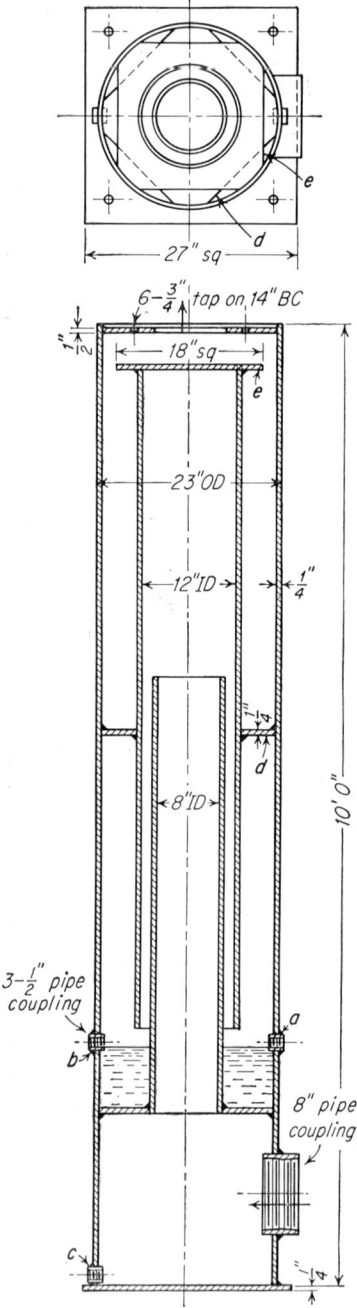

Fig. 16-13. Welded wet-type muffler for a 200- to 250 hp engine.

While wet-type mufflers are very effective in respect to silencing the exhaust noise, they have certain disadvantages because of which they are used only when a very quiet exhaust is required, such as on submarines or on stationary engines installed near residential districts. These disadvantages are: a slightly higher back pressure and a shorter service life, owing to corrosion.

16-6. Tail Pipe. Generally speaking, the exhaust gases leave the muffler, or silencer, not in a steady stream but in a pulsating one. Each wave, as it reaches the outlet, is reflected back into the pipe, and it is desirable that on its way back it reach the muffler not at the moment of a fresh discharge from it but between two discharges. This can be obtained by adjusting the length of the tail pipe, either shortening or lengthening it. The best length can be determined only by experiment. The same is true also of the length of the pipe between engine and muffler. This length, however, is usually determined by the location of the engine. Not much can be done in this respect.

The outlet from the tail pipe should be cut on the bias, as shown on Fig. 16-14. As the flow of gas is gradually diverted from a straight path by the opening on the side, it gradually loses velocity and noise is reduced.

If the exhaust pipe goes up vertically, it is called a *stack*. If a stack goes through the roof, it should be allowed to expand freely. This is best attained by putting on the roof a sheet-iron collar c (Fig. 16-15), which protects the area beneath it from the water flowing down the slope of the roof, and by attaching to the stack a sheet-iron hood h.

16-7. Exhaust-gas Turbines. In order to prevent overlapping of the exhaust-pressure fluctuations, groups of two or three cylinders which do not exhaust simultaneously are given separate manifolds which are connected to separate groups of nozzles in the turbine. Figure 16-16 shows a typical turbine-driven supercharger installation in a six-cylinder engine. The exhausts from cylinders 1, 2, and 3 discharge into one manifold and those from cylinders 4, 5, and 6 into the other, lower manifold. The flow of the exhaust gases after the turbine is so steady and, hence, noiseless that it does not require a silencer.

The exhaust-gas turbine is a delicate piece of machinery; it should be handled strictly in accordance with the detailed instructions furnished by the engine builder.

16-8. Recovery of Waste Heat. Much of the heat in the exhaust gases may be recovered by the application of a waste-heat exchanger.

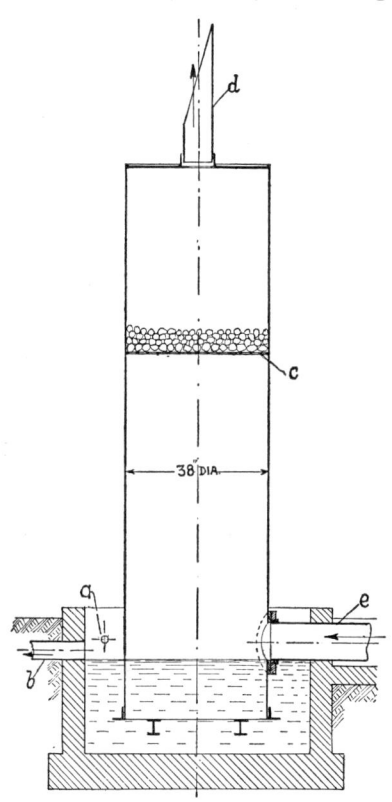

FIG. 16-14. Wet expansion muffler for a 200-hp engine.

Heat exchangers of this kind are of either the fire-tube or water-tube type. In the former, the water is in the shell and the gases flow through the tubes. In the water-tube design, on the other hand, the water is inside the tubes and the gases flow across the outer tube surfaces. The efficiencies are about the same but the water-tube heat exchanger needs less space.

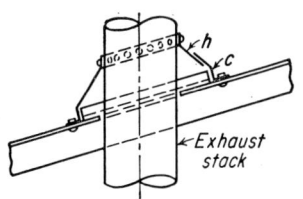

FIG. 16-15. Connection at the roof.

The amount of heating surface should be designed to give the exhaust gases, as they leave the heat exchanger, a temperature not below 300 F, for the heat recovery from gases which have a temperature lower than that drops very rapidly owing to a decrease in the temperature difference between the gases and water. The latter has a temperature of about 200 F if used for heating, and not over 225 F in low-pressure boilers.

Naturally, in order to increase the heat recovery, water is admitted after it is preheated as jacket water. The amount of heat that can be recovered depends upon: (1) the temperatures of the exhaust gases, t_1 before and t_2 after the heat exchanger; (2) the weight w of air per pound of fuel used; (3) the specific fuel consumption of the engine f,

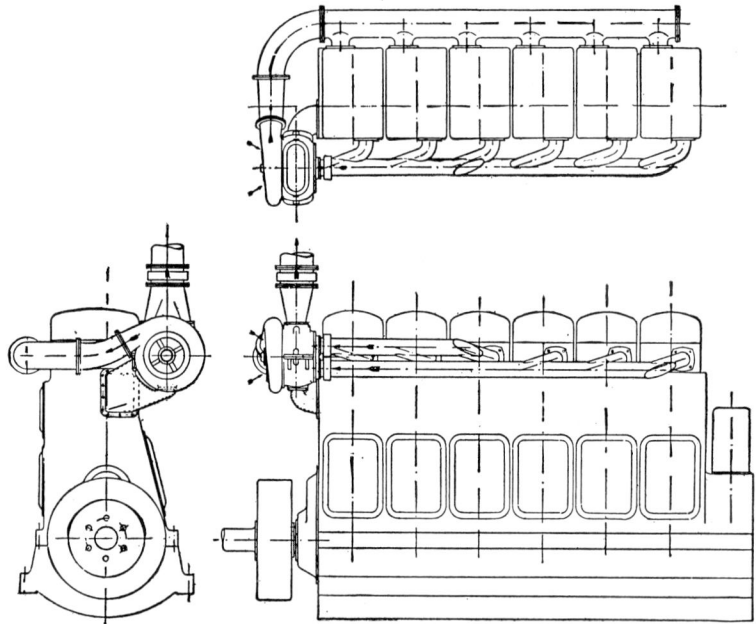

FIG. 16-16. Exhaust temperatures of engines of two different types.

pounds per horsepower-hour; (4) the power N developed by the engine, horsepower; and (5) the efficiency of the heat exchanger, e, as a fraction of 1.

Designating the amount of heat recoverable in Btu per hour by Q and the specific heat of the gases by c_p, one can write

$$Q = (t_1 - t_2)(w + 1)f \times N \times e \times c_p \qquad (16\text{-}1)$$

where the specific heat may be assumed, with sufficient accuracy, as $c_p = 0.25$ Btu per lb and the average efficiency e of the heat exchanger about 0.75.

Assuming that at full load $t_1 = 750$ F and $t_2 = 325$ F, that the engine operates with 60 per cent excess air, or uses $w = 24$ lb of air per pound of fuel, and has a fuel consumption $f = 0.42$ lb per hp-hr, the amount of heat available for recovery may be found by expression (16-1), for 1 hp,

$$Q = (750 - 325)(24 + 1) \times 0.42 \times 0.75 \times 0.25 = 835 \text{ Btu per hp-hr}$$

The amount of jacket water W that can be heated from 145 to 200 F is found as

$$W = \frac{835}{200 - 145} = 15.2 \text{ lb per hp-hr}$$

Thus, from each horsepower of the engine 15.2 lb of hot water are available for use in some process or in a hot-water heating system.

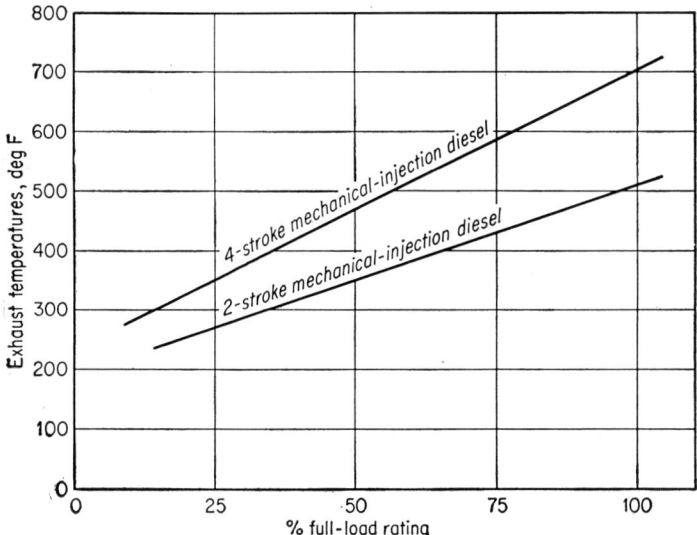

FIG. 16-17. Engine with turbine-driven supercharger.

If steam, say of 5 psig, is needed, the amount of it that can be generated in a waste-heat boiler will be found as follows: the heat content of dry saturated steam of a pressure of $5 + 14.7 = 19.7$ psi abs is 1156 Btu per lb; the heat content of the jacket water being 145 Btu per lb, the amount of steam W_s is found from $Q = 835$ Btu per hp-hr as

$$W_s = \frac{835}{1156 - 145} = 0.825 \text{ lb per hp-hr}$$

Figure 16-17 shows average exhaust-gas temperatures in four- and two-stroke airless-injection diesel engines at different load conditions. It shows that the exhaust temperatures of two-stroke engines are considerably lower than the corresponding temperatures of four-stroke engines. This is due, first, to the lower mean effective pressures in two-stroke engines and second, to the cooling effect of the scavenge air. For this reason the exhaust from two-stroke engines is not so suitable for waste-heat recovery as that from four-stroke engines.

Figure 16-18 shows a Foster water-tube waste-heat boiler with fins shrunk on steel tubes to increase the heat-absorbing surface and in Fig. 16-19 a vertical fire-tube waste-heat boiler. This boiler can produce steam up to 40 psig pressure.

Evaporators. In marine installations exhaust heat may be utilized to produce distilled water by evaporating sea water; the steam is then

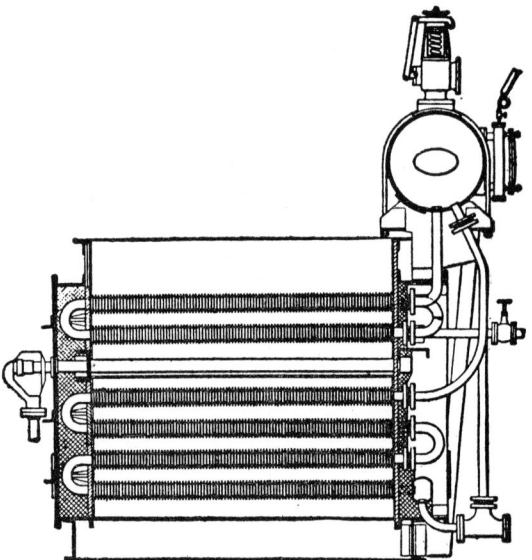

FIG. 16-18. Water-tube waste-heat boiler.

condensed, and since salt is not evaporated, the condensate is fresh water. Such an evaporator is shown diagrammatically in Fig. 16-20.

Exhaust By-pass. When a waste-heat boiler is installed, it is advisable to have a by-pass line so that at low loads the gases may be by-passed around the boiler. Otherwise the heat in the gases may be so low that the temperature of the gases leaving the boiler may drop to below 200 F, at which temperature the water vapor in the exhaust gases will begin to condense and cause rapid corrosion of the boiler.

In Fig. 16-21 is shown a by-pass swing valve that is built in sizes to go with exhaust pipes from 4 to 12 in. in diameter.

Spark Arresters. Spark arresters are devices used in dry-type exhaust systems, mostly in high-speed naval boats, in which the engines are often operated at and even over the rated power. Spark arresters separate and trap sparks and other solid particles in the exhaust gases through an abrupt change of the direction of flow of the gases. Most spark arresters have also some device for creating a

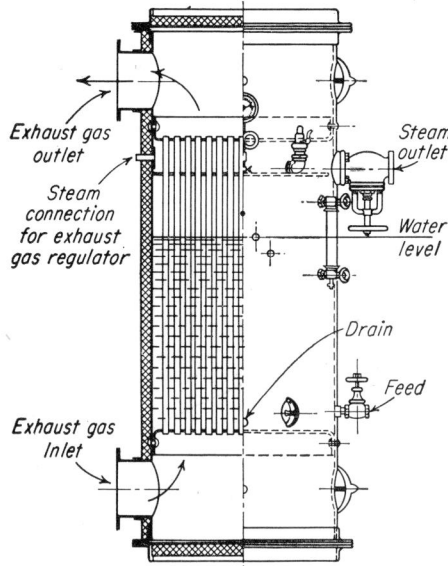

Fig. 16-19. Fire-tube waste-heat boiler.

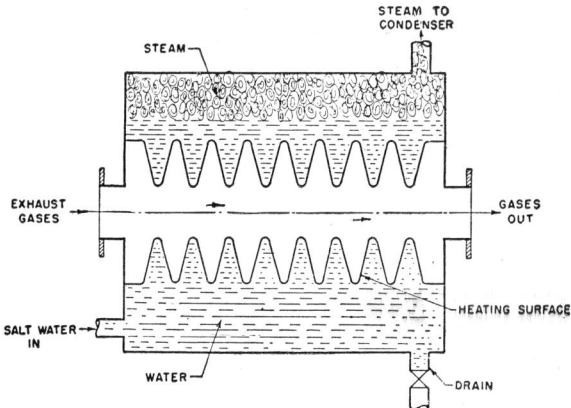

Fig. 16-20. Exhaust-gas heated evaporator.

whirling motion of the exhaust gases, such as the spiral vanes shown in Fig. 16-22. The centrifugal force due to the whirling motion throws solid particles, after the gases have entered the inner casing, toward the inside wall of this casing. From there they slide down and fall into a conical chamber from which they are removed periodically. The exhaust gases are usually admitted off center but are discharged through a central passage. Spark arresters are sometimes incorporated in a dry muffler.

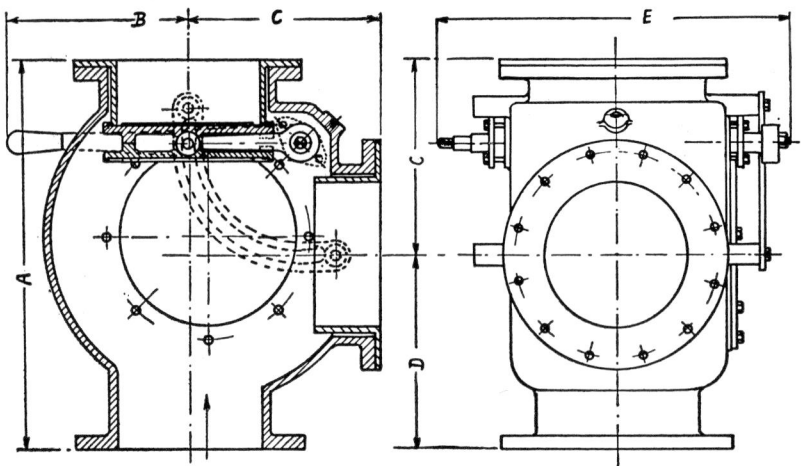

Fig. 16-21. By-pass valve for exhaust line.

16-9. Exhaust and Scavenging in Two-stroke Engines. *General Remarks.* The performance of a two-stroke-cycle engine, meaning the maximum sustained mean effective pressure and the specific fuel consumption, depends upon the type of scavenging used (Sec. 2-6) and also to a great extent upon the size, shape, and arrangement of the exhaust and scavenge ports. While very little, if anything, can be done to improve the performance of a given two-stroke engine, two engines of the same general type, but designed and built by two different engine manufacturers, may show considerably different performance characteristics, depending on how skillfully they are designed and manufactured. The difference between two-stroke engines of different make is much greater than between corresponding four-stroke engines. Therefore, before deciding to order and install a two-stroke

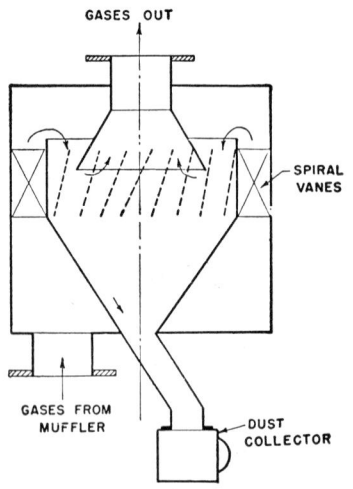

Fig. 16-22. Spark arrester.

engine, the specifications and guarantees of the engine builders should be carefully studied.

Pressure Waves. In order to permit the pressure in the cylinder to drop below the scavenge-air pressure in the shortest possible time, the opening of the exhaust ports or valves occurs much earlier and more suddenly than in four-stroke engines. Therefore, the pressure wave created in the exhaust line is much more pronounced and has a considerably higher amplitude.

Figure 16-23 shows an indicator diagram of a medium-sized single-cylinder crankcase-scavenging diesel engine, and Fig. 16-24 an offset diagram taken with the same indicator, only using a very soft spring, from the exhaust line of the same engine. Figure 16-24 shows that at the moment of pressure release point a—opening of the exhaust

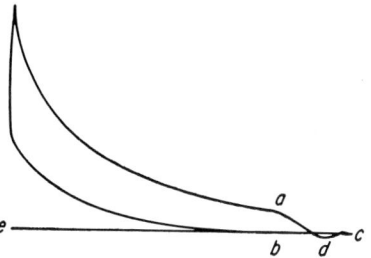

FIG. 16-23. Indicator diagram of a two-stroke diesel engine.

port—the pressure in the exhaust line is slightly below atmospheric; while at point d, at which a vacuum is created in the cylinder, the peak of the first pressure wave reaches the exhaust line; a second pressure peak occurs slightly after dead center, point c; after closing of the exhaust ports, point b, the pressure in the exhaust line gradually falls, owing to the inertia of the gases flowing out. The frequency of the pressure waves, meaning the reciprocal of the time lapse between peaks, depends upon the gas velocity and length of the exhaust pipe and can materially help or hinder the scavenging of the cylinder.

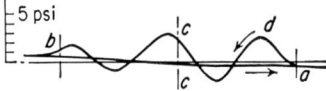

FIG. 16-24. Offset diagram from the exhaust line of a two-stroke diesel engine.

Exhaust Line. Length. While for a four-stroke engine the best pipe length is the shortest, for a two-stroke engine this is not so. There is always a certain length which will give the lowest back pressure in the line when the gases start to discharge on the next cycle. For a 14- by 17-in. two-stroke crankcase-scavenging engine turning at 300 rpm, the best length is around 45 ft. A length greater or less than this increases the pressure at the exhaust ports at the instant these are uncovered in the next engine cycle. When more than one cylinder connects to the common line, the best length can be ascertained only by experiment.

This sensitiveness to the exhaust-pipe length explains why two-stroke engines often do not have an exhaust manifold and discharge the gases by separate pipes to a common pit or muffler. Such a down pipe with a water jacket partly welded on (elbow b) is shown in Fig. 16-26; the jacket d is bolted to the elbow flange and at the bottom held by a packing e and gland f.

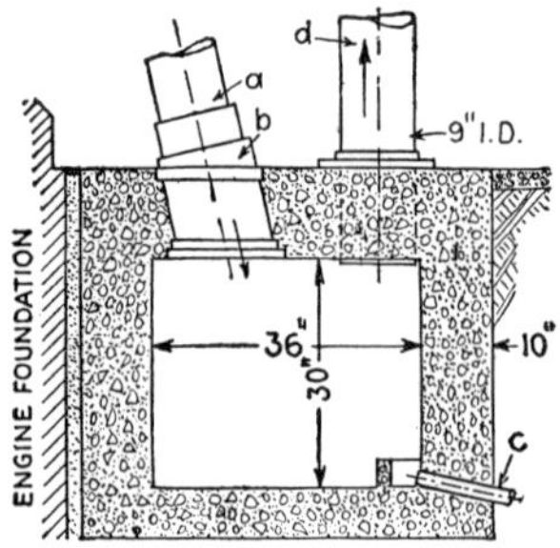

Exhaust Mufflers. The general design of mufflers is the same as in four-stroke engines. However, in two-stroke engines a low back pressure is more important and the selection of a suitable muffler is therefore much more critical than with four-stroke engines. Large volume is an important requirement.

FIG. 16-25. Exhaust pit for a two-stroke engine.

Concrete Mufflers. In Fig. 16-25 is shown the concrete pit which serves as muffler for the engine without a manifold and with separate exhaust pipes a (shown in Fig. 16-26). The pit has a separate intake casting b for each cylinder and one exhaust stack d; c is the drain pipe for condensate.

Figure 16-27 shows a muffler for a 200-hp two-cylinder horizontal engine. The upper part is tapered inward to permit the use of a smaller, easier to handle, cast-iron cover plate; noteworthy is the unusually large volume—about sixty times the piston displacement. The exhaust stack is reduced to 6 in. since in this case the discharge of the gases to the atmosphere is very steady.

Concrete mufflers along the lines of Fig. 16-11 are also satisfactory if their volume is sufficiently large.

16-10. Exhaust Temperatures. In a multicylinder engine, the temperature of the exhaust gases as they leave an engine cylinder is the best means of detecting engine trouble. If a cylinder is not receiving its share of fuel, the exhaust temperature of this cylinder will be lower than the temperatures from the other cylinders, if all other conditions are the same. Conversely, a high exhaust temperature may be caused by faulty combustion, which has continued so long during the stroke that the exhaust gases are too hot. This may be traced to a faulty injection valve or to low compression. The latter may be caused by gummed piston rings, a worn piston, or leaky valves.

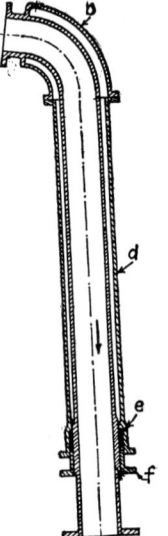

FIG. 16-26. Two stroke water-cooled exhaust pipe.

A well-trained engine operator reads the exhaust-gas temperatures hourly, and at a difference of over 25 F between readings of the cylinders of an engine, at once starts an investigation.

Pyrometers. Universally, gas temperatures are read by electric pyrometers. These instruments are based on the principle that when pieces of two dissimilar metals are joined at both ends, heating of one juncture, while the other juncture is kept at a lower temperature, will cause a flow of electric current. A pair of wires of two such metals, connected at one end and insulated electrically, is called a *thermocouple.* The thermocouple is incased in a tube which is screwed into the exhaust connection of the engine cylinder. Two wires lead from the thermocouple to a millivoltmeter, which indicates the very small electromotive force, sometimes designated as emf and measured in millivolts,

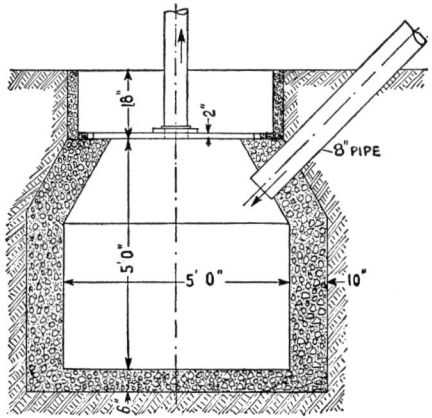

FIG. 16-27. Concrete muffler for a 2-cylinder 200-hp two-stroke engine.

that is produced by the flow of the electric current. The dial of the pyrometer is marked directly in degrees Fahrenheit instead of in millivolts, having been calibrated by the instrument maker. Of course, the warmer the gases, the greater will be the emf (millivolts) and the pyrometer dial will read a higher temperature.

Pyrometers are rugged and reliable, but if carbon settles on one of the thermocouples it will insulate the latter from the hot gases and the pyrometer will read low. The thermocouples should be inspected and carbon deposit removed at six-month intervals.

The pyrometer has a switch with a series of contacts, which enables one instrument to be connected to all the thermocouples on one, or even two, engines.

Exhaust Temperature Readings. Each engine design will show different temperatures at a given percentage of load rating, depending upon several factors. The system of fuel injection and the degree of atomization given by the fuel spray nozzle will affect combustion and, consequently, exhaust temperatures. The weight of air used per pound of fuel burned will also influence exhaust temperatures. Since the weight of air taken per stroke is approximately the same at all loads but the amount of fuel decreases with a decrease of load, the exhaust temperature decreases with a decrease of the load.

Four-stroke engines have exhaust temperatures at full load of around 750 F. Figure 16-17 shows temperatures observed at different loads on engines of two types.

It is advisable to require the engine builder to give normal temperatures at full, $3/4$, and $1/2$ load for each diesel engine.

QUESTIONS

1. Explain the pressure fluctuation in the exhaust pipe line.
2. Explain how an increase of the back pressure affects the mean effective pressure of an engine.
3. What is exhaust-valve overlapping in a multicylinder diesel engine and to how many degrees can it amount?
4. Enumerate the functions that the exhaust system of a diesel engine may perform.
5. Enumerate the parts of the exhaust system which are usually present.
6. Indicate what parts are sometimes added to the exhaust system for special purposes.
7. State the conditions necessary for a proper functioning of exhaust valves and explain how they may be lost in operation.
8. What does a short flame coming out of an exhaust test-cock indicate?
9. Enumerate the objects of watercooling exhaust manifolds.
10. Show a sketch of a coupling which is recommended for inclusion in the exhaust piping to allow for its heat expansion.
11. Explain the reasons for insulating exhaust pipes and the methods used.
12. What are the two main types of engine mufflers?
13. Draw a sketch of a dry muffler as used with small engines.
14. Draw a sketch of a welded-steel muffler and state two methods for reducing the ringing sound of the inrushing gases.
15. Draw a sketch of a so-called *snubber* and explain the principle of its action.
16. Explain the action of an absorbent silencer.
17. Show a sketch of a concrete muffler and state its advantage for larger engines.
18. Show a sketch of a wet-type muffler and explain the principle of its action.
19. Draw a sketch of a vertical wet muffler without a metal bottom and explain its action.
20. State the reason why wet mufflers, in spite of their effectiveness, are used only in special cases.
21. Draw a sketch of the recommended shape of the tail-pipe end and explain its action.
22. What two basic types of heat exchangers are used for waste-heat recovery from a diesel-engine?
23. Enumerate the factors upon which the amount of waste-heat recovery depends.

EXHAUST SYSTEM

24. Express the amount of recoverable heat Q in the form of an equation that connects the various factors upon which it depends.

25. State how much, on the average, can be recovered from a diesel engine per horsepower per hour: (*a*) heat, Btu, (*b*) water at 200 F, pounds, and (*c*) low-pressure steam, pounds.

26. Draw a diagrammatic sketch of an evaporator for obtaining distilled water from sea water.

27. Draw a sketch of a by-pass valve for use with a waste-heat boiler.

28. Draw a diagrammatic sketch of a spark arrester.

29. What determines the proper length of exhaust piping for a two-stroke engine?

30. What is the main requirement for a muffler for a two-stroke engine?

31. What information can be obtained from the exhaust-gas temperatures of a multicylinder engine?

32. What kind of pyrometers are used for exhaust-temperature measurements?

33. Draw a diagram of typical exhaust-gas temperatures for a four- and a two-stroke engine.

CHAPTER 17

STARTING AND REVERSING METHODS

17-1. Requirements. When a diesel engine is started, its crankshaft must be turned over by some outside means so that the air in the cylinder is compressed at top dead center to such a pressure and resulting temperature that, when fuel is injected, it will ignite and produce a power stroke.

There are two important requirements to be fulfilled for sure and quick starting: sufficient speed and a correct compression ratio.

Sufficient Speed. If the engine is turned over very slowly, the unavoidable small leaks past the piston rings and possibly through the intake and exhaust valves will permit part of the air to escape from the cylinder during the compression stroke. This may lower the compression pressure and temperature at the end of the stroke below those necessary to ignite the injected fuel. The heat loss from the compressed air to the metal walls of the cylinder and compression space is also greater at low speeds—when the duration of the compression stroke is longer—and this further lowers the temperature of the air. There is, therefore, a minimum speed which an engine must attain before ignition can occur and the engine can start firing. The starting speed depends upon the type and size of the engine, its condition, and the temperature of the surrounding air. In some engines the starting speed is around 70 to 75 rpm, whereas in some small engines it may be as high as 250 or even 300 rpm. There is no definite relation between the starting and operating speed of an engine. However, all other conditions being equal, an engine starts at a lower speed when it is in its best operating condition, has well-seated piston rings and valves, has correct timing, and when there is no excessive friction in the engine or in the engine-driven auxiliaries.

Sometimes, if an engine cannot be started owing to excessive leakage past the piston rings, pouring a little lubricating oil on the cylinder walls may sufficiently seal the piston rings to start the engine. Naturally, at the first opportunity the engine must be overhauled and proper compression restored.

Correct Compression Ratio. If the compression ratio is not sufficiently high, the final temperature of the compressed-air charge will again be too low for ignition. A new engine, naturally, has the correct compression ratio. However, wear of bearings may lower the piston position somewhat and, hence, the compression ratio. A retarded closing of the intake valves, caused by incorrect take-up of wear in the valve mechanism, or some other error in the valve timing, also may lower the effective compression ratio.

17-2. Starting Methods. Diesel engines, generally speaking, are started by hand, by an electric motor, by an auxiliary gasoline engine, by compressed air, or by an explosive cartridge.

Hand starting can be used only for small engines, having a bore not over 4 in. and one or two cylinders.

Electric Starting. The power required to turn the crankshaft of a cold diesel engine and then to bring it up to starting speed generally is a little less than 10 per cent of its rated output but in some cases may amount to as much as 20 per cent, especially with small engines. With large engines it is smaller, down to about 3 or 4 per cent of the rated output.

Electric starting systems use direct current because the electrical energy in this form can be stored in batteries and drawn upon when needed for starting. After starting, the battery is recharged with an engine-driven generator.

The electric starting system consists of (1) a storage battery, (2) a direct-current (d-c) electric motor, (3) a mechanical engagement between the motor and the engine crankshaft, (4) an auxiliary electric generator to charge the battery, if the engine does not drive a d-c generator as its main load, and (5) the cables, wires, and switches necessary to complete the electrical system.

Storage Battery. For small engines the current is supplied by a storage battery, usually of 6, 12, or 16 volts with a 110- to 150-amp-hr capacity. Locomotive and marine diesel engines use storage batteries of 24 or 32 volts, sometimes of 64 and even 112 volts with a capacity from about 175 to 400 amp-hr.

Electric Motor. The motor is usually of the series-wound heavy-duty type, which, for a short time—a few seconds—can carry a 100 per cent overload.

To prevent overheating of the starter motor from the heavy current drawn, some electric starting systems have a timing device in conjunction with the magnetic switch to disconnect it after about 15 sec if the engine does not start. It should be noted that electric starters are designed to produce maximum power for very short periods and should not be operated for more than 30 sec at a time.

The current required to start a small diesel engine, such as a truck or tractor engine, is quite large, over 500 amp, as shown in Fig. 17-1, and therefore is not put through the ammeter. The resistance to turning the engine increases with the viscosity of the lubricating oil, and the latter increases very rapidly with the decrease of temperature.

Mechanical Engagement. The electric motor is connected to the engine by means of spur gears; the pinion on the motor shaft engages the large gear which is part of the flywheel. Most mobile and other

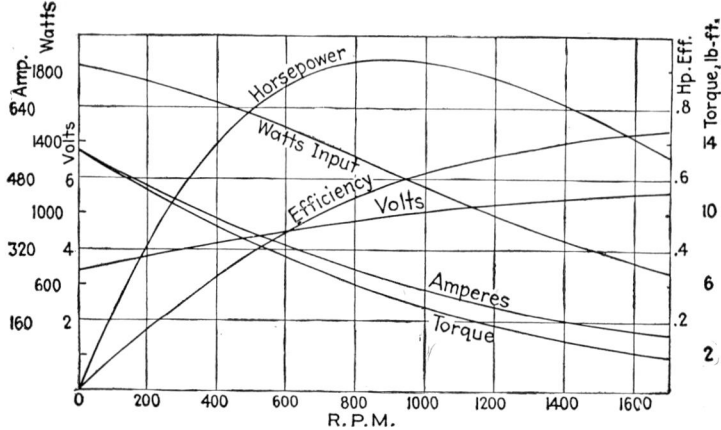

FIG. 17-1. Typical characteristics of an automobile-starting motor.

smaller engines use the Bendix or similar automatic pinion-shift mechanism, originally developed for passenger-car gasoline engines. The mechanism consists of a shaft fastened to the motor shaft by means of a flexible coupling. This shaft carries the pinion and has the form of a screw, or helix, with a steep rectangular thread. The pinion is connected to the shaft by a corresponding helical spline that forces the pinion, while turning freely on the screw shaft, to move axially approximately 1½ in. When the starter pedal is depressed, it closes a switch between the battery and the motor, whose armature begins to spin. The inertia of the pinion, assisted by a special counterweight, causes the pinion to lag from the speed of the motor shaft. This turns the pinion on the spline, moves it away from the motor, and brings it in mesh with the teeth cut in the rim of the engine flywheel, which it begins to turn. When, after a few turns, the engine starts to fire, the peripheral speed of the flywheel immediately becomes greater than the peripheral speed of the pinion, and the latter is pushed back on its helical spline and disengaged from the flywheel teeth.

Generator. The battery is recharged by an electric generator of the constant-voltage type, which operates when the engine is running. To prevent the discharge of the battery when the engine is running slowly or not at all, the generator is equipped with an automatic cutout. If a generator is operated at a greatly varying speed, as in a truck or locomotive engine, its average efficiency will be rather low, varying at best from 30 to 60 per cent. In conjunction with the comparatively low efficiency of the starting motor, as shown in Fig. 17-1, and the battery losses, the energy which must be supplied by the automotive diesel engine for each starting operation is rather high. If the engine is often stopped and started, the additional fuel consumption for charging the battery may amount to an appreciable part—2 to 3 per cent—of the total fuel consumption.

The charging characteristics of a 150-amp-hr 32-cell 64-volt battery for cranking a locomotive engine are shown in Fig. 17-2. The auxiliary generator for charging the battery is equipped with a device that gradually reduces the ampere input after the battery has reached its full voltage, which is limited by another device. This takes place when the battery is about 80 per cent charged, and after this the current input gradually tapers down to about 5 per cent.

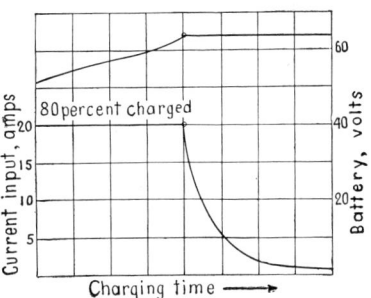

FIG. 17-2. Charging diagram of a storage battery.

Cables and Switches. Because the voltage used for starting motors is comparatively low, the electric current drawn during starting may rise to values of 500 amp or more, so that heavy cables and switches must be used. In order to keep the electric losses as low as possible, the main cables are made as short as possible. The starting switches, in the larger engines, are of the magnetic type which can be remotely operated by a push-button switch and require less current to operate.

17-3. Auxiliary Gasoline Engine. Instead of using an electric motor and battery, a diesel engine can be cranked by a small 3- to 5-hp gasoline engine which operates a Bendix drive similar to the one used with an electric motor. This drive is connected to the diesel engine by V belts or gears. The gasoline engine is connected to the V belt or gear drive by means of a friction clutch and is started by hand cranking. When it attains the proper speed, the clutch is let in and the diesel engine is turned over until it begins to fire. When the diesel engine gets under way, the peripheral speed of the flywheel turns the Bendix screw and disengages the drive from the flywheel. After this the gasoline engine is shut down. This method is used for engines with a bore up to 8 in.

It may also be mentioned that the warm water from the gasoline-engine jacket can be used in cold weather to preheat the jackets of the diesel engine and thus facilitate its starting.

Some diesel engines have the starting gasoline engine incorporated in the main engine (Fig. 17-3). Turning the crank 1 opens valve 3 leading to the auxiliary chamber 4. This added space reduces the compression ratio to about 6:1, which is suitable for a gasoline engine.

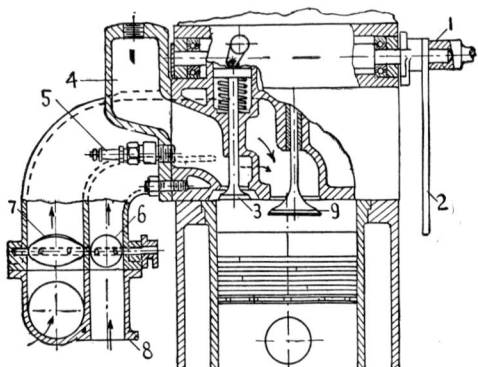

Fig. 17-3. Gasoline starter of a diesel engine.

As the crankshaft is turned, usually by hand, butterfly valve 6 permits the drawing of an air-gasoline mixture from a carburetor attached to flange 8, through the intake valve 9. A high-tension magneto supplying the current to the spark plug 5 is also engaged at the same time. When the engine has made about 700 revolutions, rod 2 automatically releases the shaft turned by the crank 1. This cuts out the auxiliary chamber and magneto and turns the double butterfly valve, shutting off the flow of the air-gasoline mixture, and opens wide butterfly valve 7, admitting air. The engine begins to operate as a compression-ignition oil engine.

17-4. Air Starting. Stationary and marine engines of over 25 hp in most cases use compressed air for starting purposes. The reasons are that compressed air is cheap to produce and easy to store and, being a gas, behaves during expansion similarly to the combustion gases in the cylinder.

Compressed-air starting is particularly suitable for large diesel engines which require the use of considerable energy in a short time. Compressing air into tanks and using the air from the tanks can furnish the required energy in any amount needed. The compressed-air tanks and the whole starting valve gear do not add much extra bulk or weight to the engine, and this method is generally used on large engines.

The usual air pressure is 150 to 300 psi. Air-injection engines have high-pressure air compressors and, in order to reduce the size of air tanks, use air pressures from 500 to 750 psi.

The volume of the air tanks necessary to start an engine may be taken as fifteen to twenty-five times the total piston displacement for small engines, decreasing to about seven to ten times the piston displacement for large engines. As a minimum requirement this volume may be taken as 30 times the displacement of one cylinder of the engine.

The compressed air used for starting can be restored over a comparatively long period of time, after the engine has been started; the air compressor can thus be small and does not require much power. The compressor may be driven directly from the engine or from a separate power source, such as a small hand-started combustion engine or an electric motor. A separately driven air compressor is always installed with marine diesel engines, in addition to an engine-driven compressor, to insure a positive supply of starting air.

Engine-driven compressors are used in marine engines, in which compressed air is needed for other purposes than starting the engine, and in truck engines where the truck and trailers are equipped with air brakes.

Starting Valves. The admission valves for starting air are of the poppet type and are mounted in the cylinder head.

Depending upon the method of operating these valves, air starters fall into three general groups: (1) hand-operated, for single-cylinder engines, (2) cam-operated, for medium-sized engines, and (3) compressed-air-operated by means of cam-operated pilot valves, for all larger engines.

Hand-operated Valves. Single-cylinder engines frequently are fitted with a simple lever-operated globe or gate valve. This is done in order not to increase the price of the engine by the cost of a more elaborate system.

Before starting such an engine, it must be turned over until the piston is just past dead center on the power stroke. Air is admitted through a port in the cylinder head by pulling the lever of the air valve. The valve is held open while the piston makes about one-half of its stroke. The piston is moved at first by the pressure of the air in the tank—about 150 to 250 psi—and from the closing of the air valve to the end of the stroke by the pressure of expanding air. The momentum given the flywheel stores up enough energy in it to enable it to continue to turn the crankshaft, bringing the piston back and

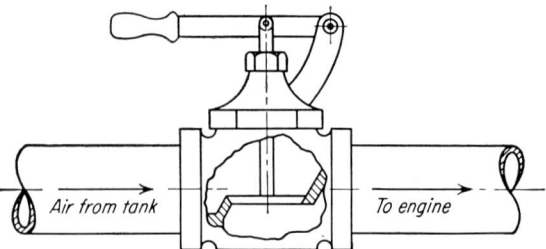

FIG. 17-4. Manually-operated air-starter valve.

over dead center. The engine operator continues to open and close the valve until the engine starts to fire the fuel.

The drawback of this system is that it requires from the operator a certain skill in opening and closing the air valve. If this is opened too late, the engine will not come up to speed quickly and, if the valve

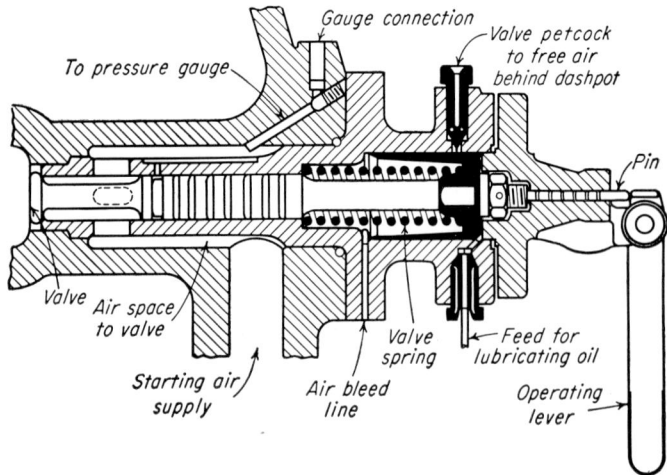

FIG. 17-5. Manually-operated air-starter valve combined with check valve.

is opened too early, the engine may start to rotate backwards. A hand-operated valve of this type is shown in Fig. 17-4; it requires a poppet check valve in the cylinder head.

Figure 17-5 shows a hand-operated valve combined with a check valve in one unit. The valve is held to its seat in the cylinder head by a heavy spring. Since the valve stem is enlarged to the inside diameter of the valve seat, the valve is balanced and the air pressure does not affect it. The operator opens it by pushing in the pin when he pulls the operating lever to the right. The valve spring closes the valve as soon as the lever is released.

Hand-operated air valves are used mostly on two-stroke engines. They are more difficult to operate on a four-stroke engine because of the danger of opening it on the suction stroke and thus wasting compressed air.

17-5. Cam-operated Valves. Engines with two and more cylinders are started by valves similar to the one shown in Fig. 17-5 but operated by cams, located on the engine camshaft, through valve-actuating gear, push rods, and rocker arms, in the same manner as the intake and exhaust valves. The cams have a contour to open the air-admission valves 3 to 8 deg after top dead center and to close them about 55 to 45 deg before bottom dead center, in four-stroke engines. To make the turning over of the engine easier, in some four-stroke engines the compression is released during starting by slightly opening the exhaust valve near the middle of the stroke.

Two-stroke engines require less air for starting because of the absence of the exhaust and suction strokes which are made at the expense of the flywheel energy. In these engines the air-admission valves are closed 85 to 75 deg before bottom dead center.

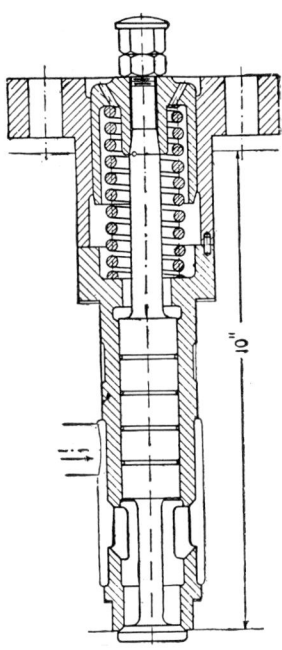

Fig. 17-6. Mechanically-operated air-starter valve.

In stationary multicylinder engines, starting valves are put on only two or three, and not more than one-half of the cylinders. The cylinders without starting valves fire first, because they are not chilled by the expansion of the starting air. In marine engines with ten cylinders or less, all cylinders are equipped with air-starter valves. In engines with twelve and sixteen cylinders only one-half the cylinders usually have air-starter valves. In double-acting engines air-starting valves are located in each cylinder, but only in one head, usually in the upper one.

A mechanically operated air-starting valve, as used in smaller and medium-sized engines, is shown in Fig. 17-6. No stuffing boxes are used on the stem to avoid the danger of causing the valve to stick. With a good ground fit of the stem and labyrinth grooves in it, which fill up with lubricating oil, the leakage of compressed air to the atmosphere is very small. The size of the pipe connecting compressed air to the valve must be equal to or slightly larger than the valve diameter. The force exerted by the spring when the valve is closed must be slightly larger than the projected area of the valve seat times the air pressure.

The air valve may be engaged to the valve-actuating rocker arm by two methods: by moving the rocker-arm or by lifting the valve top.

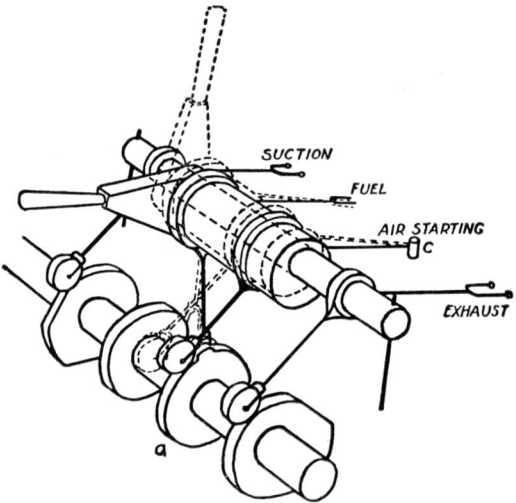

FIG. 17-7. Diagram of shifting air-starter control.

In the first method the rocker-arm shaft has the fulcrum for the air valve rocker-arm in the form of an eccentric. When the engine is in regular operation, the eccentric lifts the rocker-arm end with the roller so that the air-starter cam does not touch the roller. For starting, the rocker-arm shaft is turned by a lever about 90 deg, the eccentricity of the fulcrum points down, the rocker-arm is lowered, and its roller comes in contact with the cam. If the piston of this cylinder is set just beyond top dead center, the nose of this air-starter cam will hold the starting cam open and air will be admitted into the cylinder. The engine begins to turn. The camshaft now turns and eventually removes the nose of the cam from underneath the rocker-arm roller, whereupon the air valve is closed by its spring.

Figure 17-7 shows a diagram of the eccentric fulcrum-shaft and rocker-arm arrangement. Cam *a* and rocker *c* control the air-starting valve. The lever is shown in the *starting* position; the air rocker-arm roller rests on cam *a* and the fuel cam is pulled away from its cam. When the engine begins to fire on the other cylinders, the handle is turned to the position shown by broken lines.

To simplify the air-starting mechanism in larger engines, valve *d* is engaged by compressed air admitted to a small air cylinder *a* (Fig. 17-8). The force of the spring *b* is such that it keeps valve *d* closed during the suction stroke. When, before starting, compressed air is turned on, it pushes the piston *a* up and prevents valve *d* from opening and admitting air into the engine cylinder. When, in turning, the air-starter cam pushes down the rocker arm whose end is in contact with the top of piston *a*, the air valve *d* is permitted to open and admit air into the cylinder until the cam turns to a position in which the rocker-arm end moves away. Since with such a valve the engine is started without shutting off the fuel injection, the engine is relatively easy to start. When the engine begins to fire, the globe valve in the pipe line from the tank is closed and a small valve in the line near the engine is opened to release the air trapped in this portion of piping; piston *a* comes down and a spring removes the rocker-arm end from the air-starter cam.

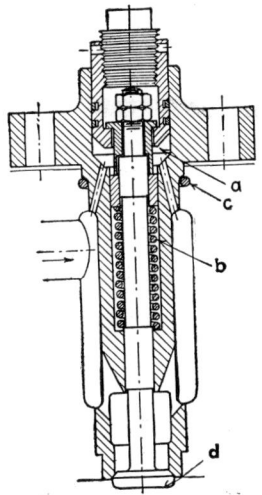

Fig. 17-8. Automatic air-starter valve.

By far the majority of engines up to 1,000 hp employ this general type of starting equipment. The arrangement is simple and functions reliably. The operator should see that the valve does not remain cocked open when the cam releases it. If it does remain open, high-pressure air will flow into the cylinder during the power stroke, which will mean the loss of much of the air in the storage bottles. Occasionally a little kerosene should be applied to the valve stem to free it of gum.

Leaks at the air valve are exceedingly objectionable, for they will allow air to leak into the cylinder during the compression stroke when the engine is started, and, as a result, the compression pressure may rise to a dangerous point.

Ordinarily, air leakage can be detected by a gradual loss in air pressure. The operator should turn on the air into the lines before he shifts the control to the starting position, and then close the valve at the air bottle. If the starting valve leaks, the pressure gauge will start to fall.

17-6. Air-operated Valves. In the case of starting valves operated by compressed air, the valve in the cylinder head is simply a check valve (Fig. 17-9a) and the air admission is controlled by a separate poppet valve operated by a cam c (Fig. 17-9b). When the engine begins to fire, the roller between the cam and valve stem is pushed down, out of contact with the cam.

In multicylinder engines, the air admission is controlled either by a separate poppet-valve and cam for each cylinder or by a rotating distributor. A slot in the distributor disk or sleeve admits air successively to a number of ports, connected by pipes to the individual cylinders.

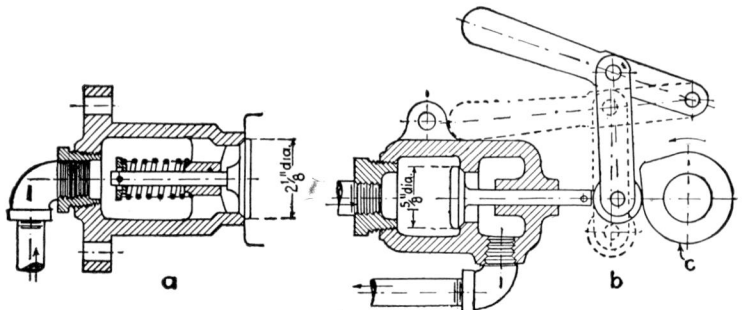

FIG. 17-9. Air-operated starter valve.

A more elaborate system of air starting, employed where two, three, or four cylinders of a multicylinder engine have air starting equipment, is shown in Fig. 17-10.

Air is stored in a tank at 400 psi pressure, and when the gate valve is opened, it flows to the valve box. From this a pipe line leads some of the air to the starting valve.

After the operator has barred over the engine to the starting position, he moves the throttle handle. This closes the bleed port of the starting valve and lifts the latter from its seat. This allows air to flow through pipe E to the ram of the starting air valve. The ram moves upward and opens both the starting air valve and the lifting valve. Opening the starting air valve permits the air from the tank to flow through the valve box into the air-starting manifold and then to the main air-starting valve.

When the lifting valve is lifted, air is admitted through pipe K to the pilot valve. This valve is held closed by a spring in the case until the cam, driven by the camshaft of the engine, opens it to the position shown in the drawing. When the pilot valve is opened by the cam, air passes from pipe *K* through the pilot valve and pipe *F* to the piston on the main air-starting valve, which then opens. Air flows through the manifold and the main air-starting valve into the cylinder as long as the cam holds the pilot valve open. Air flowing into the engine cylinder causes the piston to move and the crankshaft to rotate.

When the rotation of the engine causes the cam to release the pilot valve, the latter snaps closed and the air in pipe *F* blows out through the release port of the pilot valve. This then causes the main air-starting valve to close, thereby shutting off the supply of air to the

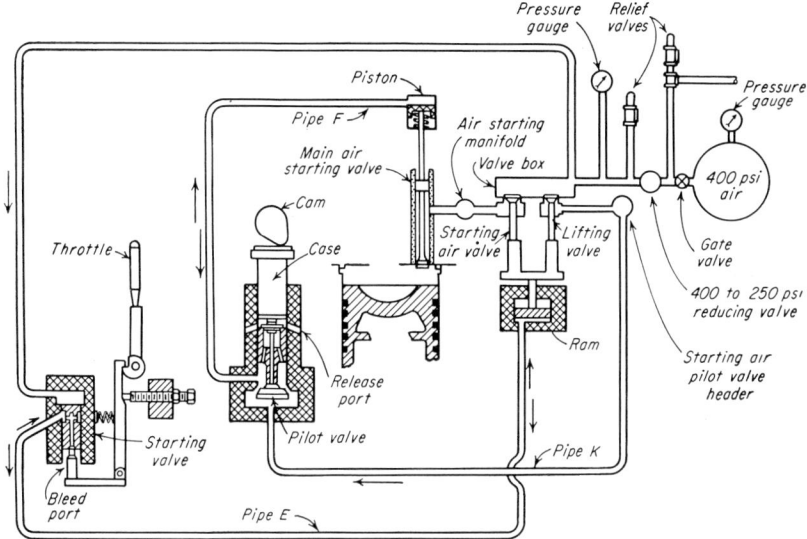

Fig. 17-10. Air-starting system of a Cooper-Bessemer diesel engine.

cylinder. This cycle of opening and closing of the main air-starting valve by the action of the cam on the pilot valve continues until the operator shifts the throttle to permit the starting valve to seat. The bleed port then lets the air in pipe *E* leak out, and with this pipe pressure removed, the ram is pushed down by a spring (not shown), closing the lifting valve and the starting air valve. Closing the lifting valve stops the flow of air to the pilot valve and puts the piston out of action. Closing the starting air valve stops the flow of air to the main air-starting valve.

The pilot valve and the body are made of bronze and fit together closely. In normal operation, with clean starting air, they will require no attention other than routine inspection. If, however, the starting air contains grit, scale, or any other abrasive that cuts the seats, they should be ground in. A good grade of grinding compound should be used sparingly and care should be taken that none of it finds its way to parts other than the seat.

If the oil becomes gummy, the valve will operate sluggishly. Since it will not close properly, it will act as if it were out of time. When the starting air is opened to the engine, all the pilot valves should jump toward the camshaft snappily, and come away readily when it is shut off. They will not be so quick in the latter operation as the former, since the air in the starting header or manifold tends to hold them toward the camshaft until it all leaks out.

If the pilot valves persist in following the cam after the starting air is shut off, this shows that pressure remains in the pilot headers and is coming from the pilot air throttle.

If the pilot air throttle is leaking, the stop valve to the engine can be closed to prevent loss of air and operation of the pilot valves.

Moisture in the starting air, together with a little oil carried over from the air compressor, may cause a gummy deposit to form on the inner surfaces between balance piston and valve. During operation the lower part of the valve that is exposed to combustion gases gets warm and bakes this gum into a scale, which in time builds up sufficiently to tend to cause the balance piston to stick. This scale cracks off and is sometimes caught between valve and its seat when the engine is started. This permits combustion gases to enter the air-starting manifold; if the engine is allowed to operate in this manner more than a few minutes, it will heat up and paint will be burned from the manifold. If it is necessary to keep the engine running, fuel should be shut off from this particular cylinder.

17-7. Starting Position. A four-stroke engine can be started from any position of the crankshaft only if it has five or more cylinders, each equipped with an air-starting valve. All other four-stroke engines must first be turned to a position such that one of the pistons is on the downward stroke, slightly past dead center and ready to take the starting air. On the other hand, a two-stroke engine can be started from any position of the crankshaft if it has three or more cylinders, all with air-starting valves.

In small engines the crankshaft is turned to the necessary position by turning the flywheel, either directly by hand or by means of a bar inserted into holes on the flywheel rim. Larger engines are turned either by a worm which is engaged during this operation with teeth cut in the flywheel rim or by a ratchet mechanism, or jack (Fig. 17-11). In large engines the worm may be operated by an electrical or compressed-air motor and the ratchet mechanism by a reciprocating compressed-air motor.

The operation of the ratchet jack is very simple: When the lever a is pushed down, the pawl b turns the flywheel in the direction of the arrow and pawl c goes down, sliding over the teeth in the rim. When lever a assumes a horizontal position, the upper end of pawl c drops into a recess in the rim; when the lever is pulled up, pawl c is turning the flywheel and pawl b is going down until its upper end drops into a recess, and the pumping cycle with lever a is repeated. It is important to remove the pawls from the rim before starting the engine.

17-8. Cartridge Starting. A cartridge engine starter consists of two separate devices, the breech and the starter proper. In the breech, a cartridge ignited by an electric current produces high-pressure combustion gases. These gases are admitted through a pipe connection to a piston which, by means of a series of helically splined shafts, first engages the starter jaw with the clutch jaw fastened to the end of the engine crankshaft. When the two jaws are engaged, the continuing motion of the piston produces a rotary motion of the jaws and turns the engine crankshaft, thus starting the engine. When the piston reaches its extreme position, the gases are released through an exhaust valve, the piston is returned to its initial position by a spring, and the starter jaw is disengaged from the crankshaft jaw.

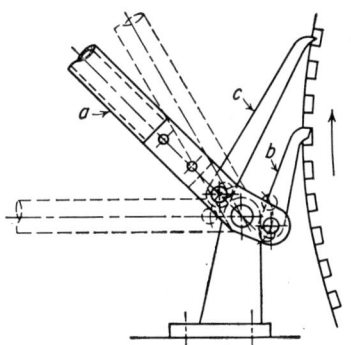

FIG. 17-11. Flywheel-barring mechanism.

17-9. Low-temperature Starting. Diesel engines are, generally speaking, built to operate and be started at an average atmospheric temperature around 70 F. When a diesel engine must be started after it has been shut down for several hours at a temperature considerably lower than 70 F, two kinds of difficulties may occur: (1) the starting device may not turn the engine fast enough to produce the necessary compression temperature and pressure, or (2) the compression temperature may be below the required minimum for igniting the fuel, even though the engine turns fast enough.

Slow turning results from an increase of the viscosity of the lubricating oil. The frictional resistance to turning of an engine is directly proportional to the lubricating oil viscosity, and can therefore be decreased by lowering the viscosity of the oil. The main friction being in the oil film between the piston and cylinder walls, the most effective remedy is to heat the cylinder walls by raising the temperature of the water in the jacket. Some small engines have an electric heater built into the cylinder jacket when the engines are intended for operation in a cold climate. Another procedure is to admit to the lubricating-oil pressure pump oil with a low viscosity just before stopping the engine in order to supply a low-viscosity oil-film between the pistons and cylinder walls. Sometimes even the lightest diesel lubricating oil is still too heavy and it may be advisable to dilute it slightly with gasoline, which will evaporate during regular engine operation, or with about 10 per cent kerosene.

Low Air Temperature. If other factors remain constant, the final temperature of compression is directly proportional to the initial temperature of the air charge. How much a drop in temperature affects the final temperature of compression may be illustrated by the following example.

If the initial air temperature is 70 F, the final compression temperature in a cold small diesel engine with a compression ratio of 14:1 that is being turned over will be about 700 F. The question is to find the final compression temperature if the initial air temperature drops, say to 30 F.

Such calculations must be made with all temperatures expressed in absolute degrees or degrees Rankine. The temperature in degrees Rankine, R, is equal to the temperature in degrees Fahrenheit plus 460 degrees.

Thus the normal room temperature will be $70 + 460 = 530$ R, and compression temperature will be $700 + 460 = 1160$ R, or a ratio of $1160/530 = 2.19$.

Since for any compression ratio the final temperature after compression is proportional to the absolute temperature at the beginning of the compression, a lower intake temperature of $30 + 460 = 490$ F will yield a compression temperature of $490 \times 2.19 = 1073$ R, or $1073 - 460 = 613$ F.

Thus a drop of the intake temperature of 40 F lowers the final compression temperature by $700 - 613 = 87$ F; this may, under certain conditions, prevent the ignition of the injected fuel.

Methods and devices used to overcome low intake-temperature effects in diesel engines may be listed as follows: (1) electric intake-air heaters, (2) electric flame primers, (3) glow plugs, (4) punks, (5) fuel-pump overload devices, (6) injection of ether or gasoline into the intake, (7) increase of the compression ratio; and (8) heating of the jacket water.

All these devices are used only in smaller engines cranked by an electric motor and batteries. Large engines are kept in engine rooms, where the temperature is not allowed to drop much below 70 F, and do not need devices for low-temperature starting.

Electric heaters for intake air consist of coils of resistance wire installed in the intake manifold and heated by current drawn from the starting storage battery. This is a good scheme that is used rather extensively.

Flame primers consist of a hand-operated fuel-oil pump, a fuel-spray nozzle, and an electric spark plug with a continuous vibrating spark

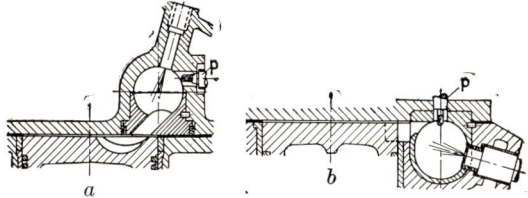

FIG. 17-12. Combustion chambers with glow plugs.

coil, with the nozzle and spark plug installed in the air-intake manifold of the engine to give a continuous series of sparks. The finely atomized fuel spray is directed against the spark-plug electrodes; when it is ignited, it heats the air drawn into the engine during the starting. This method is satisfactory and is used in some marine engines.

Glow plugs are located in the cylinder heads of the engine so that the incoming air passes by them, as shown by p, p in Fig. 17-12. They are heated by current from a regular or low-voltage storage battery (2 to 3 volts) and their action is satisfactory. However, being in the combustion space and subject to hot gases all the time that the engine is in operation, their thin wire deteriorates rather rapidly. In some engines they are used simultaneously with an electric resistance-wire air heater.

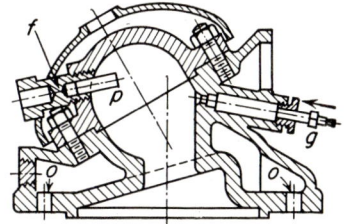

FIG. 17-13. Cylinder head of a two-stroke hot-ball oil engine.

Punks. Some engines have a steel plug f, Fig. 17-13, reaching into the combustion space opposite the fuel injection spray; the plug has an axial hole drilled into its inner end and terminates in a screw thread on the outer end, which holds it in place. Before the engine is started, this plug is screwed out, a piece of punk p is inserted into its cavity and ignited, and the plug is screwed back into the cylinder head. When the engine is turned over, the fuel spray from the injector g strikes the glowing punk and is ignited. To prevent overheating of this cylinder head, called a *hot ball*, the lower part is water-cooled through the openings o, o.

Fuel-pump Overload Devices. Regardless of the method by which a diesel oil is produced, it is always a mixture of different hydrocarbons, some more, some less volatile. By increasing the amount of fuel injected per stroke, the amount of the more volatile, more easily ignitable fuel particles can also be increased. This method requires certain changes in the engine construction and is not used often.

Ether or gasoline may be used for priming if the engine does not have an adequate cold-weather starting device or in conjunction with an air heater in extremely cold weather. A very small amount of either highly volatile fuel may be poured by hand into the air-intake opening; since it will have evaporated by the time it reaches the cylinder, it will assist in the ignition of the regular fuel injected into the cylinders. This method is effective but dangerous, as even a small amount of ether or gasoline is apt to produce excessive combustion pressures, overstrain the cylinder-head studs and, in extreme cases, may wreck the engine. Also, the storage of ether creates a great hazard.

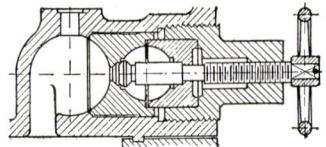

FIG. 17-14. Dual-compression chamber of Lister engine.

Increased Compression Ratio. As mentioned before, the higher the compression ratio, the higher will be the final compression temperature. Therefore, an effective means of ensuring the starting of a cold engine is to increase the compression for the starting and then to reduce it back to the normal amount during operation of the engine. This is accomplished by dividing the combustion space into two unequal parts so that a small space can be cut off from the main combustion chamber during starting.

In the Lister engine the combustion chamber is divided into two spaces by a wall with an opening closed by a valve (Fig. 17-14). When the engine is running, the valve is in its extreme right-hand position and the compression ratio is 15:1. Before starting, the valve is screwed to the left by means of a handwheel. The compression space is reduced and the compression ratio increased to 19:1, which results in a compression temperature sufficient to start the engine from cold in any weather. For starting a single-cylinder engine, the exhaust valve is kept open by means of a lifter while the engine is being spun. When the lifter is released, the engine is kept going by the flywheel, and the high compression ignites the very first fuel charge. After a few minutes of operation the valve is moved to the running position. The same method is used in the National-Superior oil engines (Fig.

17-15) by turning the plug e 90 deg. In this engine the compression ratio is also raised 4 points, from 13:1 to 17:1.

17-10. Reversing. From about 200 hp up, marine diesel engines with six and more cylinders are usually direct-coupled to the propeller and must therefore be provided with a means of reversing the direction of engine rotation for astern operation of the vessel. For safe maneuvering, the reversing mechanism must (1) quickly slow down the engine to a complete stop and (2) start it up in the opposite direction. Furthermore, this must be done against the action of the ship propeller which always turns the engine in the original direction owing to the continued motion of the ship through the water.

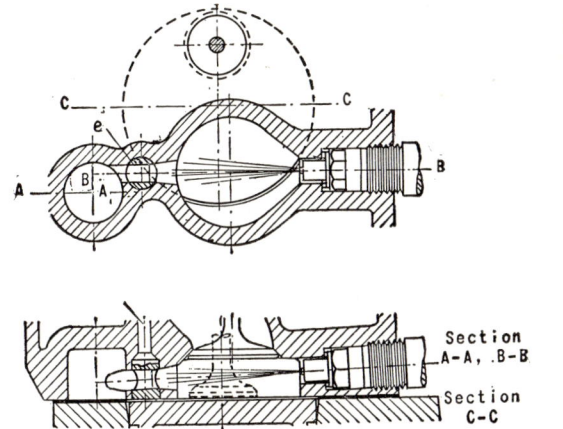

FIG. 17-15. Combustion chamber with a Ramsey energy cell.

The only satisfactory method of reversing a diesel engine direct-coupled to a ship propeller is by the use of compressed air in the air-starting system. This is done first by changing the timing of the air-starting valves so as to admit compressed air to the cylinders to oppose the original direction of rotation, to stop the engine, and immediately after that to start it in the opposite direction. While the engine is being stopped, the timing of all other valves—intake, exhaust, and fuel injection—is changed to correspond to the new direction of rotation.

The gears used for reversing the engines fall into two well-defined classes: (1) gears with a sliding camshaft and (2) gears with a movable-link system.

Although either type of gears can be used on both four- and two-stroke engines, some schemes are better for two-stroke and others for four-stroke engines.

Four-stroke Engines. When a four-stroke engine is reversed, the order of the operating strokes is also reversed. The reversing system, therefore, must change both the valve timing with respect to the crankshaft position and the sequence of valve operations. This may be done by using a separate set of cams to operate the valves in each direction of engine rotation. Separate ahead and astern cams for each valve are placed side-by-side on the camshaft, and one or the other set is brought into operation, depending upon the direction of rotation desired. Two methods may be used to bring the desired set of cams into position to operate the valves, namely: (1) sliding the camshaft with the cams axially or (2) shifting the cam followers.

In smaller engines, up to 500 or 600 hp, when a sliding camshaft is used, the cam lobes are beveled (Fig. 17-16) and the rollers have

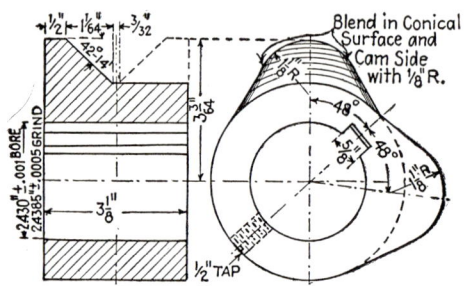

Fig. 17-16. Intake and exhaust cam of direct reversible engine.

rounded edges to permit them to slide up and down between the two cams when the camshaft moves endwise. The force to slide the camshaft must be sufficient to overcome the pressure of the valve springs which are compressed when the followers are lifted as the ramps of the cams slide under them. The force required for sliding the camshaft is usually provided by compressed air acting on a piston connected to the shifting lever.

In order to reduce the force necessary for sliding the camshaft, on larger engines, the rollers of the cam followers are lifted clear of the cams while the camshaft with the cams is moved endwise into place. The cam followers are then returned to their operating position in contact with the second set of cams.

With the second method, the engines have cam followers which are shifted from contact with one set of cams to contact with the other while the camshaft remains in place. One scheme of this type uses double-roller followers mounted opposite each set of cams on movable links in the valve gear, as shown in Fig. 17-17. The valve gear is designed so that one set of rollers is in contact with the corresponding set of cams for valve operation in one direction, roller *a* with cam *c*. To provide valve operation in the opposite direction, links are moved in the position shown in dotted lines and roller *b* is brought in contact with the second cam *d*.

When the cams which control the air-starting valves and fuel injectors are mounted on the same camshaft, the reversing process can be accomplished by a second complete set of cams according to one of the systems previously outlined. If, however, the air-starting and fuel-injection systems have two separate camshafts, other methods must be used to reverse the timing.

One method is to provide symmetrical cams on the separate camshafts which operate these systems. By rotating these camshafts a small angle with respect to the engine crankshaft, when the engine is

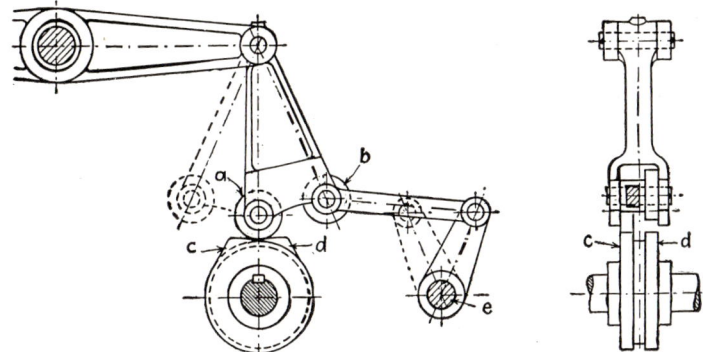

Fig. 17-17. Roller-type reversing gear.

reversed, the opposite side of the cam will operate the starting air and fuel injection in the correct timing for reverse engine rotation.

Another method is to use symmetrical cams for inlet and exhaust and separate reverse cams for starting air and fuel injection.

When the engine has separate camshafts driven by helical gears, a simple method of rotating the camshaft with respect to the crankshaft is to slide one of the helical gears axially. The angle of the gear teeth and the distance the gear moves determine the number of degrees that the separate camshafts are rotated. This method may be used where a sliding main camshaft has a helical gear to drive the two separate camshafts. When the main camshaft is moved endwise for reversing, its gear turns the separate camshafts a certain number of degrees and changes the timing of the air-starting and fuel-injection systems.

Two-stroke Engines. While the methods described above for changing the valve timing may be used with two-stroke engines, a simpler type of reverse gear can be used in which the expensive duplication of cams is avoided. Taking as an example an engine with exhaust valves and scavenge ports with separate camshafts for the exhaust valves and fuel pumps, and omitting temporarily the starting-air valves, the timing diagrams for ahead and astern operation will be along the lines of Fig. 17-18. All that is necessary to effect reversal, for both fuel pump and exhaust valves, is to turn the fuel-pump camshaft through an angle α and the exhaust-valve camshaft through an angle β. However, it is more convenient to reverse only the exhaust valves by turning the camshaft and to use independent means, such as duplicate cams, for the fuel pumps and starting valves. Of course, the effect of turning the camshaft on its timing is taken into consideration in fixing the angular positions of their cams.

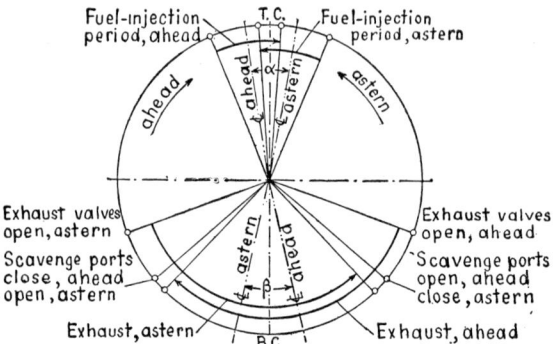

Fig. 17-18. Timing diagram of a direct-reversible two-stroke engine.

Instead of turning the camshaft through a certain angle relative to the cam roller, the same effect may be obtained by turning the roller relative to the camshaft. One such arrangement is shown in Fig. 17-19. In this scheme, the valve lift in the astern position is smaller than in the ahead one on account of the changed leverage, but that is not important. It is convenient to reduce the reversing angle by one-half by using a half-speed camshaft and double-nosed cams as shown in Fig. 17-19.

Another gear based on the same principle is shown in Fig. 17-20. By turning the eccentric fulcrum a through an angle β, the fuel-cam roller is displaced from the ahead position h to the astern position s. This corresponds to an angle α of change in the timing. When the roller is in the neutral position n, the cam does not touch the roller.

When the camshafts are driven by gears from the crankshaft, they may be rotated with respect to the drive gear by sliding the helical gears or splines, as explained before. When they are chain-driven, the timing of the camshafts with respect to the crankshaft is changed by movable idler sprockets which shorten the chain length on one side of the drive sprocket and lengthen it on the other, and vice versa.

When the camshaft is driven by a train of gears, as in large Nordberg engines, the timing of the camshaft may be changed by changing the position of one of the idler gears.

In an opposed-piston engine, the crankshaft connected to the pistons which control the exhaust ports leads the other crankshaft connected to the pistons controlling the scavenge ports by several degrees of rotation when operating in the normal ahead direction. The crankshafts are geared together in this fixed relative position, and when the engine is reversed, the crankshaft controlling the exhaust ports will lag

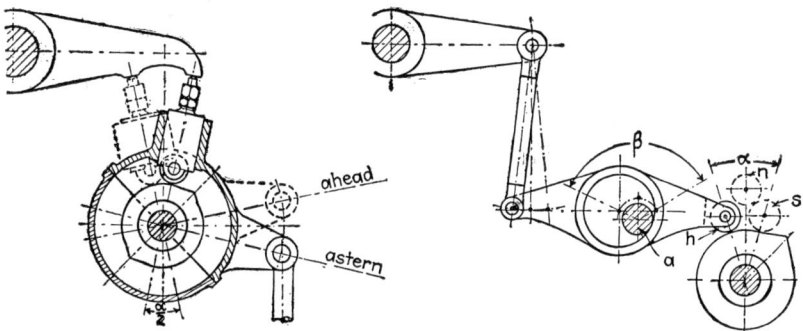

Fig. 17-19. Reverse gear with movable roller.

Fig. 17-20. Reverse gear with eccentric motion.

behind the crankshaft controlling the scavenging air ports. This condition is not desirable for proper performance. However, it can be tolerated for operation astern, when full power is not required. For astern operation the reverse gear in such an engine changes only the timing of the air-starting valves and of the fuel injection.

Pumps. When the direction of rotation of an engine is reversed, all engine-driven pumps, for water, oil, and air, will begin to operate in the opposite direction. In the case of reciprocating pumps, this does not make any difference. Gear pumps have automatic reversing check valves, which, by the pressure of the fluid discharged from the pump, keep the discharge piping connected with that side of the pump from which the fluid is being delivered. Positive-displacement air blowers have either a similar arrangement or change-over valves operated mechanically. Centrifugal pumps for reversible engines are usually built with straight radial vanes, as shown in Fig. 14-20, and the discharge opening remains the same regardless of the direction of rotation of the impeller.

QUESTIONS

1. State the two important requirements for quick starting of a diesel engine.
2. Explain why a certain minimum speed is required for successful starting of a diesel engine.
3. Enumerate the methods used for starting diesel engines.
4. What is the power required from an electric motor to start various diesel engines?
5. Enumerate the component parts of an electric starting system.
6. What is the voltage and ampere-hour capacity of storage batteries used with various diesel engines?
7. State briefly the characteristics of an electric starting motor.
8. Describe briefly the construction and operation of the mechanical connection between the starting motor and the engine flywheel.
9. State briefly the characteristics of a battery-charging generator.
10. What is the size of compressed-air tanks used in air-starting systems?
11. Enumerate the different methods of operating air-starting valves.
12. Explain briefly, with necessary sketches, the construction and action of a cam-operated air-starting mechanism.
13. Explain briefly, with necessary sketches, the construction and action of an air-operated air-starting valve.
14. State what engines can be started by air from any position of the crankshaft and explain why.
15. Explain methods used to turn a diesel engine to a starting position.
16. Explain briefly the arrangement and operation of a cartridge starter.
17. State the difficulties that may occur in starting a diesel engine at low temperatures.
18. Enumerate the methods and devices used to overcome difficulties of starting a diesel engine at low temperatures.
19. Draw a sketch of a Lister engine combustion chamber with a variable compression.
20. Draw a sketch of a variable-compression combustion chamber used in National-Superior engines.
21. What method is used for reversing marine diesel engines of 200-hp and larger rating?
22. State what two types of reversing gears are used in direct-reversible engines.
23. Explain the methods used for reducing the force necessary for sliding the camshaft in (*a*) smaller engines, up to 500 or 600 hp, and (*b*) in larger engines.
24. Sketch a reversing gear with a movable linkage.
25. State what arrangements are used for a large two-stroke diesel engine; draw the necessary sketches.
26. What steps must be taken in respect to engine-driven pumps in direct-reversing engines?

CHAPTER 18

VIBRATIONS AND BALANCING

18-1. Causes of Vibration. If a force displaces an elastic body from its position, the body will develop a restoring force which tends to return it to that state. Therefore, when the displacing force is removed, the body will move toward its original position. Owing to its mass—or rather, to its inertia—the body, on its return movement, will not stop at its previous position of equilibrium but will be carried beyond, causing a displacement in the opposite direction. The restoring force thus developed in the opposite direction will reverse the action and the process will continue until these consecutive movements called *oscillations* or *vibrations* are dissipated or damped out by friction or other resisting forces. The number of vibrations per second or the rate at which they occur is known as the *frequency of vibration*. If an elastic body is allowed to vibrate freely, it will do so at a certain rate known as its *natural frequency of vibration*, which depends upon its shape and the material of which it is made.

When displacing forces act repeatedly on an elastic body, they cause vibrations known as *forced vibrations*. When these forced vibrations occur at the same rate as the natural frequency of vibration of the body, or some multiple of it called a *harmonic*, the free vibrations will be reinforced or amplified by the forced vibrations. Under this condition, known as *resonance*, the displacement, or *amplitude*, of the resultant vibrations will be greatly magnified and may produce excessive stresses in the body.

Vibrations in engines are due to displacing forces resulting from various unbalanced forces acting in the engine. If all the forces in an engine were constant in magnitude and direction, they could easily be balanced. Actually, however, the forces within an engine are constantly changing in both magnitude and direction and are, therefore, difficult to balance. The problem of balancing these changing forces is also made more difficult by the reciprocating motion of certain parts.

Engine vibrations may thus occur due to unbalanced rotating forces, unbalanced reciprocating forces, and variations in gas pressure, inertia forces, and torque.

If these fluctuating forces in an engine happen to occur at the same rate as the natural frequency of vibration of the engine structure or one of its parts, the resulting condition of resonance may increase the amplitude of the vibrations to such an extent that serious damage will be done. Generally, the natural frequencies of the engine structure and engine parts are considerably higher than the frequency at which the unbalanced forces in the engine are likely to occur under normal operating conditions. However, if these forced vibrations occur so that they reinforce every second, third, fourth (and so forth) natural vibration of a particular engine part, a condition of resonance may occur at one of these higher harmonics of the natural frequency of vibration. While resonance with these harmonics of higher order will not produce as great a magnification of the vibrations as resonance with low-order harmonics, they are more likely to occur and are responsible for the so-called vibration points or *critical speeds* in respect to a certain engine part, within the normal operating speed range of the engine. Engine parts that are apt to develop synchronous vibration, "pick-up vibrations," are push rods, gears, beams, and brackets attached to an engine, and engine supports. Valve springs also can develop deflections interfering with their proper functioning. Crankshafts also pick up vibrations but more frequently of the torsional type.

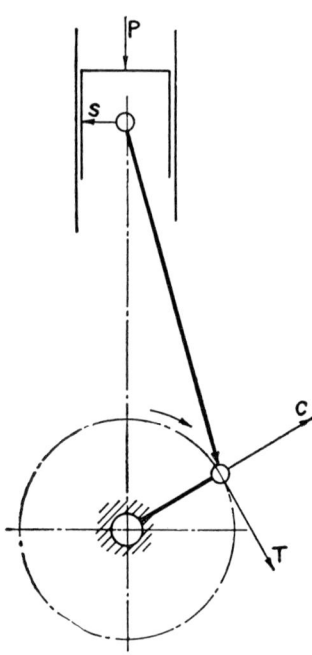

Fig. 18-1. Forces acting in an engine.

The gas pressure P (Fig. 18-1) acting on the piston of a diesel engine is resolved into two types of forces: a tangential force T turning the crankshaft and a side thrust S on the cylinder walls. To these is added the centrifugal force C produced by the rotation of the crankshaft with all parts attached to it. All these forces vary in size, direction, and point of application, and all can cause some kind of vibration.

18-2. Engine Vibrations. Vibrations of the engine as a whole, noticeable only if the structure supporting the engine is flexible, may be classified by the type of displacement which they cause, as shown in Fig. 18-2:

Shaking—caused by fluctuating vertical or horizontal forces which tend to move the engine up and down or sideways.

Rocking—caused by fluctuating horizontal forces acting above the center of gravity of the engine which tend to rock the engine about a line passing through its center of gravity.

Pitching—caused by fluctuating vertical couples which tend to make the ends of the engine rise and fall.

Yawing—caused by fluctuating horizontal couples which tend to turn the engine crossways or move the ends to the left and right.

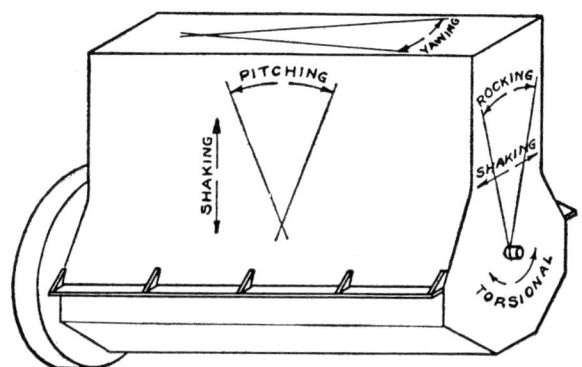

FIG. 18-2. Types of engine vibration.

Torsional vibration—caused by fluctuating torque reactions which tend to twist the crankshaft as it rotates.

From the above definitions it will be seen that shaking is due to the unbalanced reciprocating forces and vertical or horizontal components of the centrifugal forces, while pitching is due to the unbalanced couples produced by these forces. Rocking is caused primarily by variations in the horizontal component of the piston reaction or side thrust S as shown in Fig. 18-1 and is caused by changes in gas pressures, inertia forces, and load reactions. Yawing is caused chiefly by unbalanced couples produced by the horizontal components of the centrifugal forces in a vertical engine and the horizontal components of the reciprocating forces in a V-type engine. Torsional vibrations are caused primarily by variations in torque as a result of changes in gas pressures, inertia forces, and torque load reactions.

While all these vibrations are to some extent interrelated, the gas pressures and load reaction will have little effect on the engine vibrations other than torsional vibration and rocking. Shaking, pitching, and yawing vibrations are caused by unbalanced reciprocating and rotating forces and couples which occur at all engine loads unless the engine is equipped with means to balance them.

In addition there are internal vibrations in the engine structure itself caused by fluctuations in the gas pressures and inertia forces. This is evidenced by the so-called *engine roughness*, which occurs at certain vibration points when the frequency of the fluctuating forces coincides with the natural frequency of the engine structure or some multiple of this frequency. In order to prevent resonance with these vibrations, the engine frame is made as rigid as possible to increase its natural frequency of vibration.

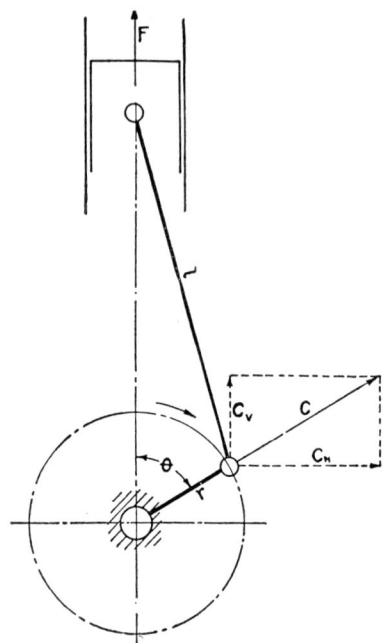

FIG. 18-3. Action of centrifugal force.

18-3. Unbalanced Engine Forces.
Inertia forces are proportional to the square of the engine speed and become very large as the speed of the engine increases. These reciprocating forces are considered as acting only along the line parallel to the center line of the cylinder.

In order to determine the magnitude of unbalance in a reciprocating engine, the moving parts may be considered as divided into those which are reciprocating and those which are rotating. The small end and the adjacent part of the connecting rod are considered as reciprocating, while the large end and the rest of the connecting rod are considered as rotating. The reciprocating weight thus consists of the piston, piston rings, piston pin, and the weight of the upper end of the connecting rod. The rotating weight consists of a weight, assumed to be concentrated at the radius of the crank which will produce a centrifugal force equivalent to that of the entire crank structure, plus the weight of the lower end of the connecting rod.

The magnitude of the inertia force F of the reciprocating parts (Fig. 18-3) varies throughout each stroke and may be computed by the approximate expression

VIBRATIONS AND BALANCING

$$F = \frac{Wrn^2}{35,200}\left(\cos\theta + \frac{r}{l}\cos 2\theta\right) \qquad (18\text{-}1)$$

where: W = the weight of all reciprocating parts, lb
n = the crank speed, rpm
r = the crank radius, in.
l = the connecting rod length, in.
θ = the crank angle, deg, from top center

The first term in expression (18-1) is known as the *primary force*, and the second term is known as the *secondary force*. The primary force depends upon the weight of the reciprocating parts, as enumerated above, and upon the crank radius and engine speed. The primary force is equal to the inertia force which would be produced if the connecting rod were infinitely long. The secondary force represents the influence of the angularity of the connecting rod. It depends upon the same factors—weight of reciprocating parts, crank radius, and engine speed—and in addition upon the ratio of the crank radius to the length of the connecting rod. Owing to their natures, the primary force alternates in magnitude and direction once per revolution while the secondary force alternates twice per revolution.

The centrifugal force of the rotating parts always acts radially outward along the line of the crank and thus moves around the shaft at crankshaft speed. The magnitude of the centrifugal force C may be computed from the expression

$$C = \frac{Wrn^2}{35,200} \qquad (18\text{-}2)$$

where W = the weight of the rotating parts, lb
r = the radius of rotation, in.
n = the number of revolutions per minute

As indicated in Fig. 18-3, this centrifugal force may be resolved into two perpendicular components: one acting along the cylinder center line and the other acting across the center line. As the crank rotates, the magnitudes of these two component forces change with each position of the crank, even though the resultant centrifugal force remains constant at a given speed.

The magnitude of the vertical and horizontal components of the centrifugal force, which vary throughout each revolution, may be expressed in terms of the angle of the crank from top center position as the product of the centrifugal force and the cosine and sine of the crank angle, respectively. Since these forces alternate in magnitude and direction once per revolution, they also are considered part of the primary forces. By combining the components of the primary rotating and reciprocating forces, the resultant unbalanced primary forces are obtained.

18-4. Crankshaft Balance. A crankshaft whose center of gravity coincides with its center line is said to be *statically balanced*. With its ends supported on two horizontal knife edges, the shaft will be stable at any position of the cranks, that is, it will have no tendency to roll. The centrifugal forces from the rotating parts of each crank structure of a statically balanced crankshaft, if considered to act in the same plane, will all balance each other. Actually, however, the centrifugal forces of the rotating parts of a crankshaft do not act in the same plane but at various distances from the middle point of the shaft. These forces will therefore form couples which, unless balanced, will tend to cause pitching and yawing as the crankshaft rotates. When these couples are completely balanced, the crankshaft is said to be in *dynamic balance*.

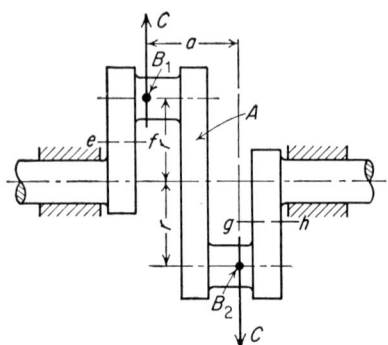

FIG. 18-4. Two-throw crankshaft.

Before the other moving parts of an engine can be balanced, its crankshaft must be in dynamic balance. Some crankshafts, by virtue of their crank arrangement, are always in dynamic balance while others must be dynamically balanced by means of weights, known as *counterbalances* or *counterweights*, placed opposite their cranks. By the proper choice of the weight of the counterbalance and the point at which it is placed, its centrifugal force may be made exactly equal to and acting in the opposite direction from the centrifugal force of the other rotating parts of the crankshaft.

A two-throw crankshaft with 180 deg between the cranks (Fig. 18-4) may serve as an illustration. When this shaft is properly machined, it will be balanced statically but not dynamically. The central double crank A and the center parts of the end cranks up to the planes e–f and g–h are in perfect balance; however, the crankpins B_1 and B_2 with the adjacent crank webs are not balanced and will produce unbalanced centrifugal forces in the planes of rotation, each located at the center of gravity of the corresponding unbalanced rotating mass. Each of these two centrifugal forces C may be computed from Eq. (18-2) above. The two centrifugal forces act in different planes of rotation and in opposite directions and therefore give a couple Ca, where a is the distance between the centers of gravity of the unbalanced masses.

If an equal but opposite couple lying in the same plane is provided, the system will become dynamically balanced. This additional couple may have a different arm a' instead of a and the centrifugal force may act at a different radius of rotation r' instead of r. However, according to Eq. (18-2), the relation between the two new rotating masses, or counterweights W', and the unbalanced weights W must evidently be

$$W'r'a' = Wra \qquad (18\text{-}3)$$

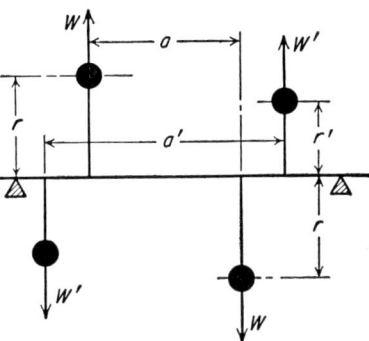

This condition is represented diagrammatically by Fig. 18-5. A crankshaft constructed with counterbalances opposite every crankthrow to balance all the centrifugal forces will thus also be in dynamic balance.

Fig. 18-5. Dynamic balance.

Counterbalances, however, are not always necessary and, since they add extra weight to the crankshaft, are not always used. Thus, crankshafts which have their cranks arranged in symmetrical pairs, each of which acts at the same angle and distance on either side of the middle

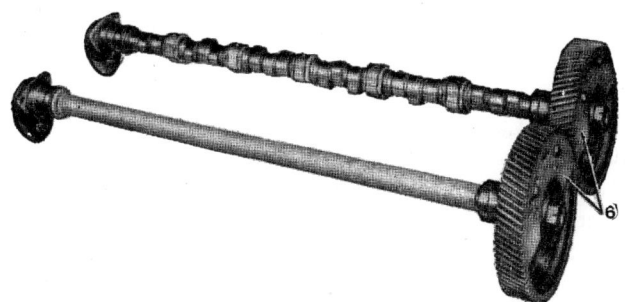

Fig. 18-6. Cam- and balance-shaft assembly.

point of the shaft, will be in dynamic balance because the couples acting about the center will be equal and opposite.

In the case of multi-throw crankshafts in two-stroke cycle engines, the centrifugal forces of the symmetrical cranks on each side of the center of the shaft act at equal distances but in different directions. These forces will form couples which are not balanced, and consequently the shaft will not be in dynamic balance. In order to balance such a shaft dynamically, counterweights must be added to produce couples which are exactly equal and opposite to the resultant unbalanced couples. While this could be accomplished with counterbalances opposite each crank throw, some crankshafts are made with counterweights only opposite the outer cranks as shown in Fig. 18-6. By placing these counterweights at a certain angle with the outer crank throw instead of directly opposite it, the unbalanced couples from the inner cranks will also be balanced. In this way, while the centrifugal forces from all cranks are not individually balanced, the resultant couples produced by them, which would tend to cause pitching or yawing, are balanced by the counterbalances at each end and the crankshaft is in dynamic balance.

18-5. Balancers. Comparing Eqs. (18-1) and (18-2) and the discussion concerning the components of the centrifugal force in connection with Fig. 18-3, it will be noted that the expression for the primary reciprocating force is the same as the expression for the vertical component of the centrifugal force. Thus if a counterbalance with a weight equal to the weight of the reciprocating parts is placed at a radius equal and opposite to the crank, the primary reciprocating forces will be balanced. The horizontal components of this added centrifugal force, however, would be unbalanced.

This principle is employed in single-cylinder engines to balance part of the reciprocating forces. A weight that produces a centrifugal force equivalent to one-half the reciprocating forces is added to the counterbalance that is computed to balance the centrifugal force of the rotating masses. At the top and bottom of the stroke this excess centrifugal force opposes the reciprocating force and thus reduces it by one-half. At positions midway in the stroke the excess centrifugal force acts across the axis of the cylinder, since there is no reciprocating force to balance it. The net result is a balancing of one-half the reciprocating force at the expense of creating an equal unbalance in the horizontal plane.

In engines with two or more cylinders, all primary forces are generally balanced by the use of symmetrically spaced cranks or by counterbalances, as previously explained.

Primary Couple Balancers. In engines which have unbalanced primary couples, particularly two-stroke engines of two, three, four, and six cylinders, an equal and opposite couple must be provided which has the same frequency as the crankshaft. One method of providing this type of balance is the use of two balance shafts driven in opposite directions at the same speed as the crankshaft and provided with counterweights at each end, as shown in Fig. 18-6. As will be seen from the illustration, the centrifugal forces due to the counterweights when they are in a horizontal position are always balanced. When the weights are in a vertical position, the centrifugal forces will act upward at one end of the engine and downward at the other end and will produce a definite couple. The magnitude of these counterbalances on the balance shafts are so chosen that the resulting couple will be equal and opposite to the unbalanced couple of the crank mechanism and will thus balance it. One of the counterbalance shafts also acts as the valve camshaft, and the counterweights at one end are built into the drive gears of the camshafts. This arrangement will balance the primary couple, which would otherwise tend to produce pitching vibrations.

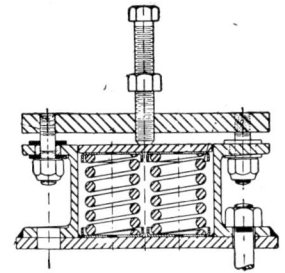

Fig. 18-7. Vibration isolator.

18-6. Vibration Absorbers. In spite of all the precautions taken, it is not possible to balance all forces and couples which occur in a reciprocating engine. In order to keep under control the vibrations which cannot be balanced, the engine must be provided with means that will absorb these vibrations without injury to the engine or its supports. Engine-support vibration absorbers consist essentially of a flexible engine mounting made of rubber, cork, or steel springs. In order to absorb all types of vibrations such a mounting should allow a small deflection of the engine in all directions and absorb the vibrations without transmitting them to the foundation. An example of an engine-vibration absorber with coil springs, also called a *vibration isolator*, is shown in Fig. 18-7. However, the fitted pins in this absorber restrict its effectiveness to the absorption of vibrations caused by vertical shaking, pitching, and rocking. Figure 18-8 shows a vibration absorber in which heavy rubber pads are interposed between the engine bedplate and the floor plate of a boat. This arrangement allows small displacements in all directions.

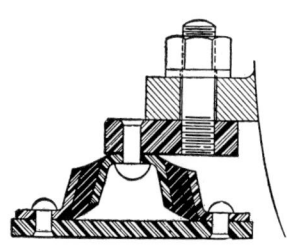

Fig. 18-8. Vibration absorber with rubber.

Finally, it may be of interest to describe a method used to overcome vibrations transmitted to a building by two four-cylinder diesel engines installed to drive alternating-current generators in parallel. It was found that when each of these engines was running alone, the vibrations were not pronounced but that when they were running in parallel, the vibrations sometimes became really dangerous.

An investigation of the conditions disclosed that the vibrations were worst when all eight pistons passed the dead centers simultaneously. The vibrations were smallest when the cranks of the two engines were at right angles.

In order to ensure the operation of the engines with the minimum of vibration a special synchronizing arrangement was installed. On the end of each crankshaft was mounted a contact finger that slides over a stationary commutator ring with a short contact as shown in Fig. 18-9. Thus one crankshaft would make contact when it was in

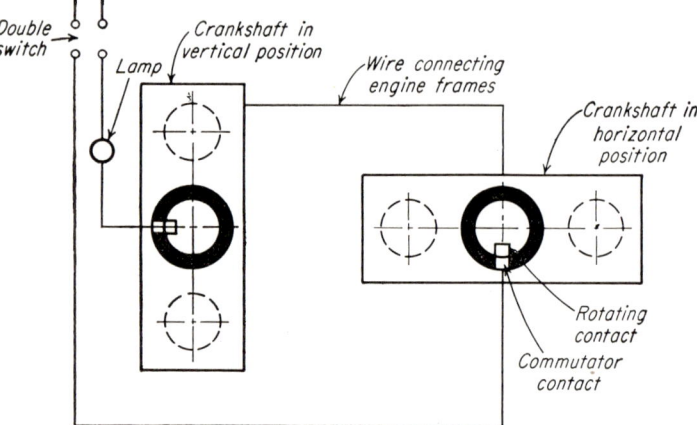

Fig. 18-9. Diagram of timing contact connections.

the vertical plane and the other one when it was in the horizontal plane. These contacts were connected to a d-c source, such as a storage battery, through a lamp and switch.

The method of putting the two engines together is to have the incoming engine running a little fast, as indicated by the synchronoscope, then to wait for the shafts to come in a position which causes the lamp to flash, and at this instant to close the switch of the incoming engine. The rotors of the two machines will be held in the same relative positions as long as the generators are connected together, and the vibrations will remain at a minimum.

18-7. Torsional Vibration. If a long steel shaft with a flywheel fastened at one end is held firmly at the other end and the flywheel is given a slight turn, the shaft will be twisted and the straight line OO on the surface of the shaft (Fig. 18-10) will become a right-handed helix OA. When the flywheel is released, the shaft will untwist itself and will turn the flywheel in the opposite direction. Because of its inertia, however, the flywheel will not stop turning when the shaft is fully untwisted, but will continue to turn until the shaft is twisted in the opposite direction. The straight line OO will become a left-handed helix OR. Since steel is elastic, this shaft will continue to twist and untwist in opposite directions until the internal friction between the steel fibres slows it down or *damps out* the angular vibration. Only one flywheel arm is shown in Fig. 18-10 in order not to obscure the picture.

Such a twisting and untwisting of a shaft is known as *torsional vibration*. Every shaft has a certain natural frequency of torsional vibration depending on its size, shape, elasticity, and attached weights. For instance, a long, thin steel shaft with one end firmly secured and the other end fastened to a heavy flywheel will vibrate, or twist and untwist, much more slowly than will a short, thick steel shaft to the free end of which the same flywheel is attached. Therefore the long, thin shaft together with the flywheel is said to have a lower *natural frequency of torsional vibration*. The shorter and thicker an engine shaft is, the higher will be the natural frequency of its torsional vibration.

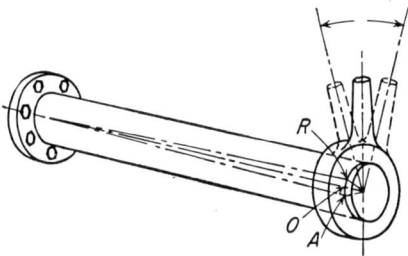

FIG. 18-10. Torsional vibration of a shaft with a flywheel on its end.

In an engine crankshaft the forces exerted by firing in the cylinders tend to twist the crankshaft while it is rotating. This gives rise to periodic vibrations which in normal operation are so small that they are not evident and are not ordinarily harmful. If, however, these periodic impulses happen to coincide with the natural frequency of torsional vibration of the engine crankshaft and attachments, these impulses will reinforce each natural vibration and the resulting vibrations may become so great that the stresses in the crankshaft will exceed the safe limit and the shaft will finally break. However, even before the crankshaft breaks, an excessive torsional vibration may cause other serious damage, such as the pitting or breaking of the teeth in the gears that are driven by the crankshaft. The existence of dangerous torsional vibrations in a crankshaft is usually revealed by a periodic noise, especially a grinding noise made by the teeth of gears keyed to the crankshaft.

Critical Speeds. The speed of crankshaft rotation at which these twisting impulses occur at the same rate as the natural frequency of torsional vibration is called the *critical speed of the first order*. The crankshafts of most diesel engines have high natural frequencies so that the critical speed of the crankshaft system is not reached in normal operation.

If an engine is run at a speed at which the twisting impulses occur at a rate which is one-half the crankshaft's natural frequency of torsional vibration, the twisting impulses will reinforce every second natural vibration. The speed at which this vibration point occurs is called the *critical speed of the second order*. Similarly, depending upon the engine characteristics, critical speeds may occur at which these impulses reinforce every third or fourth natural vibration, and so on; this is called the *critical speed of the third or the fourth order*, etc. When, in bringing an engine up to speed, a crankshaft vibration point is reached, it is usually due to a critical speed of a higher order which, although not so destructive as the first-order critical speed, should be run past as quickly as possible to avoid possible detrimental effects. Where an engine is to be run at a constant speed other than the rated speed for which it was designed, it must not be at or near a critical speed.

High-speed multicylinder engines are more likely to run through critical speeds than low-speed engines because the size and weight limitations prevent the use of large-diameter crankshafts and yet a long shaft is necessary. A large number of cylinders and a high operating speed result in a great number of firing impulses tending to twist the crankshaft. These factors tend to bring the low-order critical speeds within the operating speed range of the engine.

In order to obtain high natural frequencies, crankshafts are made as short and as large in diameter as possible. In order to keep the weight down and yet assure sufficient rigidity to avoid critical speeds within the operating range of the engine, some crankshafts are made large in diameter but hollow. For the same reason, the flywheels are made as light in weight as possible.

18-8. Torsional-vibration Dampers and Absorbers. It is not always possible to make a crankshaft so rigid that resonance with some of the higher-order harmonics of its natural frequency will not occur within the operating speed range. Therefore, crankshafts of multicylinder high-speed engines are generally equipped with devices which prevent vibrations from building up to dangerous amplitudes when the engine is operated at or near a critical speed.

The devices in use are based on two different principles: the damping of the vibrations or their absorption. Damping involves friction, as between two bodies pressed one against the other, or internal friction of the vibrating body, as in a cast-iron structure. Absorption is accomplished by creating a force equal and opposite to the disturbing force.

Torsional-vibration dampers limit the amplitude of the vibrations by providing frictional resisting forces which oppose them. The energy of vibration is thus converted into heat, which is dissipated without harmful effects on the engine.

One type of torsional-vibration damper consists of a small flexibly driven flywheel mounted at the opposite end of the crankshaft from the main flywheel. In some cases, the damper flywheel is driven by a spring-loaded friction clutch which tends to keep it rotating at crankshaft speed. If the crankshaft starts vibrating torsionally it will tend to turn the damper flywheel either faster or slower as it twists first one way and then the other; but the inertia of the damper flywheel will tend to keep it rotating at constant speed and will thus oppose the vibrating forces. By adjusting the tension of the springs on the friction-driven clutch, it can be made to slip when the vibrating forces reach a certain magnitude. The energy of vibration will then be converted into heat by doing work against the clutch friction, and this heat is dissipated to the surroundings.

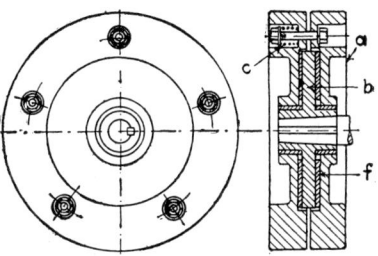

Fig. 18-11. Vibration damper.

Such a vibration damper used in automotive engines is shown in Fig. 18-11: a two-piece flywheel *a* floats pressed by springs *c, c* to the disk *b* which is lined with friction disks *f* and keyed to the crankshaft. When the crankshaft and with it disk *b* begin to oscillate, the friction between disks *f* and *a* damps out the vibration and prevents it from becoming dangerous.

Sometimes torsional vibrations are *absorbed* by driving the flywheel through a rubber coupling. In this case, the vibrating forces will tend to deflect the rubber and thus will expend their energy against its internal resistance. In order to absorb vibrations occurring at different speeds, some engines are equipped with a vibration absorber consisting of two small flywheels of unequal weight so that their resisting forces will be most effective at different speeds.

Another type of torsional vibration damper, sometimes called a *harmonic balancer*, consists of an auxiliary flywheel driven by flexible leaf springs. The vibrating energy in this case is dissipated by the friction between the leaves of the springs sliding over each other. The same principle is also employed in elastic couplings between the crankshaft and the main flywheel, which serve to damp out some of the torsional vibrations. This type of damper usually operates in lubricating oil which assists in damping the vibrations by its viscous friction and also carries away the heat generated. In a similar manner, hydraulic couplings or fluid drives also damp out and prevent the transmission of torsional vibrations between the crankshaft and the driven load.

Pendulum-type Absorber. This device, often called a *vibration damper*, consists of two or more symmetrically located heavy steel segments or balances m, m (Fig. 18-12) suspended by links b, b to a disk c so that they can swing in the plane of rotation, like a pendulum.

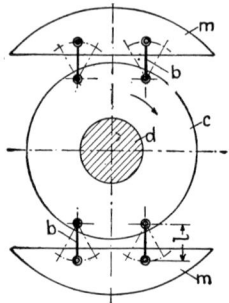

Fig. 18-12. Schematic arrangement of a tortional-vibration pendulum absorber.

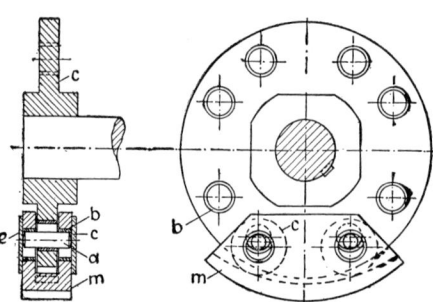

Fig. 18-13. Torsional-vibration pendulum absorber.

The length l of the links b, b is so selected that the natural frequency of swing of the segments is equal to the frequency of vibration of the shaft system which must be absorbed. The disk c is fastened to the free end of the crankshaft, the end that has the maximum amplitude of torsional vibration. During undisturbed rotation of the shaft, the centrifugal force keeps the pendulum weights at the greatest distance from the axis of rotation. When the shaft starts to vibrate, the weights begin to swing and thus are periodically drawn nearer to the shaft axis (Fig. 18-12). The work of bringing the pendulums closer to the axis against the resistance of the centrifugal force is obtained from the energy put into the shaft when it starts to vibrate and thus reduces this energy and damps out, or at least reduces considerably, the angle of torsional vibration of the shaft.

The masses of the pendulums do not influence their frequency but influence only the amount of energy that they can absorb. The heavier the pendulums, the greater is the energy absorbed and the more effective is the damper. The frequency of swing of the pendulums m, m can be adjusted to make it equal to the frequency of the dangerous harmonic by changing the length l of the links. The weights are used in pairs in order not to disturb the shaft balance.

In an actual absorber the weights are suspended not on links but are held by large-diameter pins a, a of hardened steel inserted in holes with hardened bushings b, b in the disk c and in the weights m, m (Fig. 18-13). The pins are held in place by retainer plates e, e. Figure 18-13 shows a vibration absorber with four weights, of which three have been taken off. Using pins of a slightly different diameter is equivalent to changing the length of the links. Thus the absorber can be *tuned* to a desired frequency. If the shaft has two critical speeds, one pair of weights is tuned to one critical speed, and the second pair, at right angles to the first, is tuned to the second critical speed. The damping effect can be intensified by increasing the weight of the swinging masses and the distance of their center of gravity from the axis of rotation. These vibration absorbers are used rather extensively in high-speed multicylinder engines.

Counterweights attached to the cheeks of a crankshaft can be made to act also as absorbers by suspending them on pins, as in Fig. 18-13, instead of fastening them rigidly.

Detuner Flywheel. A special type of damper is used in the Doxford opposed-piston marine oil engines. The crankshaft has two medium-weight flywheels, one on each end; the flywheel on the free end is connected to the shaft by means of a flexible spring coupling, with a spring rigidity increasing with the angle of deflection. This gives a variable natural frequency of vibration to the whole shaft system and prevents exact synchronism from being established at any speed; hence the name *detuner*.

QUESTIONS

1. Explain how vibrations are caused in an elastic body.
2. What is frequency of vibration?
3. What is the natural frequency of vibration?
4. What, if any, is the difference between vibrations and oscillations?
5. What is forced vibration?
6. What is a harmonic vibration?
7. What is a condition of resonance?
8. What is the amplitude of a vibration?
9. What, in general, causes vibration in an engine?
10. Enumerate the forces that may cause vibration in an engine.
11. What is the critical speed of an engine?
12. What engine parts may develop or pick up vibrations?
13. Enumerate the five types of vibration of the engine as a whole and illustrate them by a diagrammatic sketch.
14. What two types of engine vibration are caused by gas pressures and load reaction?

15. What types of engine vibrations are caused by unbalanced reciprocating and rotating forces and couples?

16. What is engine roughness?

17. Into what two groups are engine parts divided when the balancing of an engine as a whole is being considered?

18. What term represents the primary force in the expression (18-1) of inertia of reciprocating parts?

19. What term represents the secondary force in the expression (18-1) of inertia of reciprocating parts?

20. Into what two components is the centrifugal force of rotating parts resolved?

21. What is a statically balanced crankshaft?

22. What is a dynamically balanced crankshaft?

23. Explain how counterweights are used to obtain the dynamic balance of a crankshaft.

24. Explain how a counterweight fastened to a crankshaft can balance all or part of the inertia of reciprocating parts.

25. In computing the weight of a counterweight, why is it desirable to balance only one-half the inertia forces of reciprocating parts and not their full value?

26. What are primary-couple balancers? In what engines are they used?

27. What is a vibration damper?

28. Draw a diagrammatic sketch of a so-called vibration isolator.

29. What is torsional vibration of a shaft?

30. How is the presence of torsional vibration of an engine crankshaft revealed?

31. What is the critical speed of the first order of an engine in regard to the torsional vibration of its crankshaft?

32. What is the critical speed of a higher order of an engine in respect to torsional vibration of its crankshaft?

33. What types of devices are used to control torsional vibrations?

34. What is the purpose of torsional-vibration dampers?

35. What is a harmonic balancer?

36. Explain the principle of the pendulum-type torsional-vibration absorber.

37. What is the detuner flywheel?

38. What is the main principle of action of a detuner flywheel?

PART 4

ENGINE INSTALLATION AND OPERATION

CHAPTER 19

ENGINE INSTALLATION

19-1. Selection of Engine. In selecting a diesel engine for a new installation the following basic characteristics must be determined first: (1) the power necessary, (2) the speed factor that is desirable, and (3) the type of construction that would be most advantageous.

Power. When the engine to be installed will drive machinery whose power requirement is fixed, such as a pump or compressor or machinery in a metal or woodworking shop, the problem is rather clear: the rated power of the engine must be equal to or slightly greater than the maximum demand of energy.

When the engine drives an electric generator, d-c or a-c, and the current is used for lighting and power, the problem becomes more involved. Experience shows that even if the number of lights and their wattage and the number of electric motors and their power are originally determined very carefully, as time goes by there will be more lights and additional motors installed and the load on the generator and diesel engine will increase. It should be remembered that the worst thing that can be done in respect to successful operation and proper maintenance of a diesel engine is to overload it. Therefore, if the diesel engine is intended to generate electric energy, a good practical rule is to compute carefully and rather liberally the maximum amount of energy required and then to add on top of the resulting figure another 20 or 25 per cent. Also, if the energy thus computed is expressed in kilowatts, the diesel-engine horsepower required should be taken as 50 per cent higher. Thus the rated power of an engine connected to a 240-kw generator should be

$$240 \times 1.5 = 360 \text{ bhp}$$

This coefficient of 1.5 allows for both the conversion of kilowatts to horsepower units and the efficiency of the electric generator.

Altitude. At sea level the average, or standard, atmospheric pressure is 14.7 psia. With an increase of the altitude above sea level, the atmospheric pressure decreases and as a result the engine receives air of less density. This means that the engine receives less air by weight per stroke and therefore can burn less fuel, and as a result the available power drops in the same proportion. The decrease of pressure, air charge, and power is about 3.5 per cent for every 1,000 ft of altitude.

Thus a 400-hp engine installed somewhere in a mine at an altitude of 6,000 ft can develop, without overloading, only

$$400\left(\frac{1 - 0.035 \times 6000}{1000}\right) = 400(1 - 0.21) = 316 \text{ hp}$$

Therefore, if an output of 400 hp is required at an altitude of 6,000 ft, the rated horsepower of the engine at sea level must be

$$\frac{400}{1 - 0.21} = 507 \text{ hp}$$

Thus if an engine is installed at a certain altitude and is expected to deliver the power at which it is rated at sea level, it will be overloaded. In the example above discussed a 400-hp engine would be overloaded by

$$\frac{400 - 316}{316} \times 100 = 26.5\%$$

This, evidently, would be inviting trouble as soon as the engine is put in operation.

Speed Factor. The next question is whether the engine should be in the low-, medium-, high-, or super-high-speed classification, as outlined in Sec. 2-9. While it is true that at present engines with a speed factor of 6 and even higher are built for heavy-duty service, it should be noted that if an engine is installed for continuous operation and if a long life with small maintenance cost is desired, a low- or medium-speed engine with a speed factor not over 6 would be the proper selection. Engines with a speed factor $c_s > 6$ are lighter in weight, occupy less space, require lighter foundations. These are their main advantages.

High-speed engines, with a speed factor of 27 and up, are suitable for portable or temporary installations.

Engine Type. Under this title should come first the question of cycle, whether to use a two-stroke or a four-stroke engine. The difference between the two types is purely mechanical.

The chief advantages of two-stroke engines, at least theoretically, are: one-half smaller piston displacement for a given power, meaning an engine practically one-half as heavy and therefore less expensive; and a flywheel only about one-half as heavy for the same uniformity of rotation because of twice the number of working strokes.

However, the actual performance of two-stroke engines is greatly influenced by the scavenging of the cylinder. If scavenging is not very good, the cylinder charge will be contaminated by residual exhaust gases, less fuel can be burned, and the mean effective pressure will be lowered, sometimes to the same value as that obtained in a four-stroke engine.

Only engines with separate scavenge pumps and operating on one of the more perfect scavenge schemes give really good results. On the other hand, for small and medium power, crankcase scavenging engines have the advantage of simplicity, fewer parts to get out of order, no valves to grind, and low initial cost.

The final decision, whether to select a two-stroke or a four-stroke engine, is usually influenced more by the availability of an engine of the proper power and speed factor than by a personal preference for one type or the other.

Cylinder Number. The question of how many cylinders the engine should have may be solved as follows: the greater the number of cylinders for a given engine power, the greater is the uniformity of rotation and, generally speaking, the better is the engine balance. Thus to drive an electric generator a four-stroke engine should have at least four, preferably five or six, cylinders. A three-cylinder two-stroke engine is as good in this respect as a six-cylinder four-stroke engine. A number of cylinders greater than six is used chiefly to increase the power of an engine without increasing its height and width. Also the weight of an engine per horsepower decreases with an increase in the number of its cylinders. On the other hand, the greater the cylinder number, the greater is the number of moving parts, the more numerous the places subject to wear, the greater the amount of maintenance work needed, and the greater the chance of failure of some part.

When an engine is installed as an addition in an existing power plant, it is very convenient to have it of the same make and same bore and stroke as the other engines. First, because it cuts down the number of spare parts which must be kept on hand to avoid prolonged shutdowns, and second, because it simplifies the operation and maintenance for the personnel. Naturally, these considerations should not influence the selection of a new engine if the older engines are not of a satisfactory type or make.

19-2. Foundations. The foundation of a diesel engine must perform three functions: (1) support the weight of the engine, (2) maintain the engine in proper alignment with the driven machinery, and (3) absorb the vibrations produced by unbalanced forces created by reciprocating and revolving masses.

The third function is the most important one; however, if an engine is properly balanced, as are automotive-type and boat engines, it does not need a foundation, *i.e.*, not a heavy one of concrete.

The first object of a foundation is achieved by making its supporting area sufficiently large. The safe loads vary from about 1,000 lb per sq ft, for alluvial soil or wet clay, to 4,000 lb per sq ft, for gravel, coarse sand, or dry clay. In computing the area necessary to support the engine, the weight of the foundation itself must be added to the weight of the engine. In these computations 135 lb per cu ft may be used for the weight of concrete.

If the soil is marshy, it will not support even 1,000 lb per sq ft. In such a case, it will be safer to reduce the load to about one-half that figure and to extend the base of the foundation to form a mat or raft. The concrete mat should be reinforced by steel bars both lengthwise and crosswise and must be at least 8 in. thick for a small engine and up to 12 in. for a large one.

The second object of a foundation is attained by making it strong and rigid and, in larger sizes, by reinforcing it with steel bars at the bottom and at the top. In the case of a belt drive, there must be a sufficient side area and weight to resist the outside forces. A good practical rule is to make the foundation depth from 3.2 to 4.2 times the engine stroke, the lower factor for well-balanced multicylinder engines on a firm soil and increased factors for engines with fewer cylinders or on less firm soil.

The third object is achieved by giving the foundation a sufficient weight or mass. Theoretically, the minimum weight required to absorb vibrations could be expressed as a function of the weight of the reciprocating masses and of the speed of the engine. However, for practical purposes it is simpler to use the empirical formula

$$W_f = C W_e \sqrt{n} \qquad (19\text{-}1)$$

where W_f = the weight of the foundation, lb
W_e = the weight of the engine, lb
C = an empirical coefficient
n = the engine speed, rpm

The coefficient C is given in Table 19-1.

If the weight of the engine and its speed are not yet known, as in preparing the estimated cost of a new installation, the volume of the concrete for the foundation may be estimated with sufficient accuracy from the data of Table 19-2.

TABLE 19-1. VALUES OF COEFFICIENT C IN FOUNDATION FORMULA (19-1)

Type of engine	Cylinder arrangement	Number of cylinders	Coefficient C
Single-acting................	Vertical	1	0.15
Single-acting................	Vertical	2	0.14
Single-acting................	Vertical	3	0.12
Single-acting................	Vertical	4, 6, 8	0.11
Single-acting................	Horizontal	1	0.25
Single-acting................	Horizontal, duplex	2	0.24
Single-acting................	Horizontal, twin duplex	4	0.23
Double-acting................	Horizontal	1, 2	0.32
Double-acting................	Horizontal, twin tandem	4	0.20

TABLE 19-2. VOLUME OF CONCRETE FOUNDATION, CU FT PER HP

Number of Cylinders	1	2	3	4	5–8
High-speed engines...............	4.0	2.5	2.0	1.7	1.5
Medium-speed engines............	5.0	3.1	2.5	2.1	1.9
Low-speed engines................	6.0	4.0	3.0	2.6	2.3

Thus, from the data of Table 19-2, a 200-hp medium-speed four-cylinder engine requires a foundation having a volume

$$V = 200 \times 2.1/27 = 15.5 \text{ cu yd}$$

Foundation Material. While the early diesel engines had foundations made of stone or brick with mortar for bond, present-day machinery is set on concrete foundations exclusively. The concrete mixture varies with the kind of material, but when crushed rock, ranging up to $2\frac{1}{2}$ in. across, is available, along with clean, sharp sand, a mixture of one part cement, two parts sand, and four parts rock, designated as a 1:2:4 mixture, gives a reliable foundation. For good firm soil, reinforced concrete foundations for large engines may use a leaner mixture, down to 1:3:6.

19-3. Vibrations. Most persons can detect vibrations of 0.004 in., and more sensitive persons may notice vibrations as small as 0.002 in.

In building an engine foundation the problem is not so much to prevent the vibration of the engine or its foundation as to prevent these unavoidable vibrations from being transmitted to the power-plant building or to other buildings in the neighborhood. Many parts of a building, such as roof trusses or floor beams, have a low natural frequency of vibration and are apt to develop resonance.

There exist several measures of precaution against the transmission of vibrations. First, the engine foundation must not come in direct contact with any part of the building foundation; in places where the engine foundation comes within a few inches of the building foundation, the intervening space should be filled with dry sand. Second, the engine foundation may be isolated from the subsoil by putting under it a layer of cork. Specially prepared slabs made of compressed granulated cork can support a load up to 3,500 lb per sq ft. A third measure is to isolate the engine from the concrete foundation by putting it on special flexible supports, as shown in Figs. 18-7 and 18-8. Naturally, if the engine-driven machine is direct-connected to the engine by a flange or flexible coupling, the whole unit must have a common subbase, usually of welded construction, and the vibration isolators are put between this subbase and the concrete foundation, which in this case can be made much lighter.

19-4. Excavation for the Foundation. If the soil is firm, no retaining walls will be needed. It is necessary merely to make the excavation wider and longer than the foundation is to be, and then to erect the concrete forms, supporting them by braces from the dirt walls. For small engines the excavation may be made to the foundation dimensions, so that the concrete may be poured directly, without wooden forms. When wooden forms are employed, they must be removed after the concrete hardens, and dirt filled in all along the concrete.

Template. Except for small engines, no diesel engine should be installed without a template being constructed to insure that the foundation bolts are properly located.

This template is made up of 1- by 8-in. or 1- by 6-in. yellow-pine or douglas-fir boards, with the positions of the foundation bolt holes in the engine bed located on the boards. In some instances the engine builder is willing to supply the template; otherwise, it must be built locally and the holes for the foundation bolts laid out from the engine blueprints. In Fig. 19-1 is shown such a template for a small engine.

Setting the Template. With the foundation excavation completed and the side forms in place, the next step is to place the template over the opening and hang the foundation bolts in place.

The template is best suspended from or attached to the under sides of two 2- by 4-in. beams set on edge as shown in Fig. 19-2. The beams are heavy enough to resist any accidental shove of a workman.

The template should be set above the top of the foundation form a distance equal to the height of the engine base; that is, a distance equal to the amount the foundation bolts must rise above the foundation top to enable them to pass up through the bolt openings of the engine and, in addition, enough to enable the erector to screw on the top nuts when the engine is leveled up.

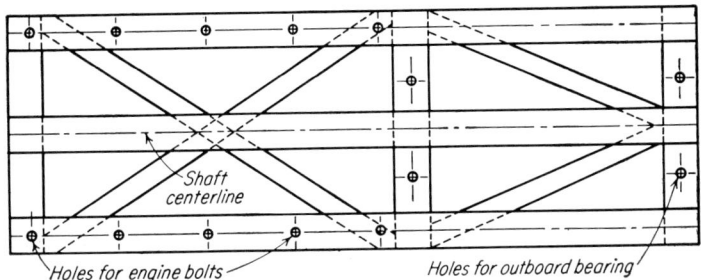

Fig. 19-1. Template for a small engine.

The template top has marked on it the center line of the crankshaft; it must be shifted with the supporting beams until the template center line coincides with the desired location of the crankshaft. As for the foundation bolts, also called *anchor-bolts*, it is necessary to decide what bolt type is to be adopted.

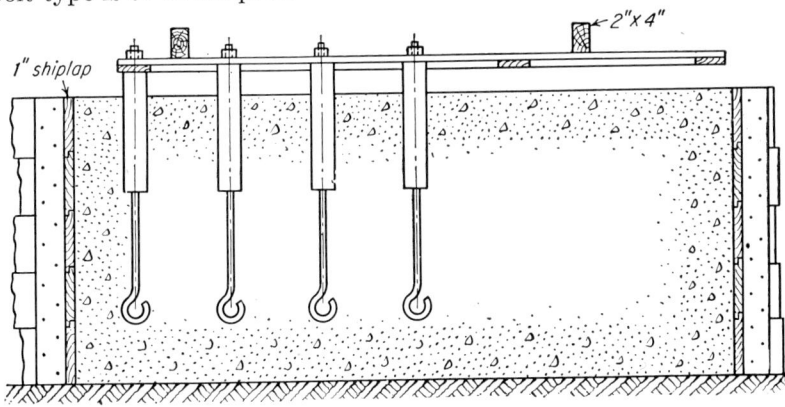

Fig. 19-2. Hanging of template and bolts.

With large engines, it is common practice to make the bolts with nuts on both ends and to insert them after the engine is in place. To do this, collapsible wooden braces *a, a* are placed at the bottom of the excavation and against the side forms, as shown in Fig. 19-3a and *b*. From the top of each box, a piece of 3-in. sheet-metal tubing, such as a roof down pipe, is placed so that each pipe will form a conduit for the foundation bolt. Better still, the bolts may be hung in place inside the tube and then removed when the foundation has been finished. After the engine base is in place, the bolts are dropped back into place and, after removal of the collapsible braces, the lower nuts are screwed on. The lower nuts should rest against cast-iron anchor plates b, Fig. 19-3c, round or square, about six to eight times the anchor-bolt diameter.

The advantage of this arrangement is that the engine can be shifted upon the foundation without being raised to clear the bolts as is necessary when the bolts are fixed in place.

For engines under 1,000 hp, the anchor plate is placed in advance at the lower end of each bolt and the bolt is slipped inside the sheet-metal pipes and suspended from the template. The space between

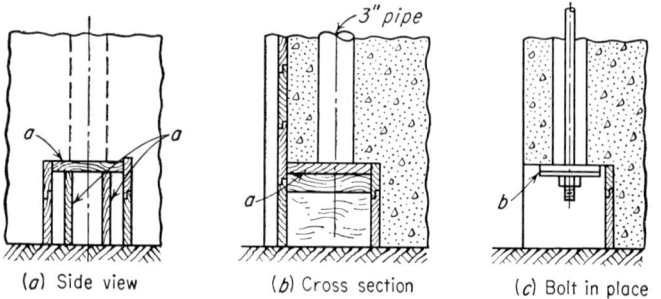

(*a*) Side view   (*b*) Cross section   (*c*) Bolt in place

FIG. 19-3. Placing of wooden braces at bottom of excavation and against side forms.

the bolt and the pipe enables the bolt to accommodate any inaccuracies in bolt location in the engine bed. The space may be filled with cement grouting after the engine is in place.

Smaller engines, up to about 500 hp, usually have anchor bolts with the lower ends bent to form a hook or ring and embedded when the concrete is poured. The upper ends of the bolts are surrounded by a 2- or 3-in. sheet-metal pipe, 18 to 24 in. long, to permit them to be bent slightly to fit the holes in the bedplate. To prevent pulling out of the bolts when the nuts are tightened, the length embedded in concrete should be equal to at least 30 bolt diameters.

Setting of Concrete. Before the engine is put on the foundation, the concrete must be allowed to set, or harden, properly; the time required is at least one week for a small foundation and up to two weeks for a large one.

19-5. Leveling the Engine. Larger engines are often shipped partly dismantled; the bedplate and crankshaft form one piece, the crankcase or frame forms another piece; the cylinders are also separate. In erecting such an engine, the first step is to move the bedplate onto the foundation. If the anchor bolts are inserted later, the bedplate is moved by using pieces of 2-in. pipe as rollers. If the foundation bolts are placed in advance, it is necessary to build up a cribbing sufficiently high to enable the bottom of the bedplate to clear the upper ends of the anchor bolts. If the bedplate is shipped mounted on wooden skids, pieces of gas pipe are inserted between the cribbing and the skids to act as rollers after the bedplate has been raised sufficiently by jacks. Two or three crowbars properly applied will force the bedplate along until it is in place over the bolts. The bedplate is then slightly raised using four railroad jacks or hydraulic jacks and the skids are removed. After this the cribbing is gradually removed and the bedplate lowered within about 1 to $1\frac{1}{2}$ in. from the top of the concrete foundation. In this space are slipped steel wedges which are driven in when the bedplate is brought to level. Care must be taken to use a sufficient number of wedges—not less than 4 or 5 on each side, and even more on long bedplates—to prevent sagging of the bedplate under the influence of its own weight. A more convenient method is used by several engine manufacturers: the lower flange of the bedplate has a number of tapped holes, usually near the foundation-bolt holes. The holes are tapped for 1 to $1\frac{1}{2}$-in. screws, depending on the engine size.

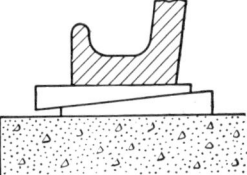

Fig. 19-4. Wedges for leveling engine.

Screws with hex or square heads inserted in these tapped holes are used as jack screws, permitting easy and accurate leveling. To prevent chewing up of the relatively soft concrete surface, a flat piece of steel about 2 by 2 by $\frac{1}{4}$ in. thick should be put under each screw. When leveling a heavy engine with wedges, it is advisable to use double wedges, as shown in Fig. 19-4, to prevent the concrete from being chewed up and the edges of the bedplate from being damaged.

The top of the bedplate is always machined exactly parallel with the crankshaft center line. If the top of the bed is leveled, the crankshaft is also made level.

To level the bedplate, a spirit or machinist's level is placed lengthwise along the top of the bed. The wedges are then moved until the bubble of the level is at center. The same is done for both sides of the bed.

Levelness crosswise is checked by placing the spirit level crosswise of the machined ends. With large engines the erector should check the alignment by measuring with a strain gauge the distance between the two webs of each crank when the shaft is at top and bottom center of a particular crank. The gauge should show no variation over 0.002 in. for a medium-sized and 0.003 in. for a very large engine. If the gauge shows a greater difference, the bed is distorted and the leveling must be rechecked.

With the care to be expected of any erector, the bed is finally leveled in all directions. The nuts of the foundation bolts should then be run down by hand, without the use of a wrench, or at best, with only a short wrench.

19-6. Grouting the Bed. A mixture of cement and sand, a 1:2 ratio, should be watered until it is free-flowing. This mushy grout is then poured under the base sometimes so that the spaces between the bolts and sheet-metal pipes are filled. However, a better practice is to stuff the spaces between the bolts and sheet-metal pipes with waste to prevent the grout from filling these spaces. The greater free length of the foundation bolts gives more resilience, prevents unscrewing of the nuts due to vibration, and reduces the stress in the bolts due to tightening of the nuts. A stiffer grout is then troweled under the base and made smooth all along the exposed joint.

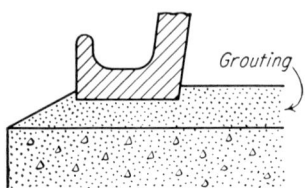

FIG. 19-5. Grouting of the bedplate.

From the outside the grouting should not extend above the bedplate bottom and should be sloped, as shown in Fig. 19-5, to prevent lubricating oil from accumulating on top of the foundation. Oil has a tendency to penetrate into concrete and to deteriorate it. The penetration of oil can be definitely prevented by applying to the top and sides of the foundation a coat of concrete-floor primer and two coats of concrete-floor enamel. This should be done as soon as possible after the grouting has hardened. One must be sure to use only oil-proof enamel made specially for engine-room floors.

While some erectors do not remove the wedges, good practice requires that they be removed after the grouting hardens and before the foundation-bolt nuts are tightened by wrenches. Otherwise the engine is supported by the small surface of the wedges. Also later one of the wedges may be driven in further by error, which would distort the cast-iron bed and throw the crankshaft out of alignment.

Erecting the Frame. The engine frame, or crankcase and cylinder assembly, whichever may be the design, is then shifted onto the bedplate and located by the dowel-pin holes which have been drilled in assembling the engine at the factory. Other parts are set in place in the natural process of engine assembly.

19-7. Aligning the Shaft Extension. In many instances an engine is supplied with a generator whose shaft is coupled to the engine shaft.

In this case it is necessary first to set the engine in place and then to slip in the rotor and its shaft. To locate the outboard bearing, which supports the outer end of the generator shaft, the two coupling halves are shifted together and the extension shaft moved until the fitted coupling bolts enter the reamed holes in each coupling half. This gives a rough alignment, and the finer adjustments are made with the aid of a feeler gauge like the one shown in Fig. 19-6.

The outer end of the shaft is moved by shifting the outboard bearing until the feeler gauge shows the same clearance all around the coupling gap. The coupling bolts are then drawn up tight.

A similar procedure is used if the engine is coupled to an extension shaft with a belt pulley.

Outboard Bearing. A mistake sometimes made when the outboard bearing is not furnished with the engine is to use a pillow block with a ball or roller bearing, instead of a plain bearing similar to the main bearings of the diesel engine. The latter will gradually wear and the crankshaft center line will go down. A plain, babbitted outboard bearing, if selected properly,

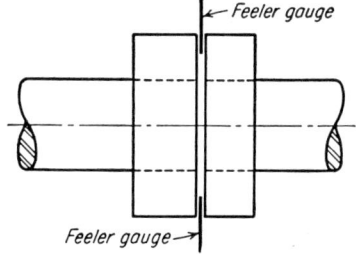

Fig. 19-6. Aligning shafts with a feeler gauge.

will show about the same wear and the center line of the engine crankshaft and extension shaft will remain a straight line. A ball bearing or roller bearing does not wear down and the center of the outboard bearing will become higher than the center line of the crankshaft. This will cause bending stresses and may lead to breakage of the crankshaft or of the extension shaft.

Boat Engines. The crankshaft center line of a boat engine driving a propeller is not horizontal but slopes down toward the stern. The stern bearing and comparatively long propeller shaft and intermediate shaft are installed and aligned first, and the engine is lined up with them by the use of a feeler gauge at the flange couplings (Fig. 19-6) by means of steel wedges or jack screws if the latter are provided. Instead of grouting, proper shims of steel strips or hardwood are prepared and inserted between the engine base and the ship girders, the wedges removed, and the holddown-bolt nuts tightened. Since the foundation is not so rigid as a concrete foundation of a stationary engine, it is good practice to use a nut and a locknut to prevent unscrewing of the nut. If the locknut is thinner than the main nut, it should be put on first. In either case, when screwing up the top nut, the lower nut should be held by a wrench and thus lock the two nuts through friction.

19-8. Protection of Parts. It is important to remember that during the entire work of erection all parts should be protected from dirt. Each crankpin should be covered with cloth and wrapping paper. Each of the main bearings should be covered with cloth and all oil holes plugged with wads of waste. The generator, as soon as it is in place, should be covered with a tarpaulin, or in the absence of this, with building paper. All tapped holes should be closed by plugs and threaded pipe ends by caps or at least with pieces of oil rags tied by wire.

QUESTIONS

1. What three basic considerations must be taken into account in selecting a new engine?
2. How does altitude affect the power of a diesel engine?
3. What factors should influence the selection of an engine from the viewpoint of speed?
4. When is a two-stroke engine used in preference to a four-stroke one and vice versa?
5. What are the factors governing the selection of the number of cylinders in an engine?
6. What three functions must the foundation of a diesel engine perform?
7. What are the safe loads for an engine foundation with different soils?
8. What is the empirical formula for determining the foundation weight sufficient to absorb vibrations produced by a diesel engine?
9. What material is used for a diesel-engine foundation, and to what specifications?
10. What is the purpose of a template for an engine foundation?
11. Explain with a sketch how a foundation template is set.
12. What type of bolts is used for the anchor-bolts for engines (*a*) up to 500 hp, (*b*) up to 1,000 hp, and (*c*) over 1,000 hp?
13. How long does it take a concrete foundation to set before the engine can be placed on it?
14. Describe briefly how an engine is placed on the foundation.
15. What methods are used for leveling the engine bedplate on its foundation?
16. What is *grouting?* How is it made up, and what is its purpose?
17. Describe how a shaft extension connected to the crankshaft by a flange is aligned with the engine crankshaft.
18. Explain why it is wrong to use a pillow block with ball or roller bearings for an outboard bearing.
19. Why is it recommended to use lock nuts with the foundation bolts of a boat engine?
20. Describe how various engine parts should be protected from dirt during engine installation.

CHAPTER 20

OPERATING A DIESEL ENGINE

20-1. Before Starting. There are several steps to be taken before starting a diesel engine, especially the first time, and it is good practice to work out a certain routine to be followed always:

1. All moving parts of the engine must be examined for proper adjustment, alignment, and lubrication. This includes valves, cams, valve gear, fuel pumps, the fuel-injection system, the governor, lubricators, oil and water pumps, and the main driven machinery.

2. The whole engine and machinery must be examined for loose nuts, broken bolts, loose connections, and leaky jackets, joints, or valves. It is well to remember that nothing that must be tight should be loose and nothing that must be free should be tight.

3. All tools from the tool board should be checked to make sure that none is missing. They may be needed in a hurry when the engine is running or, if misplaced and left on the engine, may drop off from vibration and damage some moving parts.

4. All pipes and valves for fuel, lubricating oil, water, and air, as well as ducts, must be checked for clogging up, lack of adjustment, cleanliness, etc.; absence of foreign matter in the piping systems must be checked especially carefully if the engine has been idle for some time or is just being put into service. In the latter case it is advisable to blow out the entire piping system with compressed air.

5. A complete checkup must be given to the lubricating system to make sure that oil is present in every place required, that the lubricator and all bearings that are individually oiled have an ample supply of clean oil, that all grease cups are filled. The lubricator should be checked for proper functioning of the pumps and for the amount of oil delivery, and filled with oil to the proper level; the lubricator should be turned by hand and the points to which it delivers oil should be well lubricated. Make sure that the engine will receive proper lubrication the very moment it starts to turn.

6. The cooling system must be checked, and if the pumps are driven by electric motors, they must be started; the suction line must be opened to have water in the engine before starting. The correct amount of water circulation should be adjusted later, while the engine is being warmed up.

If the engine has oil-cooled pistons with oil delivered by a special pump, start the oil pump and adjust the pressures to the amount stated on the name plate or given in the engine builder's instruction book.

7. The fuel-oil system must be checked in every respect, to make sure that pipes are clean, pumps are working, and that a supply of fuel is in the tanks. The fuel-injection pumps should then be primed, and air or water removed from the discharge line, valves, or nozzles. One or two strokes on the fuel-injection pump is usually sufficient. Care should be taken not to force too much fuel into the combustion chamber or cylinder in order not to obtain an excessively high pressure with the first firing—causing the safety valves to pop—and not to get the fuel oil into the crankcase sump. However, the fuel pumps must be primed sufficiently so that each discharge line is filled clear to the nozzles. The fuel-control lever is set wide open so that injection will start at once. The fuel-pump control is put in the FUEL-ON position.

8. The safety valves, usually installed on each cylinder head, should be checked. These valves are set to pop off at about 750 to 1,250 psi, depending upon the maximum pressure allowed in the engine. The valves are exposed to high-temperature gases and have a tendency to stick. The checking may be done either by compressing the spring with a crowbar or by unscrewing the cap and taking the valve out for inspection.

9. The engine should be turned over one or two times if it has not been operated for some time. To do this it is necessary to open the indicator cocks or compressor-relief valves and to turn the engine over, either by hand with a bar inserted in the holes in the flywheel rim, or with a jack or air motor, as the case may be. Then the indicator cocks should be closed with the engine in proper position for starting—one cylinder having the starting-air valve open and the piston about 10 deg past top center.

10. The starting-air in the tanks must be checked to see that it is up to the required pressure. If not, it must be pumped up. The starting-air system from the tanks to the starting main control valve must be opened, after it has been checked that the main control valve is closed. With an air-injection engine the bottle with injection air must be checked and if necessary pumped up to the required pressure.

OPERATING A DIESEL ENGINE 339

11. The engine load should be off: the switch should be open if the engine drives a generator, or the clutch should be in neutral position if the drive is through a friction clutch. If the engine drives a pump or compressor, the by-pass should be open.

20-2. Starting. If all eleven points of the preparatory program have been observed, starting with compressed air is very simple.

First, the main starting-air valve is opened and the starting lever manipulated according to the instructions given in the engine instruction book.

Second, the engine is watched; no unnecessary air should be used. At the first indication of combustion, air should be cut off and the ventilating valve opened. An engine in good condition usually begins to fire between the second and fourth revolution of the crankshaft.

Third, if the engine fails to start after four or five revolutions, there is something wrong. Useless turning of the engine should be stopped and the cause of trouble investigated, as outlined in Sec. 20-7.

Low Air Pressure. If the starting-air pressure is too low either from a slow loss of air through some leaky joints or from failure of the engine to start at the first attempt, and there is no air compressor to pump up air, several methods may be used for securing the necessary starting pressure but *never should pure oxygen be used* for starting purposes.

Flasks of compressed air may be obtained and the contents equalized into the engine receivers, or a flask of carbon dioxide may be obtained from some local soda fountain and piped to the starting bottles. This gas is a liquid at ordinary temperatures and at about 800 psi pressure. It is therefore necessary to apply some heat in order to evaporate this liquid carbon dioxide. This heat may be applied by pouring hot water over the bottle or by applying rags soaked in hot water.

20-3. Warming Up. After the engine is started, before putting on the load, it should be allowed to idle for a few minutes (up to five) and warm up. During these 5 min. the following observations must be made:

1. Listen to find out if combustion is regular and firing order correct, check all cylinders for combustion, and note the working of the fuel-injection pumps to see whether they all operate properly.

2. Observe the cooling water system throughout to see whether the pumps are working and there is sufficient water; watch to see if the water temperature is building up properly; and regulate the water flow accordingly.

3. Observe lubrication pressures and the working of the lubricator, and count the number of drops for correct operation. Feel whether any of the cylinders is warming up too fast—indicating an unlubricated

piston—and listen for an unlubricated piston-pin or crankpin bearing. If any moving part receives an insufficient amount of lubricating oil, serious trouble may result.

4. Observe the exhaust, color and sound, to note proper conditions. These observations should be repeated after the load is put on. The color of the exhaust can tell many things, as will be shown later.

The making of these observations during the first five minutes after starting should be a regular habit with the engine operator. This procedure is the best and most reliable method of preventing improper operation. It is based upon the fact that a diesel engine requires neither much attention nor continuous attention, but it requires proper attention at the proper time. It is also based on the known fact that a diesel engine should be operating properly in five minutes or there is something wrong which should be detected in those five minutes.

However, it should be noted that certain observations should be carried on even after the 5-min warming-up period. Thus, if there are any leaky water jackets, injection valves, air valves, etc., they may not show up until full expansion of the corresponding part has taken place after the engine has been in operation a longer time at normal load. No leaks of any kind should be allowed; if they cannot be stopped while the engine is running, the engine should be stopped and not restarted until the trouble is corrected.

20-4. Running. In general the attention which an operator must give to the engine in regular operation is along the same lines as during the warming-up period. The difference is that the corresponding observations should be made periodically every 15 or 20 min. and at least every half hour, even if the engine is equipped with a sufficient number of automatic danger-warning signals, and second, that all observations must be entered in the engine log.

The entrances in a complete engine log are the following:

1. *Time* of entering the readings, or rather the first reading in each series.

2. *Engine load* or, in the case of electric load, volts and ampere readings.

3. *Engine speed* from the tachometer, or if the engine has an adding revolution counter, the counter reading; in this case it is essential to have in the engine room a large electric clock with a hand indicating seconds, to enable the operator to read the revolution counter at exact intervals.

4. *Fuel consumption,* either the instantaneous reading of a Rotameter or the reading of a fuel meter in which case it is also important to make the readings at exact intervals.

5. *Exhaust:* (*a*) readings of the temperature of exhaust from each cylinder; (*b*) exhaust temperature in the exhaust line close to the exhaust manifold; (*c*) color of exhaust—either by simple description such as "clear, little haze, light gray, gray, dark gray, and very dark gray," or better, by a number according to a standardized smoke scale, such as Ringelman's scale.

6. *Lubricating oil:* (*a*) pressure as discharged from the oil-pressure pump; (*b*) temperature of the oil before the oil cooler; (*c*) temperature of the oil after the oil cooler.

7. *Cooling water:* (*a*) temperature of the water as delivered to the water-cooling manifold; (*b*) temperature as discharged from each cylinder, or in the water-outlet line; (*c*) flow, gallons per minute, either from a Rotameter or a water meter.

8*a*. *Scavenge air:* (1) temperature after blower; (2) pressure after blower, usually in inches of mercury.

8*b*. *Supercharger conditions:* (1) temperature of air after booster pump; and (2) pressure of air after booster pump, psi or inches of mercury.

9. *Barometric pressure*, inches of mercury.

10. *Temperature of the air* at the air intake, before the air filter.

11. *Remarks* about what happened at a certain moment during operation of the engine, such as: put second engine on line or stopped it; found lubricating-oil filter clogged by dirt as indicated by excessive pressure drop, switched to the second filter, or by-passed filter and exchanged filter element, etc.

Between taking readings and entering them in the engine log, the operator should listen to find out if the engine is running uniformly, without unusual sounds or knocks; he should feel whether the bearings are running warmer than usual, and particularly watch that the engine as a whole does not become overloaded or some of the cylinders become overloaded because the combustion in one or two cylinders does not proceed correctly, as indicated by a considerably lower or higher temperature of the exhaust from that particular cylinder or cylinders.

Naturally, the operator must also see that the day fuel tank is not depleted and, if the engine has hand-lubricated places, that they are oiled at regular intervals.

The camshaft valve levers, valves, fuel-pump levers, etc., should be oiled every two hours. The exhaust valve stems should receive a few drops of kerosene instead of oil every three or four hours in order to keep them in good working condition. The circular groove around the valves and the whole top of the cylinder head must be wiped clean at all times. Oil must not be allowed to accumulate on the cylinder head and run down the sides of the engine, as it could easily work into the joints between the cylinder and heads and decompose the rubber gaskets which form the water joint.

If the flow of cooling water or oil should stop for any reason, the engine or any one cylinder will become overheated. The engine must be stopped at once and permitted to cool off gradually. It is extremely dangerous to admit water to a hot engine as a sudden change in temperature may cause the pistons to seize or one of the cylinder heads, liners, or the exhaust manifold to crack.

Exhaust Conditions. Normally, the exhaust from the engine should be perfectly clear. However, if the engine is operating under an overload, the exhaust may become visible, with a light-grayish smoke. If the engine exhaust is visible under other than overload conditions, the cause should be found and immediately remedied. An engine should under no condition be operated for any length of time with a visible, smoky exhaust.

If a pyrometer with thermocouples is installed on the engine, the cylinder that yields a smoky exhaust may be found by noting the exhaust temperatures of the various cylinders. If an abnormal condition exists in any one cylinder, this condition will usually be accompanied by an increase in the temperature of the exhaust from this cylinder. However, smoky exhaust may be caused by one or two cylinders with abnormally *low* exhaust temperatures, which indicates that these cylinders do not get their share of fuel, and as a result, the other cylinders are overloaded. If possible, the engine should be stopped and the cause found and remedied as indicated in Sec. 20-13.

20-5. Stopping the Engine. To stop the engine, proceed as follows: Move the fuel-pump control to STOP position and shut the fuel supply valve.

The cooling water and piston-cooling oil should be left running after the engine is shut down until the outlet temperatures are not more than 5 to 10 F higher than the inlet temperature. This prevents local overheating which would cause scale deposits in the jackets. If hard water is used and the engine is supplied with direct-connected pumps, it will be necessary to start the auxiliary pumps to cool the engine down as indicated above.

If the engine is to be shut down for a considerable length of time, the water jackets must be completely drained so as to prevent rust and in cold weather also to protect the jackets from bursting if the water in the engine room should freeze. Naturally, all drop oilers must be stopped, all switches cut out, and friction clutches put in neutral position.

20-6. Types of Trouble. The term *trouble* covers a great number of conditions that can be classified into three separate groups.

A. Troubles *interfering* with proper operation of the engine:

1. Failure of the engine to start.
2. Failure of the engine to come up to speed.
3. Failure of the engine to develop full power.
4. Irregular engine speed.
5. Overspeeding of the engine.
6. Sudden stopping of the engine.

B. Troubles *noticeable* while the engine is running:

7. Smoky exhaust.
8. Abnormal cylinder pressure.
9. Abnormal exhaust-gas temperature.
10. Incorrect cooling-water temperature.
11. Excessive piston-cooling oil temperature.
12. Overheating of the engine.
13. Noisiness of the engine.
14. Vibration.
15. Hot starting-air pipe.

C. Troubles *found* when the engine is partly dismantled and inspected:

16. Piston and rings gummed up.
17. Fuel injector or exhaust valve carbonized.
18. Water in crankcase.

With the large number of different diesel engine makes in the field, it is impossible to cover the ground so thoroughly as to anticipate all possible causes of trouble and to prescribe remedies for all of them. The object of the following pages is to show the general relations between various symptoms, actual causes, and remedies. If the engine operator obtains from these pages a general picture of what happens, he ought to be able to find in this picture the proper place for his engine and with the aid of the instructions furnished by the manufacturer of his engine to rectify, or, as it is said, to *shoot*, his *engine troubles* without too much difficulty.

20-7. Failure to Start.

1. The flywheel, having turned ahead during the expansion, exhaust, and suction strokes, stops during the compression stroke in the starting cylinder and turns back.

The starting valve either sticks or leaks; it must be taken out, examined, and made to operate properly.

2. Engine turns over but does not come up to a sufficient speed to start firing. This may be due to one of several reasons: (a) starting-air pressure in the tank is too low; (b) the valve between the air tank and engine is not opened properly; (c) air-starter valve is out of order; the valve must be taken out, examined, and fixed; (d) air-starter timing is not correct. The timing should be checked and corrected if necessary; and (e) excessive friction in the bearings or pistons. This may happen after an overhaul in which insufficient care was taken. The clearances should be checked carefully and adjusted to standard.

3. The engine turns over freely but does not fire. This may be from any of three main reasons: (a) no fuel oil is injected into the cylinder; (b) fuel oil is injected too late; (c) compression pressure and temperature are too low.

Lack of Fuel. There are various reasons to account for lack of fuel injection.

a. The fuel supply tank may be empty or a hand-operated valve in the supply pipe closed.

b. The fuel pump, filter, or injection line may be air-bound. Disconnect the injection line at the nozzle or header and operate the fuel pump by hand until the fuel discharge becomes regular, without air bubbles.

c. The fuel-injection pipe fittings may be leaking badly. The pipe must be inspected for breaks about the fittings and for poor surface joints.

d. The suction or delivery valve of the fuel pump may leak or stick. The valves must be removed, inspected for a broken spring, and seats examined, cleaned and ground, if necessary.

e. The fuel-pump plunger may be leaking from excessive wear and not building up enough pressure for injection. Both the plunger and the barrel must be replaced by a new barrel-and-plunger unit.

f. Fuel lines may be stopped with scale, dirt, or other matter sometimes found in fuel. All impurities must be removed, the strainers inspected and cleaned.

g. Fuel-oil filter may be clogged up. Inspection of the filter and the maintenance log should show the cause and suggest the remedy.

h. Water may be present in the fuel. If water has been found, the system must be drained until there is assurance that all water and dirt are removed.

i. Fuel oil may not flow freely because of cold weather. The fuel will have to be heated or a lighter oil used for starting the engine.

Late Injection. If the fuel enters the cylinder when the piston has started downward, the expansion of the air may cause its pressure and temperature to drop to such an extent that the fuel cannot ignite.

j. Injection timing must be checked and possibly advanced.

Low Compression. The compression pressure may be too low from any of the following causes:

k. The air filter or the slits in the intake pipe may be clogged; these parts should be inspected and cleaned.

l. The exhaust or intake valve may not seat properly. The valve lifts should be checked, the valves reground, and the valve stems freed with kerosene.

m. The valve cages may leak. The gaskets must be inspected and, if necessary, renewed.

n. The gasket between cylinder and cylinder head may leak. The gasket must be tightened or renewed.

o. Piston rings may be stuck, broken, or worn, causing blow-by. To check blow-by, the crankcase cover is removed, the piston is set in the air-starting position, the crankshaft secured by wooden blocks so that it cannot turn, the safety valve in the cylinder head replaced by an air-pressure gauge, and air admitted through the air-starter valve. A hissing sound into the crankcase and quick drop of the pressure in the compression space will indicate a blow-by, necessitating pulling of the piston. All piston rings must be freed and any broken or excessively worn ones renewed.

p. The mechanical clearance between the piston crown and cylinder head may be excessive. The clearance must be measured and adjusted to the normal value by inserting shims between the crankpin bearing and connecting-rod leg.

20-8. Failure to Come Up to Speed. If the engine starts firing but does not come up to normal speed, without load or under a small load, the possible causes may be:

1. Fuel supply is not sufficient because fuel pump valves stick or leak or a pump plunger is excessively worn.

2. Fuel suction pipe or fuel filter clogged.

3. The governor does not act properly, needs adjustment.

4. One or more of the cylinders are not firing properly or missing, regularly or occasionally. The trouble may be in their injection equipment or timing. If the engine is not equipped with a pyrometer and thermocouples at the exhaust from each cylinder, the missing cylinder may be located by opening the indicator cocks or test pet cocks.

5. Misfiring may be caused also by insufficient compression in one cylinder because of leaky valves or gaskets, or stuck or broken piston

rings. Locating of the exact cause and its elimination are very important for satisfactory operation of the engine in the future.

6. Misfiring at light loads may be caused by excessive cylinder cooling, if the normal compression ratio of the engine is not very high. This cause is easily found and eliminated by reducing the flow of jacket water and thus raising its temperature.

7. Water leaks into one of the cylinders. The cylinder head must be removed and tested by hydraulic pressure to about 100 psi, and repaired if a leak is found. The cylinder head gasket should be examined and possibly renewed. The cylinder liner should be examined for cracks or leaks. If these cannot be fixed, a new liner must be inserted.

8. Poor combustion. In an air-injection engine this may be caused by an excessively low injection-air pressure. In an airless-injection engine poor combustion may be caused by poor atomization or incorrect timing.

9. Exhaust back pressure is high. The exhaust system may be clogged, the exhaust ports may be filled up with carbon, or the lift of the exhaust valves may be below normal.

10. High internal friction of the engine caused by a hot bearing. Due warning is generally given by the rise of the lubricating-oil temperature and a gradually increasing knocking in one of the cylinders. Large units are often and should be always equipped with some kind of thermometer giving the temperature of each main bearing. When the temperature of one of the bearings begins to rise, the trouble should be at once investigated in order to determine the cause and to try to eliminate it. If the bearing temperature or lubricating-oil temperature continues to rise, the engine should be shut down as quickly as possible.

11. High internal friction of the engine caused by a piston which is ready to seize. Piston seizure is usually caused by one of three causes: (a) Insufficient or poor cooling of the liner walls, mostly because of scale deposits from hard or dirty water. (b) Improper cooling of the piston itself. (c) Insufficient or poor piston lubrication.

The operator usually has a warning of an incipient seizure in the form of a gradually increasing knocking in one of the cylinders. If the knocking is not too loud, it may be possible to prevent seizure and nurse the piston back to normal by shutting off the fuel from the specific cylinder and by feeding to it a greatly increased amount of lubricating oil. With splash lubrication of the pistons the engine should be shut off at once and the actual cause of the trouble investigated and remedied.

If the engine operates normally, but gradually or suddenly the speed drops and stays below normal, the most common cause is that the engine has become overloaded. This is easily checked by reducing the load or from exhaust temperatures. At the same time any one of conditions of items 1 to 9 may have occurred and require attention.

20-9. Failure to Develop Full Power. Generally the causes are the same as when the engine starts firing but does not come up to normal speed when carrying a small load, only possibly they are not quite as pronounced. Causes and remedies are given in Sec. 20-8, items 1 to 9.

20-10. Irregular Engine Speed. The possible reasons are the following:

1. Water in fuel pipes. The action to be taken is indicated under $3h$ in Sec. 20-7.

2. Air in the fuel system. Some types of booster fuel pumps with vacuum in the suction line are apt to get air into the fuel system. Repack the packing glands, or better yet, change the day-tank location so that the fuel will flow to the pump by gravity.

3. Irregular operation of the fuel-injection system caused by sticking fuel-pump valves or nozzle parts causing dribbling. The fuel-pump and nozzle should be disassembled, carefully examined, and fixed.

4. Fuel oil is too heavy—actually, too viscous. This happens in air-injection engines. The trouble is eliminated by installing a fuel-heating device to make it more fluid or by changing to a lighter oil.

5. Governor is not functioning properly. The governor sleeve may have too much friction or the dashpot is not adjusted. The governor should be removed and inspected for worn or broken parts. The dashpot should be adjusted.

20-11. Engine Overspeeds. This condition is very dangerous, regardless of whether the engine has an overspeed governor or not and must be corrected at once. There are two possible causes:

1. The governor sticks in the full-load position. The engine must be shut down at once and the governor mechanism inspected for broken parts.

2. The fuel by-pass may be clogged or the mechanism not properly adjusted. The engine must be shut down at once and the fuel by-pass examined and fixed.

20-12. Engine Stops Suddenly. Several different causes are possible:

1. Lack of fuel. The various actual causes and remedies are the same as those discussed in Sec. 20-7, items $3a$ to h.

2. The fuel pump is out of order. Examine for broken drive, slipping coupling, and worn parts.

3. Intake and exhaust valves not working. The valves should be checked for sticking stems. The operating mechanisms should be checked for broken parts, particularly for broken valve springs.

4. Transfer or booster fuel pump does not function. Check for broken drive or worn parts.

5. Governor out of order. Check for sticking, broken, or worn parts.

6. Bearing seizure. Remove crankcase access covers as quickly as possible, find out whether bearings are hot or discolored. If indications are that bearing seizure took place, let engine cool off, remove suspected bearing shell, determine how badly it is damaged. If it can be reconditioned, proceed with the work, checking clearances and the lubrication of all bearings carefully.

7. Piston seizure. When crankcase access covers are removed, a check of the temperature of the skirt and appearance of the cylinder surface will usually reveal which of the pistons has seized. The piston must be pulled and inspected for further damage, such as cracks. If the damage is not excessive, scored areas both in the cylinder and on the piston surface should be smoothed down by stoning; the clearance between the piston and liner should be measured to make sure that it is not too small as well as the gap clearances in the piston rings. The piston lubrication and cooling should be checked. Improper cooling of the liner caused by scale deposits may impair lubrication and cause piston seizure.

20-13. Smoky Exhaust. A smoky exhaust, aside from that observed occasionally during starting, indicates improper operating conditions that will increase the operating expenses for fuel and for needless extra maintenance. There are several possible causes of smoky exhaust:

1. Engine is overloaded. This is the most common and one of the most dangerous causes. As mentioned repeatedly before, overloading a diesel engine for any length of time is a very poor practice. It not only increases the maintenance cost but shortens the life of the engine.

2. Fuel nozzle check valve is leaking. This allows the air charge to pass this valve during compression and thus retards the injection. The check valve must be examined and reground.

3. Fuel-nozzle tip holes are plugged up or worn on one side. This may interfere with proper atomization and formation of a proper cone-shaped fuel spray and permit fuel to impinge on relatively cold surfaces. The fuel nozzle must be taken out, tested in a fuel-nozzle tester, and the tip either cleaned out or exchanged.

4. Compression pressure in one of the cylinders is too low. This may cause this cylinder to misfire. This is discovered either from the

exhaust pyrometer readings or by measuring the firing pressures with a special indicator, as discussed in detail in Sec. 20-7, items $3k$ to p.

5. Fuel-injection pressure too low. A leaky suction or by-pass valve in the fuel-injection pump in a common-rail system may cause a reduction of fuel-injection pressure and poor atomization. The pump must be examined and repaired.

6. Injection-air pressure is low. In an air-injection engine the injection-air pressure must be adjusted, depending upon the load and speed of the engine.

7. Injection is late. If the exhaust temperature is near normal but the firing pressure low, injection is probably too late. The fuel timing must be checked and corrected.

8. Slow igniting fuel. A high-speed engine requires a fuel of a certain minimum ignition quality, measured by the cetane or aniline number or some other index. If the fuel has a lower ignition quality, the engine will smoke at full load, regardless of the adjustments made. The only remedy is to change to a more suitable fuel.

9. Too much lubricating oil gets into the combustion space. In low-speed engines with special feeds to the cylinder sometimes too much oil is being fed, too many drops per minute. The oil feeds must be checked and adjusted.

In high-speed engines the pistons are lubricated by splash and oil mist in the crankcase. The object of oil-control piston rings is to scrape off excess oil which drains back into the sump through oil holes drilled in the piston skirt. The oil-scraper rings must be inspected, as well as the oil-drain holes in the piston. If the holes are plugged up, they must be cleaned. Under certain conditions it may be advisable to increase the size of the drain holes slightly, if they plug up easily, or to drill additional drain holes.

Color of Smoke. After presenting an over-all picture of the causes of a smoky exhaust, it may be helpful to mention the relation between the color of the exhaust smoke and the conditions responsible for it.

There are three basic colors encountered in diesel-engine exhaust: white, light gray to black, and bluish.

White smoke is the result of a low combustion temperature which occurs with a low compression pressure as discussed in Sec. 20-7, items $3k$ to p, and Sec. 20-13, item 4.

White smoke or steam will also appear if water is leaking into a cylinder, as discussed in Sec. 20-8, item 7.

Gray smoke, varying in intensity from light gray to dark gray or even black, is the result of poor combustion as it occurs under conditions discussed in Sec. 20-13, items 1 to 3, 7, and 8.

Late injection, discussed in Sec. 20-13, item 7, often gives very dark or black smoke.

Bluish smoke usually indicates burning of lubrication oil, as discussed in Sec. 20-13, item 9. However, bluish smoke may also be caused by fuel oil being impinged on the walls of the combustion chamber due to plugged-up fuel nozzle holes.

20-14. Abnormal Cylinder Pressure and Exhaust Temperature. These two factors are so closely connected that they should be discussed together. However, since the number of operating conditions that influence the cylinder pressure and exhaust temperature is very great, it is simpler to present the whole picture in the form of Table 20-1. The application of Table 20-1 is very simple: when during the engine operation either the pressure or exhaust temperature or both happen to be below or above normal values, the line in Table 20-1 which contains these two values will indicate the probable operating conditions that have caused the corresponding pressure and temperature.

TABLE 20-1. CYLINDER PRESSURE, EXHAUST TEMPERATURE, AND OPERATING CONDITIONS

Cylinder pressure	Exhaust temperature	Corresponding operating conditions
Low compression	Low	Mechanical clearance too great; compression ratio too low
	Normal	Air intake clogged or air delivery by blower insufficient
	High	Air-charge loss through leaky valves, piston rings, or worn or scored liners
High compression	Low	Mechanical clearance too small; high compression ratio
Low firing	Low	Fuel rack too far out
	Normal	Air intake clogged or low-cetane fuel
	High	Injection timing late, injection nozzle dirty or leaky; high back pressure
Normal firing	Low	Light load
	High	Overload; high back pressure
High firing	Low	Injection timing early
	Normal	Worn orifices in injection nozzle
	High	Fuel rack too far in

EXAMPLE: the firing pressure is low, exhaust temperature is high; this combination is given in line 7 and the operating conditions causing it are given as: injection timing too late, injection nozzle dirty or leaky, or high back pressure.

20-15. Incorrect Cooling-water Temperature. With the exception of engines equipped with a thermostatic control of the cooling water circulation, most engines require that the flow of water be regulated by hand in accordance with the load. If with a more or less constant load the jacket-water temperature begins to climb, the operator must locate the cause at once. There are only two possibilities: (1) either the supply of water has been decreased or shut off by an inadvertent closing of a valve or stopping of the water-circulating pump or (2) a piston is dragging, ready to seize. This condition was already discussed in Sec. 20-12, item 7.

20-16. Excessive Piston Cooling-oil Temperature. The cause may be

1. In the piston as explained in Sec. 20-8, item 11.
2. That the pump circulating the oil does not deliver enough oil.
3. That the oil cooler is not working properly, being either scaled up on the water surface or clogged up by dirty oil.
4. That the oil cooler does not receive sufficient cooling water.

The cause of trouble is usually located without difficulty and must be eliminated as quickly as possible.

20-17. Overheating of the Engine. The possible causes are

1. Flow of cooling water is insufficient. In this case flow must be increased. If water-circulating pump is belt-driven, the belt may be slipping.
2. Scale deposits on cylinder and cylinder-head water jackets. Water jackets must be cleaned as explained in Chap. 25.
3. Water-cooling tower by-pass is open. The flow of water over the tower must be checked.
4. Insufficient lubrication of the pistons. Cylinder feeds must be checked and adjusted.
5. Lubricating oil is poor, dirty, or diluted with fuel. Oil must be renewed; only oil recommended by the engine maker or previously tested and found to be satisfactory should be used.
6. Clogged lubricating-oil filters. Filters must be cleaned and refills replaced where needed.
7. Worn lubricating-oil pump. The pump must be examined and worn parts exchanged or repaired.
8. Incorrect fuel-injection timing. The timing must be checked and corrected according to the engine builder's specifications.

9. Carbonized fuel nozzle. The fuel nozzle must be cleaned and its cooling checked.

10. Afterdribble. The fuel-nozzle valve must be checked for sticking, for a weak spring, and worn nozzle parts.

20-18. Engine Is Noisy. Unpleasant noise, usually in the form of knocking, may have two basic causes: one mechanical, when some engine part hits another part, the other combustion, usually called *fuel knock*.

Knocking from *mechanical causes* may come from several sources among which are the following:

1. Badly worn piston pin or piston-pin bearing. Piston pin must be reconditioned or renewed; piston-pin bearing probably needs to be renewed.

2. Excessive clearance in crankpin bearing. Bearing clearance must be adjusted or bearing renewed.

3. Badly worn pistons or liners or both causing piston slap. A check of the clearances and the appearance of the surfaces will reveal this condition. Liners must be reconditioned or renewed; the same applies to pistons.

4. Piston hits inlet and exhaust valves. This happens in smaller engines with a very small mechanical clearance when a very thin cylinder-head gasket is inserted. The gasket must be exchanged.

5. Loose flywheel key. The key must be tightened. If the keyway in the shaft or hub is worn, it should be cleaned and a new, wider taper key fitted. If the key becomes loose periodically, especially in a large engine, a second key should be installed or two tangential keys used, as explained in Sec. 6-6. A knocking flywheel cannot be tolerated as it may result in a broken crankshaft.

6. There are other mechanical causes of knocks which must be found and remedied. Sometimes it is impossible for the operator to determine what the trouble is even after a thorough investigation. In such a case it is best to have a factory-trained expert investigate and remedy the trouble.

Fuel knock may be caused by several conditions:

7. Injection is too early. The timing should be checked and adjusted. In a large low- or medium-speed engine, it is advisable to take indicator diagrams, regular and offset dead-center, from each cylinder and to adjust the fuel-injection timing accordingly.

8. Fuel-injection system out of order. Examine fuel pumps and injection nozzles for worn and broken parts and replace them if necessary.

9. Improper fuel oil. Fuel knock caused by fuel oil with a low igni-

tion quality will be noticed in more than one cylinder and usually is erratic and intermittent. If equal parts or more, if needed, of fuel oil with good ignition quality are added, knocking should disappear. If possible, the fuel supply should be changed to a brand having good ignition and burning qualities.

10. Too high an injection-air pressure. In an air-injection engine the injection pressure must be adjusted to the load of the engine. If the engine has an injection-air regulator, the latter may become gummy and sticky. Periodical inspection and cleaning is essential.

20-19. Vibration. If the engine begins to vibrate, the causes may be either of the following:

1. Loose anchor bolts. The nuts of the foundation bolts must be tightened; this should be done periodically.

2. One cylinder is misfiring. The missing cylinder must be located and the cause remedied.

20-20. Starting-air Pipe Hot. The air-starting check valve probably does not seat properly. The stem of the check valve may be gummed up and should be loosened by a few drops of kerosene. If the trouble continues, the valve cage should be removed and the valve examined and fixed.

20-21. Gumming Up of Pistons and Rings. This may be the result of one of several causes:

1. Poor lubricating oil. Only oil recommended by the engine builder or tested in the power plant and found satisfactory should be used. Changing to a cheaper oil, even if the salesman claims that his oil is just as good or better than a proven brand, does not pay.

2. Excessive use of lubricating oil. Even the best lubricating oil, if admitted in excessive amount to the combustion space, is apt to give gummy deposits. This condition is discussed in Sec. 20-13, item 9.

3. Incomplete combustion. This condition is discussed in Sec. 20-13, items 2 to 8.

4. Excessive cooling of the engine.

20-22. Carbon Deposits on Fuel Injector and Exhaust Valve. The possible causes are the following:

1. Incomplete combustion. This condition is discussed in Sec. 20-13, items 2 to 8.

2. Incorrect fuel oil. This condition is discussed in Sec. 20-18, item 9.

3. Excessive back pressure. Exhaust passages in cylinder head and exhaust pipe may get clogged by carbon; they should be periodically inspected and cleaned. Excessive back pressure may also be caused

by an excessively long pipe or by a pipe of too small a diameter; the installation must be changed.

20-23. Water in Crankcase. Possible causes are the following:
1. Cracked cylinder head.
2. Leaky cylinder-head gasket.
3. Cracked or leaky cylinder liner.

All these conditions are discussed in Sec. 20-8, item 7.

4. Lower seal of liner is leaking. Liner must be pulled and rubber rings at the bottom renewed.

20-24. General Procedure. If a certain undesirable condition in the operation of an engine is noticed and the above list shows several possible causes of trouble, the operator should start with the simplest one and gradually go over to the other ones until the actual cause of the trouble is determined and eliminated.

One should also remember that due to differences in type, design, and operating characteristics of various diesel engines, not all causes of trouble are encountered in every engine.

QUESTIONS

1. Enumerate the routine steps that must be taken before starting a diesel engine.
2. Describe the procedure of starting a diesel-engine by compressed air.
3. Explain what should be done if the starting air pressure is too low.
4. Enumerate what the engine operator should do after having started a diesel engine during the warming-up period, before putting on the load.
5. Enumerate the items that a diesel-engine operator should read or note and enter in an engine log while operating an engine.
6. What do exhaust-gas temperature readings indicate in respect to load distribution between the cylinders?
7. Describe the procedure of stopping a diesel engine.
8. Indicate the three main groups into which engine troubles may generally be divided.
9. Enumerate the six main troubles that interfere with proper engine operation.
10. Enumerate, if you can, all nine main troubles that may occur while the engine is running and carrying the load.
11. Enumerate the three most common troubles found when the engine is partly dismantled.
12. What does shooting trouble mean?
13. What are some of the causes for an engine's not starting?
14. Indicate why an engine could begin to turn over, but not begin to fire.
15. What could lead an engine to start firing but not to come up to speed?
16. Enumerate the causes of an engine's misfiring, regularly in one cylinder or occasionally.
17. What may be the causes for an engine's not developing full power?
18. What may be the causes for an engine's speed to be irregular?
19. What may be the causes of overspeeding?
20. What may cause an engine to stop suddenly?

OPERATING A DIESEL ENGINE

21. Enumerate the possible causes of a smoky exhaust.
22. What are the three basic colors of a visible diesel-engine exhaust?
23. What may cause a diesel engine to show white exhaust?
24. What may cause gray to black exhaust smoke?
25. What are the two main causes of a bluish exhaust smoke?
26. What does a rising cooling-water temperature indicate if the load is constant? What must the operator do in such a case?
27. What may cause excessive piston-cooling oil temperature?
28. What are the possible causes of engine overheating?
29. Enumerate the possible causes of mechanical knocks in a diesel engine.
30. Enumerate the possible causes of fuel knocking.
31. What may cause an engine to vibrate?
32. What may cause heating of the air-starting pipe?
33. Indicate the possible causes of gumming up of pistons and piston rings.
34. Indicate the possible causes of carbon deposits on the fuel injector or exhaust valve.
35. What are the possible causes of water appearing in the crankcase?
36. If there are several possible causes for a certain engine trouble, how should the operator proceed in locating the actual cause?

CHAPTER 21

ENGINE PERFORMANCE

21-1. Indicator Diagrams. The understanding of what takes place in a diesel-engine cylinder, particularly in respect to combustion, can be greatly helped by a graphical presentation called a *pressure-volume* or *p-v diagram*. In such a diagram, the ordinates, or vertical distances, show the pressure of the gases and the abscissas, or horizontal distances, show the corresponding volume occupied by the gases, *i.e.*, the volume between the cylinder head and the piston at that instant. Figure 21-1

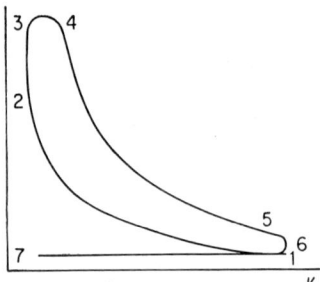

Fig. 21-1. Pressure-volume, or indicator, diagram of a four-stroke engine.

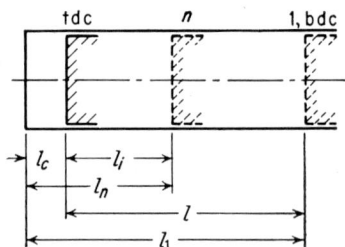

Fig. 21-2. Piston positions in an engine cylinder.

shows a diagram of a four-stroke diesel engine: curve 1-2 shows the gradual pressure rise during the compression stroke; line 2-3 indicates the very rapid pressure rise during the first part of combustion; 3-4 shows the second part of combustion at a more or less constant pressure; 4-5 shows the expansion of the burning gases; release occurs at point 5 with a pressure drop, at first very gradual, then faster and extends slightly after point 6 when the piston starts the exhaust stroke. Because of the small pressure scale, the exhaust line, the greater part of 6-7, and suction, line 7-1, practically coincide with the atmospheric line.

The volume V_n occupied by the gases in the cylinder when the piston is at point n (Fig. 21-2) may be presented as $V_n = al_n$, where a is the piston area in square inches and l_n is the distance in inches from the cylinder head to the piston crown assuming that the combustion space V_c has a cylindrical shape and is l_c in. long, or $V_c = al_c$. If at a point 1, bottom dead center, the total volume of the gases $V_1 = al_1$, the piston displacement, designated by V_d, is

$$V_d = V_1 - V_c = a(l_1 - l_c) \tag{21-1}$$

and the piston stroke $l = l_1 - l_c$. Since a is constant, therefore the volume of gases at any point n of the piston position may be expressed by the length $(l_c + l_i)$, where l_i is the distance traveled by the piston from top center. At top center l_i is zero and the volume of the combustion space is expressed by the length l_c. At the bottom center $l_i = l = l_1 - l_c$ and the total volume of the gases is expressed by the length l_1.

Based on this relation a p-v diagram may be obtained from an engine in operation using an instrument called an *indicator*. Basically an engine indicator consists of a cylinder, one end of which is connected to the combustion space of the engine and the other is closed by a gas-tight but free-moving piston. The piston is fastened to a spring that is compressed or extended, depending upon the construction of the indicator, when the piston is moved by the increasing gas pressure. When the gas pressure begins to decrease, the spring moves the piston in the opposite direction. The motion of the piston is increased several times by a lever arrangement similar to a pantograph and is traced on a drum. The drum turns back and forth imparting to a piece of paper fastened to it a motion proportionate to the engine piston motion. A pencil, fastened to the end of the lever mechanism, is moved up and down by the variable gas pressure and traces a continuous line on the paper, in this way giving a record of pressures at consecutive positions of the engine piston and thus giving a complete picture of the engine cycle. The traced outline is called an *indicator diagram* or *indicator card*. In accordance with this designation, a p-v diagram showing the events in an engine cylinder is called an *indicator diagram*, even if it is drawn not by an indicator but from theoretical computations. In this latter case the diagram is more correctly called a *theoretical* indicator diagram.

An indicator card serves two purposes. First, it shows what happens in the engine cylinder and thus indicates whether the engine is operating properly or whether adjustments should be made to improve its performance. Second, it permits computation of the power developed by the engine. The procedure is based on the fact that the area enclosed by the outlines of an indicator diagram represents the work done during one cycle.

Figure 21-3 shows an indicator diagram of a two-stroke engine: curve 1-2 is compression, 2-3 and 3-4 are combustion, and 4-5 is the expansion curve; release occurs at point 5 and after the pressure in the cylinder drops to the scavenge-air pressure, the burned gases are pushed out by the incoming air; after closing of the exhaust and scavenge-air ports, somewhere near point 1', compression of the fresh charge starts.

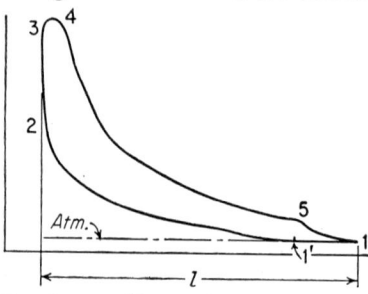

FIG. 21-3. Indicator diagram of a two-stroke diesel engine.

21-2. Indicators. The instrument used to take indicator diagrams is called an *indicator*. In its most usual form an indicator consists of a cylinder c (Fig. 21-4) fitted with a piston p that is leakproof but moves with a very small friction. The piston is moved by the variable gas pressure inside the engine cylinder and this motion is resisted by

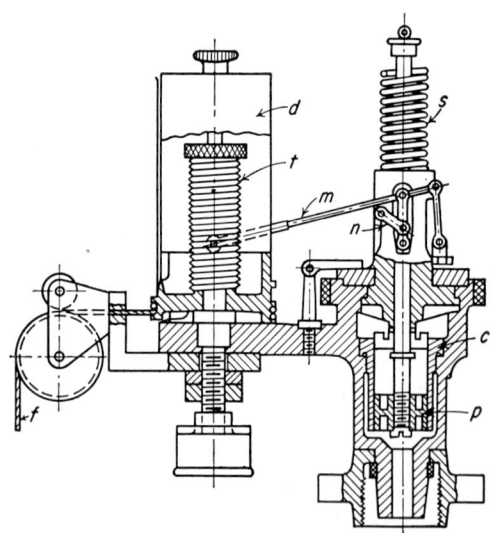

FIG. 21-4. Engine indicator.

a helical spring s of known scale. The motion of the piston is recorded on a paper-covered drum d by means of a magnifying parallel motion m, n connected to the piston rod. The indicator cord f is connected to a reducing motion and its pull turns the drum d. This pull tightens the spiral torsion spring t which, on the next stroke of the reducing mechanism, turns drum d in the opposite direction. Spring s is located outside the cylinder, where it is easy to exchange and is not subjected to the hot gases acting upon the piston.

The normal-size piston has an area of one-half square inch. Pistons of smaller sizes are referred to as one-half normal size, with an area of one-quarter of a square inch, or one-fifth normal size, with one-tenth of a square inch area.

Springs are rated by the pressure corresponding to one inch of the diagram height when used with the normal-size piston. When used with a smaller piston, the spring scale is increased inversely to the ratio of the piston areas. Thus an 80-lb spring will give a diagram of 80 psi pressure per inch of height with a normal piston and a diagram of $80 \times 5 = 400$ psi per inch of height with a piston one-fifth normal size.

The selection of the piston size and spring depends upon the maximum cylinder pressure and engine speed. The best, the least distorted diagrams are obtained by using the largest size of piston possible with a spring that is capable of recording the maximum pressure and produce a diagram sufficiently high to be measured accurately. A diesel-engine indicator diagram should have a height of $1\frac{1}{4}$ to $1\frac{3}{4}$ in. A good piston and spring combination to obtain such a diagram is a piston one-fifth normal size and an 80-lb spring.

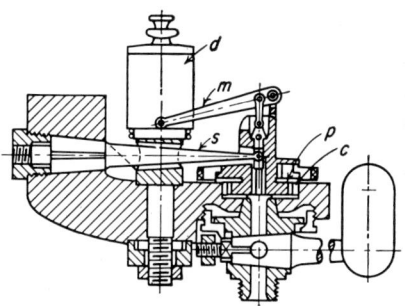

FIG. 21-5. High-speed engine indicator.

High-speed Indicator. Acceleration increases as the square of the engine speed and thus diagram errors that are due to the inertia of moving indicator parts increase in the same proportion. These errors can be reduced by decreasing the weight of the moving parts or by using a stiffer spring and increasing the rate of magnification of the pencil movement in order to obtain a diagram with a sufficient height. Figure 21-5 shows a high-speed piston-type indicator with a cantilever-bar spring s. Such a spring permits one to reduce the size and weight of the recording mechanism. The other letters designate the same parts as in Fig. 21-4.

Indicator Drum Motion. The drum movement must be strictly proportionate to the engine-piston motion. In present fully enclosed diesel engines the simplest way to obtain this motion is by means of a crank-gear reduction mechanism (Fig. 21-6). To obtain accurate diagrams, the ratio of the connecting-rod length l to the crank throw r, r/l, must be the same as that of the engine. The crank is fastened to a disk held by screws to the engine shaft in such a position that the crank lies in the same plane with the engine crank; this is checked by a level placed on the cut-off edge of the disk.

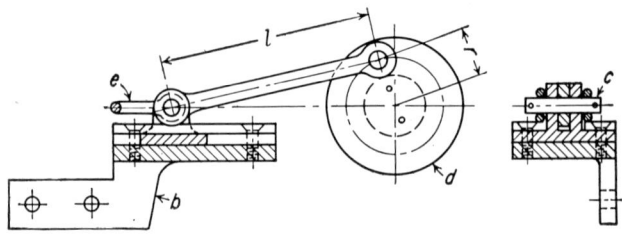

Fig. 21-6. Crank-gear type reducing motion.

Instead of the accurate crank-gear an approximate but easier to make arrangement of Fig. 21-7 may be used. It consists of a crank c, a pair of guide rollers o, o, and a piano wire w for a connecting rod. The crank is extended and its end is bent to the center in order to permit easy slipping on and off of the ring b when the engine is running.

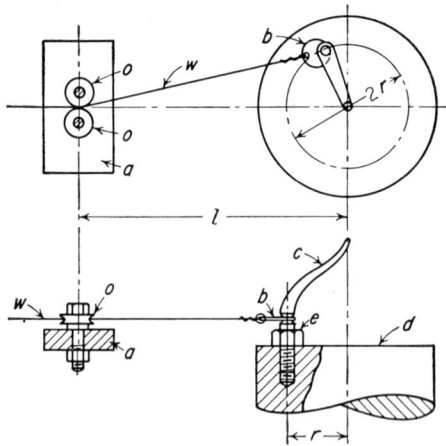

Fig. 21-7. Simple reducing motion.

The crank is screwed into a hole drilled and tapped in the end of the engine crankshaft d or of the key holding the flywheel, as the case may be, and held in position by a lock nut e. An alternate arrangement is shown in Fig. 21-8: the indicator crank c is screwed into a block b that can be adjusted on a steel bar a. This bar is clamped to the end of the engine shaft by a heavy set screw d. The piece f is held by a pin inserted in one of the holes e, e, depending upon the diameter of the engine shaft. The piece b with the crank c can slide along a and is held in position, so that its end point is on the shaft centerline, by a set screw g.

21-3. Offset Diagrams. A regular indicator diagram gives the cylinder events referred to the piston stroke. As explained in Sec. 2-8, the piston motion is not uniform: its speed is zero at one dead center and increases, at first slowly, then faster, and reaches a maximum about mid-stroke; after this it decreases to zero at the other dead center. Because of these conditions a small motion of the piston near one of the dead centers corresponds to a considerably larger time lapse than a similar motion farther away from the dead center. However, in all engines, the important events of ignition and combustion take place near the dead centers, as does the scavenging in a two-stroke engine. Hence these events are not shown clearly enough on a regular indicator

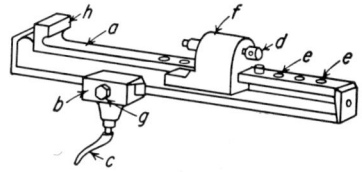

FIG. 21-8. Attachable reducing motion.

card. These events can be investigated much better by means of a so-called *offset diagram*. This diagram is taken with the indicator motion offset by 90 deg to the engine crank. On an offset diagram the events occurring near the dead centers are recorded near the middle of the card where the velocity of the drum is highest and therefore are represented by greater horizontal distances. The action of offsetting may be explained using Fig. 21-9: in a regular indicator diagram, point a corresponds to the moment when the engine crank is 45 deg before dead center on the compression stroke, point b corresponds to the piston being at dead center or when the crank is a few, say $7\frac{1}{2}$ deg, after it. If the pair of guide rollers on Fig. 21-7 is moved 90 deg ahead, keeping the same distance from the engine-shaft center line, a diagram as shown on the right side of Fig. 21-9 is obtained: compression starts at point c; during the compression stroke, when the engine crank is at 90 deg, the drum motion is reversed, point e, and at point a the crank is 45 deg before upper dead center; ignition occurs near dead center, point b, and the $7\frac{1}{2}$-deg motion of the engine crank is distinctly visible, showing that ignition is too late, after dead center. Combustion is practically completed, point f on both diagrams, and pressure release starts at point g on both diagrams. A slight variation of the combustion process, caused by irregular injection, that would give only a heavier line on the regular diagram will give several distinct lines from d to f on the offset diagram.

21-4. Draw Cards. In the absence of a mechanism to take regular and offset diagrams, valuable information may be derived from records made by an indicator with the drum motion obtained by pulling the indicator cord by hand. If the cord is pulled slowly, the pressures in the cylinder are recorded as vertical lines. Such a diagram is useful to check compression and firing pressures, especially in multicylinder engines, to make sure that both compression and firing pressures are reasonably the same, respectively, in all cylinders.

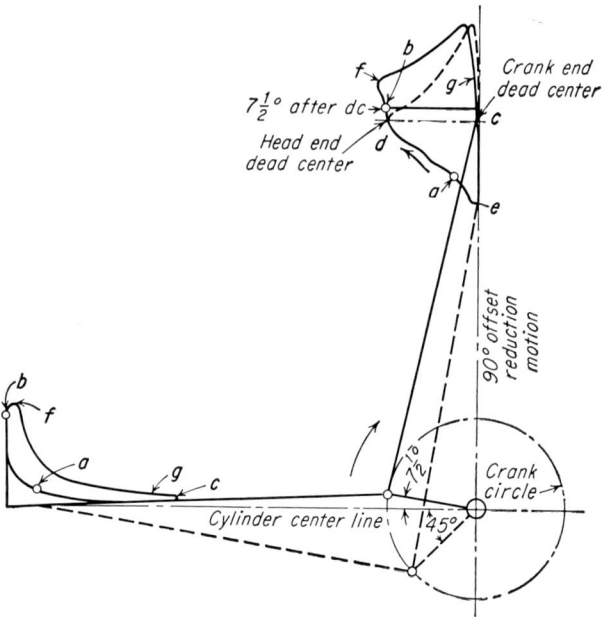

Fig. 21-9. Relation between regular and 90-deg offset indicator cards.

It is recommended to keep a continuous record of cylinder-compression readings in order to notice any falling off of the compression caused by wear or sticking of piston rings or leaking valves. Tests for compression should always be made under the same operating conditions, especially as to speed and load—preferably at full load.

The usual procedure is to take consecutive records, first of the compression and then of the firing pressures of each cylinder, all on the same card, the indicator being attached to each cylinder in succession. In order to avoid stopping the engine to move the indicator cock, every cylinder should be equipped with an indicator cock. However, during normal operation of the engine the indicator cocks should be removed, since prolonged stay on a cylinder in operation burns them out. When taking a draw or pull card the indicator pencil is pressed lightly against the paper for a few strokes (six to ten) and the drum moved slightly, about $\frac{1}{32}$ in., after each stroke.

Naturally, when measuring the compression, the fuel injection to the cylinder under test must be cut off carefully, either by stopping the pump plunger or by opening an air-bleed cock in the fuel line between pump and injection nozzle, or by closing the fuel cock to the particular cylinder in a common-rail injection system. In air-injection engines the injection air should also be shut off, if accuracy is desired.

Another way of testing the compression is to operate the engine on starting air. It may be of interest to use both methods in order to compare the compressions in hot and cold cylinders. Also it should

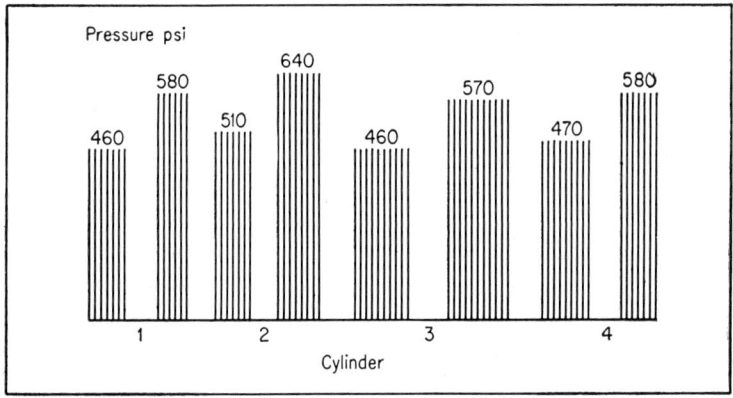

FIG. 21-10. Composite pull diagram of compressions and firing pressures of a four-cylinder diesel engine.

be remembered that the unavoidable leakage past piston rings and valve seats is greater at lower engines speeds, where it lowers the compression pressure.

Figure 21-10 shows a sample of a pull card of a four-cylinder engine. Good practice is to see that the compression in the cylinders does not vary more than 15 psi between the lowest and the highest pressures; in no case should it be allowed to drop more than 15 psi below the pressure recommended by the engine manufacturer.

21-5. Firing Diagrams. If the indicator cord is sharply and evenly pulled by hand just before the engine piston reaches top position on the compression stroke, a diagram can be obtained similar to an offset diagram. However, taking such a diagram requires considerable experience and skill, and the difficulty of obtaining a usable diagram increases with engine speed.

Figure 21-11 shows a hand-pulled diagram of a four-stroke air-injection engine that shows good engine performance. Combustion starts at dead center, the pressure rise is very gradual and small, only 75 psi above the compression. The regular indicator diagram in Fig. 21-11 shows the same condition, but not so clearly.

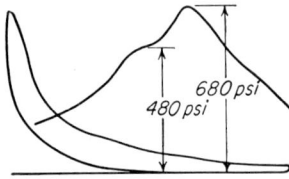

Fig. 21-11. Four-stroke air-injection engine.

Figure 21-12 shows a similar double diagram of an airless-injection engine; Fig. 21-13 and 21-14 show typical diagrams of good performance of two 2-stroke engines. Figure 21-15 shows a diagram of a late injection with ignition occurring considerably after dead center. Figure 21-16 illustrates a case of a defective fuel nozzle: clogging of the holes in the nozzle with carbon or dirt interferes with the timing of injection and results in poor atomization, late ignition, uneven burning, and

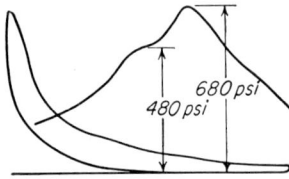

Fig. 21-12. Four-stroke airless-injection engine.

Fig. 21-13. Two-stroke air-injection diesel engine.

Fig. 21-14. Two-stroke airless injection engine.

Fig. 21-15. Late injection.

Fig. 21-16. Late injection and slow burning—afterburning.

Fig. 21-17. Low-cetane fuel, fuel knock.

a considerable amount of afterburning. Another case of a defective fuel nozzle is shown in Fig. 21-17: the holes in the nozzle tip were enlarged by erosion, or the cutting action of the high-velocity streams of fuel passing through them; the larger holes give coarser fuel sprays and therefore poorer atomization; as a result, a comparatively large portion of the fuel charge has entered the cylinder when ignition takes place; this accumulated fuel burns with a violent pressure rise. The insufficient atomization is also indicated by the jags in the second part of the combustion line.

Finally, Fig. 21-18 shows a combined regular and pull diagram from an engine with a great resistance in the intake system—a long and not sufficiently large pipe from the air filter to the intake manifold of the engine. Because of the low cylinder pressure at the beginning of the

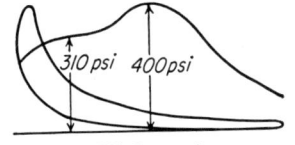

FIG. 21-18. High-suction vacuum.

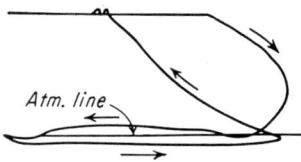

FIG. 21-19. Weak-spring diagram, four-stroke engine.

compression stroke, the final compression pressure is abnormally low and so is the maximum combustion pressure—only 460 psi.

21-6. Intake and Exhaust. For checking the low-pressure events of the engine cycle, the normal piston, one-half of a square inch in area, and a light spring should be used. A 40-lb spring is suitable for checking the exhaust-valve timing, whereas a 20- or 10-lb spring may be used for checking the intake-valve timing and air-inlet and exhaust back pressures.

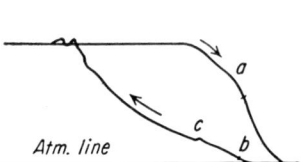

FIG. 21-20. Weak-spring diagram, two-stroke engine.

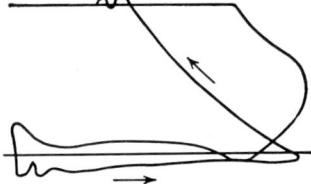

FIG. 21-21. Exhaust-valve closing too early.

Figure 21-19 is a light-spring diagram of a normally operating four-stroke engine. The pressure waves at the beginning and end of the exhaust stroke are due to the inertia of gases in a rather long exhaust line but have no ill effect on the performance of the engine.

Figure 21-20 shows a light-spring diagram of a two-stroke engine: at a is the pressure release when the piston uncovers the exhaust ports, at b they are again covered, and at c the piston covers the upper scavenge ports, line b-c showing a slight supercharging. Figure 21-21 shows a diagram with too early a closing of the exhaust valve; Fig.

21-22, too early an opening of the exhaust valve. Figure 21-23 is a diagram taken with a very light spring showing a high back pressure due to some obstruction in the exhaust system. Finally, Fig. 21-24 shows a diagram of an engine with a late opening of the intake valve and Fig. 21-25, one with a very high vacuum during the suction stroke due to excessive resistance in the intake-air system.

The above diagrams are only samples that show various abnormal conditions in the operation of diesel engines which can be discovered

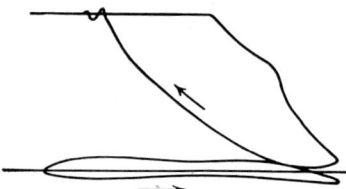

FIG. 21-22. Exhaust-valve opening too early.

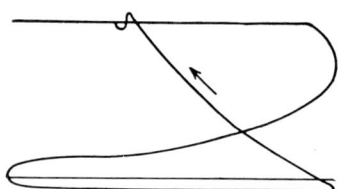

FIG. 21-23. Back pressure high.

by means of an engine indicator. A man in charge of a diesel engine should use the indicator and find out for himself what causes the different deviations from a normal diagram. If he does this and files the diagrams of various faulty operating conditions with the correct explanations, he will gradually build up a most valuable reference library and will be in a position to check his engine easily and simply and thereby to maintain it in proper operating conditions.

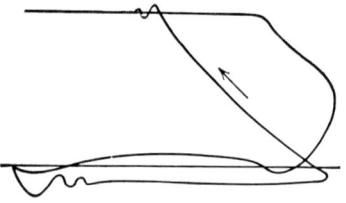

FIG. 21-24. Intake-valve opening late.

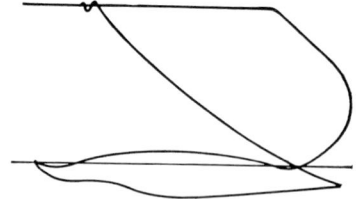

FIG. 21-25. Intake-system resistance high.

21-7. Pressure-Time Diagrams. Another method to investigate events occurring near dead centers is to use special indicators that produce diagrams on which the horizontal distances are proportional to time. Such an indicator is built with the paper drum revolved by a clock mechanism. Since the rotation of the crankshaft is rather uniform, the paper-drum motion may be obtained also from the rotation of the shaft with a suitable speed reduction. The indicator card obtained by such an instrument will have the abscissas proportional to time or to crank-angle degrees, which is practically the same. A sample of such a diagram is shown in Fig. 10-4.

Indicators giving pressure-time diagrams are rather expensive and are used only in research work. They are especially useful for an investigation of high-speed engines.

21-8. Mean Indicated Pressure. Mean indicated pressure is the average constant pressure which, if acting for one stroke, would develop the same power as the variable pressure recorded on an indicator card. Mean indicated pressure, mip for short, is found as follows: the area of the diagram, such as area (1-2-3-4-5-1) in Fig. 21-3, is measured and found to be a square inches; a is divided by the length l, inches, of the diagram and gives the mean height h_m; the mean indicated pressure p_m, pounds per square inch, is found by multiplying h_m by the scale of the ordinates m, called *indicator-spring scale* or *spring number*, and showing how many pounds per square inch correspond to 1 in. Because the mean pressure p_m, found in this manner, would do the same work as the variable pressure of the diagram, i.e., have the same effect, the mean indicated pressure is often referred to as indicated mean effective pressure or simply *mean indicated pressure*.

21-9. Indicated Horsepower. Indicated horsepower is computed from the general equation that states that work done is equal to the force multiplied by the distance through which it acts. The force F, pounds, is equal to the piston area a, square inches, multiplied by the mean indicated pressure p_m, pounds per square inch; the distance is equal to the piston stroke l, inches. Thus the work per one cycle $W = ap_m l$ in.-lb. If this work is multiplied by the number of power strokes per minute, which for a two-stroke engine is equal to the number of revolutions n per minute, the product will give work done per minute, or power P, inch-pounds per minute. Dividing P by 12 will convert it to foot-pounds per minute and again dividing by 33,000, the number of foot-pounds per minute equivalent to 1 hp, will give the indicated horsepower N_i developed by one cylinder:

$$N_i = \frac{ap_m ln}{12 \times 33{,}000} \tag{21-2}$$

Changing the designation p_m to p and rearranging the order of the multipliers gives the standard expression

$$N_i = \frac{plan}{12 \times 33{,}000} \tag{21-3}$$

In a four-stroke engine one power stroke occurs for every two revolutions and n in expressions (21-2) and (21-3) must be divided by two.

The indicated horsepower N_i developed by a multicylinder engine is evidently equal to the sum of N_i for each cylinder, or if the mean effective pressure p is the same in all cylinders, the total horsepower is equal to the N_i of one cylinder multiplied by the number of cylinders. In a double-acting engine the piston area upon which the gas acts from the underside is reduced by the area of the piston rod and the N_i of the engine is equal to the sum of the N_i developed by each of the two combustion spaces times the number of cylinders.

21-10. Brake Horsepower. Not all of the power N_i developed by combustion of the fuel is available as useful power to do outside work. Part of it is absorbed by the engine itself and is referred to as *mechanical losses* or *friction horsepower* N_f.

The mechanical losses of a diesel engine may be divided into four main groups: (1) friction—between the cylinder surfaces and pistons and piston rings, in bearings, gears, and valve mechanisms including opening of the valves; (2) the work of charging—absorbed during the exhaust and suction strokes and by the superchargers in four-stroke engines or by scavenge pumps or blowers in two-stroke engines; (3) resistance of the air to rotation of the flywheel; and (4) the power required to drive auxiliaries—fuel pumps, governor, lubricating-oil and water-circulating pumps, and air compressors in air-injection engines.

As a result the power delivered to the crankshaft and available for doing useful work is appreciably, 15 to 30 per cent, smaller than the indicated power. The available power is called *shaft* or *brake horsepower* and designated N_b. The name brake horsepower is derived from the usual method of its determination—the engine is tested with the power absorbed and measured by a brake. Evidently the difference between N_i and N_b is N_f:

$$N_b = N_i - N_f \qquad (21\text{-}4)$$

It is well to recall that friction losses are really not a loss of energy, since energy cannot be lost or destroyed, but is only a transformation of work into heat. This heat is dissipated to the atmosphere and becomes useless or unavailable for the purpose for which the engine is operated.

Brake mean effective pressure, designated as *bmep,* is a fictitious but very useful concept; bmep may be calculated from expression (21-3), substituting N_b bhp instead of N_i ihp, and solving for the mean pressure p, now designated as bmep,

$$\text{bmep} = 33{,}000 \times 12 \times \frac{N_b}{lan} \qquad (21\text{-}5)$$

The value of bmep gives an indication of what duty an engine carries, what its output is for its displacement. An engine which develops continuously a higher bmep carries a heavier duty, develops a greater horsepower per cubic inch of displacement and per pound of its weight than an engine with a lower bmep, all other things being equal. For a given engine, bmep changes in direct proportion with the load.

The following figures may give an idea of bmep found in present engines: in four-stroke medium-speed continuous-operation engines bmep = 75 to 85 psi; in four-stroke high-output unsupercharged engines bmep = 95 to 100 psi; in supercharged engines bmep of 110 to 125 psi is found and for intermittent operation the bmep may be raised to 140 psi. In two-stroke engines, depending upon the method of scavenging, bmep varies from 30 to 65 psi; in heavy-duty high-output two-stroke engines, such as those used in the Navy, bmep may be found as high as 90 to 100 psi.

21-11. Engine Efficiencies. Efficiency in general is defined as the ratio of the output to the input. In the case of a heat engine the output is the work delivered and the input is the heat supplied in the fuel. Efficiencies are expressed as decimal fractions and are always less than unity. Another way to express efficiency is by percentage, in which case efficiency is less than 100 per cent. The difference between the efficiency fraction and unity or percentage efficiency and 100 is the loss incurred during the process discussed.

In regard to diesel engines a number of efficiencies may be discussed, the principal ones being mechanical efficiency, thermal efficiency, over-all efficiency, volumetric, or charge, efficiency, and scavenge efficiency.

Mechanical efficiency is the ratio of brake horsepower to indicated horsepower:

$$e_m = \frac{N_b}{N_i} \tag{21-6}$$

Mechanical efficiency depends on the design of the engine, on the piston and rotary speeds, on cooling conditions, on the method, quality, and quantity of lubrication, and on the accuracy used in manufacturing, fitting, and aligning various engine parts when assembling the engine.

Two-stroke engines, generally speaking, have a slightly lower mechanical efficiency because of the power absorbed by the scavenge pump. However, certain well designed and built engines may have a high efficiency (curve 4a, Fig. 21-26). Truck engines, both four-

and two-stroke ones, have a low mechanical efficiency, curve 5, because of power absorbed to drive the generator and radiator fan.

The mechanical efficiency of an engine varies with the load, falling off with a decrease of the load. As a first approximation, the losses N_f may be considered to be independent of the load, which means that the friction horsepower N_f is equal to the indicated horsepower of the engine when it is running idle. This gives for the mechanical efficiency the simple formula

$$e_m = \frac{N_b}{N_b + N_f} \qquad (21\text{-}7)$$

However, reliable values for e_m may be obtained only from actual tests, by determining both N_b and N_i.

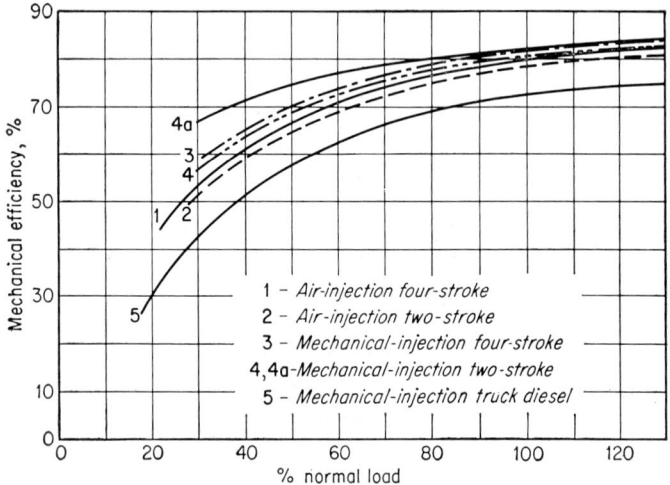

FIG. 21-26. Typical curves of mechanical efficiency of diesel engines.

Figure 21-26 gives a few typical curves of mechanical efficiency for different engines at various loads.

Thermal efficiency is the ratio of the output, or work N_i, horsepower, done by the working substance in the cylinder in a given time to the input, or heat energy of the fuel supplied during the same time. Since the work done by the gases in the cylinder is called *indicated work*, the thermal efficiency thus determined is often called *indicated thermal efficiency*. The expression for computing it is

$$e_{ith} = \frac{2{,}545 N_i}{WQ} \qquad (21\text{-}8)$$

where 2545 Btu is the amount of heat that corresponds to the work of one horsepower during one hour; W, pounds, is the weight of fuel supplied during one hour; and Q, Btu per pound, is the heat value of the fuel, or amount of heat that one pound of fuel develops with complete combustion.

Over-all efficiency is a ratio similar to indicated thermal efficiency, only using as output of the engine its useful, or shaft, work expressed in brake horsepower, N_b. Therefore over-all efficiency is often also called *brake thermal efficiency*. Thus the expression for computing it is

$$e_{bth} = \frac{2{,}545 N_b}{WQ} \qquad (21\text{-}9)$$

Volumetric efficiency of a four-stroke engine is the ratio of the volume V_a of the air admitted to the engine cylinder during the suction stroke, referred to normal pressure and temperature conditions, to the suction volume of the piston that is equal to the piston displacement V_d. Thus

$$e_v = \frac{V_a}{V_d} \qquad (21\text{-}10)$$

Volumetric efficiency shows the amount of air admitted compared with the maximum possible amount of air represented by the piston displacement.

A given volume of gas having a certain pressure and temperature is equivalent to a certain weight. Therefore volumetric efficiency is often formulated also as the ratio of the weight W_a of air admitted to the cylinder during the suction stroke to the weight, at normal temperature and pressure, of a volume of air equal to the piston displacement. With this definition volumetric efficiency is also called *charge efficiency*,

$$e_{ch} = \frac{W_a}{V_d w} \qquad (21\text{-}11)$$

where w is the specific weight of air at standard conditions, pounds per cubic feet.

The actual amount of air admitted is smaller than the amount corresponding to the piston displacement for two main reasons: the temperature of the air in the cylinder is higher and the pressure is lower than the normal, or standard, values. The temperature of the fresh air drawn into the cylinder is raised when the air comes in contact with the hot cylinder walls, cylinder head, and piston crown. The pressure in the cylinder at the end of the suction stroke is lower than that of the atmospheric air outside, because of the resistance to air flow in the air intake system, particularly through the intake valve.

The lower the volumetric efficiency, the less air is admitted and the less fuel can be burned, and the less power is developed by the engine. Taking air from a point in the engine room where the air is cooler and having all valves properly timed will increase the volumetric efficiency and through it the maximum power which the engine can develop. In present diesel engines volumetric efficiency is of the order of 80 to 87 per cent, decreasing chiefly with an increase in engine speed.

Scavenge Efficiency. For two-stroke engines the concept of volumetric efficiency does not apply. Instead, the term scavenge efficiency is used; this rating shows how thoroughly the burned gases are removed and the cylinder filled with fresh air,

$$e_{sc} = \frac{W_a}{(V_d + V_c)w} \qquad (21\text{-}12)$$

where W_a = the weight of air retained in the cylinder, lb
V_d = the volume of the piston displacement, cu ft
V_c = the volume of the combustion space, cu ft
w = the specific weight of air, lb per cu ft, at standard conditions or at the pressure of scavenge air, depending upon the basis of comparison

As with four-stroke engines it is also desirable to take the air to the scavenge blower at a point where the air is cooler. Scavenge efficiency depends greatly upon the arrangement of exhaust and scavenge-air ports and valves. The best scavenge efficiency is obtained by uniflow scavenging, as explained in Sec. 2-6.

21-12. Limitations of Performance. The power which can be developed by a given cylinder and piston stroke is limited by the engine speed and by the mean effective pressure developed in the cylinder.

The maximum engine speed is determined by the inertia forces produced by the moving reciprocating parts, which increase very rapidly with an increase of speed, and also by lubrication difficulties caused by heat which is developed by friction. The maximum engine speed as indicated by the engine builder should never be exceeded.

Mean effective pressure is limited by one of the following factors:

1. Heat losses and efficiency of combustion.
2. Charge, or volumetric, efficiency in four-stroke engines and scavenge efficiency in two-stroke engines. The amount of air charged into the cylinder and available for combustion depends upon these efficiencies. Supercharging increases the amount of air and thus helps to increase the mean effective pressure.
3. Completeness of mixing of the fuel and air. This depends upon the fineness of the fuel atomization, sufficient penetration, and good distribution of the fuel.

For each engine the limiting, or maximum permissible, brake mean effective pressure is prescribed by the manufacturer and must never be exceeded. Going beyond the recommended maximum brake mean effective pressure would result in higher combustion temperatures and hence higher temperature stresses and greater wear of many parts, increased maintenance costs, and even cracking of cylinder heads.

In an engine driving a propeller or a centrifugal pump, the mean effective pressure developed by the engine is determined by the propeller or pump speed; it increases very rapidly even with a small increase over the maximum rated speed. In an engine driving an electric generator the mean effective pressure will be within the safe limit as long as the rating of the generator is not exceeded.

Overloading a diesel engine means exceeding the limiting mean effective pressure and should therefore never be allowed.

QUESTIONS

1. What is a pressure-volume, or p-v, diagram?
2. Draw a p-v diagram of a four-stroke diesel engine.
3. Describe briefly an engine indicator.
4. What is an indicator diagram or card?
5. Draw a theoretical indicator diagram of a two-stroke engine.
6. What is the area of a normal-size indicator piston?
7. What other sizes of indicator piston exist?
8. How are indicator springs rated?
9. What is the recommended height of a diesel-engine indicator diagram?
10. What is a good piston and spring combination to obtain a suitable diesel-engine diagram?
11. Give a sketch of a simplified reduction motion for taking indicator diagrams.
12. What is an offset indicator diagram and what is its purpose?
13. Explain how an offset indicator diagram is taken.
14. What is a draw, or pull, card?
15. How are draw cards taken from a multicylinder engine?
16. What is the permissible difference in the compression pressure of the cylinders of a multicylinder engine?
17. By how much may the compression pressure be permitted to drop below the pressure recommended by the engine manufacturer?
18. What is a firing diagram? How is one obtained?
19. Show the approximate shape of a firing diagram (*a*) of good engine performance and (*b*) of an engine with late injection.
20. By means of what indicator diagrams is it convenient to check the low-pressure events of the engine cycle?
21. What is a pressure-time diagram?
22. What is mean indicated pressure?
23. By means of what expression is the indicated horsepower calculated?
24. How is indicated horsepower designated?
25. How is the expression for calculating the indicated horsepower adapted to the difference between a four- and two-stroke engine?

26. Enumerate the four main groups into which the mechanical losses of a diesel engine may be divided.

27. Enumerate the various places in a diesel engine where mechanical losses due to friction between metal surfaces do occur.

28. What approximate percentage of the indicated power developed in the engine cylinder is absorbed by mechanical losses?

29. What is the name given to the available useful power and how is it designated?

30. What is brake mean effective pressure and how is it designated?

31. Give approximate values of brake mean effective pressures found in various types of diesel engines.

32. What is the general definition of efficiency?

33. Enumerate the five principal efficiencies used to rate diesel-engine performance.

34. What is the mechanical efficiency of a diesel engine?

35. How does the mechanical efficiency of an engine vary with a decrease or increase of the load?

36. What is the indicated thermal efficiency of a diesel engine?

37. What is the over-all efficiency of a diesel engine? What other name is used for this efficiency?

38. What is the volumetric efficiency of a diesel engine?

39. What is the charge efficiency of a diesel engine?

40. What factors reduce the charge efficiency?

41. What is scavenge efficiency? To what engines does it refer?

42. What are the two main factors that limit the power output of a given engine?

43. Why should the maximum engine speed indicated by the engine builder never be exceeded?

44. What are the factors that limit the brake mean effective pressure?

45. Why is it dangerous to exceed the brake mean effective pressure recommended by the engine manufacturer?

46. What happens when a diesel engine is overloaded?

CHAPTER 22

RATING AND TESTING

22-1. Power Rating. *Methods.* There are three main methods of defining the rated power, or the horsepower, by which a diesel engine is designated: (1) by the maximum load which an engine can carry continuously; (2) by the maximum power which an engine can develop, and (3) by a figure computed from a conventional arbitrary formula.

Maximum continuous load is determined in several ways. At present most engine builders base this load, called *normal load*, on the mean effective pressure, mep, or p_e, expressed in pounds per square inch. In stationary diesel engines it is commonly considered that at the rated load, mep should not exceed 80 psi in four-stroke engines. In supercharged engines the rated mep is considerably higher, up to about 125 psi and for intermittent operation it may be raised as high as 145 psi. In two-stroke engines, it depends upon the scavenge method used and upon the scavenge efficiency obtained and varies from about 30 to 90 psi.

Naturally, it must be remembered that in four-stroke engines, p_e is referred to the compression and expansion strokes only, so that when comparing them with two-stroke engines, p_e must be divided by 2.

Only very few low-speed engines, such as large marine oil engines, are sometimes rated on the basis of the mean indicated pressure, mip or p_i, and their rating given in indicated horsepower, ihp; the rating of most engines is given in brake horsepower, bhp.

Every engine should be able to carry a certain overload, 10 to 20 per cent, but only in an emergency and not for more than a few hours at a time. Engines with a great excess of air, such as mechanical-injection diesel engines, are often capable of carrying an overload of from 30 to 40 per cent. However, under such a load the combustion becomes incomplete, with smoky exhaust, and there is a danger of injuring the engine through high temperatures and heat stresses.

The *maximum power* which an engine can develop is sometimes used as its rating, even though the engine cannot develop it continuously.

This is done in automotive engines, because, by the nature of the load carried, they do not have to develop their maximum power for any length of time.

Small stationary and tractor engines are sometimes rated on a similar basis, with the difference that their rated horsepower is given as 90 per cent of the maximum horsepower, leaving 10 per cent for eventual overload.

The rated horsepower of engines used in the oil industry, according to the API (American Petroleum Institute) rules, is considered as 65 per cent of the maximum power, leaving an unusually large margin, about 50 per cent, for temporary overloading.

Nominal horsepower is a term used for rating automobile and truck engines, usually for purposes of taxation. The nominal horsepower N of a four-stroke engine in this country is given by the NACC (National Automobile Chamber of Commerce) formula—also called SAE rating, or sometimes AMA (Automobile Manufacturers Association) rating—

$$N = 0.4id^2 \qquad (22\text{-}1)$$

where d is the piston diameter, inches, and i is the number of cylinders.

This formula is based on a piston speed of 1,000 ft per min and a mean effective pressure $p_e = 67.2$ psi. Therefore the approximate actual horsepower N_e can be found by multiplying the above expression by the ratio of actual piston speed c to 1,000:

$$N_e = 0.0004icd^2 \qquad (22\text{-}2)$$

This would differ from the correct figure by the extent to which the mean effective pressure differs from 67.2. Long strokes, large valves, and light reciprocating parts have raised the piston speed at maximum engine output to 2,500 feet per minute and higher and the mean effective pressure to 80 psi and higher.

The question of rating automotive engines is illustrated by the data of Table 22-1, which refers to three modern truck diesel engines of different make, a 4-cylinder engine classified as a high-speed engine, and two 6-cylinder engines classified as super-high-speed engines. Table 22-1 shows how much the maximum horsepower obtained on the test stand is higher than the nominal power by the SAE formula or adjusted to a mean effective pressure of 80 psi. The two last lines show the actual mean effective pressure and piston speeds of the three engines which may be considered as representative for this type of service.

22-2. Torque. The relation between horsepower N_b and torque T in pound-feet is

RATING AND TESTING

$$N_b = \frac{Tn}{5,252} \qquad (22\text{-}3)$$

where n is the engine speed, revolutions per minute. On the other hand, the expression of horsepower for a four-stroke engine with i cylinders is

$$N_b = \frac{p_e l \pi d^2 n i}{(12 \times 4 \times 2 \times 33,000)} \qquad (22\text{-}4)$$

Equating expressions (22-3) and (22-4) and solving for T gives

$$T = 0.0052 p_e i l d^2 \qquad (22\text{-}5)$$

or

$$p_e = \frac{192 T}{i l d^2} \qquad (22\text{-}6)$$

Since for a given engine ild^2 is constant, equations (22-5) and (22-6) show that T is directly proportional to p_e. Thus for variable-speed engines both p_e and T may be represented by the same curve, only using different scales.

TABLE 22-1. RATING OF FOUR-STROKE TRUCK DIESEL ENGINES

Number of cylinders.	4	6	6
Bore, in.	3¾	4	4¾
Stroke, in.	5	4⅜	5⅜
Displacement, cu in.	221	330	572
Maximum governed speed, rpm.	1,650	2,400	2,200
Nominal hp by SAE, or AMA, rating.	22.5	38.4	54.2
Hp for 80 psi mep at 1000 ft per min.	26.8	45.7	64.5
Maximum hp at maximum rpm, observed.	34.8	86.2	149.6
Mep at observed maximum hp, psi.	75.5	86.3	94.2
Piston speed at maximum hp, ft per min.	1,375	1,750	1,970

Noticing that $l\pi d^2 i/4$ in equation (22-4) is the piston displacement V, cubic inches, the mean effective pressure may be calculated from equation (22-4) as

$$p_e = \frac{792,000 N_b}{V n} \qquad (22\text{-}7)$$

For two-stroke engines the corresponding expressions become

$$T = 0.0104 p_e i l d^2 \qquad (22\text{-}8)$$

$$p_e = \frac{96 T}{i l d^2} \qquad (22\text{-}9)$$

or

$$p_e = \frac{396{,}000 N_b}{Vn} \tag{22-10}$$

22-3. Performance Curves. *Variable Speed.* For engines that operate at a variable speed, the rated horsepower at a certain speed does not give enough information. The necessary information is obtained from performance curves which give the horsepower as a function of the speed used. In the same diagram some additional

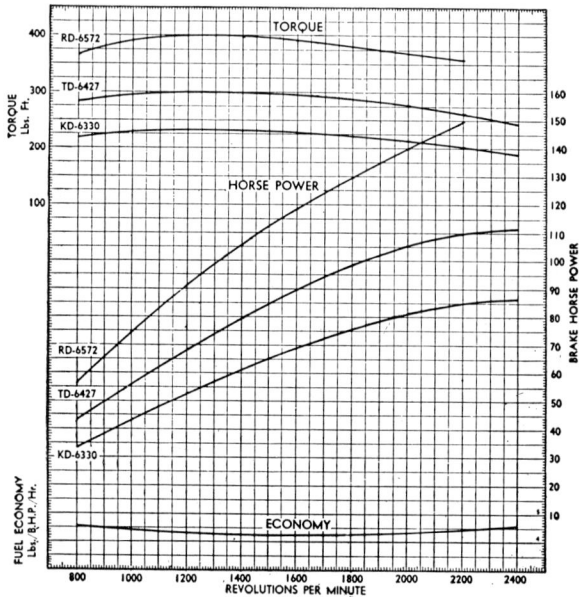

Fig. 22-1. Performance curves of a truck diesel engine. KD-6330 4 in. × 4¾ in. × 6 cyl; TD-6427 4 7/16 in. × 4 5/8 in. × 6 cyl; RD-6572 4¾ in. × 5 3/8 in. × 6 cyl.

characteristic curves are usually plotted, such as brake mean effective pressure p_e, torque T, fuel consumption in pounds per horsepower-hour, and sometimes mechanical efficiency and indicated horsepower. All these data must be obtained from an actual test of the engine.

Performance Curves. A typical set of performance curves for a series of three 6-cylinder automotive diesel engines is shown in Fig. 22-1. The engines are of identical design and differ only in bore and stroke as indicated in the caption. The fuel economy for all three engines is the same, about 0.45 lb per hp-hr at low and high speeds and a minimum of 0.425 lb per hp-hr at medium speeds, between 1,500 and 1,700 rpm.

Figure 22-2 shows detailed performance curves of an industrial four-stroke diesel engine. The engine is a 5¾ × 8 in. × 4 cyl with a displacement of 831 cu in.

The maximum-output curves show the maximum output in brake horsepower which the engine can develop with a clear exhaust. These ratings are usable only in intermittent-service applications such as excavators, locomotives, hoists, etc.

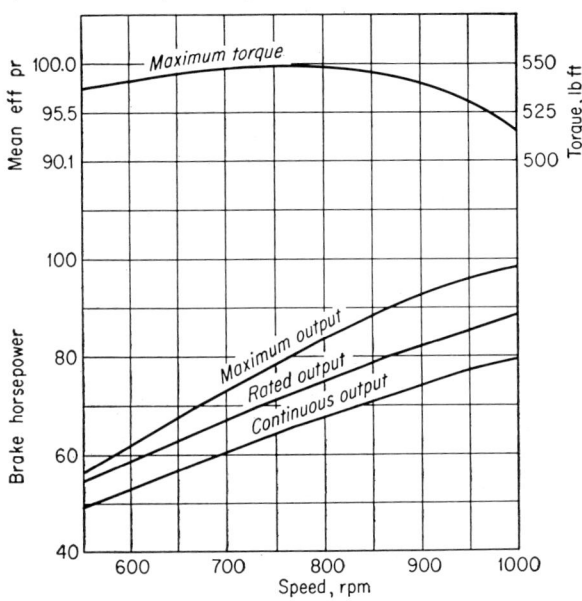

FIG. 22-2. Performance of a power unit with 5¾-in. × 8-in. × 4-cyl. diesel engine with full equipment including radiator fan.

The rated-output curves show the load in brake horsepower which the engine is capable of carrying for a period of 12 hr from each cold start, generally for 12 hr per day service.

The continuous-output curves show the maximum recommended load in brake horsepower for continuous day-and-night runs for more than 24 hr. The continuous output does not exceed 90 per cent of the rated output.

Margin of capacity is defined as follows: During a 12-hr run at rated output and after the jacket-water and lubricating oil temperatures have become constant, the engine is capable for periods of one hour of carrying a load of 10 per cent above its rated output without undue heating or other mechanical trouble and then to continue with its 12-hr rated load.

All output figures and references thereto are corrected to an air density corresponding to sea-level barometric pressure and a standard temperature of 60 F. Full equipment includes an air cleaner, a lubricating-oil pump, a fuel-transfer pump, fuel-injection pumps, a jacket-water pump and a governor.

Instead of presenting the fuel economy only for the full rated power at different speeds, a more complete picture is obtained by giving fuel-consumption curves at different speeds for loads varying from about ¼ to full load. Such curves are shown in Fig. 22-3 for the engine whose performance is shown in Fig. 22-2.

The fuel consumptions shown were obtained with fuel oils having 45 to 50 cetane number, 30 to 34 API gravity at 60 F, and 33 to 42 SSU viscosity at 100 F, but are recalculated to a fuel having a heat value of 19,000 Btu per pound.

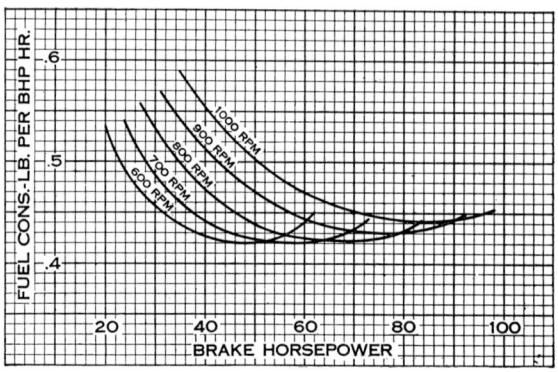

Fig. 22-3. Fuel consumption of diesel power units with full equipment including radiator fan.

Finally, Fig. 22-4 shows performance curves of a two-stroke engine. The general data for this engine—the most powerful single-acting engine built in this country—are: 29 × 40 in. × 9 cyl, normal shaft horsepower—6,000 shp at 160 rpm; bmep = 62 psi; overload capacity 10 per cent continuous, and 25 per cent for 2 hr. Since the engine is direct-connected to a ship propeller, the 110 per cent load was obtained during the tests on the test stand by increasing the speed to 165 rpm at 6,450 shp, and the 125 per cent load, by increasing it to 172 rpm at 7,300 shp. The performance curves of Fig. 22-4 are plotted using the bmep as abscissas, rather than the engine speed conventionally used. This is done because the output of such an engine and its bmep are in a fixed relation to the engine and propeller speed.

A comparison of the performance curves of diesel engines with those of carburetor spark-ignition engines shows that the torque curves of diesel engines are flatter and considerably higher at low speeds; also the peak load is beyond the range of regular engine use. These characteristics permit the use of a smaller diesel engine as a substitute for a gasoline engine in trucks, road-building, machinery, or other automotive service.

22-4. Testing. There exist a number of books which give detailed information how to test combustion engines. A person conducting engine tests should read such a book and be familiar with it. However, the tests which may be required in the operation of a diesel

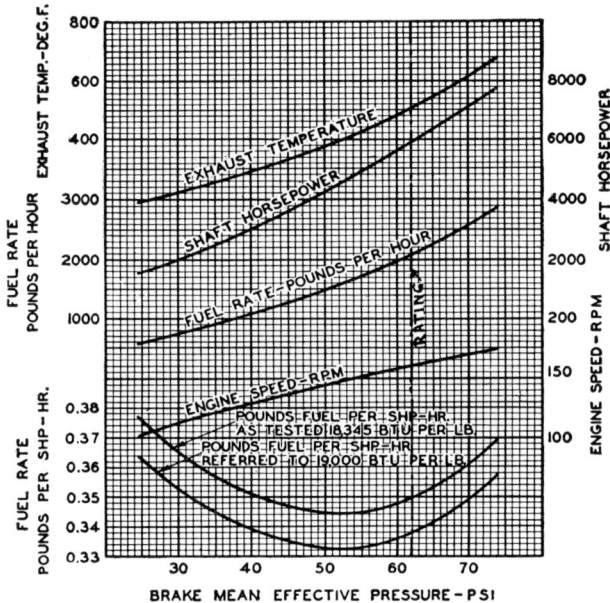

Fig. 22-4. Test results on 6,000-shp engine—power to drive scavenge blower not deducted.

power plant are rather simple and a brief outline may be of value to the operator.

Objects of Tests. Both acceptance tests of a new engine installation and routine tests of engines in a power plant are conducted with two main objects: (1) to find the maximum power which the engine can develop under certain conditions and thus to determine the safe engine load, and (2) to find the specific fuel consumption, in pounds per horsepower-hour, under the load conditions of the plant.

In addition to the two main objects, a properly conducted test can give additional information valuable for best operation of an engine, such as (3) exhaust temperatures at different loads, (4) the lubricating-oil pressure necessary for desirable oil flow and oil temperature, and

(5) the jacket-water flow needed to maintain the proper jacket temperature.

Regardless of the object of a test, it is always good to follow the instructions of the standard Power Test Codes laid down and published by the ASME (American Society of Mechanical Engineers). When testing an automotive-type engine it is advisable also to follow the SAE Rules, which differ only slightly from the ASME Codes.

22-5. Brake Horsepower. The simplest way to determine the useful horsepower of a diesel engine is by means of a brake, hence the term *brake horsepower*. A simple prony brake for a small engine is shown in Fig. 22-5. It consists of two wooden blocks b, b, a lever l made of a piece of 2- × 4-in. plank and fastened to the upper block

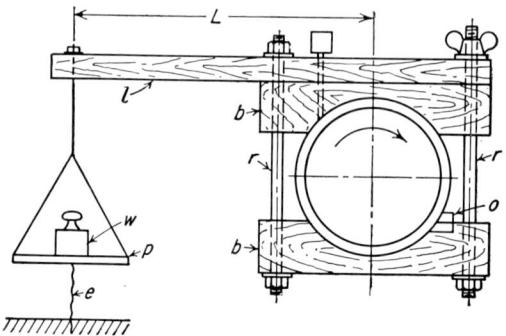

FIG. 22-5. Prony brake for a small engine.

by a couple of sixteenpenny spikes, two threaded rods, r, r with nuts on their ends, and a square board p, suspended on wires, on which the weight W is placed. The adjustment to carry the load determined by the weight W is made by turning the nut on the right, less-stressed rod r. It is convenient to equip this nut with a handle. The wooden blocks should be lubricated at the points where the rotating pulley or flywheel to which the brake is applied will spread the lubrication over the whole surface of the blocks. The upper block may be lubricated by a sight-feed oiler, the lower one by a groove filled with waste soaked in oil. A piece of rope or chain e, fastened with one end to the ground and with the other one to the platform p, serves as safety device to prevent the brake from swinging from a sudden increase of friction between the pulley and the wooden blocks.

A brake construction suitable for larger engines, up to 300 hp, is shown in Fig. 22-6. Instead of the lower single block it uses a number of comparatively small blocks fastened by screws to a steel band; the ends of the band are riveted or welded to the flattened ends of the adjustment rods r, r; the friction torque is transmitted by the lever l and the support i to a platform scale s. The brake runs more smoothly if it is lubricated by pieces of animal fat inserted between the wooden blocks, particularly by strips of pork lard.

A brake which can be made up easily for temporary use and is suitable for engines developing up to 200 bhp is shown in Fig. 22-7. It

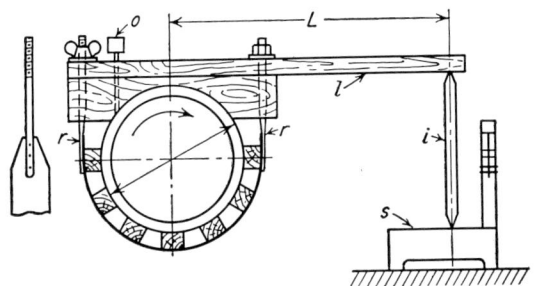

Fig. 22-6. Prony brake for medium-size engine.

consists of a double rope, to one end of which is suspended the weight or weights W, the other end being connected to a spring scale s fastened to the ground. The two ropes are held in place on the flywheel rim by wooden blocks f, f (Fig. 22-7b). In this case the rope rubs on the flywheel rim and must be well soaked in oil or lard to prevent rapid wear. A better scheme is to use a greater number of blocks made so

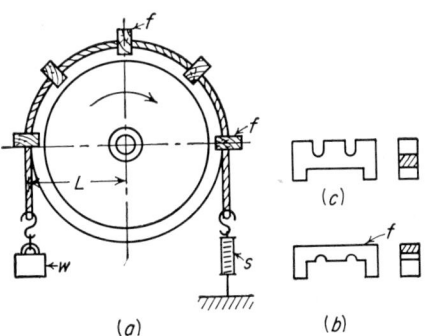

Fig. 22-7. Rope brake for small engine.

as to raise the ropes from contact with the rim (Fig. 22-7c). If the blocks should wear down, they can be replaced easily and there is no danger of wearing and breaking the ropes.

By increasing the arc of contact from 180 to 360 deg (Fig. 22-8) and using a block sheave q, the rope brake can be used for absorbing large torques without exerting a heavy pressure downward on the bearing adjacent to the flywheel.

In designing a prony brake with wooden blocks, it is advisable to use pulleys with a peripheral speed not over 2,500 ft per min to avoid the danger of the blocks' taking fire.

The power N_b absorbed by a prony brake, regardless of the type of construction, may be computed by the formula

$$N_b = \frac{2\pi LFn}{33,000} \tag{22-11}$$

where L = the length of the beam, ft
F = the net brake load, lb
n = the engine speed, rpm

The factor $2\pi L/33,000$ is usually designated as C and called the *brake constant*. The formula (22-11) then becomes

$$N_b = CFn \tag{22-12}$$

It is convenient to make the beam length $L = 5.25$ ft, as then $C = 0.001$. For small engines, L may be made 31.5 in., giving $C = 0.0005$; and for large engines L is made 10.5 or 15.75 ft, which corresponds to $C = 0.002$ or 0.003, respectively.

In a rope brake L is equal to the distance from the center of the shaft to the center of the rope.

In computing the net brake load F one must add to the weight W (Fig. 22-5) the weight of the platform p and of the unbalanced part of the lever l, the latter referred to distance L from the shaft line. When using a platform scale (Fig. 22-6), the additional weight W_2 which must be added to the scale reading to obtain F is found simply by weighing the lever l and support i. In order to eliminate friction between the brake and the brake pulley, the nuts must be loosened and a triangular file inserted between the upper block and the pulley rim to act as a pivot.

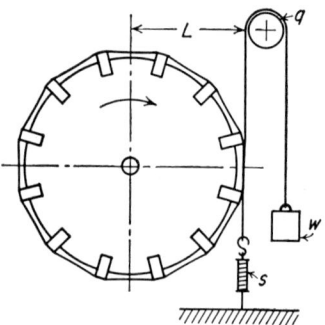

Fig. 22-8. Rope brake for larger engine, up to 200 hp.

With a rope brake the net force F is the difference between the weights W and the indication of the spring scale s.

Hydraulic Dynamometers. A hydraulic brake is often called a *dynamometer* because it permits accurate and easy measurement of the engine horsepower. In shops that recondition automotive diesel engines such a brake is often used to run in and test each engine after it has been overhauled.

The basic principle of operation is the same as in an ordinary prony brake only with the introduction of the friction of water instead of the friction between a cast-iron brake pulley and stationary friction blocks.

Figure 22-9 shows diagrammatically the scheme of operation: The rotor has pockets at the periphery and is keyed to a shaft connected by a flange coupling to the engine crankshaft; the rotor shaft passes through packing glands in the casing and is supported on two outboard

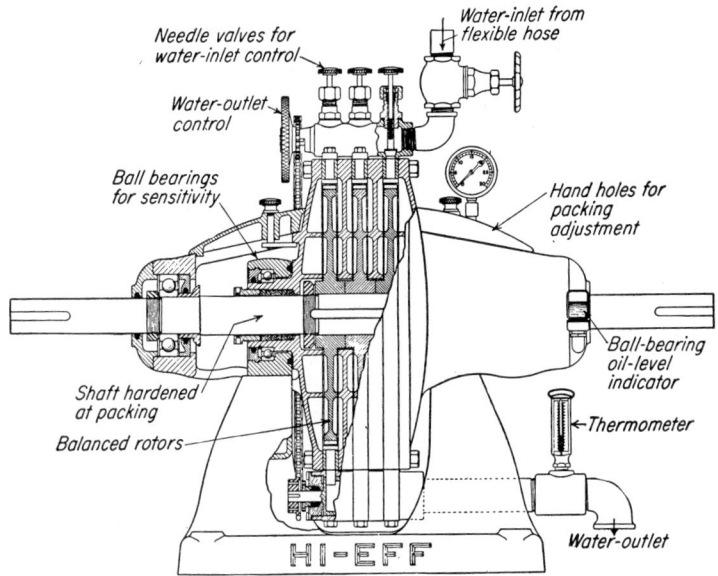

Fig. 22-9. Hydraulic brake or *dynamometer*. *Courtesy of Taylor Dynamometer & Machine Co.*

bearings; the casing has vanes (Fig. 22-9b) that oppose the movement of the water entrained by the rotor pockets; it is supported by the shaft. When the rotor begins to turn, water is thrown against the circumference of the casing by the centrifugal force and tends to turn it. An arm is fastened to the casing and the turning moment is transmitted, as with a prony brake, to scales. The same formula (22-11) is used for computing the horsepower developed. The horsepower absorbed by a water brake varies approximately as the cube of the rpm and as the fifth power of the rotor diameter. With a given brake and constant speed the power absorbed is regulated by the amount of water kept in the casing. The power absorbed is transformed into heat and is carried away by a continuous flow of water through the casing. The flow is regulated so as to maintain a constant level and constant temperature of the water. Several models of hydraulic brakes are available commercially.

22-6. Electric Loading. The *electric-cradle dynamometer* (Fig. 22-10) is a d-c generator with rotor and stator mounted on concentric ball bearings, so that the stator is free to turn. The torque is measured by suitable scales. This dynamometer is used also for testing overhauled diesel engines. Its advantages are that it can be used over a wide range of operating conditions, both as a generator for testing and as a motor for running in an engine cold.

Electric Generators. If the engine drives an electric generator, either directly or by a belt drive, the brake horsepower can be obtained from the ammeter and voltmeter readings.

The power of an engine connected to a d-c generator is expressed as

$$N_b = \frac{EI}{746 e_g} \qquad (22\text{-}13)$$

where E = the voltmeter reading, volts
I = the ammeter reading, amp
e_g = the efficiency of the electric generator

The efficiency of the generator may be taken from the efficiency curve usually furnished with the generator. If no curve is available, e_g may be assumed for larger machines (about 500 kva) as 0.93 for full load, 0.92 for ¾ load, and 0.91 for ½ load. In small machines, about 50 kva, the efficiencies are about 4 to 5 per cent lower.

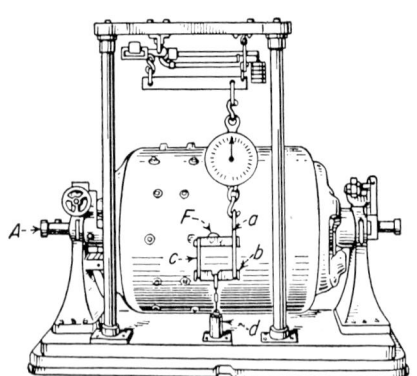

Fig. 22-10. Electric-cradle dynamometer. From Marks, *Mechanical Engineers' Handbook, Fifth Edition.*

The power developed by an engine connected to a single-phase a-c generator is

$$N_b = \frac{E \times I \times pf}{746 \times e_g} \qquad (22\text{-}14)$$

where pf is the power factor, which depends upon the characteristics of the electric system.

In a three-phase system

$$N_b = \frac{\sqrt{3} \times E \times I \times p.f}{746 \times e_g} \qquad (22\text{-}15)$$

If the electric generator is driven by a belt, the efficiency of the belt drive e_b must be taken into account. The value computed by expression (22-13) or (22-14), as the case may be, must be divided by e_b. It may be assumed with sufficient accuracy that for a flat-belt drive $e_b = 0.95$, and that for a V-belt drive $e_b = 0.96$.

22-7. Fuel Measurements. Liquid fuels can be measured either by actual weighing or by a flowmeter.

Measurement by Weight. Figure 22-11 shows a good arrangement: The day tank a has a throat b with a hook c; while the engine is running ready for the test, a is filled over the point of c, and a stop watch is started when the fuel level, in coming down, bares the point of c. The engine continues to run, using the fuel from tank a. Toward the end of the test run, a weight of fuel sufficient for the duration of the test is poured into the day tank so that its level rises again above the hook point, and, when the latter appears, the stop watch is stopped. The comparatively small cross-sectional area of b decreases the percentage of error in the fuel-rate determination.

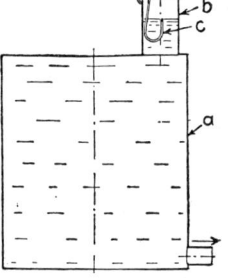

Fig. 22-11. Fuel-oil measuring device.

A fuel-economy test should be started only after the engine has attained stationary thermal conditions. This means that the engine should be allowed to run under the load for which the fuel economy is being found until the temperature of the cooling water and lubricating oil has attained a constant level. This usually takes about 30 min.

The duration of the test should also be about 30 min. To obtain a reliable figure the test should be run three times and an average taken of the three results, provided any single test does not differ more than ±5 per cent from the average value. Otherwise the test which gave a greater discrepancy should be discarded and one or two more tests run to determine the correct or, rather, the more probable figure.

Volumetric Fuel Measurement. When fuel economy is determined by means of a flowmeter, such as a Rota-meter, the reading is usually taken in gallons per minute or per hour and the specific weight of the fuel must be measured with a hydrometer to convert the volume to weight units.

22-8. Exhaust Temperature. In many instances, especially if the engine operates at different speeds, it is of interest to determine quickly with sufficient accuracy the load carried by the engine at a given time. This may be done first by running several tests with different loads and speeds and from these data plotting curves giving the interrelation between exhaust temperature, engine load, and speed.

An example of such a test is shown in Figs. 22-12 and 22-13. The 9- × 12-in. × 4-cyl engine was run with a prony brake. The first test was run with a brake load of 120.5 lb, which corresponds to a mean effective pressure of 40 psi, and a speed of about 240 rpm until stationary thermal conditions were obtained. This was indicated when

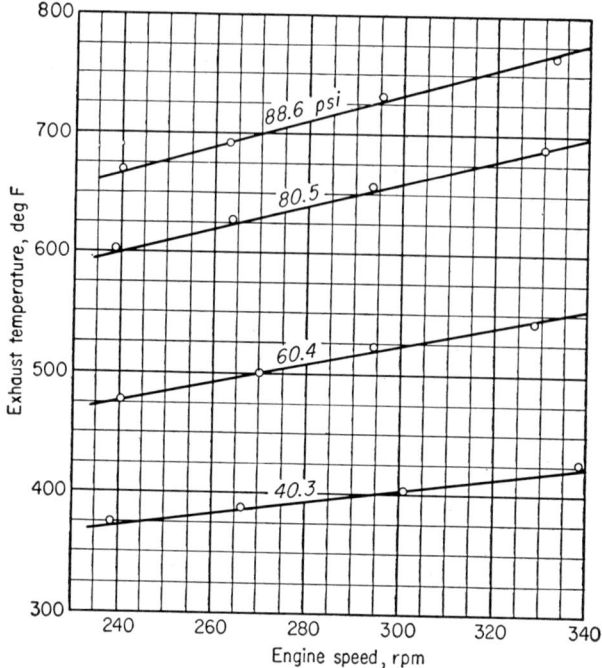

Fig. 22-12. Exhaust temperatures at different mep and speeds of a 9 × 12-in. diesel engine.

the cooling water and exhaust-gas temperatures reached a constant level that was noted. The test was then repeated with the same brake load but with the speed increased to about 270 rpm, then to 300 rpm, and finally to 330 rpm. After these tests, another series of four tests was run with a brake load increased to 180.75 lb., which corresponds to a mean effective pressure of 60 psi. Two more series of tests were run with mep = 80 psi, which is the full-load rating of the engine, and with mep = 88 psi, which corresponds to 10 per cent overload. The exhaust-gas temperatures are plotted against engine speed and through the points corresponding to the same mean effective pressure are drawn curves, which are practically straight lines

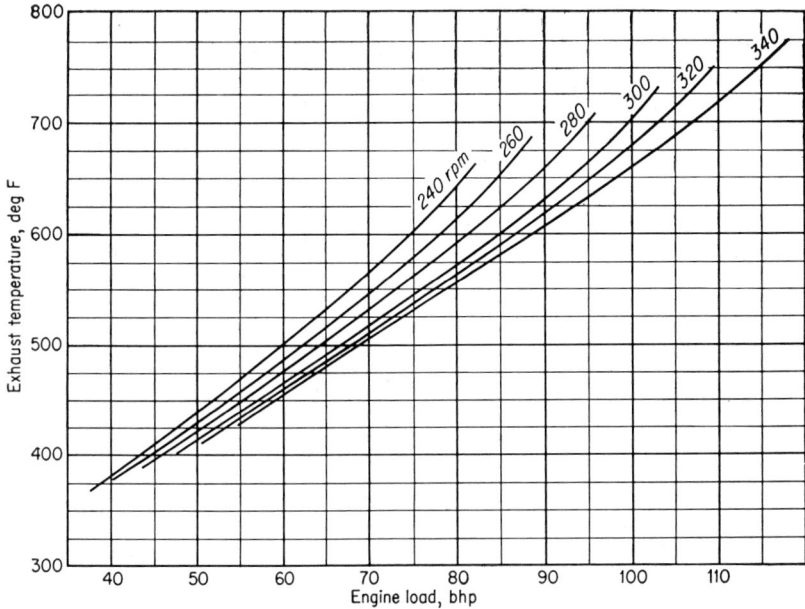

Fig. 22-13. Exhaust temperatures at different loads and speeds of a 9 × 12-in. × 4-cyl diesel engine.

(Fig. 22-12). The next step was to enter in Table 22-2 the horsepower N_b computed from expression (22-4) or (22-7) for the various mean effective pressures and speeds and the corresponding exhaust temperatures t_e found from Fig. 22-12 at the intersections of the four

TABLE 22-2. EXHAUST-GAS TEMPERATURES OF A DIESEL ENGINE AT DIFFERENT SPEEDS AND LOADS

Mep psi		Engine speeds, rpm					
		240	260	280	300	320	340
40	N_b	37.0	40.0	43.1	46.2	49.3	52.5
40	t_e	372	383	394	405	415	426
60	N_b	55.5	60.0	64.6	69.3	74.0	78.8
60	t_e	477	492	507	523	539	555
80	N_b	74.0	80.0	86.2	92.4	98.6	105.0
80	t_e	600	619	639	659	679	698
88	N_b	81.4	88.0	94.8	101.6	108.5	115.5
88	t_e	665	686	707	728	749	770

mep-curves with the ordinates corresponding to the desired speeds. The last step was to plot the temperature data from Table 22-2 against corresponding brake horsepower and to connect by smooth curves the points referring to the same speeds. The results are shown in Fig. 22-13. It goes without saying that before such a test is conducted, the engine must be tuned up properly, so that each cylinder carries the same load; the load should be divided evenly between the cylinders at all speeds, and the exhaust from each cylinder should have the same temperature. The same is true for the use of Fig. 22-13 in checking engine load: the point of intersection of the observed exhaust temperature with the curve corresponding to the engine speed will determine more or less accurately the brake horsepower developed by the engine at these conditions only if the fuel adjustment is in good shape and each cylinder carries its share of the total load.

While such tests can usually be conducted in the plant after the engine is installed, it is easier to conduct them on the test stand of the engine builder where the engine is run under a prony brake with a water-cooled brake pulley. Therefore, it is advisable, in ordering the engine, to specify that a chart such as that shown in Fig. 22-13 be furnished by the engine builder, or at least data, as shown in Fig. 22-12, from which the power-plant engineer can himself construct the rating curves of Fig. 22-13.

Back Pressure. It should be remembered that an increase of the back pressure raises the exhaust-gas temperature. However, under normal conditions, when the back pressure does not exceed 1.5 psig, the influence is rather small.

QUESTIONS

1. State the three different methods used for rating diesel engines.
2. How much overload should a diesel engine be permitted to carry, and under what conditions?
3. What is considered the rated horsepower of combustion engines in the oil industry?
4. What is nominal horsepower and how is it determined?
5. How does one give the power rating of an engine which operates at a variable speed?
6. Write the expression giving the relation between the brake horsepower, speed, and torque of an engine.
7. What are the two main objects of testing an engine?
8. What are other valuable data that may be obtained from a diesel-engine test?
9. Why is the useful horsepower of a diesel engine called brake horsepower?
10. Draw a sketch of a simple small prony brake and explain its action.
11. Draw a sketch of a prony brake suitable for a diesel engine up to 300 bhp.
12. Draw a sketch of a rope brake for temporary use.

RATING AND TESTING

13. Write the expression for horsepower as it is determined by means of a prony brake.

14. What is the brake constant?

15. Explain the action of a hydraulic dynamometer.

16. Explain how the load produced by a hydraulic brake is regulated.

17. Explain the action of an electric-cradle dynamometer.

18. Explain the determination of the horsepower of a diesel engine connected to an electric generator.

19. What are the approximate figures for the efficiency of a d-c generator for full, three-quarter, and half loads?

20. Explain the differences in the expressions used to compute horsepower from ampere and volt readings with (*a*) a d-c generator, (*b*) a single-phase a-c generator, and (*c*) a three-phase a-c generator.

21. What is the *power factor* and how can it be determined?

22. If the electric generator is driven by a belt, what efficiencies may be assumed for (*a*) a flat-belt drive and (*b*) a V-belt drive?

23. By what methods may liquid fuel consumption be measured?

24. Draw a sketch of an arrangement to measure fuel-oil consumption by weighing and explain the procedure.

25. State what instruments are necessary for measuring the fuel consumption volumetrically.

26. When should a fuel-consumption test be begun?

27. What should be the duration of a fuel-consumption test?

28. If the fuel economy is measured by means of a flowmeter, in what units is the reading taken and how is the specific fuel consumption, in pounds per horsepower-hour, determined from this reading?

29. What is the object of measuring exhaust gas temperatures at different loads and speeds?

30. Explain how the engine load may be determined from exhaust-gas temperature measurements.

31. Why is it advisable to ask the engine builder to furnish exhaust-temperature data?

32. How does an increase of the back pressure influence the exhaust-gas temperature?

33. What is the amount of back pressure that does not influence the exhaust-gas temperature appreciably?

PART 5

MAINTENANCE AND REPAIRS

CHAPTER 23

MAINTENANCE RECORDS AND DATA

23-1. Importance of Maintenance. Modern diesel engines are products of painstaking research, skillful engineering, and precision manufacture. Properly installed and given care and attention comparable to that given other types of prime movers, they will give satisfactory and dependable service for many years. There is no danger of sudden or high depreciation due to obsolescence. The ability to maintain a high thermal efficiency is characteristic for all diesel engines but depends to a great extent on a rather strict adherence to a reasonable maintenance schedule. Successful operation of a power plant is possible only with adequate maintenance of the engines and all the other equipment.

23-2. Qualifications of Operators. Owners of diesel engines should realize that only fully qualified operators can be entrusted with the operation of diesel engines. Starting and stopping of engines is a comparatively simple procedure. Even an inexperienced man can do it by following the printed instructions of the engine builder. However, judgment supported by proper experience will tell a man *when* to start and *when* to stop an engine and in general *how* to service this expensive equipment in order to keep it in proper running condition.

The right kind of operator is alert, resourceful, and understands the operating principles underlying the equipment in his charge. Although it is desirable to keep labor costs at a minimum, it is a poor policy to entrust the safety of an expensive power plant to incompetent or irresponsible hands. The increased cost of repairs, if inexperienced help is employed, and decreased dependability of the installation will more than absorb the savings in labor cost. In the long run, properly qualified and skilled operators are always cheaper.

23-3. Combustion. An important condition for the successful operation of a diesel engine is proper combustion of the injected fuel. At the same time, fuel injection and combustion are things which can easily get out of order and therefore require constant attention. As already mentioned, measuring the exhaust temperature is necessary to check for proper combustion. However, the exhaust temperature alone does not give complete information, as several engine troubles affect the exhaust temperature in the same way. By checking exhaust temperature and cylinder pressure simultaneously, a better diagnosis can be made. The determination of the causes of trouble due to faulty fuel injection and combustion and their elimination are discussed in Secs. 20-8 to 20-22. Here it is sufficient to state that proper maintenance of a diesel engine requires that the operator be familiar with the taking of indicator cards and their interpretation in connection with exhaust-temperature readings. He must make this double check periodically and also every time that the pyrometer shows a deviation from normal temperatures. The taking of indicator cards is discussed in Secs. 21-2 to 21-7.

23-4. Inspection Schedules. Maintenance work, in order to be effective, must be carried on thoroughly and regularly. A schedule of maintenance service should be laid out in every plant and this schedule should be followed as closely as the operating conditions will permit.

A typical schedule form for diesel engine inspection and maintenance is shown in Table 23-1. In this form must be inserted the maximum allowable number of operating hours between inspections of the listed part. This number of hours will vary according to the size of engine, type of load, and character of service. Its correct determination is the responsibility of the plant manager who must have proper experience and good judgment. In a new plant it is good policy first to set a short period between inspections of different parts and later, when experience of the plant shows that within the prescribed hours a certain part does not show much dirt, wear, or loss of adjustment, gradually to increase the duration between inspections. However, one should remember that accumulation of dirt, increase of wear, and loss of adjustment in general follow not a straight-line course with time, but a convex curve which gets steeper and steeper. Therefore, to prevent greater damage, it is much better to have the next inspection a little too soon than a little too late. This consideration applies not only to the schedule form shown in Table 23-1, but to the forms in Tables 23-4 to 23-6 as well. Also, it may be stated that the forms in Tables 23-1 to 23-6 are made out not for a specific engine or type of engine but for any engine—air- or airless-injection, four- or two-stroke cycle, using a cooling tower or a spray pond, etc. Therefore, when applied to a certain engine, certain lines in the schedules may not be applicable and probably some new ones may

MAINTENANCE RECORDS AND DATA

TABLE 23-1. SCHEDULE OF DIESEL-ENGINE INSPECTION AND MAINTENANCE

Engine make_____ Type_____ Rating_____bhp

Item No.	Engine part to be inspected	Recommended maximum time between inspections	
		Operating hours	Months
1	Engine cylinders or liners and pistons............	6,000	9
2	Air-intake valves...............................	3,000	6
3	Exhaust valves.................................	1,500	3
4	Starting-air valves.............................	4,000	6
5	Safety or relief valves..........................	100	1
6	Air-compressor cylinders and pistons.............	3,000	6
7	Compressor valves; suction and discharge........	1,500	2
8	Scavenge-pump cylinder and piston or rotor......	3,000	6
9	Scavenge-pump valves; suction and discharge.....	3,000	6
10	Scavenge ports and automatic valves............	3,000	6
11	Exhaust-gas flow regulators.....................	2,000	6
12	Exhaust muffler and ducts......................	6,000	12
13	Main bearings and journals.....................	6,000	12
14	Outboard bearing..............................	6,000	12
15	Thrust bearing................................	6,000	12
16	Crankpins and bearings.........................	3,000	6
17	Piston pins or crosshead pins and bearings........	6,000	12
18	Crosshead guides and shoes.....................	6,000	12
19	Compressor-piston pins and bearings.............	3,000	6
20	Vertical-shaft bearings..........................	4,000	6
21	Camshaft bearings.............................	4,000	6
22	Camshaft drive................................	2,000	2
23	Fuel pumps...................................	4,000	8
24	Fuel-pump drive...............................	2,000	3
25	Fuel nozzles or valves and fuel timing............	500	1
26	Governor links, bearings, springs................	4,000	6
27	Governor drive................................	4,000	6
28	Water- or oil-cooled pistons: packings, bearings, ball-and-hinge joints.........................	3,000	6
	Scale and sediment deposits:		
29	Cylinder heads and jackets......................	1,000	2
30	Cooling passages in pistons.....................	2,000	4
31	Compressor heads and jackets...................	3,000	6

have to be added. The object of these schedules is mainly to call the attention of the reader to the great number of different inspections and checks which must be made at regular though different intervals if an engine and power plant are to be kept in good operating condition. When an engine is operated intermittently, the number of hours of operation may be low, but certain unfavorable phenomena, such as corrosion, oxidation of oil, settling of foundations, will still take place. Therefore, in the second column are given time limits in months and the inspection and servicing should be done after the expiration of one of the limits—operating hours or time elapsed—whichever takes place first.

Naturally, whenever a part is inspected and found to be worn near or beyond the prescribed tolerance, it should be reconditioned or so adjusted that the prescribed clearance is restored. Therefore, in connection with maintenance schedules, it is necessary to provide the plant manager, or engine operator in a smaller plant, with a specific instruction sheet regarding clearances and settings for each engine or each type of engine in the plant.

Typical forms for such instruction sheets are given in Tables 23-2 and 23-3. These forms must be filled out, taking information from instruction books furnished with each engine. If the instruction book of any specific engine should not have all the desired data, the latter should be obtained by writing to the engine builder. If this cannot be done for some reason, average data may be used as indicated at the end of this chapter.

Tables 23-4 to 23-6 contain inspection and maintenance schedules for engine and plant equipment down to fire protection and the plant building itself. In this respect it should be remembered that a power plant kept clean and neat has an important psychological effect upon the attitude of people working in the plant. It makes them proud of the whole place, induces them to keep the engines clean-looking, properly adjusted, running without undue noise or smoke, in general increases the efficiency of the whole plant. It also keeps down the rate of industrial accidents. In short, money spent for upkeep of the plant is well spent and gives high returns in many respects.

23-5. Maintenance Log Sheets. To be sure that maintenance work is done properly and according to the schedule similar to the one given in Table 23-1, the operator should have maintenance log sheets made out for each engine for each month of the current year. In these log sheets he must enter as a permanent record all maintenance work done. A sample of such a log sheet for work on the engine proper is shown in Table 23-7 which is prepared for a six-cylinder air-injection

MAINTENANCE RECORDS AND DATA

TABLE 23-2. ENGINE OPERATING DATA

Engine make_____Type_____Rating_____bhp

Item	Engine builder's data	Insert data
1	Compression ratio..	
	Valve timing:	
	Air-intake valve	
2	Opens before t.d.c., crank-angle deg.........................	
3	Closes after t.d.c., crank-angle deg..........................	
	Exhaust valve	
4	Opens before t.d.c., crank-angle deg.........................	
5	Closes after t.d.c., crank-angle deg..........................	
	Fuel valve	
6	Opens before t.d.c., crank-angle deg.........................	
7	Closes after t.d.c., crank-angle deg..........................	
	Starting-air valve	
8	Opens after t.d.c., crank-angle deg..........................	
9	Closes after t.d.c., crank-angle deg..........................	
	Valve lifts:	
10	Scavenge valve, plate or strip type, in......................	
11	Fuel pump: suction valve, in.....................................	
12	discharge valve, in..................................	
13	Air compressor: low pressure—suction and discharge valves, in...	
14	intermediate—suction and discharge valves, in...	
15	high pressure—suction and discharge valves, in...	
	Pressures:	
16	Compression in cylinders, cold, psi............................	
17	Compression, hot, full-speed, psi...............................	
18	Combustion at full rated load, psi.............................	
19	Lubricating oil at main bearings, psi.........................	
20	Starting air, psi..	
	Sight-feed lubricator, drops of oil per min:	
21	Each lead to power cylinders....................................	
22	Each lead to air compressor: low-pressure cylinder...........	
23	intermediate cylinder.............	
24	high pressure cylinder............	
	Valve springs:	
25	Intake: valve closed, force exerted, lb........................	
26	valve open, force exerted, lb...........................	
27	free length, in...	
28	Exhaust: valve closed, force exerted, lb.....................	
29	valve open, force exerted, lb........................	
30	free length, in..	

TABLE 23-3. STANDARD CLEARANCES AS INDICATED BY ENGINE BUILDER

Engine make_____ Type_____ Rating_____ bhp

Item	Engine part	In. min	In. max
1	Piston in cylinder liner, skirt..........................		
	Piston-ring gap:		
2	Top ring...		
3	Compression rings.....................................		
4	Oil-control rings.......................................		
	Piston-ring land clearance, axial:		
5	Top ring...		
6	All other rings...		
7	Piston crown to cylinder head.........................		
	Compressor piston to cylinder:		
8	Low-pressure...		
9	Intermediate...		
10	High-pressure...		
	Compressor piston to head:		
11	Low-pressure...		
12	Intermediate...		
13	High-pressure...		
	Scavenge piston:		
14	To cylinder...		
15	End to head, top......................................		
16	End to head, bottom..................................		
	Bearings:		
17	Main (crankshaft).....................................		
18	Crankpin...		
19	Piston pin..		
20	Camshaft...		
21	Fuel-pump drive shaft................................		
22	Lubricating oil-pump shaft...........................		
23	Rocker-arm bushing...................................		
24	Water-pump drive shaft..............................		
	End plays:		
25	Crankshaft, thrust....................................		
26	Connecting-rod side clearance.......................		
27	Camshaft...		
	Valves:		
28	Stem to valve guide...................................		
29	Valve head to cylinder head, valve open............		
30	Valve tappet in guide.................................		
	Backlash:		
31	Crankshaft gear to camshaft gear....................		
32	Lubricating-oil pump gears...........................		
33	Water-pump gear to camshaft gear..................		
34	Timing chain, total movement.......................		

TABLE 23-4. SCHEDULE OF ENGINE-EQUIPMENT INSPECTION

Engine make_____Type_____Rating_____bhp

Item	Equipment to be inspected and serviced	Recommended maximum time between inspections	
		Operating hours	Months
	Fuel system:		
1	Filters and strainers..........................	200	1
2	Fuel booster and transfer pumps...............	2,000	4
3	Auxiliary storage tanks.......................	1,000	3
4	Supply lines.................................	1,000	3
5	Heaters for heavy fuel oil....................	3,000	6
	Lubricating system:		
6	Lubricating-oil pumps, complete................	3,000	6
7	Lubricating-pump drive.......................	3,000	6
8	Oil-supply lines.............................	1,000	2
9	Oil strainers and filters......................	200	1
10	Oil tanks....................................	2,000	4
11	Oil coolers, tightness and scale deposits.........	3,000	6
12	Pressure-feed lubricators and check valves.......	3,000	6
13	Crankcase, sediment and surface...............	2,000	4
	Air-intake System:		
14	Air filters...................................	300	1
15	Air-suction ducts............................	2,000	3
16	Air-intake silencers...........................	2,000	3
17	Air coolers..................................	3,000	4
18	Exhaust muffler, sediment and tightness.........	2,000	4
	Pressure gauges—check with standard gauges:		
19	Lubricating oil...............................	3,000	6
20	Cooling water...............................	3,000	6
21	Compressed air..............................	3,000	6
22	Exhaust-gas pyrometer, check with standard......	3,000	6
23	Pyrometer lead wires, check insulation...........	3,000	6
	Pressure-relief valve:		
24	Fuel oil.....................................	3,000	6
25	Lubricating oil...............................	3,000	6
26	Compressed air..............................	3,000	6
27	Cooling water...............................	3,000	6

TABLE 23-5. MAINTENANCE SCHEDULE OF COOLING, FUEL AND AIR EQUIPMENT

Item	Equipment to be inspected and serviced	Months between inspections
	Cooling tower:	
1	Clean, adjust, and level troughs.	2
2	Clean distribution racks.	1
3	Clean and inspect screens.	1
4	Drain and clean basin.	6
5	Copper-sulfate treatment for algae.	6
6	Spray pond, clean and adjust spray nozzles.	1
	Jacket-water heat exchangers:	
7	Descale and clean tubes.	3
8	Inspect for leaks and seal them.	3
	Water wells:	
9	Check static level.	6
10	Check pumping level.	6
11	Check flow.	6
	Water pumps:	
12	Check suction pressure with gauge.	6
13	Check discharge pressure with gauge.	6
14	Check delivery.	6
15	Check power input to each pump.	3
16	Check speed of pumps.	3
17	Pull and inspect pumps for wear.	6
18	Check thrust bearings and clearances.	6
19	Drain and renew bearing oil.	4
	Water piping:	
20	Inspect for leaks.	3
21	Clean and paint exposed pipe.	12
	Fuel-oil storage tanks:	
22	Drain off water.	6
23	Inspect for leaks.	6
24	Drain off and clean out.	12
25	Clean and paint outside.	12
	Fuel-oil piping:	
26	Inspect for leaks.	6
27	Clean and paint exposed pipe.	12
	Air compressors:	
28	Drain and renew oil.	3
29	Inspect valves and bearings.	3
30	General overhaul.	12
	Air-storage tanks:	
31	Drain off water and oil.	2
32	Hydrostatic safety test.	12
33	Check pressure gauges.	12

TABLE 23-6. MAINTENANCE SCHEDULE OF ELECTRIC EQUIPMENT AND BUILDING

Equipment to be inspected and serviced	Months between inspections
Switchboard:	
Clean with large vacuum cleaner	1
Check for ground, unintentional grounds, and ground connections	1
Inspect oil circuit breakers, oil and contacts	3
Test watthour meters and indicating instruments	3
Check, adjust, and record relay settings	1
Clean and inspect bushings and potheads	2
Generators:	
Clean with vacuum cleaner	1
Clean and paint windings	6
Inspect collector rings and brushes	½
Check air gaps, four sides	1
Test frame grounds	1
Test insulation	1
Exciters and motors:	
Clean with vacuum cleaner	1
Clean and paint windings	6
Check air gaps, four sides	½
Check coupling, pulley or pinion alignment	1
Inspect commutator and brushes	½
Test insulation	1
Electrical starting equipment:	
Clean with vacuum cleaner	1
Inspect contacts and oil	3
Check, adjust, and record undervoltage and overload-relay settings	2
Inspect and test switch, conduit, and motor-frame ground	6
Storage battery:	
Check water level and charge	½
Fire extinguishers:	
Inspect and list	6
Discharge and recharge, soda, acid, and foam type	12
Test and fill tetrachloride type	12
Power-plant building:	
Clean all windows, inside and out	6
Replace all broken glass panes	3
Inspect roof and fix all leaks	12
Inspect and clean gutters and spouts	12
Inspect electric wiring	6
Inspect plumbing	6
Inspect paint, inside and out, and repaint where necessary	12

TABLE 23-7. MAINTENANCE LOG SHEET

No._____

Plant_____ Engine No.__ Hours of running for the month_____

Month_____Year 19____.

Item No.	Parts to be inspected and serviced	Date work was done on cylinder					
		1	2	3	4	5	6
1	Main pistons........................						
2	Piston pins.........................						
3	Crankpin bearings...................						
4	Oil hole in crankpin.................						
5	Cylinder or liner wear measured......						
6	Cylinder jackets and heads scale cleaned...						
7	Main bearings, wear and clearance....						
8	Crankshaft and extension shaft alignment..						
9	Exhaust valves......................						
10	Intake valves.......................						
11	Exhaust-cam roller setting and timing.....						
12	Intake-cam roller setting and timing.......						
13	Starting-air valves..................						
14	Safety valves.......................						
15	Fuel-injection nozzles...............						
16	Fuel-pump valves...................						
17	Fuel-injection timing................						
18	Spray checks.......................						
19	Fuel-pump drive....................						
20	Governor..........................						
21	Camshaft drive.....................						
22	Lubricating-oil filter................						
23	Lubricating-oil piping...............						
24	Lubricating-oil pump................						
25	Lubricating-oil cooler...............						
26	Lubricating-oil day and storage tanks......						
27	Low- and intermediate-pressure air-compressor valves....................						
28	High-pressure air-compressor valves.......						
29	Low- and intermediate-pressure compressor piston............................						
30	High-pressure air-compressor piston.......						
31	Air cooler..........................						
32	Air receivers.......................						
33	Air-injection piping.................						
34	Air tanks drained and cleaned...........						
35	Crankcase flushed out................						

engine. With several engines in the power plant, it is convenient to have log-sheet forms printed to cover all engines and then the operator simply leaves blank the lines or columns which do not pertain to a particular engine.

Similar maintenance log sheets should be used for the maintenance of engines and electric equipment in conformity with schedules similar to Tables 23-5 and 23-6.

23-6. Maintenance Requirements. All diesel engines require essentially the same maintenance operations. In most operations the main work to be done is to dismantle and replace parts. Therefore, it is very desirable that suitable means be provided in the plant for handling the parts and that sufficient floor and bench space be available on which to put them while the work is being done.

An overhead crane or hoist with a suitable capacity to handle all parts is of great importance. The operator should not be required to use a 5-ton crane to lift a 200-lb. piston, nor should he be handicapped by trying to lift a two-ton load with a one-ton chain hoist. Two men with the right kind of equipment and tools can accomplish far more in a shorter time than four men can do if they have to depend on man strength and inadequate tools.

The procedures of maintenance may be grouped under four headings: cleaning, adjusting, reconditioning, and replacing parts.

The nature and amount of equipment required to carry out these procedures depend upon the system of maintenance adopted in respect to the last two procedures. In a large plant or a plant with large engines, reconditioning of worn parts in the plant itself will usually save money and time. In this case a certain number of machine tools and precision gauges will be required, as well as at least one skilled mechanic. In small plants, or plants with small engines, it is more economical to use replacement parts from stock and to send the worn parts for reconditioning to the engine builder or shops specializing in this kind of work.

23-7. Compression and Combustion Pressures. The compression pressure in each cylinder of an engine should be maintained within ± 4 per cent of the recommended pressure. Thus in an engine with a recommended compression pressure of 500 psi, the maximum variation may be $500 \times 0.04 = 20$ psi either way. The cylinder with the lowest pressure may have a compression of 480 psi, but not less, and the cylinder with the highest pressure may have a compression of 520 psi, but not more.

The usual causes for decreased compression are leaky valves, leaky piston rings, obstructions in the suction lines or low scavenging pressure and increased piston end clearance caused by bearing wear. The procedure of adjusting the compression ratio is given in Sec. 24-3 in the paragraph on Piston-end Clearance.

The maximum combustion pressure should be kept within ±5 per cent of the recommended full-load pressure. The combustion pressure is affected by the compression pressure, the conditions and characteristics of the fuel or spray valve, the injection-air pressure in air-injection engines, the injection timing and fuel-oil characteristics in mechanical-injection engines.

The combustion should be complete at all times. Incomplete combustion is a direct fuel loss; it also leads to undue wear and increased cost of maintenance. There is no excuse for a smoky exhaust in the operation of a diesel engine.

23-8. Details of Maintenance. In the next chapter are given practical and proven procedures and helpful hints on how an operator, using schedules and data similar to those given in Tables 23-1 to 23-7, can keep his engine or engines and the power plant as a whole in tip-top condition with the minimum expenditure of time and effort. To an experienced operator some of the details may seem obvious and almost superfluous. However, in the first place, all instructions are intended not for a man who knows all about a certain subject, but for one who wants to know all he can learn about that subject, and second, the efficiency even of an experienced man is increased and strengthened when he sees in print in a systemized form rules and recipes which he has come across in his practice.

Finally, it should be understood that all instructions given in this or any other book, by necessity are only of a general character and are not intended to replace the instruction book furnished by the engine builder with his engine. The operator must read the engine builder's instruction book, become thoroughly familiar with it, and keep it on hand, so as to be able to consult it at any moment when a question comes up. This is a condition without which he cannot be a good engine operator.

23-9. Safety Precautions. Safety precautions are the result of casualties that have occurred in some plants. In order to prevent the same thing from happening in another plant, safety precautions have been drawn up and pertinent warnings should be posted near every piece of equipment. The operator, seeing these warnings all the time, is helped to remember and follow them at the proper occasions. They are a substantial aid in maintaining the efficiency of the plant.

As a sample of such safety rules, below are tabulated some rules concerning bearings:

1. Never use an engine when it is known that the bearings are in bad shape, unless absolutely necessary.

2. The rapid heating of a bearing is a danger sign. If it takes a bearing an hour or more to reach a steady temperature (although one that is uncomfortably hot for the hand), the bearing is probably safe. But if the same temperature is reached in 10 or 15 min, it will not stop there and there is apt to be trouble.

3. Never permit a bearing to begin to smoke. When a bearing gets hot, take steps to cool it or stop the engine.

4. Use clean rags, not waste, to wipe a bearing clean after opening it.

5. Never attempt to nurse a wiped bearing to shape. Remove the cap and fit the bearing to the journal by scraping.

6. Don't squirt oil into an oil hole before seeing that it is open and not clogged with dirt.

7. When using an oil can, make sure that it contains the proper oil and that the spout opening is not stopped up.

8. Do not forget to service a bearing which is in an inaccessible or out-of-the-way place.

9. All self-oiling bearings with oil wells must be drained, washed out, and refilled at regular intervals; all rotating parts must be thoroughly hand-oiled before starting the engine.

10. Do not file or machine bearing-shell joints. Either rebabbitt them or renew the shells.

11. When taking leads, make sure that the bearing halves are set up metal to metal.

12. When rebabbitting a bearing, never pour babbitt into a shell which is not perfectly dry. Failure to observe this rule may cause serious burns or even a fatal accident.

13. After rebabbitting a bearing, do not forget to open the oil holes.

14. When cutting oil grooves, do not extend them to the ends of the bearing.

Similar safety rules should be worked out and posted near other equipment.

23-10. Average Clearances. Good practice requires about the same relative clearances between moving parts in engines of the same general type, such as all diesel engines with water-cooled cylinder jackets, regardless of size of the bore and engine speed. In the absence of engine builders' data for a particular engine, the following figures may be used:

Piston to Cylinder. Designating the nominal bore of the cylinder by D, the recommended clearance between the piston and cylinder may be computed as follows:

Cast-iron pistons: skirt clearance c with new pistons and liners:

$D = 3\frac{1}{2}$ to $6\frac{1}{2}$ in.: $c = 0.001$ to $0.0012\ D$
$D = 7$ to 12 in.: $c = 0.0009$ to $0.0011\ D$
$D = 13$ to 20 in.: $c = 0.0007$ to $0.0009\ D$

Piston-pin section, when the piston is tapered down from bottom to crown:

$D = 13$ to 20 in.: $c = 0.001$ to $0.0012\ D$

Aluminum pistons: skirt clearance c with new pistons and liners, at bottom:

$D = 3\frac{1}{2}$ to $6\frac{1}{2}$ in.: $c = 0.0012$ to $0.0019\ D$
$D = 7$ to 20 in.: $c = 0.0012$ to $0.0015\ D$

From $D = 8$ in. and up these pistons are usually made with a taper toward the crown and before the first piston ring above the wrist pin $c = 0.0014$ to $0.002\ D$, without any definite relation to the relative size.

Piston Rings. The gap clearance, whether bias-cut, butt, or lap joint, measured with the ring compressed, in the cylinder, may vary from $c = 0.003$ to $0.005\ D$, and in some high-speed heavy-duty engines even up to $0.008\ D$.

The axial clearance, between ring and land, according to good practice, is given in Table 23-8.

Piston Pin. The necessary running clearance c of the piston pin, if d is the piston-pin diameter in inches, may be computed from the expression

$$c = 0.001 \text{ to } 0.0015\ d$$

Bearing Clearances. Main bearing: if the journal diameter is d, the suitable running clearance c is

$$c = 0.0015 \text{ to } 0.0025\ d$$

Crankpin bearings: same clearance as main bearings.

Camshaft bearings: these bearings are usually run with the same rather small clearance as the piston pins.

Fuel-pump Drive Shaft. This and all other auxiliary drive shafts—same clearance as the piston pins.

End Clearances. There are no set rules for the end clearances, such as between main bearing shells and crank webs. These clearances in most cases are not important, with the exception of those clearances between the main bearing shells and crank webs that locate the crankshaft axially and those between the crankpin bearings and crank webs, since these clearances locate the connecting rods. The end clearances are governed by design and manufacturing considerations rather than by operating conditions. For each engine they can be readily measured and for future reference should be entered in the permanent record of the engine.

Valves. Valve-guide clearance around stems is almost independent of the size, and varies from 0.002 to 0.003 in. in small engines with a 3-in. bore, to 0.003 to 0.005 in. in large ones, with a bore up to 20 or 22 in.

Valve-tappet clearance in the guide varies from 0.001 to 0.0015 in. in small engines and up to 0.002 to 0.003 in. in large ones.

TABLE 23-8. PROPER SIDE CLEARANCES OF PISTON RINGS, INCHES

Cylinder diameter, in.	Two-stroke engines		Four-stroke engines	
	Top ring	Other rings	Top ring	Other rings
Below 6	0.005	0.004	0.003	0.002
6–12	0.006	0.005	0.004	0.003
Over 12	0.008	0.006	0.005	0.004

Valve-stem clearance to the rocker arm should be measured when the engine is hot and should be for both intake and exhaust valves, for all engine sizes, $c = 0.010$ to 0.015 in.

If this clearance is measured when the engine is cold, it should be $c = 0.012$ to 0.017 in. for intake valves and $c = 0.015$ to 0.030 in. for exhaust valves.

However, the clearance for the exhaust valves should always be checked when the engine is hot.

Backlash. Backlash from the *crankshaft gear* to the camshaft gear should be as small as possible, only so much as to avoid binding at high speeds. In small engines it may vary from 0.0005 to 0.0015 in., increasing to about twice these values in large engines.

Governor-drive gear backlash may vary from 0.001 to 0.002 in. in small engines, increasing to about 0.003 to 0.005 in. in large engines.

Oil-pump gear backlash may vary from 0.005 to 0.008 in., regardless of engine size.

Water-pump gear backlash may vary from 0.002 in. to 0.004 in. in small engines; it should be only slightly greater in large engines.

QUESTIONS

1. Enumerate the requirements for being a qualified diesel-engine operator.
2. What is the most important condition for successful operation of a diesel engine and what are the means of checking it?

3. Give a sample of a typical schedule form for diesel-engine inspection.

4. Explain how the maximum number of operating hours between inspection of various parts is determined and upon what factors it depends.

5. What is the purpose of maintenance log sheets? Why should they be made out for each engine for each month of the year?

6. Give a sample of a typical maintenance log sheet.

7. Enumerate the separate kinds of maintenance work required by a diesel engine.

8. Within what limits must the compression pressure in each cylinder be maintained in respect to the recommended pressure?

9. State the usual causes for a decrease of compression.

10. Enumerate the items that influence the maximum combustion pressure.

11. How are safety precautions established in the first place?

12. What is the object of posting safety precautions regarding a certain piece of equipment near it?

13. State some of the safety precautions listed in Sec. 23-9.

14. State the average clearance between a cast-iron piston skirt and liner recommended for (a) small-bore engines, (b) medium-size engines, and (c) big engines.

15. Give the same information for aluminum pistons.

16. What is the recommended gap clearance of piston rings in respect to engine bore?

17. What is the recommended clearance between the piston pin and its bushing?

18. What is the recommended running clearance for main bearings?

19. What is the recommended clearance for crankpin bearings?

20. What is the recommended clearance for camshaft bearings?

21. What is the recommended clearance between the valve stems and valve guides in various sizes of engines?

22. How must the clearance between the exhaust-valve stem and the rocker arm or other part actuating it be checked?

23. What are the recommended backlashes for different gears in a diesel engine?

CHAPTER 24

MAINTENANCE AND REPAIRS

In this chapter are discussed some of the special procedures which must be followed in the servicing and repairing of the more vital parts of a diesel engine.

24-1. Cylinder Heads. In removing cylinder heads particular care must be taken not to damage or bend the cylinder-head studs.

First, all attached manifolds, pipes, rocker arms, etc., must be disconnected from the cylinder head and removed.

Second, the stud nuts must be loosened and removed. On large engines the stud nuts must be loosened by special wrenches with long handles furnished with the engines. One should never try to loosen the nuts by hammering, nor to tighten them by using an additional extension in the shape of a piece of pipe.

Next, the lifting eye bolts are screwed into the tapped holes in the top of the cylinder head. With a large engine and a heavy cylinder head, the eye bolts must be braced one against another by a piece of 2- × 4-in. wood cut to fit between the bolts to prevent the bolts from being bent or by a stayboard as shown in Fig. 24-1.

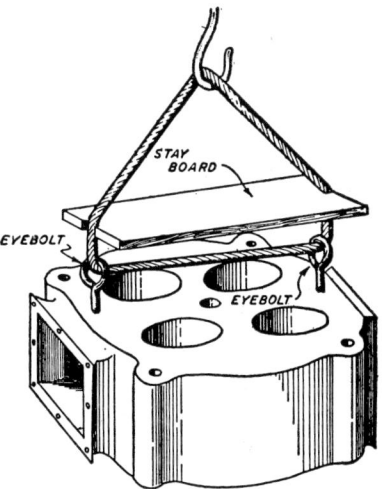

FIG. 24-1. Correct method of lifting a cylinder head.

Finally, hooks or slings are attached to the eye bolts in such a manner as to engage the overhead hoist centrally, and the head is lifted, guiding it to avoid damage to the studs.

Cylinder heads should be cleaned periodically, usually when they are taken off for pulling the pistons. All holes in the top of the head must be plugged up by wooden plugs, the head turned upside down and the water space filled with muriatic acid solution. When the gas bubbles cease to rise, if steam is available, a steam hose is stuck into the solution. This will stimulate the chemical action of the acid. The solution is kept near the boiling point until all signs of effervescence stop. After that the solution is removed and the water space thoroughly washed, first with cold and then with hot water. Finally, the plugs and hand-hole covers are removed and all remaining scale thoroughly scraped and washed out.

24-2. Cylinders and Liners. Whenever a piston is pulled, the cylinder or liner should be cleaned; scored or burned places must be polished by stoning. The bore should be checked for wear. Unavoidable liner wear is always greatest at the top. When the wear reaches 0.005 in. per inch of cylinder diameter, the liner must be replaced. If the cylinder has no removable liner, it must be rebored and new, oversize pistons installed.

In low-speed engines, with cylinder lubrication by special feeds, when the pistons are out for inspection, the cylinder lubricators should be worked by hand until oil comes freely from each of the feed holes.

Cylinder Jackets. Cylinders and cylinder blocks cast in one piece with the water jackets should be inspected periodically for scale and, if scale is found, cleaned. It is well to remember that scale, even $\frac{1}{32}$ in. thick, affects heat transfer from the cylinder to the water considerably and $\frac{1}{16}$-in.-thick scale is an effective heat insulator.

The cleaning procedure is as follows: first the jackets are washed out with water to remove mud and sand, and then cleaned with a solution of muriatic acid, one part of acid to about ten parts of water. A good procedure is to mix up the solution in a barrel, connect the discharge of a hand pump to the cooling-water manifold, the suction of the pump to the barrel and to circulate the acid mixture through the cylinder jackets until no signs of effervescence show in the return flow, indicating that all hard scale has been dissolved and loosened. The cylinders should then be washed out first with cold and then with hot water until the least trace of acid has been removed. The hand-hole covers should then be taken off and all loose scale scraped out and removed.

24-3. Pistons. The pistons should be pulled at regular intervals, as given in the schedule, the rings removed and pistons and rings thoroughly cleaned and washed with kerosene. All scored places on a piston should be smoothed down by stoning.

The condition of the pistons and liners gives a good indication of the quality and quantity of lubricating oil used. If the liners and piston rings have a bright surface and the rings are free, it indicates that the oil is good and supplied in the proper amount. If an inferior oil is used, the wearing surfaces will have a dull grayish appearance and the rings will show a tendency to stick. An excessive amount of lubricating oil, even of good quality, will result in the accumulation of a gummy deposit on the piston and rings.

If an engine is operated with poor combustion, as shown by a smoky exhaust, a gummy deposit will be found on the rings even if a suitable lubricating oil and in a proper amount has been used.

The heads of the pistons should be examined for possible cracks and burned spots. If the burn is deep, it may be advisable to cut out the burned metal with a sharp chisel and to weld up the depression to restore the crown to its original shape.

The oil hole in the piston pin should be cleaned out.

In pistons with fitted and fixed piston pins, when the latter has been removed, care must be exercised, in placing it back, not to distort the piston. After assembling, the piston should be checked for roundness, using a micrometer gauge. If found to be out of round, it must be brought back to the round shape by (sometimes vigorous) tapping with a nonmetallic mallet.

When a piston is replaced after cleaning, care should be taken to put it back in the same relative position to the liner as it was before. When the connecting rod is put back in place, it also must go back in the same way as it was before. The bolts of the connecting rod must be marked before they are taken out and put back in the same holes.

Placing the piston and connecting rod back into the engine should be done very carefully, whether in a large or small engine. The general procedure is the following:

1. If the engine is provided with a piston puller, the latter is attached to the piston top. The piston and connecting rod are carefully wiped with a clean rag and lubricated all over, including the piston-pin bushings, and the liner surface as well.

2. The ring guide is placed on top of the liner. The engine is barred over until the crankpin is at top dead center. The piston assembly is lowered carefully into the liner, taking great care not to scratch the liner with the connecting rod. Proper attention must be paid to each ring as it enters the ring guide. More about ring guides will be found in Sec. 25-4.

3. If the crankpin box is a separate piece from the connecting rod, it should be assembled on the crankpin and the bolts held by special set screws instead of the nuts.

4. The upper half of the box with the bolt ends being heavier than the lower box half, it will turn down after the box is assembled on the crankpin. The box assembly must be turned around until the dowels and bolts match up with the holes in the T end of the rod.

5. The connecting rod is assembled to the crank-box assembly, the nuts are drawn uniformly and locked by cotter pins or bent washers, as the case may be.

6. The set screws are backed away from the connecting rod bolts, and the nuts are locked with jam nuts.

7. The piston puller is removed.

Piston-end Clearance. The compression pressure of a power cylinder is determined by the distance between the top of the piston and the cylinder head at top dead center. This distance is called *piston-end clearance* or also *mechanical clearance* to differentiate it from the volumetric clearance or the volume occupied by the gases at top dead center. The allowable variation for the mechanical clearance is ± 0.25 per cent of the piston stroke. Thus in a small engine with a 4-in. stroke the tolerance is only ± 0.010 in. This must be kept in mind when changing a cylinder-head gasket.

The wear of the main and crankpin bearings gradually increases the mechanical clearance and must be compensated. On large engines—with a separate crankpin bearing, often called *box*—the proper mechanical clearance is restored by inserting shims between the box and the T end of the connecting rod. Small high-speed engines usually have one-piece connecting rods and precision bearings. When such bearings are worn, they are easily replaced by new ones. In some small engines the mechanical clearance can be adjusted by decreasing the thickness of the cylinder-head gasket.

Piston-side Clearance. Whenever a liner or piston is replaced, the clearances between the piston and liner or cylinder must be checked and must compare closely with those recommended. It should be remembered that aluminum pistons require greater clearances than cast-iron ones. In checking the clearances, it should be also remembered that the top of the piston, from the top groove to the crown, is turned down smaller to allow for the higher temperature and greater expansion of that part of the piston.

Piston Rings. It is particularly important to keep the piston rings in first-class condition since they control the power developed by the engine and the lubrication of the piston and cylinder. Stuck and broken piston rings are one of the main factors contributing to wear and scoring of liners and cylinders. Ring condition should be investigated at the first signs of blow-by. When renewing piston rings, one must make sure that they are given the proper side and gap clearances as recommended by the engine builder.

24-4. Bearings. The radial clearance between a shaft and its bearing and the lubrication determine the service that can be obtained from both the shaft and the bearing. Even with proper lubrication, if a bearing is fitted too loosely, repeated pounding may flatten the shaft and crack the bearing babbitt. If the clearance is too small, the bearing may begin to heat up and burn out from excessive friction. Since the allowable tolerance in adjustment is very small, it is necessary to know the proper clearance and to use every effort in obtaining and maintaining this clearance.

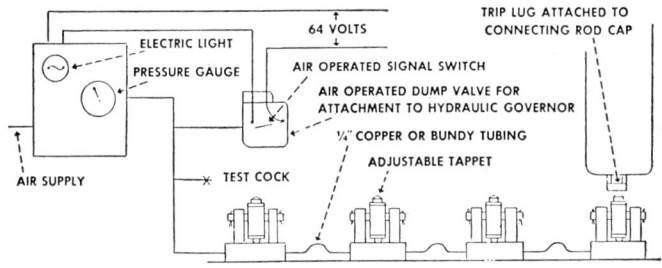

FIG. 24-2. Diagram of "bearing-watchdog" installation. *Courtesy of Paxton Diesel Engineering Co., Omaha, Nebraska.*

Crankshaft and Main Bearings. Crankshaft alignment is very important and should be checked at least once a year. Misalignment may cause a shaft to break in a few weeks of operation without any warning in the form of pounding. The differences in quantity of individual bearing lubrication, in quality of bearing metal, in fitting of the bearing shells to the bedplate, and in bearing pressures all have a tendency to destroy the original crankshaft alignment. Therefore, it is important to keep a close check on bearing wear and, if the change in clearance is not uniform on all bearings, to check the shaft alignment as explained in Sec. 25-10.

Bearing Safeguard. Figure 24-2 shows diagrammatically the so-called "bearing-watchdog" system which is designed to protect a diesel engine from damage by shutting it down automatically as soon as the crankshaft is endangered by a worn connecting-rod bearing or main bearing.

The system consists of trip levers with adjustable tappets, installed in the crankcase under each crankshaft throw, trip lugs fastened to each connecting-rod bearing cap, a dump valve attached to the servomotor of the hydraulic governor, a control box containing an air-pressure reducing valve, a telltale light, and copper tubing to connect the various parts. The clearance between the tappet of the trip lever and the lug on the connecting-rod cap is set to equal the maximum amount of wear permissible. The lower end of the trip lever keeps closed a check valve on the pressure line.

The system can be operated either pneumatically or hydraulically. With pneumatic operation, the system is connected to an air supply line and a constant air pressure of 10 psi is maintained in it. When the combined wear of a main and connecting-rod bearings reaches the clearance for which the system is adjusted, the trip lug strikes the trip lever. Its motion releases the valve which it kept closed, the air pressure in the whole system drops and this actuates the dump-valve on the governor; the engine is immediately shut down and simultaneously the electric light is lighted.

The system will also shut down the engine if a main bearing or connecting-rod bearing goes out, or if a bearing cap, connecting-rod bolt, or a connecting rod itself breaks.

In addition to protecting the engine and particularly the crankshaft, the "bearing-watchdog" system makes maintenance easier, permitting less frequent bearing inspection. The system is used successfully in locomotive, stationary, and marine installations. In engines not having a hydraulic governor the dump valve may be operated by means of a special air cylinder.

Crankpin Bearings. The crankpin bearings require considerably more attention than the main bearings: the work on the bearings is facilitated if the operator keeps on hand a special gear. The gear, however, is not complicated and consists of a good heavy wrench or socket wrench to fit the crank-bolt nuts, a skeleton wrench to back them off, a small wrench to loosen up the locking screws, as used in large engines, a pinch bar to feel the amount of clearance between the crankpin and the bearing by jumping the crank box or big end of the connecting rod, and on large engines, a set of rope tackles to lift the bolts, chain tackles to handle the crank box, and a good sledge hammer.

If the box can be jumped with the pinch bar, it is an indication that slack exists. The exact amount of slack and hence the thickness of shims which must be taken out, may be measured as follows: With the piston near top dead center, the bottom half of the bearing is lowered and two or three lead wires about $\frac{1}{32}$-in. thick are placed in a circumferential position on the bearing surface, two wires near the ends and one in the center of the bearing. Then the bottom half of the box is drawn up and the nuts on the crank bolts are screwed on tight or, on a large engine, hammered up to their original position as indicated by a scratch mark that has been made previously on the nut and bolt. After that the bottom half box is again lowered, the leads are removed and their thickness measured with a micrometer gauge. This will indicate the existing clearance. A sufficient thickness of shims is removed to reduce the clearance to the standard amount and the bottom half is tightened for operation.

The same procedure, using lead wires for determining the clearance and thickness of shims to be taken out, may be used with the main bearings. The difference is that in the main bearings, both on four- and two-stroke single-acting engines, the pressure is on the lower halves of the bearings all the time, and the clearance between the journals and upper bearing halves is not important, whereas in the crankpin bearings of four-stroke engines the pressure during each cycle is exerted first on the upper half, then the lower half. This change occurs near the top dead center on the exhaust stroke and is due to the inertia of the reciprocating parts. With an excessive clearance in the crankpin bearing, these inertia forces will pound the bearing and increase its wear and may even cause cracking of the babbitt.

24-5. Valves. All valves must be kept in good condition. It is good practice to grind the valves before they need to be remachined. If valves are allowed to leak, the seating surfaces will begin to wear rapidly and very soon it may become impossible to recondition them, making a replacement necessary.

Valve lift affects considerably the service that can be obtained from a valve and its seat. The lift recommended by the engine builder will give the best service. The valve lift should be checked carefully every time a valve is ground and corrected to the recommended height.

Valve Timing. Valve timing can be checked with a protractor put on the crank web. Another procedure is to have the flywheel marked for top dead center of one of the cylinders, usually the one nearest the flywheel, and to lay off the flywheel rim in degrees.

Corrections for exhaust- and intake-valve timing should not be made if the variation is less than five degrees from the setting given in the instruction book. Before making an adjustment, always take the readings for all valves and use the average value.

The fuel- or spray-valve timing is extremely important and great care should be exercised in its checking and adjustment. The recommended settings must be followed within ± 1 deg of crank travel.

Valve Mechanism. Clearance. The clearance in the valve-actuating mechanism, whether it is between the cam and the cam follower or between the rocker arm and the tip of the valve stem, plays an important part in engine performance and operation. This clearance must be checked at regular, relatively small time intervals, as indicated in the schedule, and kept absolutely to the standard given in the instruction book. If the clearance is smaller than that required when the engine is cold, it can entirely disappear and prevent the valve from fully closing when the engine parts reach operating temperature. On the other hand, an excessive clearance will change the timing and cause a dangerous increase in impact—a so-called *valve lash*—which may lead to undue wear or even breakage.

24-6. Fuel Pumps. On air-injection engines fuel-pump settings determine the load distribution between the cylinders of a multicylinder engine; consequently, they determine the maximum power which an engine can develop and also affect its steadiness. The adjustments should be carefully checked and all pumps set to deliver equal amounts of fuel to all cylinders.

On airless-injection engines with individual fuel pumps, the latter have two functions—namely, they measure the quantity of fuel and time its delivery to each individual cylinder. This makes it extremely important that they always be accurately adjusted and timed.

Fuel-pump Valves. When the valves of larger fuel pumps are inspected, they should be cleaned and, if necessary, ground. The lifts of the valves should be checked and readjusted if the valves are ground. The clearance of the hand lifter should be checked.

24-7. Mechanically Operated Fuel Valves. Fuel-injection valves of mechanical-injection engines generally require little maintenance beyond cleaning and perhaps a gentle lapping of the fuel valve to the nozzle-tip seat. However, they should be checked at frequent intervals, about every 500 hr, because, if they begin to leak, the whole performance of the engine will be affected.

If a valve begins to leak and persists in leaking soon after being adjusted, the simplest and probably the safest procedure is to replace it by a spare valve and to send the offender to its maker for repair. However, this plan is not always open to a plant manager in remote places, and a local solution of the problem may be desirable.

If during an inspection test a valve is found to have developed a leak, a very small amount of lapping should stop the leaking. If a considerable amount of lapping is necessary, the valve should be taken out and tested after about one hundred running hours, and if it is found that the leak has again appeared, the valve should be withdrawn from service for detailed examination. Excessive lapping should be avoided by all means as this increases the width of the needle seat, upsetting the performance of the valve.

The main cause of a persistent leak is misalignment between the long needle valve and the nozzle. This causes the needle to move sideways a small amount at every valve operation and wears the needle and its seat out of round. The cause of misalignment may be a bent fuel needle, a defective nozzle-tip seat, or a distorted valve body.

The equipment necessary to investigate these conditions consists of a surface plate, a V block, a surface gauge, and a dial indicator. The needle should be first examined by clamping the guide in a V block, setting the dial indicator as shown in Fig. 24-3, and turning the needle in the guide. Any bend in the needle will at once be discovered on the gauge and can be rectified by bending the spindle in the required direction until it is straight.

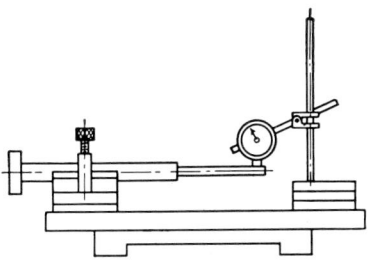

Fig. 24-3. Checking of a bent spindle.

The valve body may be checked next in the following manner. The end of a defective nozzle tip is ground away to the center line, as shown in Fig. 24-4. This tip is placed on the valve body and fastened with the hold-down nut. Next the needle is placed in its guide and the whole pushed home in the valve body. When the needle is resting in its working position in the cutaway nozzle tip, inspection with a magnifying glass will enable any misalignment to be detected. Rotation of the guide and the needle will check whether the guide is true, since the needle has already been trued up. Should the guide pass these tests, one may safely assume that the fault is in the nozzle tip. Replacing the latter will usually correct the whole assembly. However, should any misalignment be found, it must be corrected before anything further is done in the nature of fitting new parts.

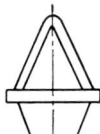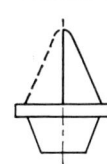

Fig. 24-4. Fuel-injector tip ground off for checking other nozzles.

While the fuel-valve body naturally was true when leaving the factory, it can be distorted by not sufficiently careful handling at the engine plant. The body may be distorted when the nozzle-tip nut is unscrewed, or by being screwed down too tight, or by being dropped. The correction of a distorted fuel-valve body is a delicate job, which requires skill and accuracy; it may be too difficult for the average operator. Nevertheless it can be done with the tools available in a good power plant, and the procedure is as follows.

A steel mandrel is first made slightly larger than the valve body, turned to fit the needle guide exactly; the other end is reduced to allow it to pass through the valve body. Next a flat cutter is made with a shank to fit the taper in the lathe-spindle nose, as shown in Fig. 24-5. This cutter is machined to the exact angle of the nozzle seat, hardened, and carefully ground away to form a cutting edge. The centers are now fitted on the head and tail stocks of the lathe, and the mandrel with the valve body on it held between them. The center line of the valve body now coincides with the center line of the lathe spindle, and the body is now securely clamped to the lathe carriage in this position. This clamping requires a certain care, the surface gauge and dial indicator being used to detect any movement of the mandrel as the lathe centers are withdrawn. This arrangement is shown in Fig. 24-6.

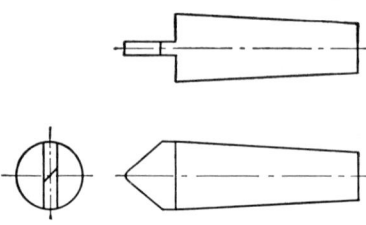

Fig. 24-5. Flat cutter for correcting a distorted fuel-valve body.

After the body is thus locked in position, the lathe tailstock is moved back, the carriage is also moved back, the mandrel is withdrawn, and the cutter (Fig. 24-5) substituted in place of the headstock center. In proceeding to clean up the nozzle seat, the cutter is run slowly—very little pressure is exerted on it, and plenty of cutting oil is used to ensure the necessary high finish. After removal from the lathe the nozzle is gently lapped into its position in the body with a fine metal polish. The valve is now assembled and tested. This

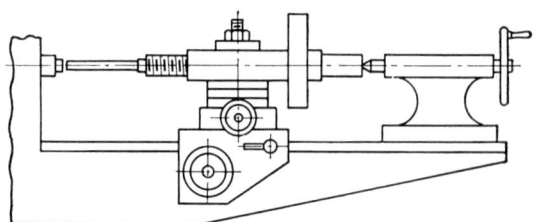

Fig. 24-6. Mounting injector body in a lathe for correcting it.

method gives more accurate results than truing up nozzle seats by means of a reamer with a pilot.

The whole procedure is described here in some detail to make clear the principles involved, which may be applied to reconditioning fuel valves of a different construction.

Lapping Fuel-valve Parts. All parts of the fuel system are of such precision that only the best, most uniform compounds should be used

TABLE 24-1. LAPPING COMPOUNDS FOR FUEL-SYSTEM PARTS

Type	Maker	Catalogue No.
Medium	American Bosch, Springfield, Mass.	TSE 7752
Fine	American Bosch, Springfield, Mass.	BM 10007
Fine	Carborundum Company, Niagara Falls, N.Y.	H-400

in servicing them. The compounds indicated in Table 24-1 have been found satisfactory for this work.

Lapping hardened steel parts with a fine compound is a rather slow process and operators ought to be glad to use a time-saving device such as that shown in Fig. 24-7. First a sturdy and accurately fitted guide is made for the protruding end of the needle. After cleaning the needle valve and seat thoroughly and placing a small amount of fine compound on the needle tip, the needle is set down on the seat. A rawhide strap is wound around the needle two or three turns. One end of the strap is fastened to a spring of sufficient tension to pull back the rawhide when it is released. The spring in turn is fastened to a bracket or wall. The other end of the strap is alternatively pulled by hand and released. The travel of the hand should be such that the valve makes at least one full turn on the seat. Another convenient way to rotate the valve stem is between the palms of the hands, with the body of the valve held in a vise.

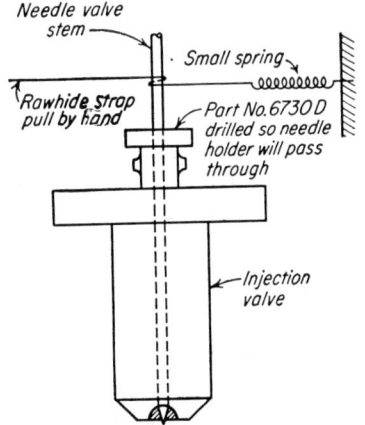

FIG. 24-7. Method of grinding injection valves.

A valve should not be lapped to its seat for longer than 5 min. Longer lapping does not add anything to obtaining a tight seat and may groove the valve. After lapping, the valve must be withdrawn and all traces of lapping compound should be rinsed away from the valve and seat with kerosene or gasoline and blown with air. After that the fuel nozzle is reassembled and checked on the test stand before being returned to service.

24-8. Hydraulic Nozzles. Nozzle trouble is indicated when (1) the engine loses power, (2) the engine smokes without being overloaded, (3) the reading of the exhaust-gas pyrometer of one cylinder is too high or too low, or (4) one of the cylinders begins to knock.

Some injectors are provided with a feeler pin with a stem in contact with the injector needle and extending over the top of the nozzle, usually covered with a removable cap. When a finger is pressed on the feeler pin while the engine is running, a sharp kick will be felt at every injection, if the injector is working properly. A feeble motion or no motion at all is an indication that the injector needle is not working properly. If the injector has no feeler pin, the work of the nozzles may be investigated by pressing the injector line between the thumb and forefinger and feeling the pulsation of the fuel pressure.

An alternative method is to cut out the nozzles, one at a time, by opening the bleeder valve or the line connection. A nozzle whose cutting out causes little or no drop in the power output is defective.

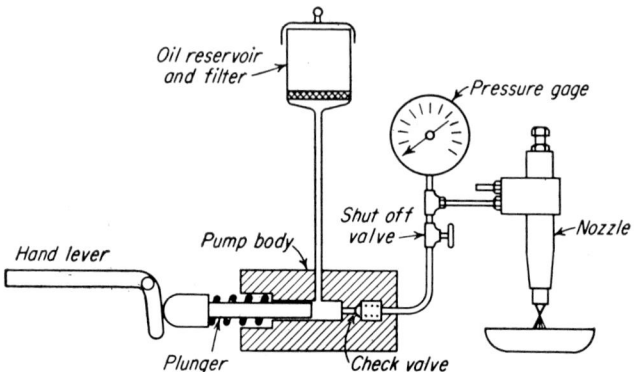

FIG. 24-8. Diagram of a nozzle tester.

When by either of these methods the defective nozzle has been located, it must be removed from the engine and examined with a nozzle tester.

Nozzle Testing. The nozzle tester consists of a hand pump with a check valve capable of producing pressures up to several thousand pounds per square inch, an attached pressure gauge, and a short line connection to which the nozzle is attached. The tester usually has a fuel reservoir with a filter and a shut-off valve between the pump and the nozzle. Figure 24-8 shows such a tester diagrammatically; it is often furnished with the diesel engine or may be built comparatively easily with the facilities usually available in a machine shop.

Fuel nozzles may be tested in a nozzle tester for (1) valve opening pressure, (2) spray appearance, (3) leakage, (4) popping, and (5) chattering.

Not all types of nozzles need to be tested for all these items. For the purpose of testing, nozzles may be classified as open, closed non-popping, closed popping, and mechanically operated nozzles.

Open nozzles may be tested only for the appearance of the spray. If the spray is asymmetrical, or lopsided, the nozzle is defective.

Closed nozzles must be checked for valve-opening pressure. While the lever of the tester is moved slowly, injection must occur within the limits specified by the manufacturer. If it does not, it must be changed, using an adjusting nut provided for this purpose or shims placed under the spring.

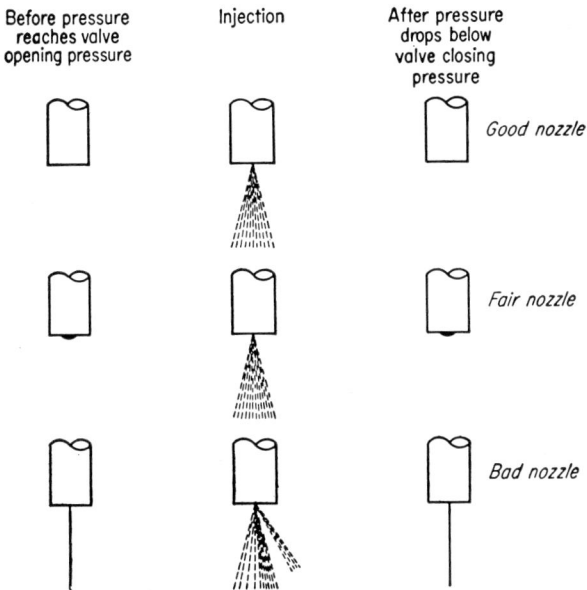

FIG. 24-9. Characteristics of sprays from spring-loaded nozzles.

Nonpopping closed nozzles must also be checked for leakage. By pressing down on the lever of the tester, a pressure is produced just below the valve-opening pressure. If no fuel appears at the spray tip, the needle valve seats properly. If drops, a dribble, or a jet appears below the valve-opening pressure, the valve is not acting properly. Thorough cleaning often corrects the trouble. If it does not, the defective valve parts must be replaced; in some types they may be lapped in, as explained later.

A very good method is to compare the suspected nozzle with others that are known to be in good condition and give satisfactory performance. Figure 24-9 illustrates the appearance of the spray in typical good, indifferent, and bad nozzles. Indifferent nozzles may, however, give satisfactory performance in the engine, especially in an engine with a high rotary speed. Mechanically actuated nozzles, as used in air-injection and common-rail-injection engines, can also be tested in the tester for leakage, but for spray observation a special fixture, including the cam mechanism, has to be provided.

Precaution. When testing a nozzle in a nozzle tester, one must be careful to keep the hands away from the end spray which has a very high velocity and penetrates the skin and flesh causing severe blood poisoning. Such injury should be examined at once by a physician, particularly if inflammation is visible.

Nozzle Defects. By testing a defective nozzle in the nozzle tester and disassembling it, it is comparatively easy to establish the trouble. The trouble may fall into one of the following categories: (1) incorrect spring tension, (2) defective needle-valve seating, (3) sticking of the

FIG. 24-10. Scraping tool for cleaning nozzle. *Courtesy General Machinery Corp.*

valve stem, (4) eroded valve stem, (5) defective orifices, (6) clogged or carbonized tips, or (7) broken spring or valve parts.

The way to correct spring tension has already been mentioned.

Defective needle-valve seating is often caused simply by dirt. A good cleaning procedure is to wash the parts in a clean glass dish with carbon tetrachloride or acetone. Petroleum ether is less expensive than the above cleaners and equally effective, though dangerous because of its inflammability. Wood sticks soaked in oil and a soft brass-wire brush may be helpful in removing sticky matter from the parts. Hard or sharp tools and abrasives should never be used. For scraping dirt from an inaccessible place, like the recess of a fuel-nozzle body, the scraper must be made of brass (Fig. 24-10).

After a thorough cleaning, the parts should be blown by air. After that they should be bathed in clean fuel oil in another glass dish and reassembled, observing maximum cleanliness. The parts should be laid on clean smooth-surfaced paper, not on tissue paper; lapped surfaces should not be touched by bare fingers since this would cause corrosion, as already explained.

If cleaning fails to correct the leakage, one may try to lap the needle valve to its seat by turning it back and forth on the seat, alternately lifting it. It is convenient to use for this the holder shown in Fig. 24-11. No abrasive should be used in lapping hydraulic nozzles—only clean lubricating oil or, in an extreme case, a little liquid metal polish. It should be noted that the apex angle of the needle valve is about one degree greater than the apex angle of the seat. This is necessary to obtain a correct contact under the high pressures existing in operation of the injector.

If the needle requires considerable lapping to make it smooth, a dummy with the correct apex angle should be used, as shown in Fig.

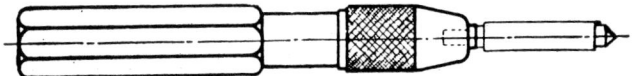

FIG. 24-11. Holder for lapping injector needle to seat.   *Courtesy General Machinery Corp.*

24-12. This lapping can be done with a very fine abrasive such as aluminum oxide lapping powder, holding the needle valve in a lathe. A dummy needle with the proper apex angle may be used to restore the smoothness of the seat, as shown in Fig. 24-13. After that the needle valve and nozzle body must be thoroughly cleaned and lapped together with clean oil.

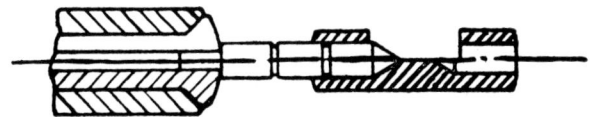

FIG. 24-12. Lapping tool for injector needle.   *Courtesy General Machinery Corp.*

If these special tools are not available and simple lapping with oil does not restore proper seating, both the needle and the valve body should be replaced, as these parts are made in pairs by selective assembling.

Freeing Stuck Needles. A needle valve occasionally sticks in its guide. If it cannot be freed by hand, do not attempt to remove it by

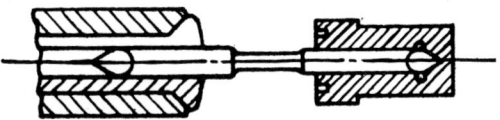

FIG. 24-13. Lapping tool for nozzle valve seat.   *Courtesy General Machinery Corp.*

force. The best method is to soak the nozzle in acetone for 10 to 12 hr. If this does not help, the nozzle is probably distorted by heat and should be replaced.

Cleaning Orifices. For cleaning clogged orifices, music wire of a size slightly smaller than the orifice should be used. It is convenient to insert a piece of such wire in a special holder similar to a drill chuck or simply in a wooden handle. The wire should be moved back and forth without turning, as turning tends to increase the size of the orifice.

Water-cooled Injector. The installation of a water-cooled injection nozzle usually presents certain difficulties, as the tip of the nozzle has to be pressed against a ring gasket to seal the water space in the jacket. The gasket has a tendency to shift as soon as the nozzle touches it. This can be overcome by inserting from underneath a cork (Fig. 22-14) which acts as a register for the gasket until it is pushed out by the lowered nozzle tip. A cork is used because it is soft and cannot damage the nozzle tip.

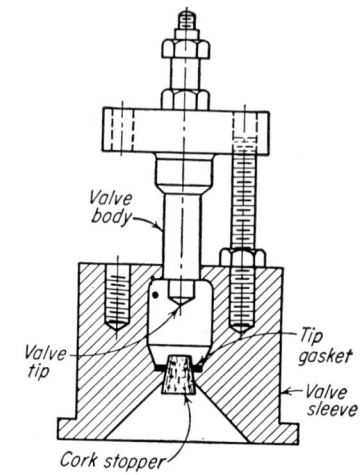

FIG. 24-14. Assembling water-cooled injection nozzle.

24-9. Camshaft Gears. Camshaft gears should be inspected for proper backlash. The thrust bearing should be examined for end play. For most engines an end play of about 0.003 in. is satisfactory. The dowels in all the gears should be examined for tightness. The lubricating arrangement should be checked to make sure that the gears are being properly lubricated.

24-10. Governor. The governor should be examined at regular intervals for loose links, bushings, or spring connections. Any bushings that are found loose should be renewed. All ball bearings should be carefully examined and those which show wear should be replaced. Drive gears should be inspected for wear.

The Woodward Governor Company has on the market a special test stand with an IR Pneumatic Tool Company air motor for testing and adjusting hydraulic governors. The air motor speed can be set between 150 and 4,200 rpm and varied in operation. Such a stand has adapters for several governor types and will pay for itself in a diesel-repair shop.

24-11. Lubricating-oil System. The filters and strainers should be thoroughly cleaned and occasionally boiled out with soda.

All lubricating-oil piping should be blown out with air, then washed with kerosene and again blown out with air. For these operations the ends of the pipes must be disconnected so as not to blow any dirt into the bearings.

All oil tanks, main storage and day tanks, must be periodically drained and cleaned with kerosene.

24-12. Air Compressor. Air-compressor valves should be cleaned with soapsuds. The carbon should be removed with a copper or brass scraper. The valves should be ground in if necessary. If the springs show the effect of heat or wear, they should be replaced. Great care should be taken in putting in the gasket under the valve seat.

In air compressors, the mechanical clearance of the pistons affects both the efficiency and output of the compressor. These clearances must be kept always at a minimum.

Air Coolers. The inside of the low-pressure cooler casing and of intermediate- and high-pressure cooler coils must be cleaned out with Oakite solution of one-half pound of Oakite Platers Cleaner in 1 gal of water. The casing is filled with this solution and the latter brought to boiling with steam. The grease and carbon will be dissolved and the dirty solution is removed through the drain in the bottom of the casing. The inside of the low-pressure coils and of the compound-cooler casing must be washed out with water to remove any mud or sand and then filled with a solution of one part muriatic (hydrochloric) acid to ten parts water to remove any accumulation of hard scale. This solution is forced through the cooler until no more effervescence shows up. After this treatment the cooler must be thoroughly washed out, first with cold and then with hot water to remove any trace of the acid.

Air Receivers. These must be cleaned with Oakite solution in the same way as the air-cooler coils are cleaned.

Air-injection Piping. The piping should be taken, a section at a time, and heated up with a blow torch to burn out any accumulation of carbon that may have formed on the inside. The pipe should then be thoroughly cleaned out and blown out with compressed air.

QUESTIONS

1. Explain the procedure of cleaning scale from cylinder heads.
2. How much wear is permitted in a cylinder or liner before it should be replaced?

3. Explain why the accumulation of scale on the outside of cylinder liners is objectionable and dangerous.

4. Explain the procedure of cleaning scale from cylinders and cylinder blocks cast in one piece with water jackets.

5. Explain what indications are obtained from inspecting the working surfaces of cylinders and pistons.

6. What precaution should be taken when assembling a piston with a fitted piston pin?

7. Enumerate the steps that should be taken when placing a piston and connecting rod back into the engine.

8. What is the permissible amount of variation in the mechanical clearance of a piston?

9. What are the effects of stuck or broken piston rings on the wear of liners?

10. Draw a sketch of the "bearing-watchdog" system and explain its action.

11. Explain the procedure of measuring crankpin- and main-bearing clearances by means of lead wires.

12. Explain the method of checking valve timing.

13. Explain what may happen if the valve-mechanism clearance is smaller or larger than necessary.

14. What is the most frequent cause of leakage of a mechanically operated fuel valve?

15. What are the possible causes of misalignment of a mechanically operated fuel valve?

16. Enumerate the indications of a poorly functioning hydraulic fuel nozzle.

17. Enumerate the items for which a fuel nozzle may be tested in a nozzle tester.

18. What personal precaution must be observed when testing a fuel nozzle in a nozzle tester?

19. Describe the procedure and precautions to be taken when cleaning fuel needles, their seats, and fuel tips.

20. Explain why the apex angle of the needle valve is made greater than the apex angle of the seat and by how much.

21. Explain how a stuck needle valve should be freed.

22. Explain how clogged tip holes are cleaned.

23. Explain, giving a sketch, how to make easier the assembling of a water-cooled fuel injector.

24. Enumerate the items that must be inspected in the crankshaft-drive gears.

25. What items should be inspected and, if necessary, exchanged in a governor?

26. Explain what parts of the lubricating system should be cleaned, and how.

27. How should air-compressor valves be cleaned?

28. Explain how the inside of an air cooler is cleaned.

29. Explain how an air receiver is cleaned.

30. Explain how the air-injection piping should be cleaned.

CHAPTER 25

ENGINE OVERHAULING

25-1. Definitions. Overhauling a machine, strictly speaking, means *going over* it. When a diesel engine is being overhauled, it is completely taken apart; every piece is inspected and whatever piece shows appreciable wear is reconditioned to its original shape and dimensions or replaced by a new one. After this the engine is reassembled, put on a test stand, started, tuned up, and carefully tested. Larger stationary engines, which are maintained in accordance with definite comprehensive schedules, as outlined in Chap. 23, do not need overhauling as they are always kept in good condition. In respect to these engines the word overhauling is sometimes used to designate major maintenance work such as the simultaneous pulling of all pistons, the cleaning of scale from cylinder jackets and heads, and the checking of the crankshaft alignment.

Marine diesel engines have to be overhauled because they have to operate continuously for many days and weeks at practically full load without stopping and without a chance for much maintenance work while at sea. The same is true of automotive diesel engines, although for slightly different reasons. In the first place, it is more difficult to keep an accurate maintenance log for an engine which is away from the garage sometimes for days or weeks. Second, the operators of trucks, tractors, road-building machinery, etc., are not necessarily first-class mechanics, and third, so long as such an engine runs and pulls the load, nobody is particularly concerned how efficiently it operates, whether the exhaust is clear or smoky, whether it runs quietly or is noisy. Therefore, depending upon the type of work, a general overhauling of these engines is done at regular intervals, every 12 or 18 months, as the case may be, or in the case of trucks after 75,000 to 250,000 miles, as found necessary for the kind of service.

In general, the methods of overhaul do not differ from maintenance procedure, except that in overhauling a more strict examination is in order as the intervals are considerably greater than in maintenance work; a part even slightly worn, if it is not subject to maintenance inspection periodically, should therefore be either reconditioned or replaced.

Therefore, in the following we will describe only procedures and special tools not usually needed in regular maintenance work. However, some of the procedures may be considered as special repair work and some of the tools may come in handy in maintenance work, too.

25-2. Dismantling. There are several points which should be rigorously observed when major inspection and overhaul work are done.

One of the most important rules is to make sure that all parts are well marked and identified as the engine is dismantled. It is particularly important to mark camshaft gears and valves if manufacturer's marks cannot be found. Center-punch markings are the most convenient to use.

The second point is to clean thoroughly every dismantled part, carefully examining it for cracks or pitting.

After that, accurate measurements must be taken of all dimensions subject to wear and also of adjustment points of the various parts as the dismantling progresses. A complete record should be kept of each and every measurement, properly entered on consecutively numbered sheets to prevent them from being overlooked or lost. The maintenance work is easier and much time is saved if this rule is followed. A special log book is even better except that it means double work and danger of mistakes in copying the figures as it is not practical to enter the measurements in a book with dirty and oily hands. In order to make these records, the maintenance crew should be well equipped with micrometers, both inside and outside, gauges, scales, trams, etc. It is impossible to obtain accurate readings with inadequate equipment. In taking measurements one must be as careful and accurate as it is humanly possible. No guesswork is permitted. Inaccurate measurements are worse than none.

Finally, in removing small parts of a subassembly, such as a fuel pump, fuel injector, pressure-relief valve, or similar mechanisms, it is well to prepare marked boxes or other containers for each piece of equipment and to place all parts of a mechanism in the same container.

25-3. Cylinder Liners. A wet-type liner is removed comparatively easily with a tool, as shown in Fig. 25-1a: a is a cast-iron disk fitted to the lower end of the liner and tapped to take the thread on the end of the rod b; c is a cast-iron or steel crossbar resting with the ends on the nuts of the cylinder studs e, e; the tie rod b is threaded on both ends and goes freely through the cross bar c. When the liner must be pulled, the nut d is gradually tightened, pulling b, a, and the liner upward. For replacing the liner (Fig. 25-1b), disk a is placed on top of the liner, the upper screw end of b screwed into it, the crossbar c is turned around, held by two nuts on the cylinder-head studs. By

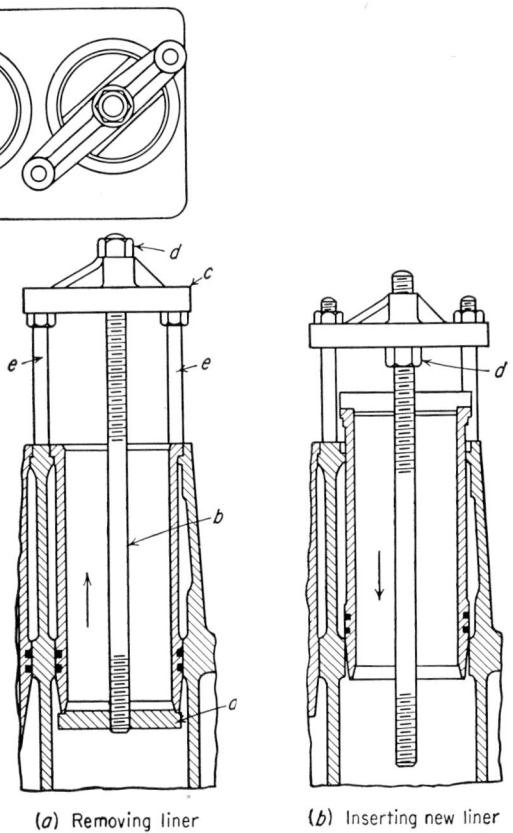

(a) Removing liner (b) Inserting new liner
Fig. 25-1. Tool for removing and inserting wet liner.

tightening nut d, the tie rod b and disk a, and with it the liner, are pushed down.

If the cylinder block does not have studs, as is found in smaller engines, the crossbar may be supported on two wooden blocks put endwise and the crossbar itself may be a piece of 2- by 4-in. block with a hole in the center for the tie rod. Replacing of the liner in such a case is done by tapping it through wooden blocks or using a lead hammer in order not to damage the upper face. The final setting is done by tightening the cap screws with the cylinder head in place but without a gasket. To reduce friction, it is advisable to coat both ends of the liner where it is in contact with the cylinder casting with a mixture of linseed oil and white lead. Lubricating oil cannot be used here as it has a tendency to rot the rubber sealing rings.

If the liner is not very tight, it may be removed by using the crank as a lever (Fig. 25-2) and turning the crankshaft by the flywheel.

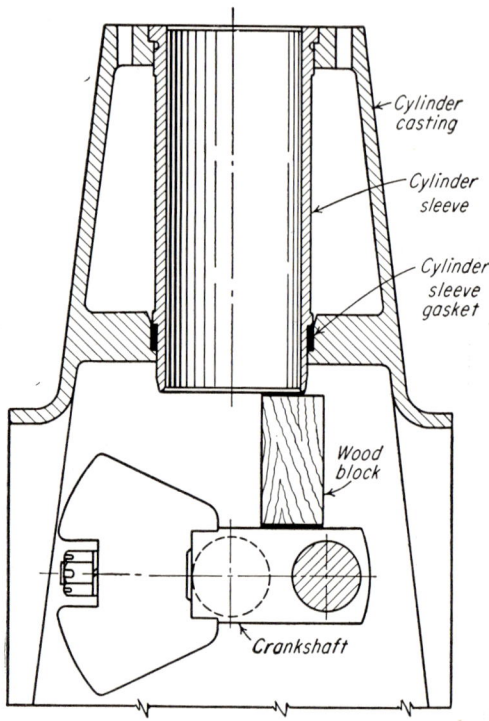

Fig. 25-2. Method of removing wet liner.

Sometimes, if the liner has been in place a long time, several bumps with the crankshaft are required to loosen the liner.

The removal of a dry-type liner, or *sleeve*, is usually more difficult because, as a rule, it is press-fitted into the cylinder. Even a strong puller may not always loosen the liner unless a cut is made in the sleeve, running from top to bottom. A cape chisel is a good tool to use in this case.

An electrically driven thin, small-diameter grinding wheel can also be used but requires the building of a special holder with an extension shaft to reach into the cylinder and cage to limit the depth of the cut.

In Fig. 25-3a is shown a tool for removal and in Fig. 25-3b for pressing-in of press-fit dry sleeves. It is similar to the tool shown in Fig. 25-1, only it is stronger and uses a thrust bearing to reduce the effort required to tighten the top nut.

When inserting a new dry liner, a good procedure is to place the liner in a paperboard box along with a small amount of dry ice. The ice will chill the liner and its shrinkage will reduce the outside diameter of the liner so that it will slip easily in the frame bore.

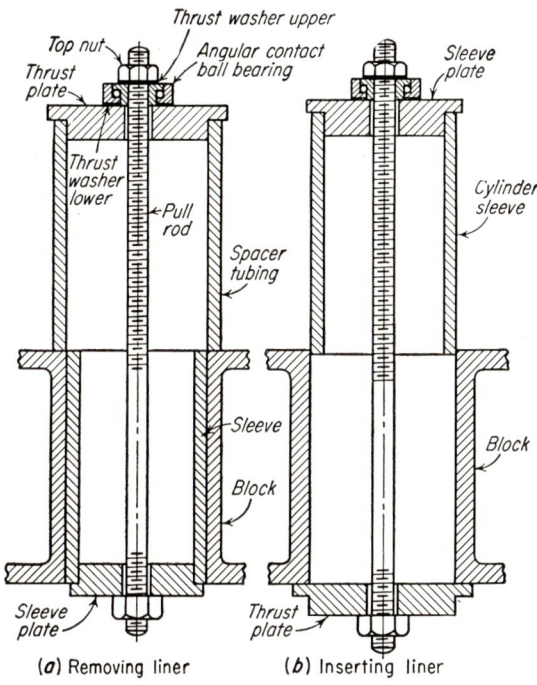

(*a*) Removing liner   (*b*) Inserting liner

FIG. 25-3. Tool for removing and inserting dry liner.

Whenever a liner is replaced, even when a press is used, the outer surface of the liner should be coated with machine or lubricating oil to make it slip more easily.

Some later small-bore diesel engines use dry sleeves with a loose fit, the sleeve being 0.001 to 0.002 in. smaller than the cylinder bore. After such a loose liner is in operation for a considerable time, oil and coke will lock it in the cylinder and the removal of the liner is made easier by the tool furnished with each engine and shown in Fig. 25-4. The tool is used as follows: (1) the lower clamp is slipped up the puller rod and off its conical seat and cocked on the rod so that it will slide down through the liner when the rod is lowered and will drop back onto its seat in a horizontal position when the rod is pulled up; (2) the

434　MAINTENANCE AND REPAIRS

upper clamp is slid down the rod and locked on it by a set collar or some other arrangement; (3) the sliding cast-iron weight is slid on the rod and the upper nut screwed on; (4) with the tool so mounted, the upper nut is struck a sharp blow with the sliding weight. The liner is usually loosened by the first blow.

Liner Measurements. In order to see whether it is time to replace the liner by a new one, the bore of the liner must be checked periodi-

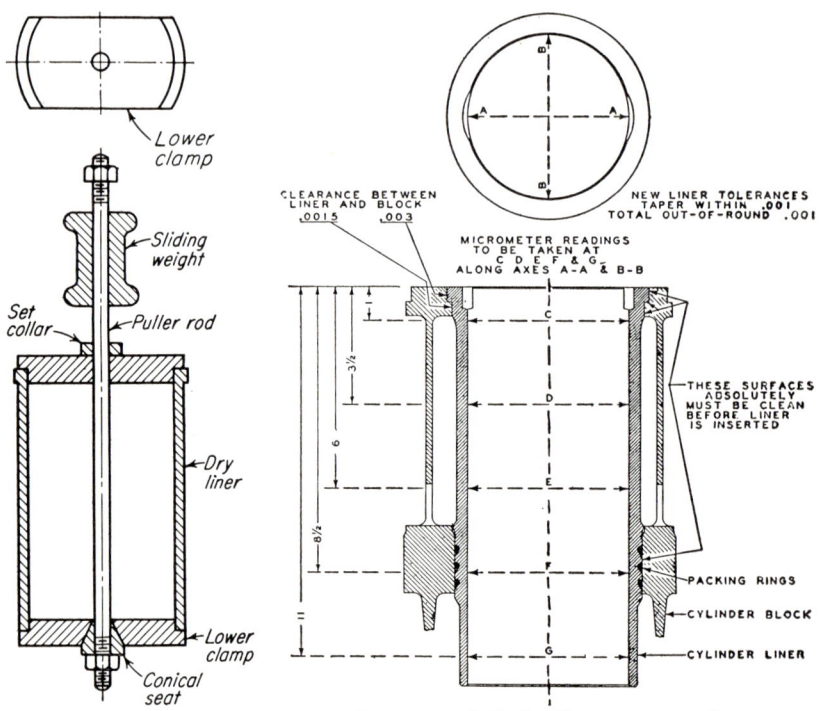

Fig. 25-4. Tool for removing dry liner.

Fig. 25-5. Cylinder-liner gauge marks.

cally by taking ten measurements, five in the plane normal to the crankshaft center line and five at right angles to them. Figure 25-5 shows the recommended location of the measurements for a $4\frac{1}{4}$-in. liner. For larger liners the same number of measurements is sufficient with a similar relative location axially. The measurements must be entered in the permanent engine records and will indicate not only when the liner must be exchanged, but also whether the wear is normal. If the wear is too great for a cast-iron liner, over 0.001 in. per 1,000 hr of operation, regardless of engine size, the cause should be investigated and eliminated. The possible causes may be (1) unsatisfactory lubricating oil, (2) unsatisfactory cleaning or filtering of the oil, (3) excessive removing of oil by the oil-control piston rings, (4) an excessive pressure exerted on the cylinder walls by compression piston rings, or (5) washing off of the oil film by fuel oil or water condensing from the exhaust gases, if the temperature of the jacket water is kept too low.

Dry-type sleeves usually have a hard inner surface and their wear should be smaller. Chromium-plated liners also should wear not over 0.001 in. in 2,000 to 3,000 hr.

Honing. All small-bore liners and many medium-sized ones, up to about 9 in., are honed to give a smooth finish and a true cylindrical surface. If a liner is worn out of round, bell- or barrel-shaped, but the total wear is not large—a few thousandths of an inch—it can be trued up by honing and used again, if the honing operation is carried out in the proper manner. The safest way to hone cylinder liners and be sure that no abrasive particles will get into the bedplate and bearings is to remove the liner from the cylinder block, place it in a honing fixture affording at least a line-to-line fit, and then hone it to the desired size. With small-bore engines a scrapped cylinder block or part of it makes an excellent fixture. After honing, the liner must be thoroughly washed in a solution of hot caustic soda and blown out with air.

If it is desired to hone the liners without taking them out, the engine must be completely dismantled and after honing the whole block must be washed out thoroughly in a solution of hot caustic soda and blown out with air to make sure that all and any abrasive particles are removed.

The selection of hone stones and of the method of honing is important. A holder with an adjustment for setting the cutting radius of the stones is better than a holder with spring-loaded stones, especially when truing distorted bores. Spring-loaded stones will follow rather than remove irregularities in the bore. The stones must be frequently dressed and brushed to prevent them from becoming loaded with metal particles. The instructions of the hone manufacturer must be followed closely regarding the use of oil or kerosene on the stones. With a dry-type hone, cutting fluids should not be used.

For honing cast-iron wet-type liners ordinary medium honing stones are satisfactory. The dry-type liners or sleeves are usually made of much harder material. For rough honing, coarse stones of 46 grit are satisfactory and for smooth mirror finish required for the inside of the liners, stones of 150 grit should be used.

When liners have been in service for a certain length of time, the inside surface often becomes very smooth, or glazed. This glaze lengthens the time required to seat new piston rings. Therefore, it should be removed or broken by working the 150-grit stone up and down in the liner a few times.

Honing Procedure. Start the hone with the coarse stones and feel out the bore for high spots. These cause an increased drag of the hone. Move the hone up and down the bore with short overlapping strokes about 1 in. long. In the first cut concentrate on the high spots. As these are removed, the drag of the hone will become lighter and smoother. Liners of two-stroke engines with scavenge ports should not be honed at the ports as long as in the rest of the bore, since this area cuts away more rapidly. When the drag of the hone becomes quite light and smooth, the feed on the stones should be increased. The feed increase must be very light to avoid an oversize bore. Roughing stones cut rapidly even under low pressure.

When the bore is fairly clean, the hone is removed to inspect the stones and measure the bore. These measurements will indicate which spots must be honed most. Moving the hone from top to bottom in the liner will not correct an out-of-round bore. If the hone is left in one spot too long, this may cause a taper. Where and how much to hone can be judged by feel. A heavy cut produces a more steady drag of the hone than a light cut and so makes it difficult to feel the high spots in a distorted bore. Therefore, it pays to use a light cut with frequent adjustments.

The rough hone should bring the bore within 0.0005 in. of the final dimension. In finish-honing with the 150 grit, the procedure is along the same lines.

After honing the bore should be thoroughly scrubbed with soap and water and dried. A clean bore should show no dirt when rubbed with a white rag.

Reboring Cylinders. When a cylinder is worn out of round or to a bell shape, usually near the top, it must be trued up by reboring. Any good standard boring bar may be used. The cylinder is bored to within 0.001 to 0.0005 in. under the desired measure and the surplus metal left after boring should be removed by honing.

This procedure may be used for gray-iron and alloy-iron cylinders and blocks and for blocks with dry sleeves, provided they will not turn in the block and their surface hardness does not exceed 450 Brinell hardness number. Sleeves with a hardness in excess of 450 Bhn should be ground and honed with a Heald grinder.

Standard practice is to rebore cylinders and liners of automotive-type engines up to 0.060 in. oversize, not in one overhaul but in several. Cylinders and liners of larger engines usually can stand a cumulative bore increase up to 0.100 in., if pistons of so much oversize are available.

25-4. Pistons. In engine overhauling the work on the pistons consists in pulling them out, removing the piston rings if they are worn or stuck, cleaning the pistons, inspecting and measuring the outside diameters, the ring grooves, and the piston pin bushings. This inspection will help to decide whether to use the same piston after stoning it and doing other minor reconditioning work or to discard it—usually because the liner bore, in being trued up, has been increased to oversize dimensions or must be rebored.

After a piston is pulled, it must be disconnected from the connecting rod. Often it is difficult to push out a fixed piston pin without distorting the piston. The removal is made much easier and quicker by using the arrangement shown in Fig. 25-6: a heavy steel washer a is placed on one end of the piston pin; a heavy wooden block b, 2 by 4 in., 4 by 4 in., or 4 by 6 in., with a hole in the middle, is placed against the piston and tied by bolt c to the washer a. A hardwood block is put against the head of bolt c and given heavy blows with a hammer to drive the pin out. At the same time the nut on the bolt is kept tight to exert a pull on the pin. A washer d prevents block b from being chewed up by the nut. The elasticity of block b helps materially to drive the pin.

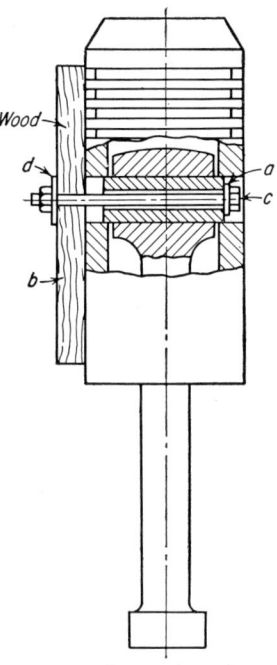

Fig. 25-6. Removing piston pin.

Piston Regrooving. The top ring groove usually wears faster than the other grooves because it is exposed to the most intense heat, receives the least lubrication, and often is subject to dust coming in with the intake air. Even a slight initial wear in the ring groove rapidly grows from the pounding of the loose ring.

The groove must be trued up and this results in a wider groove requiring a special piston ring. To avoid this the Sealed Power Corporation has developed a so-called *contracting groove insert* which is a thin but wide snap ring. The Sealed Power regrooving tool makes the worn groove $\frac{1}{32}$ in. wider than the original groove and at the same time cuts a $\frac{1}{32}$ in. recess at the bottom of the groove (Fig. 25-7).

When the contracting steel insert is put in place, the width of the groove returns to its original size and a standard piston ring may be used.

Piston Rings. Removing and replacing piston rings without breaking them is at best a delicate job which is made easier by the use of tongs, as shown on Fig. 25-8. The tongs can be made by the engine operator himself: the two side plates are riveted or brazed to one leg and have holes for the small bolt forming the pivot for the other leg.

New ring in worn groove

Groove recut

Groove insert in place

Fig. 25-7. Reconditioning top piston groove.

Separate pairs of tongs should be made for pistons of different sizes.

After the piston ring is opened with the above tongs, four or more steel strips are inserted between the piston and the ring so as to lift the ring out of the groove and to enable to slide it off the piston. For small rings the strips may be cut from sheet steel; for larger, stiffer rings they may be made from used or broken hacksaw blades. In either case the edges must be smoothed and rounded with a file or by grinding, respectively, so as not to scratch the piston and permit easy placing of the strips under the ring and easy sliding of the rings over them.

Stuck Rings. The removal of a stuck ring is facilitated by leaving the piston up-side-down in a bucket of kerosene overnight. If such soaking does not help, the piston should be heated by boiling it for several hours in water to which is added a generous amount of powder soap. Except in the severest cases, this will cause the rings to spring out of their grooves one by one without further attention. If neither soaking nor heating will loosen a ring, it becomes necessary to break it out. The use of a cape chisel for this purpose is not recommended as it is apt to damage the sides of the groove. A piece of brass should be applied to one end of the ring at the gap, and with careful use of a hammer on the brass, a small section can be broken out. The remainder of the ring is then broken out piece by piece by driving a suitably shaped brass wedge between the ring and bottom of the groove.

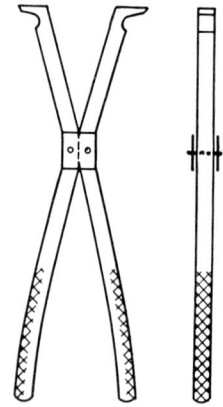
Fig. 25-8. Piston-ring tongs.

Piston Cleaning. When the rings are removed, the piston should be thoroughly cleaned, both outside and inside; the ring grooves must be cleaned with brass scrapers and all oil holes in the grooves carefully opened by passing twist drills of corresponding sizes through them. As mentioned before in connection with maintenance work, if the piston crown is burnt, it should be reconditioned by welding.

Replacing Rings. Putting piston rings on the piston is facilitated by the use of the same steel strips and spreading tongs as those used for their removal. Care should be taken, in opening the rings to slip them on the piston, not to overstress them by opening them too much as they become rather easily distorted and even broken.

Careful tests have shown that heat transfer from the piston to the rings and further to the water jacket depends greatly upon a good contact between the flat surfaces of the ring and the sides of the ring grooves. This is obtained by grinding and polishing the flat surfaces of the rings and by close fitting by scraping the sides of the grooves. The same tests have shown that when both flat sides of a ring groove are accurately fitted to polished ring surfaces, the heat transfer is not impaired if the groove is a few thousandths of an inch wider than the piston ring.

Another point not sufficiently known is that polished rings should not be touched with bare fingers. Moisture given off by human skin contains salts and acids which cause quick oxidation and rusting on finger prints. Rust acts as a heat insulator and destroys the usefulness of polishing and close fitting of the rings. Cotton gloves should be worn when handling and installing highly polished rings. Coating rings in stock with rust-preventing grease helps considerably but does not make the use of gloves unnecessary.

After piston rings are installed in the piston with both end and side clearances adjusted to the required tolerances, they should be checked for thickness to determine whether they can be pushed into the ring grooves flush with the lands. If not, the grooves must be rechecked for stuck carbon and cleaned and the inner corners of the rings slightly beveled. Rings which cannot recede flush with the lands cannot be used because they may easily score the cylinder liner or do even greater damage.

Piston-pin Bushings. With full-floating piston pins, it is advisable to have the fit of the pin in the piston bosses of aluminum pistons or in the bronze bushings of the bosses of cast-iron pistons much tighter than in the bronze bushing of the connecting-rod end. Such a difference in fit gives a slow creeping rotation of the pin in the piston bosses and this results in a desirable gradual change in the area of the piston pin and reduces the wear of the pin, which is subjected to very high bearing pressure.

Piston Installation. The most convenient means of making the piston rings enter the liner is the use of a cast-iron ring guide tapered about 3 in. per foot, as shown in Fig. 25-9. With a liberal supply of lubricating oil such a guide functions automatically, but if the rings are doweled, as in two-stroke engines, care must be taken to see that the ends of the rings are properly located and do not foul the dowel pins. If the liner has a counterbore, the shoulder between the counterbore and the working bore must be scraped to a smooth and gentle bevel, about 30 deg., to ensure that the ring will not get caught by the shoulder. The ring guide can be made conveniently from a discarded cylinder liner.

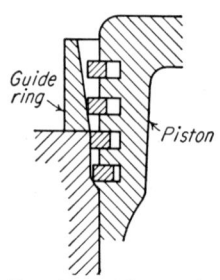

Fig. 25-9. Piston-ring guide.

For small-bore engines a so-called *compression ring*, as used in automobile garages, may be made: a sheet-metal piece sufficiently wide to cover all the rings in the piston with a strap of heavier material around it which is tightened with a screw and thumb nut (Fig. 25-10). When using such a compression ring, care must be taken to have it well lubricated inside and not to tighten the strap too much.

If no ring guide of any kind is available, the rings may be held down in their grooves by means of a sheet-metal band seized with two or three soft wires. This arrangement is not practical for large-engine bores, and the time and expense spent to make a permanent ring guide will be amply returned by time saved in reassembling the engine.

25-5. Bedplate. No matter how strong an engine foundation looks, if there are some suspect indications, such as uneven wear or heating of main bearings, it is advisable to check whether the bedplate may be distorted, thus causing misalignment of the crankshaft. If it is distorted, there is nothing else to do but to tear it loose from the foundation, chip off the old grouting and install it anew. Such conditions are rather unusual, but they have occurred before.

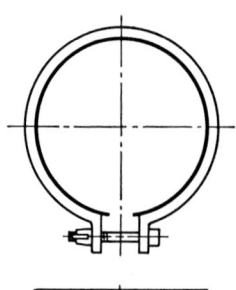
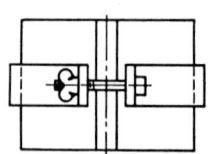

Fig. 25-10. Compression ring for sliding piston in place.

If the bedplate is not distorted, all that has to be done is to clean it thoroughly, check with compressed air all lubricating oil pipes to make sure there are no leaks, and if the bedplate was not painted inside or was painted with paint which can be scratched off, to scrape it thoroughly, wash it with kerosene and then repaint the inside with a good, oil-resistant paint or enamel of a light color—white or light ivory.

Crankcase. The crankcase also must be cleaned thoroughly inside and, if necessary, repainted with the same paint or enamel as the inside of the bedplate.

25-6. Connecting Rods. The connecting rods must be thoroughly cleaned, particularly the rifle bore through which lubricating oil is delivered to the piston pin.

When precision-type bearing shells are used, the rods must also be checked for elongation of the bearing bore and for rod alignment.

If the bore elongation in automotive-type engines exceeds 0.002 in., the rod should be exchanged. This condition may be caused by flexing in service or by filing the bearing cap, a practice left over from the

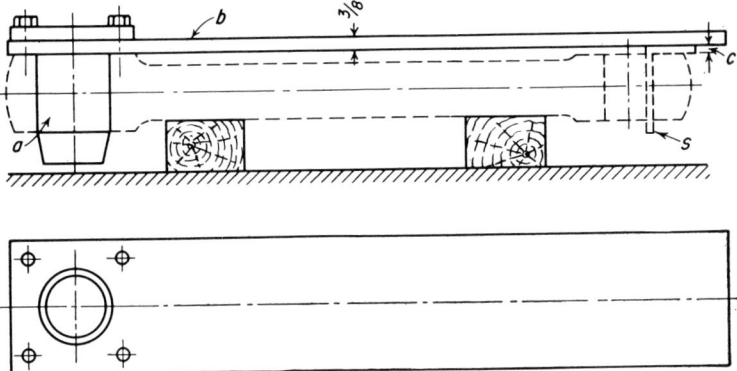

FIG. 25-11. Template for checking connecting-rod alignment.

times when the connecting rods were babbitted and did not have inserted bearings.

If the center lines of the rifle bore at the big and small ends are not parallel within 0.001 in. in 6 in., the rod is bent and must be aligned. The piston-pin center line also must be parallel with the crankpin center line.

A convenient tool for checking rod alignment is shown in Fig. 25-11: the accurately turned and polished finger a has a diameter exactly equal to the inside diameter of the rod bearing and is strictly normal to the plate b. The finger is inserted into the big end of the connecting rod and the distance c is measured with feelers and a steel block. Then the rod is turned with the other side up and the distance c is measured again. If the crankpin end alignment is correct, c will be the same in both cases, otherwise the rod must be straightened out until c is the same for both positions.

The alignment of the piston-pin bushing is checked by means of a small square s, shown by dotted lines.

Precision Bearings. Most modern medium- and high-speed diesel engines use on the big end of connecting rods so-called *precision bearings* which do not require fitting. On the other hand, if a precision bearing is worn or badly scored, it cannot be repaired and must be replaced by a new one. All remarks concerning precision bearings, spread, crush, and importance of an even tightening of the bolts, made in Sec. 25-11 in connection with precision main bearings, apply in full to precision connecting-rod bearings.

Needle-roller Bearings. There are two methods of assembling these piston-pin bearings, both requiring a cylindrical mandrel with a smooth outer surface and a diameter 0.002 to 0.003 in. smaller than the piston pin whose surface serves as the inner race for the needle rollers. For ease of handling the mandrel should be hollow, made of a piece of pipe or tubing, of such a length as to fit between the two bosses in the piston and slightly longer than the small end of the connecting rod.

The usual way of assembling a bearing of this type is to coat the mandrel with heavy cup grease and to place each row of needles on the mandrel, with retainer rings between the rows. The mandrel with the needle rollers is then slipped into the piston-pin bushing which serves as the outer roller race and was previously pressed into the rod.

If the needle rollers are big and heavy, it is convenient to hold the rollers, while assembling them on the mandrel, by heavy rubber bands. The rubber bands are removed as each row of needles is slipped into the outer race.

Another method which does not require rubber bands is as follows: the connecting rod is laid horizontally with the piston-pin bushing upward and overhanging; the mandrel is coated, as before, with cup grease and is placed inside the bushing; the mandrel is held with one hand and a retainer ring is slipped over it and down a distance about the length of a needle roller and held at this distance by a piece of sheet iron or a screwdriver held with the same hand as the mandrel. With the other hand the needles are inserted into the space between the mandrel and bushing. When all the needles of the first row are inserted, a retainer ring is placed on top of the rollers and the whole row with both retainer rings is lowered inside the bushing a distance about the length of a needle roller, using a screwdriver or similar tool to push them downward. This procedure is repeated until all rows are installed.

To prevent the mandrel and the rollers from slipping out if the connecting rod is not intended to be assembled with the piston at once, the ends of the mandrel m (Fig. 25-12) are covered with two washers w, w made of heavy-gauge sheet iron. The washers and the whole assembly are held together by the tie-bolt t and nut. Figure 25-12 shows an assembly with three rows of needle rollers n and four retainer rings r; b is the outer-race bushing.

25-7. Crankshaft. At each major overhaul the crankshaft should be carefully inspected for cracks. A simple but effective means of detecting incipient cracks is to wipe the suspect surface clean and dry and powder it with precipitated chalk. If there is a crack, capillarity will cause it to be filled with oil; the greater absorbing ability of chalk will pull the oil out and the chalk will turn yellow. In general it is not advisable to run an engine with a cracked crankshaft, regardless of how small the crack is. However, sometimes a crack may be a flaw present in the shaft from the beginning, one which will not increase with time. A careful inspection 500 hr of operation after the crack has been discovered will show whether it is growing or not. If the crack grows, the crankshaft should be changed as soon as a new one can be obtained. If the crack does not change, the shaft may be left in the engine but must be closely inspected every 500 hr and the inspections must be recorded. If the crack begins to grow, a new shaft must be installed.

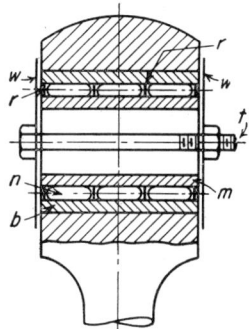

Fig. 25-12. Method of assembling needle-roller bearing.

Also, whenever a crack in the crankshaft has been discovered, the alignment of the shaft must be carefully rechecked because crankshafts of modern diesel-engines hardly ever break except from misalignment.

The crankshaft should be inspected also for the condition of the journals. Slight wear and scoring is usually not dangerous, but overheating is bad. Crankshafts of high-speed diesel engines are generally heat-treated and surface-hardened. If such a shaft has been overheated and has lost the required surface hardness, it cannot be used any more and a new crankshaft must be installed.

After the inspection the oil passages and holes must be cleaned and blown out with compressed air.

The next thing is to measure with a micrometer the diameters of the main journals and crankpins. After long operation the main journals become elliptical from the fact that the main load from the piston comes at or near top dead center. This, in itself, is not dangerous, but during an overhaul, if the journals are worn too much, the shaft should be taken to a machine shop, put in a lathe and the journals either turned, if they are not hardened, or ground true. Also, while the shaft is in the lathe, a check should be made to see whether the shaft has been sprung in any manner.

The very heavy crankshafts of large engines are sometimes turned without being removed from the bedplate. Out-of-round or tapered crankpins can be ground true in place by a special grinding machine. Shops specializing in diesel-engine reconditioning have such machines and should be entrusted with this kind of work.

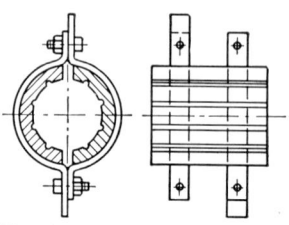

Fig. 25-13. Wooden crankpin lap with lead inserts.

The engine operator has seldom to recondition a crankpin that is out of round or tapered, but with skill and patience it can be done by hand, if the condition is not too bad, especially on a small engine. The best method is to file the pin and check it with calipers and a small square with the corner ground off so that it will not interfere with the crankpin fillet. File and check it until it is almost true, then true it up to within 0.001 or 0.002 in. After the filing is finished, make a lap of hardwood, lead, or combination of both, as shown in Fig. 25-13, and with fine carborundum and oil, work it around the pin with a back-and-forth motion until the pin surface is smooth and well polished.

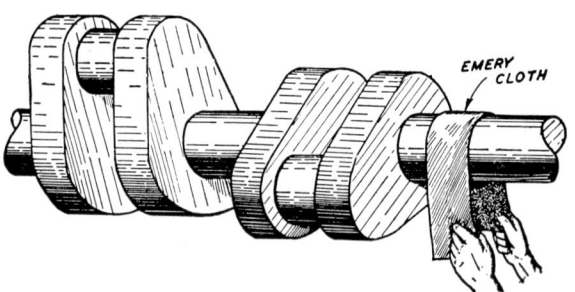

Fig. 25-14. Lapping a worn journal.

An oilstone may also be used on small pins, but the work with it is discouragingly slow.

Every precaution should be taken to prevent the abrasive from entering the oil passage in the pin or getting into the bedplate.

If a journal is slightly scored, it may be smoothed with fine emery cloth by rapidly reciprocating the sheet about the journal, as shown in Fig. 25-14. Of course, such polishing will not remove the scores, but it will remove all the burrs along the cuts which would damage the babbitt of the bearing.

25-8. Main Bearings. When the bedplate has been checked and found to be properly leveled, the first thing, particularly with a large engine, is to check the main bearings by means of the bridge gauge furnished by the manufacturer. After the bearing caps and upper bearing shells have been removed, the bridge gauge is placed over the journal as shown in Fig. 25-15. The distance from a to the journal is measured with feelers and compared with the original measurement taken when the engine was erected and usually stamped on the bridge gauge. The difference between the present and original readings will

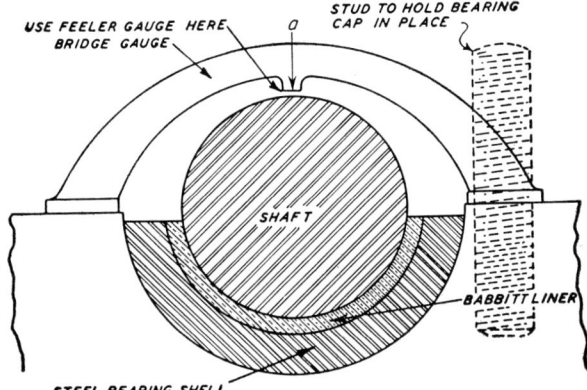

FIG. 25-15. Using bridge gauge on shaft bearing.

indicate the wear of the bearing. The wear should be uniform on all bearings; otherwise there is a danger of bending the shaft when the engine is running and possibly fracturing it. To make sure that the shaft is not sprung, bridge-gauge readings should be repeated with the shaft turned 180 deg.

The bridge gauge must be handled with care, particularly when stored away, because if it is distorted or if the faces of the two supporting legs or the face at a are burred, false readings will be obtained. Likewise, the surfaces of the bedplate on which the bridge gauge rests should be perfectly clean, but should never be scraped or filed.

In addition to the time of overhaul, bridge-gauge readings should be taken once a year, if practicable, and a record kept of the amount that each bearing has worn since the previous inspection.

Removing Bearings. The lower main-bearing shells can usually be removed by rolling them out, *i.e.*, sliding them out around the journals. In small engines the sliding motion is obtained by tapping with a hammer on a curved piece of copper or brass held against the center of the bearing edge. In large engines, placing a zinc slab against the center of the bearing edge and striking the slab with a maul may help. If the pressure of the heavy shaft is too great, the rolling out of the bearing shells may be assisted by putting a jack under one end of the crankshaft and lifting it just a few thousandths of an inch. The use of a dog fastened to the crank, as shown in Fig. 25-16, is helpful; one man should gradually turn the shaft and another man tap or push down the opposite end of the shell in order to eliminate the cocking action of the eccentric application of the dog.

Inspection and Rebabbitting. After removal, the bearing shells should be wiped clean and inspected. Bearing shells often show cracks, but if the cracks are not deep and no pieces of babbitt are loose, the condition is generally not serious. Often a bearing with small cracks will run for years without causing trouble and is sometimes better than a bearing rebabbitted and fitted by a repair shop not familiar enough with the requirements of the particular diesel engine.

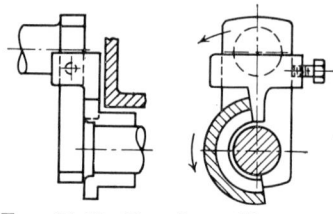

Fig. 25-16. Dog for rolling out bearing shell.

If the cracks are rather deep, it may be advisable to weld them with a fine-pointed acetylene torch, adding some new babbitt from a bar cast to about the dimensions of a bar of solder.

A piece of babbitt may be found so loose that, when the bearing is turned bottom up, it falls out. If such a piece is not over 2 or 3 sq in. in area, a patch may be puddled in, using bars of babbitt, as mentioned above, and an acetylene torch. First, the old oil is washed away with gasoline and the gasoline is blown out with air and burned away. The bottom of the patch is tinned and new metal run into the empty space, making sure that it is bonding well to the edges of the old metal and flowing freely. Some oil may remain under the old metal and come bubbling through the patch in the form of gas which leaves small flow holes that do no harm. After the puddle of babbitt cools down, the patch is peened down hard, dressed to about the height of the surrounding bearing surface and then spotted and scraped until it fits the journal.

If the babbitt is so cracked or worn that it is decided to reline a shell, the old babbitt must first be removed with a chisel or melted out. Then the shell should be cleaned with diluted muriatic (hydrochloric) acid and tinned by having a thin layer of solder run over the entire surface to be covered with the babbitt after the temperature of the shell is brought near the melting point of the solder.

If one shell is to be rebabbitted, a mold a is prepared of sheet iron (Fig. 25-17) and clamped to the bearing shell by straps b, b, and c, c and four small bolts. If two shells are rebabbitted, not necessarily both for the same bearing, a mandrel is made of a piece of pipe or tubing, about ¼ in. smaller in diameter than that of the journal.

The babbitt for relining should be of the same type as that used originally by the engine builder and, if possible, should be obtained from the engine builder or from the source indicated by him.

Compositions and properties of tin-base babbitts suitable for diesel-engine bearings, alloys SAE 10, B, and C, are given in Table 25-1. Lead-base alloys, such as SAE 14 and SAE 13, are much cheaper but are suitable only for lighter service in general machinery.

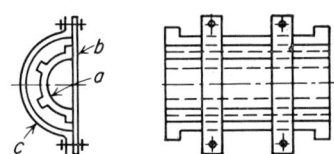

Fig. 25-17. Babbitt mold for main-bearing shell.

The shell with the mold clamped to it or the pair of shells tied together with wire and with wooden shims between them and the mandrel inside them are set vertically upon a piece of asbestos cardboard with a fine-clay dam to seal the base joint. The babbitt is heated in a ladle, over either a gas burner or a charcoal forge. When the babbitt has become fluid enough to pour, wood splinter thrust into the hot metal and then withdrawn will char but not ignite. Before pouring, the greenish-silver scum on the surface must be skimmed off for this is the dross, or impurities; if left in the ladle, it will give a porous, defective bearing. After the babbitt is poured, the bearing is allowed to cool. Later the mold or the mandrel is removed and the two halves of the bearing sawed apart with a hacksaw through the wooden shim.

TABLE 25-1. BABBITT METAL COMPOSITIONS AND PROPERTIES

Alloy grade	Composition, per cent by weight					Brinell hardness		Temperature, deg F	
	Tin	Antimony	Copper	Lead	Arsenic	70 F	212 F	Melting point	Pouring point
SAE 10	90	4–5	4–5	0.35	0.1	28.5	12.7	450	820
B	83	8.3	8.2	0.35	0.1	34.4	15.7	462	920
C	75	12	2.85	10	0.15	29.6	12.8	365	680
SAE 14	10	14–16	0.5	74	0.6	22.5	10.5	464	640
SAE 13	5	9–11	0.5	84	0.25	19.0	8.5	459	620

25-9. Fitting Rebabbitted Bearings. After the babbitting is completed, the shells are placed on a lathe, boring mill, or drill press, and the babbitt is bored to exact dimensions. Where many shells are being manufactured, the shells are rough-bored and finish-reamed.

An oil space should be relieved at the sides of the bearing halves for a distance of 30 deg. from the edge to the center, or a total of 120 deg circumferentially. The depth of the relief may be $\frac{1}{32}$ to $\frac{1}{16}$ in. at the edge, gradually tapering off to zero. Longitudinally, the relief is carried to within $\frac{1}{2}$ or $\frac{3}{4}$ in. of the ends of the bearing. The metal at the ends should fit the shaft snugly in order to form an oil seal. Even if the shaft is pinched slightly at these points, the heat generated through friction will be carried away and the wear will soon bring a nice fit and a good oil seal.

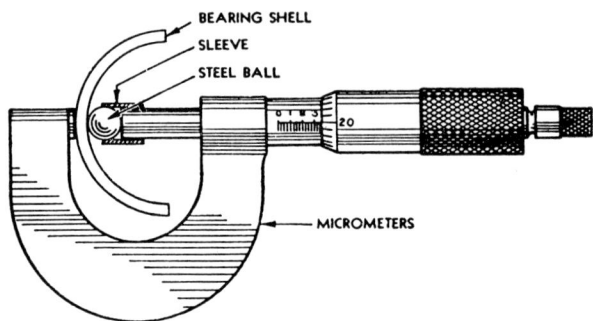

FIG. 25-18. Measuring thickness of main-bearing shell.

Shims should be carried in close to the shaft, but not allowed to touch it unless they are of brass sheets or of laminated brass material. The latter is preferable as by peeling off it is easier to reduce the shim thickness a definite amount as required by the measured wear.

Measuring Wear. The wear of a bearing shell is determined by measuring its thickness and comparing with the thickness of a new shell as given in the engine instruction book. To reach the bottom of the shell from the inside, the micrometer must be equipped with a sleeve and ball from a ball bearing, as shown in Fig. 25-18.

Spotting. If the bearing must be fitted to the journal, the procedure called *spotting* is as follows: a very light coat of prussian blue, or red lead with oil, or a mixture of lampblack and kerosene is smeared on the journal. The bearing half is placed on the coated journal and gently rotated a slight amount. Taking off the shell, it will be found that the babbitt has spots of prussian blue or lampblack. These spots are lightly scraped off with a sharp scraper. The bearing shell is again set in place and moved a few degrees around the journal. Taking off the shell, it will be found that the number of spots and even their size has increased. The spots are again scraped off, the coat on the journal evened out, and the procedure repeated until the spots begin to cover the whole surface in contact with the journal. The bearing may be considered in proper shape and the final fitting will be obtained by wearing it in while the engine is running.

Both theory and experience indicate that the behavior of a bearing is considerably improved if it is fitted not only to the journal, but also to the bedplate. This fitting helps to carry away the heat of bearing friction and thus lowers the bearing temperature and improves its lubrication. The fitting is done by the same procedure of spotting with prussian blue or red lead. If the bedplate has a fine bore, the scraping or filing is done on the outside of the bearing shell. If the

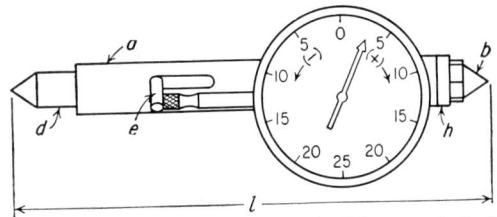

FIG. 25-19. Strain-gauge attachment for checking crankshaft alignment.

bedplate bore is markedly rougher than the surface of the bearing shell, it is advisable to scrape the bedplate bore.

25-10. Checking Bearing Alignment. The procedure is greatly simplified by an attachment using a small strain gauge as indicator. This attachment can be made in any machine shop. Figure 25-19 shows the device assembled and Fig. 25-20 shows the eight component parts. The centers, b and d, must be of steel with hardened points, the rest may be of steel, brass, or aluminum. The dimension s of adapter h should be such that the over-all length l of the whole device is a little greater than the distance between the crank cheeks. By having adapters h of different lengths, the device can be used for checking crankshafts of different sizes. The spring f should be sufficiently stiff to hold the instrument against the center-punch marks on the checks when the crankshaft is being turned.

The bearings supporting a crankshaft with all journals having the same diameter must be perfectly level. If the bottom of one of the bearings is lower than that of the rest, the crankshaft, during rotation, will be subjected to repeated bending and may break. The simplest and most accurate method of checking the alignment is by using a strain gauge, as shown in Fig. 25-19. The strain gauge is inserted between two center-punch points made near the ends of the cheeks of the crankshaft, one opposite the other, and readings are taken—first with the cranks down, as shown diagrammatically in Fig. 25-21a and then with the shaft turned 180 deg (Fig. 25-21b). If the left bearing is lower, as shown in Fig. 25-21, the first reading will be greater

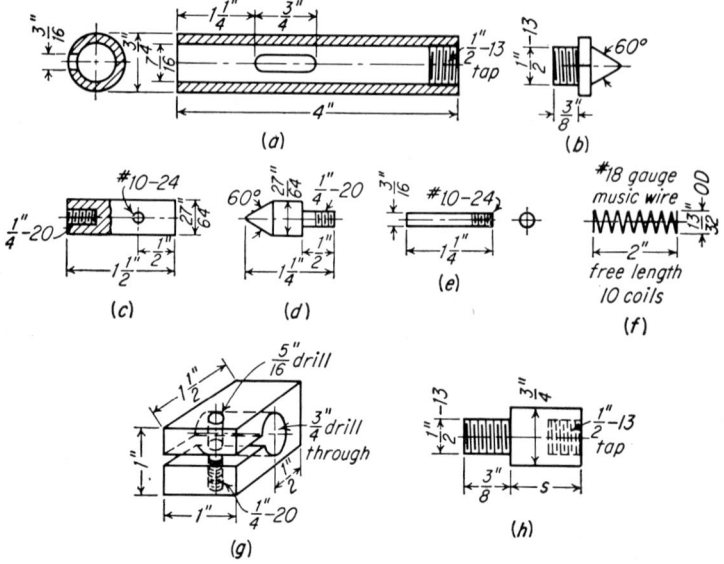

Fig. 25-20. Details of strain-gauge attachment.

than the second one. If the difference is 0.005 in. or more, the alignment must be corrected. This is done by scraping the higher bearings.

25-11. Precision Bearings. Precision bearings are generally made of a steel shell with a very thin layer of bearing metal. When wear causes the clearance between journal and bearing to increase to an

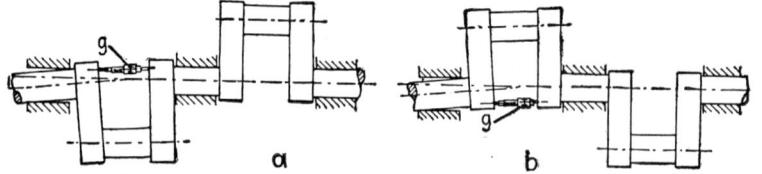

Fig. 25-21. Checking alignment with a micrometer.

extent that it exceeds the limit permissible for retaining lubrication, no adjustment can be made and the worn shells must be exchanged. In small engines the permissible wear is only a few thousandths of an inch. Thus a diesel-engine crankshaft with $2\frac{3}{16}$-in. journals has a normal bearing clearance of 0.002 in. When the clearance, from wear, increases to 0.006 in., the worn bearing shells must be exchanged. Since precision bearings are so manufactured as to be completely interchangeable, both bearing halves are identical, and only the lower half usually needs to be renewed.

However, in larger engines, even precision-type bearings need some fitting, but only from outside, to the bearing support; one may proceed in the same manner as with heavy babbitted bearings. The back must be fitted to obtain a better heat transfer. Also, if there is not an even contact over the whole area, a flexing of the bearing shell may result, causing the thin layer of bearing metal to crack and flake off. Since the back is usually made of steel, it is quicker and better to file

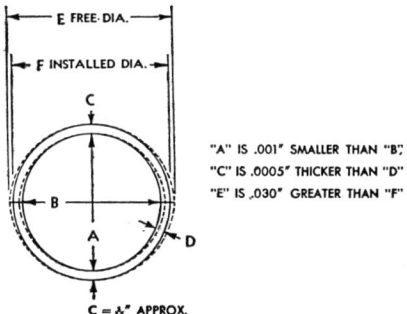

FIG. 25-22. Inside diameter of bearing shell at parting line and 90 deg. to parting line.

down the high spots, shown by the paint, with a fine file rather than to use a scraper.

In order to assist the obtaining of a good contact between the bearing shell and its support the shells are made with a *spread*—the distance from edge to edge is greater than the working diameter of the shell, as illustrated by Fig. 25-22. When the bearing is installed and the holding bolts tightened, the shells form a true circle and the spring action of the ends ensures better contact with the support. If the spread is not what it should be, it can be corrected by giving the shell a few soft blows, as shown in Fig. 25-23a, for increasing the spread, and in Fig. 25-23b, for decreasing it.

Also, the vertical height of each bearing half is made a certain very small amount greater than one half the diameter of the bore into which the two halves are assembled. When the assembly is drawn up tight, the bearing is compressed this extra height, and this assures, in conjunction with the before mentioned spread, a very good contact between the bearing back and the bore into which it is seated. This extra vertical height is called *crush* and is carefully determined by the bearing maker. To obtain proper bearing assembly, the bolts holding the bearing cap must be tightened very evenly and to a solid contact between the cap and the bearing support.

Most small high-speed diesel engines use underslung crankshafts, with the shaft supported by straps which take the pressure acting on the pistons. With this construction the bearings and crankshaft are inspected and adjusted in automotive installations, by taking off the oil pan. In stationary and semistationary engines of this type the best way to look on the crankshaft and bearings is to strip the engine and to lift and turn the block upside down.

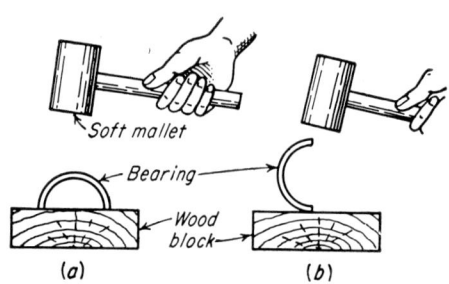

FIG. 25-23. Changing bearing-shell spread.

25-12. Checking Final Bearing Fit. After all bearings are fitted and the crankshaft alignment is checked, the crankshaft—properly oiled, with the connecting rods taken off, but all bearing caps tightened as in regular engine operation—must turn without binding. In small engines it must be so free that the counterweights, when brought out of equilibrium, will start to rotate the shaft and return it to equilibrium by force of gravity. In larger engines the turning of the crankshaft will require, naturally, a greater but still a relatively small effort. If by unscrewing the nuts on the studs holding the bearing caps the effort to turn the crankshaft is reduced, the bearings are too tight and must be rechecked and readjusted.

Naturally, before making such a check, the running clearance of every bearing should be properly adjusted by means of leads, as explained in Sec. 24-4.

25-13. Tramming. When an engine is being overhauled it is completely dismantled. In reassembling it is of the greatest importance that all timing, such as of the intake and exhaust valves, fuel-injection, and starting, should be brought back to standard, to correspond to certain piston or crankshaft positions. The exact placing of the crankshaft at various positions corresponding to certain operating events may be checked by means of a protractor. However, this requires checking the exact position of the air bubble in the level, which is not so easy to do in the crowded crankcase. A more convenient and accurate method is to determine the crankshaft position by means of a *trammel*, also called a *tram rod*, similar to the procedure described in Sec. 6-3. This is done as follows: Before the engine is dismantled, the flywheel is turned to the exact position which must be checked and deep center-punch marks are made on the bedplate or crankcase and on the flywheel rim so that the line connecting these two marks (Fig. 25-24) will be approximately at a right angle to the flywheel

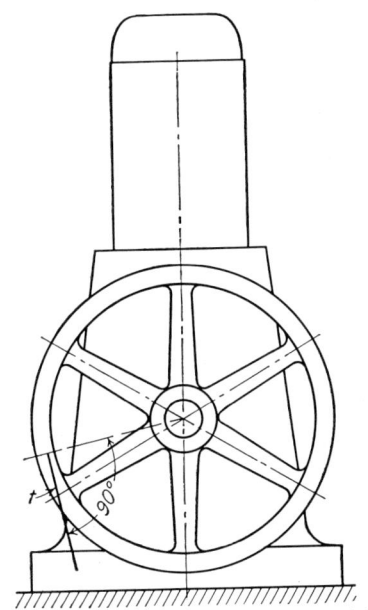

FIG. 25-24. Tramming crankshaft position.

radius. A trammel *t* is made to fit with its ends into these marks. The trammel is made of heavy steel wire or a rod, depending upon its length, with both ends sharpened to a sharp point. Depending upon the engine construction the trammel may be a straight piece, or a piece with one end bent as a hook or both ends bent as hooks, sometimes in opposite directions. When, after overhauling, the engine is reassembled and the flywheel turned so that the trammel will fit with its ends exactly into the corresponding punch marks, the crankshaft will be in the desired position and the camshaft can be set accordingly.

A tram rod may be used also for straight line movement, such as adjusting pushrods. In this case the tram rod is made in the shape of a hook; the pointed end of the hook must fit into a punch mark on the pushrod with the valve fully open, while the other end rests in a mark on the tappet guide or other fixed engine part. The use of a tram rod for exact location of a moving part is called *tramming*.

QUESTIONS

1. What does the term *overhauling* mean in respect to diesel engines in general?
2. What does overhauling mean in regard to a larger stationary engine?
3. How often should automotive diesel engines be overhauled?
4. Enumerate the main steps and rules to be observed when dismantling a diesel engine.
5. Explain, giving a sketch, the procedure of removing a wet cylinder liner.
6. Explain the procedure of removing a pressed-in dry-type liner if (a) it will be junked and (b) if it should be saved.
7. Explain the procedure of inserting a pressed-in liner.
8. Draw a sketch of a tool that helps to remove a stuck loose-fit liner.
9. Explain the procedure of taking inside measurements from a cylinder liner.
10. What is considered normal wear for the following types of liners: (a) ordinary cast-iron, (b) hardened dry-type sleeve, and (c) chromium-plated liner?
11. Enumerate the possible causes of excessive liner wear.
12. Liners of what engines are honed?
13. Describe precautions that must be taken if liners are honed in place, without being removed from the cylinder block.
14. Explain why a stone holder with an adjustment for setting of the cutting-stone radius is better than a holder with spring-loaded stones.
15. Describe the procedure for honing a cylinder liner.
16. When is it advisable to rebore an engine cylinder or liner?
17. What is the permissible oversize for rebored cylinders and liners (a) of automotive-type engines and (b) of larger engines?
18. Enumerate the work to be done on pistons when the engine is being overhauled.
19. Show a sketch of a tool that helps to remove piston rings without breaking them.
20. Describe the procedure for removing piston rings without distorting or breaking them.
21. Describe the procedure for removing a firmly stuck piston ring.
22. Describe the procedure for replacing piston rings.
23. What precaution should be observed with newly polished piston rings?
24. How is the wear of a full-floating piston pin reduced by having different fits of the pin in the piston and connecting-rod bushings?
25. Describe the procedure of installing a piston into the cylinder.
26. What are the indications that a bedplate is distorted and must be torn loose, reinstalled, and regrouted?
27. What work should be done on a bedplate at the engine overhaul?
28. What work must be done on connecting rods?
29. Draw a sketch of a tool for checking connecting-rod alignment.
30. Describe the procedures of assembling piston pins with needle-roller bearings (a) if the needles are small and (b) if they are big and heavy.
31. Describe the method of inspecting a crankshaft for cracks.
32. Describe the procedure for checking whether a small crack is only a flaw,

which is not very dangerous, or an incipient crack due to misalignment which requires replacing the shaft by a new one.

33. What other work must be done on a crankshaft at a complete overhaul?

34. Explain the procedure for checking main bearings by means of a bridge gauge.

35. Describe the procedure for removing main-bearing shells without taking out the crankshaft.

36. Explain when large babbitted bearings should be repaired and when re-babbitted anew.

37. Explain, with a sketch, the procedure of rebabbitting a large bearing.

38. Describe the procedure for finishing and fitting rebabbitted bearing shells.

39. Explain how the wear of a bearing shell is measured.

40. Explain the procedure for fitting, or spotting, main bearing shells.

41. Draw a sketch of a device that facilitates the checking of the alignment of a crankshaft.

42. Describe the procedure of checking the alignment of a crankshaft.

43. In what engines do precision bearings need fitting and in what respect?

44. What is the *spread* of a precision bearing and what is its object?

45. What is the *crush* of a precision bearing and what is its object?

46. Explain the procedure of checking main-bearing fit after overhauling an engine and reassembling it.

47. What is *tramming* and how is it done?

48. Describe the procedure of making a trammel, or tram rod, for checking the timing of an engine.

CHAPTER 26

SPECIAL REPAIRS AND SALVAGING

26-1. General Considerations. In every plant there may occur from time to time the problem of what to do about a defective engine part, major or minor, when the defect is not of a nature which is taken care of in routine maintenance or overhaul work. The repair of such defects may call for some special equipment and knowledge, but usually not beyond the facilities of an average diesel power plant, especially when the assistance of special shops may be obtained. In other instances the repair should be given to shops that specialize in diesel engine repair and salvage work. These special repairs may be in order in several different cases, namely, when (1) a part must be reconditioned in a hurry, (2) to recondition a part without dismantling the engine is cheaper and quicker than to send it to a repair shop, (3) reconditioning of a part is cheaper than replacing it, (4) the failure is caused by the design of the part, and (5) the failure, such as a crack, tends to increase if not taken care of at once.

Usually it is not difficult to determine whether the repair of the defective part comes under one of the above cases and to act accordingly.

26-2. Repair and Salvage Methods. Before going into details—how to repair or salvage certain engine parts—it is advisable to outline briefly the various methods available for this type of work. These methods are (1) welding, electric-arc and gas; (2) brazing; (3) cold-welding, formerly called *lacing;* (4) connecting by shrink links and anchors; (5) surface-welding; (6) metal spraying; (7) electroplating; (8) various machining operations, mostly in connection with one of the above methods, either as a preliminary or as a finishing operation.

26-3. Welding. Welding is used to repair cracks and to connect broken-apart pieces of an engine. Electric-arc welding is used more generally than gas welding, being more positive. Welding is applied to parts made of steel, cast iron, or aluminum. The welding of steel parts is rather simple and usually does not present special problems.

SPECIAL REPAIRS AND SALVAGING

Cast Iron. The welding of cast iron is more difficult, especially if the welded place has to be machined after welding, because the weld and the adjoining area have a much greater hardness than the original casting. If machining is necessary, there are several methods by which the weld metal and weld area in the casting can be made machinable. One is to anneal the entire casting by heating to a dark cherry red, or if expansion will cause no difficulty, to heat the weld and adjacent area only and to allow them to cool slowly by covering with asbestos or sand.

Usually a $\frac{1}{8}$-in. coated-steel electrode is used to keep the heat down. The electrode is made positive and the work negative; a current of about 80 amp is used. The welding should be done very intermittently, or gradually. Sometimes skip-welding is used with a weld not over 3 in. long made at one time. Immediately after each bead is deposited, it should be lightly peened, thoroughly cleaned with a wire brush, and allowed to cool before the next bead is applied. Care should be taken to keep the work clean and not to allow it to become too hot.

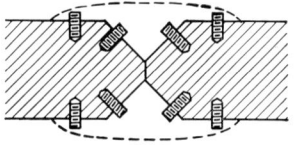

Fig. 26-1. Studding for cast-iron welding.

Some welders prefer to use a heavily coated metallic electrode, having 18 per cent chrome, 8 per cent nickel, and a low-carbon, 0.07 per cent max, steel. The procedure itself is the same as that described above.

Iron castings may be welded also using a carbon arc with a cast-iron filler rod. Proper manipulation of the arc and filler rod, when welding in a flat position, will produce a fairly machinable weld on heavy castings. The use of a dehydrated borax flux enables the operator to float out some of the undesirable impurities. The tensile strength of the weld metal is higher than that of the casting.[1]

Studding. The chilling action of the casting increases the hardness and brittleness, weakening the strength of the casting just back of the line of fusion. Welds in cast iron, if the casting is sufficiently thick, may be strengthened by the insertion of studs.

Steel studs of $\frac{1}{4}$ to $\frac{3}{8}$ in. diameter may be used. The casting should be V-ed, holes drilled, and tapped, as shown in Fig. 26-1. The studs should be screwed in to the depth of one diameter and project about three-fourths of a diameter. The cross-sectional area of all studs should be equal to about 25 to 35 per cent of the area of the

---

[1] More detailed information may be found in the booklet *Recommended Practices for Salvaging Automotive Gray Iron Castings by Welding*, 1950, American Welding Society, New York.

weld surface. It is advisable first to weld one or two beads around each stud, making sure that fusion is obtained both with the stud and the casting. Straight lines of weld metal should be avoided so far as possible. Welds should be deposited intermittently and each bead peened before it cools.

In some cases it may be more practical to shape out grooves in the casting with a round-nosed tool instead of studding, as shown in Fig. 26-2.

Aluminum. Aluminum alloys, forged or cast, can be welded with either a metallic or a carbon arc. For metallic-arc welding a heavy coated electrode of aluminum alloy with 5 per cent silicon is used. The electrode coating must be such that it will dissolve any aluminum oxide that may be formed during welding. The coating should also form a very fusible slag to cover the molten weld metal and protect it from oxidation while cooling. To supply sufficient heat and eliminate a tendency toward porosity along the lines of fusion, it is advisable to preheat the work slightly.

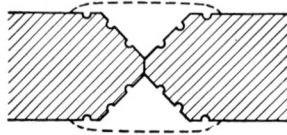

Fig. 26-2. Grooved V-joining in cast iron for welding.

Table 26-1 gives some data for butt welding of ¼- and ⅜-in. aluminum plates. The speeds given are referred to the actual welding time. The voltage given is the actual voltage across the arc while delivering the specified current. The currents designated are those used to obtain the specified speeds. Welding of butt joints of 3/16-in. sections and heavier should be done with two beads without a V cut or backing as indicated in Fig. 26-3. The data in Table 26-1 are intended only as a guide and may be replaced to fit existing conditions, such as thickness of the piece to be welded, type and size of electrode used.

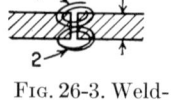

Fig. 26-3. Welding of aluminum section.

For best results, the polarity of the welding current is generally reversed. The electrode is connected to the positive, the work to the negative cable. Direct current is recommended. The arc should be short with the electrode coating almost touching the molten pool of metal. The electrode should be held approximately normal to the work. The arc should be so directed that both edges of the joint to be welded are properly and uniformly heated. Welding should advance at such a rate as to form a uniform bead. Before starting a new electrode the slag should be removed mechanically from the crater of the weld and back approximately one inch. To start a new electrode the arc should be struck in the crater of the bead, then quickly moved back along the completed weld for $\frac{1}{2}$ in., and after the crater is completely remelted the welding may proceed forward.

Final cleaning of the bead is done by first chipping off the excess slag, then soaking the weld in a hot 3 per cent nitric acid solution or in a warm 10 per cent sulfuric-acid solution for a short time and rinsing the weld thoroughly with hot water.

TABLE 26-1. BUTT WELDING OF ALUMINUM

Metal thickness, in.	Electrode diameter, in.	Electric current, amp	Minimum arc, volts	Speed, in. per min
$\frac{1}{4}$	$\frac{3}{16}$	170	20	14
$\frac{3}{8}$	$\frac{1}{4}$	250	20	12

Welding of aluminum in a vertical plane may be accomplished by proceeding either downward or upward. Either a straight-line or a weaving motion may be used in advancing the arc. Overhead welding should be made with a number of straight beads.

26-4. Brazing. Brazing, or bronze welding, is a form of gas welding which produces a strong joint in ferrous metals without actually melting the base metal. In brazing the edges of the joint are heated to a dull red heat by an oxyacetylene flame. With the base metal at the proper temperature and with the aid of a suitable flux, molten bronze from a bronze welding rod will unite with the base metal and form a strong bond. A bronze weld is comparable in strength to a fusion weld.

26-5. Cold-welding. Cold-welding was known for years under the name of *lacing* but was not used very extensively, partly because many repair men tried to keep the details of the process secret and the engine operators as a result lacked confidence in its practicability.

During the Second World War, thanks to the efforts of the SAE Maintenance Methods Coordinating Committee, the method, given the name of *cold-welding*, was brought to the attention of all those interested in repairs to such castings as cylinder heads, jackets, and engine blocks.

The method is used to seal cracks where tensile strength is not required and only a tight joint is needed.

The tools and supplies required for cold-welding are not complicated. All that is necessary is (1) a $\frac{1}{4}$-in. electric drill, (2) an assortment of chisels, (3) the usual hand tools, (4) a supply of a commercially available special-purpose sealing compound, and (5) an assortment of steel rods and tools as given in Table 26-2.

The part to be cold-welded first must be thoroughly cleaned in a degreasing tank and closely inspected to determine the number and exact position of the cracks.

TABLE 26-2. RODS AND TOOLS FOR COLD-WELDING

Special-purpose steel rod, diam, in.	Taps	Drills	Dies
$\frac{1}{8}$	No. 5–40 thr.	No. 38	No. 5–40 thr.
$\frac{1}{8}$	No. 6–32 thr.	No. 36	No. 6–32 thr.
$\frac{3}{16}$	No. 10–32 thr.	$\frac{5}{32}$	No. 10–32 thr.
$\frac{1}{4}$	$\frac{1}{4}$–28	$\frac{13}{64}$	$\frac{1}{4}$–28
$\frac{3}{8}$	$\frac{3}{8}$–26	$\frac{5}{16}$	$\frac{3}{8}$–26

The ends of each crack must be drilled to prevent further spreading of the crack. The drilled hole is tapped, the special rod is threaded, dipped in sealing compound, screwed in, cut off, and peened. Each crack is drilled and tapped for its entire length and the pins are inserted in such a manner that each pin interlocks the one next to it and each pin is peened.

After that a quantity of sealing compound is circulated through the water jacket. As a final inspection after the part is put in place, boiling or at least hot water is circulated through the cooling system and any leak that occurs is stopped by peening.

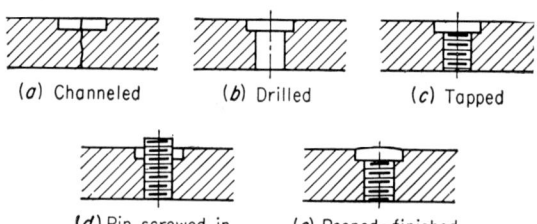

(a) Channeled (b) Drilled (c) Tapped

(d) Pin screwed in (e) Peened-finished

FIG. 26-4. Consecutive steps in cold-welding of crack.

In combustion-chamber cracks the full length of the crack must be first channeled. The width of the channel should be a shade greater than the diameter of the pin to be used. The channel should be at least $\frac{1}{16}$-in. deep, but the exact depth will depend on the thickness of the casting which is being repaired. The channel should be cut so that the crack comes in its center. Figure 26-4 shows the consecutive steps in cold-welding such a crack.

In channeling for an inlay through a valve seat, it is necessary to cut deep enough to allow for several valve-grinding jobs. The hole must be drilled as nearly normal to the seat face as possible. The side walls of the channel should be filed perfectly smooth with a small three-cornered file and slightly undercut to produce a dovetail effect. It is essential that the side wall of the channel in this case be free from scratches or cuts from the top to avoid the danger of a future valve-grinding job running into one of these and obtaining a bad line on the valve seat.

More details of the procedure to be followed when using cold-welding will be found in the description of specific repairs accomplished by this method.

26-6. Shrink Members. This method is used widely in the construction of heavy machine parts, such as large bedplates, crankcases, and flywheels. It can be employed also for repairs of similar heavy castings when the broken pieces must be held together with considerable force and welding or brazing is not practicable. The work is satisfactory if accurate machining with very close tolerances is obtainable.

26-7. Surface Welding. This method is used mostly for building up worn surfaces. The electrode should be used to give a deposit of approximately the same hardness as the hardness of the piece to be salvaged. In applying the bead, the electrode must be positive, the work negative.

The method is also used for surfacing areas subject to heavy impact, such as the seats of exhaust valves and their inserts. In the latter case the deposit is of such a hardness as to necessitate machining by grinding only. The exact hardness depends to a certain extent upon the carbon content of the supporting metal, but chiefly upon the rate of cooling of the beads. With natural cooling, hardness may be from 20 to 45 Rockwell C, or about 225 to 410 Brinell hardness number (Bhn). Peening increases the hardness, for example, from 33 to 40 Rockwell C. Quenching in cold water from 1450 F increases hardness to 50 Rockwell C.

26-8. Metal Spraying. This method is used for building up worn surfaces and consists in depositing a layer of metal on a surface, flat or curved, metallic or from any other material. The deposit is formed by a spray of molten metal obtained from a metal wire moving across a gas flame. The mechanism to feed the wire, the gas flame, and the nozzle ejecting the spray with a high velocity are combined in one tool, called the *spray gun*, that has a handle to hold it and to direct the spray in the desired direction.

When the spray of molten metal hits the relatively cold surface to be built up, it solidifies at once. There is no fusion bond formed between the piece which is being metallized and the deposited metal. The adhesion is of a purely mechanical nature. However, a fairly good bond is formed in the deposited layer itself. Therefore, the part to be built up must be prepared in a special manner: the surface must be roughened by shot blasting or, if it is round, by rough-turning so that the coarse tool marks will form a screw thread; fillets and edges usually should be dovetailed, as shown in Fig. 26-5. The deposited layer has a tendency to shrink; therefore its grip on a shaft which is metal-sprayed from the outside will be increased, but a layer deposited on the inside of a hollow cylinder would have a tendency to pull away from the cylinder.

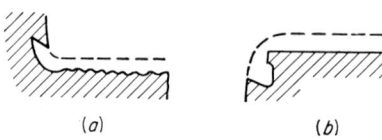

Fig. 26-5. Dovetailing for metal spraying.

Any metal or alloy, carbon or stainless steel, bronze, monel metal or aluminum, may be deposited by this method. The fuel gas used for melting the wire also may be of any kind—natural gas, acetylene, propane, butane, etc.

While metal spraying has been used successfully for many salvage jobs, such as building up worn journals and crankpins on crankshafts, building up wearing surfaces on different shafts, in many instances it has given disappointing results because improper procedure was used or because it was not suitable to the particular application. Before attempting to use this method, one should carefully analyze the problem and obtain the advice of more experienced repairmen or, directly, of representatives of the factory which builds the equipment and furnishes the wire. Metal spraying is a special trade which an interested person must learn by actually manipulating the gun under the guidance of an experienced operator-instructor before attempting to do responsible work on his own.

26-9. Electroplating. This is another method for building up worn surfaces. In using it two conditions should be kept in mind: first, that it is suitable for building up thin surfaces only—0.001 to 0.008 in. (under certain conditions up to 0.015 in.), and second, that electroplating gives a layer of rather uniform thickness. Therefore, if the wear which must be corrected was not uniform, the surface should first be trued up to the maximum wear and then built up back to the standard level.

Electroplating is used to build up by depositing any metal, but those most generally used are nickel and chromium. Lately chromium plating is being used extensively for the inside of cylinder liners and the outside of piston rings. Chromium plating reduces wear because of its hardness. It should be kept in mind that a chromium-plated surface cannot work against another chromium-plated surface. Thus chromium-plated liners must have unplated cast-iron piston rings and chromium-plated piston rings give good service only in hardened steel liners.

Electroplating is done in special plants and when used for salvaging engine parts requires special experience.

26-10. Repair Examples. *Cylinder Heads.* Cracking of cylinder heads is one of the most frequent casualties in diesel-engine plants. The reason is first, that this engine part is subjected to high stresses, and second, that there are several unrelated conditions each of which may cause failure of the cylinder head. Some of these conditions are: uneven expansion and contraction, growth of cast iron caused by alternate heating and cooling, uneven pulling down of the heads, excessive force exerted by the hold-down studs, faulty cooling-water circulation, scale deposits causing overheating, excessively high firing pressures.

Regardless of the cause of cracks, cylinder heads up to the largest ones can be successfully reclaimed by fusion welding. Some welders use electric-arc, some gas welding.

In preparing a cylinder head for gas welding, it must first be stripped of all parts, including valve-stem guides, valve-seat inserts, and studs.

Then the casting must be preheated. A casting does not necessarily have to be heated slowly—the main point is to apply the heat first to the outer surfaces and then to maintain it slightly in advance of that applied to the inner surfaces—but care should be taken in keeping the temperatures fairly close from the outside to the inside. Heating must continue throughout welding operations. At the same time it must be remembered that the length of time the casting is kept hot should be held to a minimum. The casting should be heated to a cherry red, about 1200 F, but not higher.

The way a casting is preheated depends upon its size. Small castings may be handled on a preheating table with gas firing and a hood that is drawn over the fire. Larger pieces are preheated by building a special cribwork of fire brick around them, covered with asbestos sheets to regulate the supply of air and to insulate the heat. Charcoal is used for heating and is placed inside the cribwork and around the piece to be welded.

In preparing the crack for welding, the V cut should be made with an oxyacetylene flame. This is better than chipping or grinding in advance because the burned metal is removed far enough back to reach good fusible metal as a base. The fusing temperature at the crack is raised to 2600 F by means of oxyacetylene torches.

The welding rods used must have the same coefficient of expansion as the parent metal.

If the head has no valve-seat inserts and the valve seat has a crack or is badly worn, the crack should be cut out and rewelded and the seat built up almost to proper height and then topped with a wear- and corrosion-resisting metal.

Upon completion of the welding job, the casting must be allowed to cool very gradually to leave it annealed. When it cools off, the interior must be thoroughly cleaned of all scale by using specially shaped chisels in either a pneumatic or an electric hammer; in some instances scale may be removed by using a solution of muriatic acid. After acid is used, the casting should be immersed in a hot alkaline solution, soda in water, to drive all acid from the pores of the iron, and finally washed with hot water and inspected to make sure that no scale is left.

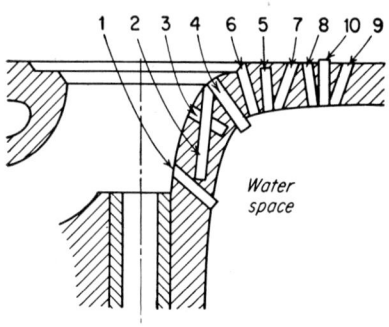

Fig. 26-6. Repairing a Caterpillar diesel cylinder head.

Cylinder heads of small diesel engines often develop cracks that may be repaired by cold-welding. In Fig. 26-6 is shown the method used to seal a typical crack from the exhaust passage into the water space in a Caterpillar diesel cylinder head. This crack, when it occurs, varies in length, sometimes running from the shoulder at the valve guide up through the seat and across the surface beside the fuel-injector tube.

The order to follow is indicated by the numbers at the pins. Before screwing in the pins, which must each be dipped in a sealing compound, a channel is cut from the shoulder just below the valve seat across the surface to the end of the crack. The cutting, where possible, should be done with a hacksaw.

After pin 1 is screwed in, a $\frac{1}{8}$-in. hole is drilled from the shoulder under the valve seat down to pin 1, following the crack all way down. A pin of a $\frac{1}{8}$-in. special rod slightly longer than the depth of the hole is cut, dipped in sealing compound, and peened straight down with a punch to swell it in the hole. If the crack does not run reasonably straight—if it has a bend that cannot be followed by a line drilling— it becomes necessary to stack pins from pin 1 on up. In some

SPECIAL REPAIRS AND SALVAGING

instances it may take two pins to bring the metal up to the proper place near the valve seat instead of one pin 2.

Pin 4 must be put as near the center of the valve seat as possible, after preparing this operation as described for valve-seat inlays in connection with Fig. 26-4.

After all the pins, up to the end of the crack, are screwed in tightly, cut off, peened, and the tops filed smooth, sealing compound is circulated through the water jacket and the head is ready to be put back in service. In Figs. 26-6 to 26-8 the cylinder heads are shown upside down as they are held during the repair operations.

Figure 26-7 shows the repairing of a crack that comes in Cummins diesel heads from the valve-seat insert over to the injector. If the crack does not run below the corner of the insert, the repair may be made according to Fig. 26-7a.

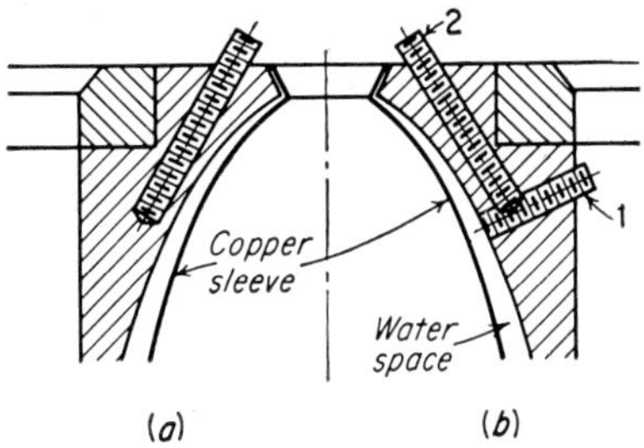

First, a slim-taper lining-up punch is inserted snugly into the injector hole and the flange of the copper tube around the crack is carefully lifted with a very sharp, thin chisel.

FIG. 26-7. Repairing a Cummins diesel cylinder head.

Then, with the punch removed, a channel is cut with a sharp $\frac{1}{8}$-in. capping chisel, from the insert over, going about half-way down in the countersink for the copper tube.

After that, a hole is drilled, starting at the edge of the countersink at a slant that will just allow the corner of the insert to be missed. This hole should be about $\frac{7}{16}$ in. deep. A No. 38 drill and No. 5-40 tap or a No. 36 drill and No. 6-32 tap may be used. A threaded pin made of a $\frac{1}{8}$-in. special-steel rod and dipped in sealing compound is inserted, the tapered punch is replaced tightly into the injector hole, and the end of the pin is then peened and finished. The lifted copper flange is laid down carefully and the punch removed.

After the repair work has been completed, sealing compound is circulated through the water jacket, and the head is ready for service.

If the crack extends down below the valve insert, it will be necessary to remove the copper injector sleeve and to make the repair using the procedure as shown on Fig. 26-7b. First is inserted through the wall pin 1 made of a $\frac{3}{16}$-in. rod. It is cut flush on the water-space side and peened and finished flush from the valve-port side. Then is inserted a $\frac{1}{8}$-in. threaded pin 2, which is then peened and finished;

the copper sleeve is replaced, and sealing compound is circulated through the water jacket.

The nature of the cracks that occur in a General Motors diesel-engine head and the method of cold-welding them are similar, as may be seen from Fig. 26-8a and b.

26-11. Cylinder Jackets. Cylinder jackets seldom have cracks; they are, however, sometimes broken by an accident, such as the breaking of a connecting-rod bolt. The repair is done by welding. If gas welding is used, the procedure is practically the same as that described above for larger cylinder heads, with preheating in a large furnace or crib built of firebricks. Whether electric or gas welding is used, the castings must be carefully annealed by being heated to about 1200 F and then cooled very slowly.

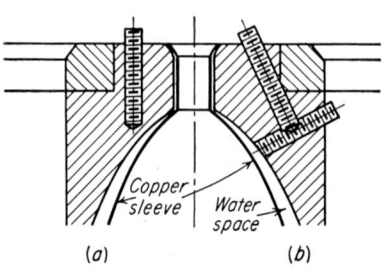

FIG. 26-8. Repairing a General Motors diesel cylinder head.

26-12. Crankcases and Bedplates. In large engines these parts also are damaged only by an accident, and the procedure for repairing castings is along the same lines as that for cylinder jackets. Welded crankcases and bedplates are usually manufactured by electric-arc

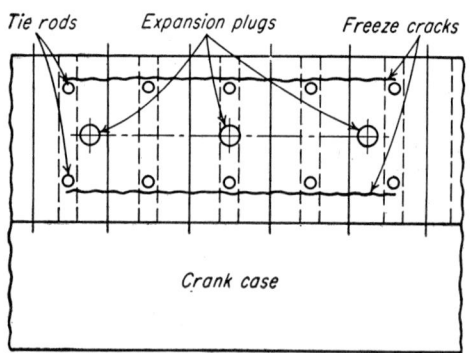

FIG. 26-9. Freeze cracks in a cylinder block repaired by tie rods.

welding and the same method is used for their repair. Stress-relieving by heating to about 1200 F is also advisable, and even necessary with larger repairs.

Crankcases and blocks of small diesel engines are occasionally cracked by freezing of water in the cooling jackets. Depending upon the nature of the cracks they may be repaired by fusion welding or

SPECIAL REPAIRS AND SALVAGING

cold-welding. The latter usually is preferred as it does not carry with it the danger of distortion of the machined surfaces.

Figure 26-9 shows diagrammatically freeze cracks on the side of a small six-cylinder diesel-engine crankcase and Fig. 26-10 (a cross-section) the method of pulling in the bulging side and reinforcing it before sealing the cracks. Depending upon the size of the crankcase and the extent to which the cracked side is bulging out, either $\frac{1}{4}$- or $\frac{5}{16}$-in. tie rods may be used. If $\frac{5}{16}$-in. rods are necessary, the procedure is as follows:

First, $\frac{9}{32}$-in. holes are drilled near the edges of the cracks in line with the spaces between the cylinders. A long drill must be used and the holes pushed through the other side. If a drill of sufficient length is not available, the holes on the opposite side have to be laid out and drilled accurately in line with those in the cracked side.

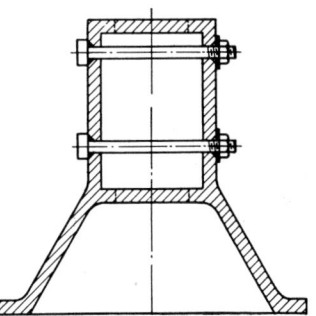

Fig. 26-10. Repairing cylinder block with freeze cracks.

The holes are reamed out to $\frac{5}{16}$-in. size and countersunk on the cracked side (Fig. 26-10) $\frac{3}{16}$ in. deep and on the opposite side $\frac{1}{16}$ in.

Long bolts with the underside of the heads turned to fit the countersink and dipped in sealing compound are inserted through the holes. Lead washers are put under the nuts and the latter are drawn up with wrenches in pairs, beginning at the ends of the cracks and tapping along the cracks with a hammer as the bulge is pulled in. The pulling in must be done very gradually. Instead of long bolts, a $\frac{5}{16}$-in. steel tie rod with threads and nuts on both ends may be used.

After the cracked side is pulled in, the ends of the bolts are cut off smooth with the nuts and peened.

Blocks that have been in service any length of time and in which the crack is at the bottom of the water space must be cleaned out thoroughly at the crack as otherwise the sealing compound may not work. This is best done by drilling a $\frac{5}{32}$-in. hole in each end of the crack and running a piece of soft wire back and forth as water is flowed in from the top.

Finally, drill, tap, and screw in $\frac{3}{16}$-in. pins along the cracks, making sure that one is placed exactly in each end of the crack to prevent its spreading and placing the rest from 1 to 2 in. apart, according to the thickness of the wall, a heavy one permitting a wider spacing.

Freeze cracks and strain cracks located near and parallel to the top of a cylinder block have a tendency to open up, as a result of the pull of cylinder-head studs or bolts.

In some types, the crack is caused by excessive pull on a stud and the crack opens up around the corner and into the exhaust ports, as shown in Fig. 26-11. First, such a crack is repaired in the usual manner by screwed-in pins dipped in sealing compound. Then the stud hole is drilled through to the bottom of the block and a hole drilled and tapped in the crankcase (Fig. 26-11) for a long stud, thus transferring the pull from the damaged block to the crankcase.

After the work has been completed, sealing compound must be circulated through the water space.

If a commercial sealing compound as used for sealing automobile-radiator leaks is not available, it can be made by mixing one quart of silicate of soda, commonly called *water glass*, with three quarts of warm, but not hot, water. This solution is poured into the water jacket, all outlets but one on top are closed, and an air pressure of 50 to 60 psi is applied through this opening for about 5 min. The surplus solution is poured out and the repaired part left to stay for 24 hr before using it. The silicate of soda forms a hard glasslike film over the surface, filling all holes and cracks. It is not affected by heat or water and neither expansion nor contraction of the metal can cause the holes or cracks to reopen.

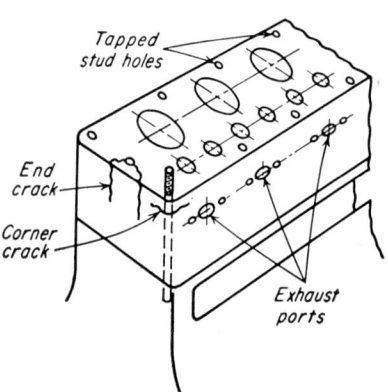

FIG. 26-11. Freeze and strain cracks in engine block.

26-13. Cylinder Liners. Maximum permissible wear of cylinder liners before reboring is about 0.007 in. per inch of bore. By that time the cylindrical surface usually becomes bell-shaped toward the combustion space and possibly barrel-shaped in the upper half of the stroke. The liner must be rebored to a true cylindrical shape and a new oversize piston fitted. In order to retard subsequent wear, it is good practice to have the liner chromium-plated by a method that gives a hard surface with small cracks. These cracks are obtained by reversing the direction of the plating current after the required thickness is obtained. These cracks act as reservoirs for lubricating oil, decreasing the wear of piston rings and reducing to a nominal amount the wear of the liner surface. Such plating is called by the trade name of *porous-chrome-plating* or *channel-type chrome-plating* and in this country is done by several plating plants under at least two different licenses. After plating, the bore should be honed.

Cylinders that are cracked or broken can be repaired by preheating and welding, but require reboring, sometimes to a slight degree only, as the result of unavoidable heat distortion. Occasionally a cylinder casting is found to be of poor or porous metal and its welding may be difficult, if at all possible.

Sometimes a cylinder or liner badly scored some distance down stroke can be repaired by soldering with hard solder and scraping the soldered surface to fit the piston rings and lapping it to a smooth condition.

26-14. Pistons. Pistons usually crack radially in the top cavity, or dome. This may be repaired by welding, as already explained. Some pistons are worn along the outside cylindrical surface and may be restored to size by metal-spraying. First, the outer surface is turned down roughly with a light undercut at the bottom of the skirt and at the edge of the first ring groove to hold the ends of the sprayed-on jacket better. Then the surface is blasted with sharp angular steel grit under 80 psi air pressure. After that it is sprayed with low-carbon steel on this surface until it is oversize, and finally ground down to the required dimensions.

The lands, or spaces between the ring grooves, are not sprayed as it is very difficult to make effective keys to hold the sprayed metal. Also there is no real necessity to build up these places to the diameter of the main piston body as the rings protrude here anyway. Occasionally, ring grooves are found worn on the sides so badly that they need to be reconditioned. This is done by welding in a dovetailed strip of metal along the sides and machining the grooves to the required width.

26-15. Piston Pins. The diameters of the piston pins, both in the center, where they are subject to a certain wear, and at the ends, where they may become loose from repeated high specific pressures, usually need to be increased only a very small amount. This may be obtained by hard-chrome plating and by grinding to the required size.

26-16. Valves. Valves usually need reconditioning of the seats and stem ends. This is done by welding on, after preheating, hard-surfacing steel layers and grinding to the finished size required. Stems of large valves generally do not wear in the diameters appreciably, but should this happen, it is usually safer and cheaper to renew the bronze bushing to fit the stem than to try to build up the stem diameter.

Valve Cages. The cages generally require reconditioning of the valve seats. Frequently the lugs for the hold-down bolts are broken off from careless, uneven tightening. Or there may be a breakdown of the partitions in the water jacketing due to clogging with sand or scale. All these repairs can be done by welding. Valve seats can be resurfaced with either cast iron or heat-resisting alloy metal. The surface metal is welded onto the base metal, usually cast iron, after preheating the cage, and is ground to the finished size required.

26-17. Crankshafts. The causes requiring salvage work on a crankshaft may be grouped under three general headings: wear, bending, and breakage.

Crankshaft journals and crankpins usually wear not only to a smaller diameter but, because of the variable pressures, also out of round. If the diameter of the shaft, after it is trued up in a lathe, is not more than 0.010 to 0.015 in. under the standard size, a simple method of reconditioning is to have the journals hard-chrome-plated to slightly oversize and then ground back to size. It seems that hard porous surfaces, as explained in touching upon plating of cylinder liners, are to be preferred.

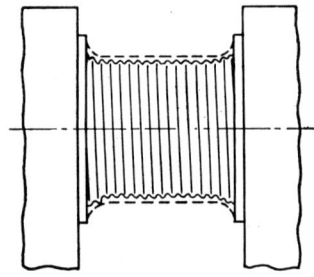

Fig. 26-12. Rough threading for metal spraying.

If the shaft, after being trued up, will have a diameter appreciably under the required standard size, and the shaft diameter is not over 6 in., metal spraying may be applied.

There are two satisfactory methods of preparing journals to be built up by metal spraying, rough threading and shot blasting. If rough threading is used, the journals should be first turned down 0.065 to 0.090 in. in diameter under the required finish size; at the fillets the cut should be run out to zero, not dovetailed, as shown on Fig. 26-12. The rough threads are cut, on small shafts up to 2-in. diameter, 20 to 24 threads per inch, on shafts over 2-in. diameter, 18 threads per inch; thus they have a depth from 0.027 to 0.036 in. To prevent metal from sticking to surfaces that must not be metal-sprayed, ordinary chalk may be rubbed on the fillets and faces of the webs. Oil holes should be closed up with carbon plugs.

With shot-blast preparation, steel grit No. 15 and not finer than No. 20 should be used and an air pressure from 60 to 100 psi. The same cut is taken as when preparing for threading. If possible, fillets and faces should be masked off. If this is not done, care should be taken not to hit the faces too much. The crank should be rotated while being blasted. The grit should hit the journal at right angles at all times and blast it thoroughly. If the faces are hit, chalk rubbed on heavily will prevent the metal from sticking.

Metal for spraying on journals should be 1.20 per cent carbon steel. The metal spraying should be done at once after the journals are prepared, to avoid oxidation of the clean surfaces.

After spraying the journals must be ground to size, as this steel is too hard for machining. Wet grinding should be used to prevent expansion of the metal layer from the heat of grinding.

Crankshafts over 6 in. in diameter are usually simply trued up and furnished with undersize bearings and metal spraying is not recommended.

Bent crankshafts are straightened out, and since it is almost impossible to straighten them out completely, a small cut is taken to true up the journals and crankpins. After truing up, they are either restored to the standard diameter by chrome plating or metal spraying, as explained above, or fitted with undersize bearings.

Broken crankshafts may in some cases be salvaged by thermit welding and machining if the break is in such a place that such a weld can be made.

Electric-arc welding seems to be the best method for repairing diesel-engine crankshafts up to the largest sizes. In general, the procedure is as follows: first the Brinell hardness of the shaft at the break is determined and, if it is made of a special material, an analysis of the crankshaft metal should be made. The material of the electrode to be used should approach as nearly as possible the material of the crankshaft after welding but should be slightly softer. The parts of the shaft affected by the break are then annealed by heating to a temperature of about 1100 to 1200 F and very slow cooling. The cracks which may appear during heating are eliminated by partial welding.

Then the openings for the welding process are prepared, leaving two resting points. These openings should not be larger than strictly necessary in order to allow easy access for the electrode, which generally does not exceed a $\frac{1}{4}$-in. diameter. The two pieces to be welded together are then carefully aligned, taking into consideration the expansion and contraction during welding.

Before starting to weld, the pieces are heated to not less than 600 F. Small shafts are welded without interruption and without letting them cool down. Large shafts are partially welded and then stress-relieved, welded some more and again stress-relieved—as many times as necessary—always maintaining the temperature, about 600 F, while filling the welds. On large shafts, two welders work simultaneously and symmetrically. This produces less distortion of the shaft and a more uniform expansion and contraction. When the operation of filling out the V cuts is completed and the reinforcements, if necessary, are in place, the shaft is again stress-relieved by being heated to approximately 1150 F and allowed to cool down slowly to the surrounding temperature.

After welding, the crankshaft must be put in a lathe for truing up.

In conjunction with machining crankshafts, an unusual but very practical method of extracting pieces of broken drills from partly drilled oil holes in crankshafts may be mentioned which may be applied to other machine parts when salvage work is being done. A tiny amount of dynamite, taken from a standard stick of blasting powder, is pushed from a small trough into the hole by means of a wooden stick. A 6-in. fuse is next inserted into a small-diameter, shoulder-free cap and dropped into the hole atop the charge. When the fuse is set afire by a match, a heavy plank or metal shield should be placed over the hole to check the flight of the piece of drill when it is blown out. As an additional precaution everybody should keep at a safe distance during the dynamiting. Drills are freed by the pressure of the blast that follows the drill flutes to the bottom of the hole and there reverses itself to drive the piece back out. If the drill breaks flush with the surface, a little funnel is built of putty around the hole and into it is put the charge and fuse.

QUESTIONS

1. Enumerate the conditions under which a diesel-engine part must be repaired without delay.
2. Enumerate the five types of special repairs that may be encountered in operating a diesel-engine plant.
3. Enumerate the various methods that may be used in repairing or salvaging engine parts.
4. State the difficulty experienced in welding a cast-iron part if it has to be machined at the weld.
5. What kind of electrodes are recommended for welding cast-iron parts?
6. Describe briefly with a sketch the procedure of strengthening the weld of a cast-iron broken part.
7. What is skip-welding? When is it used?
8. What is studding in cast-iron welding? What is its purpose?
9. What arc, metallic or carbon, is used for welding aluminum-alloy parts?
10. How is brazing done and how does its strength compare with a fusion weld?
11. What is cold-welding and when can it be used successfully?
12. Enumerate the tools and supplies necessary to perform cold-welding.
13. Describe the procedure of cold-welding a crack in a combustion chamber.
14. When are shrink members used? What is the main condition for satisfactory shrink-member repair?
15. When is surface welding used?
16. When can metal spraying be applied?

SPECIAL REPAIRS AND SALVAGING

17. How should the part be prepared for metal spraying?
18. When is electroplating used in salvaging machine parts?
19. What diesel parts are at present mostly chrome-plated?
20. State what the maximum permissible wear of cylinder liners is before they must be rebored.
21. What method is recommended for reclaiming cast-iron cylinder heads?
22. Describe the consecutive steps to be taken in reclaiming a cylinder head by electric-arc welding.
23. What tools are necessary for cold-welding?
24. To what temperature should damaged welded crankcases and bedplates be heated after being repaired by electric welding?
25. How is a crack in the water jacket sealed after it has been repaired by cold-welding? Describe the procedure in detail.
26. How are worn cylinder liners restored to proper usefulness?
27. How are pistons salvaged by metal spraying?
28. What is the usual method of reconditioning loose piston pins?
29. How are valve seats reconditioned?
30. What method is used in reconditioning valve cages of large engines?
31. What are the three types of damage requiring salvage work on crankshafts?
32. How is a worn crankshaft journal or crankpin reconditioned (a) if after truing up the shaft its diameter is not more than 0.010 to 0.015 in. under the standard size and (b) when it is more than 0.015 in. under size?
33. Describe the procedure of reconditioning a small crankshaft by metal spraying.
34. Describe the consecutive steps followed in salvaging a crankshaft that has been bent.
35. Describe the procedure of salvaging a broken crankshaft.
36. Describe a practical method of extracting a piece of a broken drill from a crankshaft or other metal part when the hole is only partly drilled.

APPENDIX

Each engine part has its own particular function to perform and in conjunction with other parts, equally as important, comprises the assembly called the *internal-combustion engine* or, for short, *combustion engine*.

An understanding of the operation or functions of the individual parts is necessary for a better understanding of the whole engine.

A person who intends to work in the diesel-engine field must know how to recognize the engine parts by sight and must learn their correct names and also their particular functions.

In Figs. A-1 and A-2 are shown cross-sectional views of a diesel engine of the heavier type. In Fig. A-1 the section of the cylinder head is taken through the injector, or spray nozzle, and starting-air valve, whereas Fig. A-2 is a section by one vertical plane through the center line of the engine and in the cylinder head shows the intake and exhaust valves.

In Fig. A-3 is shown the cross section of a diesel engine that is used in the automotive transportation field. Attention is called to the difference in the general construction of these two types of engines.

The following glossary of terms gives the student a preliminary understanding of the component parts used in the diesel engine and their functions as well as an explanation of some basic units and definitions of a more theoretical nature but indispensable to a person who wants to work intelligently in the chosen field and to rise from apprentice to engine operator and possibly to chief engineer of a large plant.

The way to use the drawings and the glossary is to study the drawings and to look up in the glossary every part name for explanation.

However, even a man familiar with diesel engines should have some use for the glossary looking up terms, units, and occasionally a definition to make sure that he remembers them correctly. Knowledge confirmed by an authoritative source is the best knowledge a person may have.

Absolute pressure The pressure in pounds per square inch, psi, above absolute zero pressure, or perfect vacuum. A Bourdon or mercury gauge registers the difference between the pressure within a receiver and that of the outside atmosphere. At sea level and under standard barometric

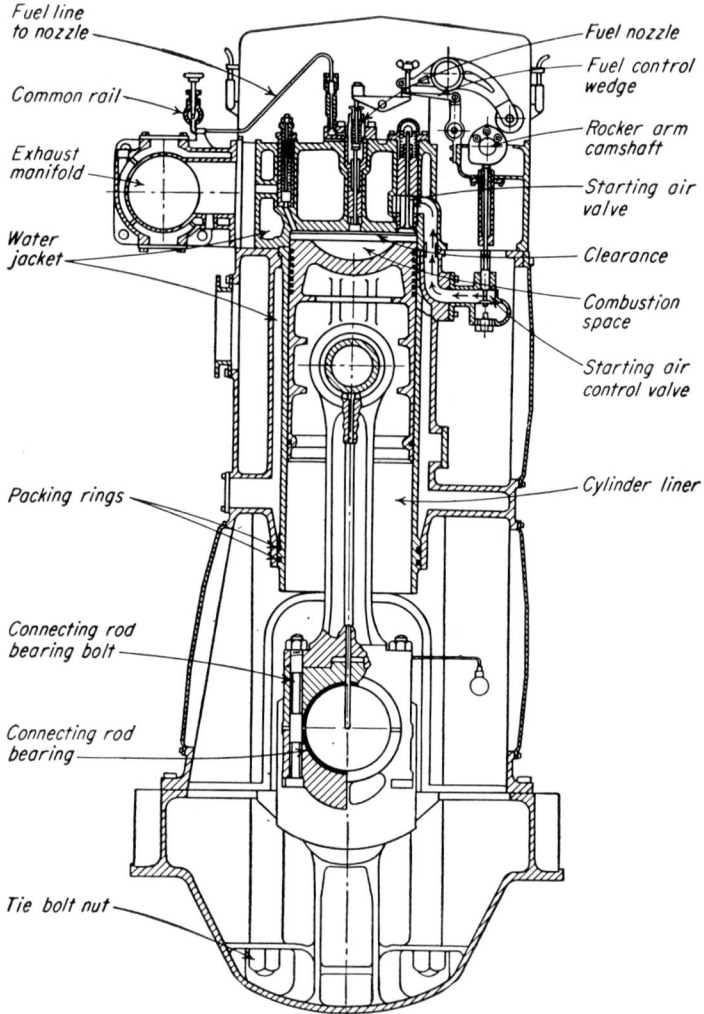

FIG. A-1. Cross section of a marine diesel engine showing fuel-injection and air-starting systems. *Courtesy of National Supply Company.*

conditions, the atmospheric pressure is 14.7 psia. To find the absolute pressure, add 14.7 to the gauge reading in psig.

Absolute temperature The temperature above absolute zero. If a Fahrenheit thermometer scale is used, absolute zero is −460 deg. To find the absolute temperature, add 460 to the Fahrenheit reading.

Accelerate To increase the speed of movement, such as increasing the speed of a piston or flywheel.

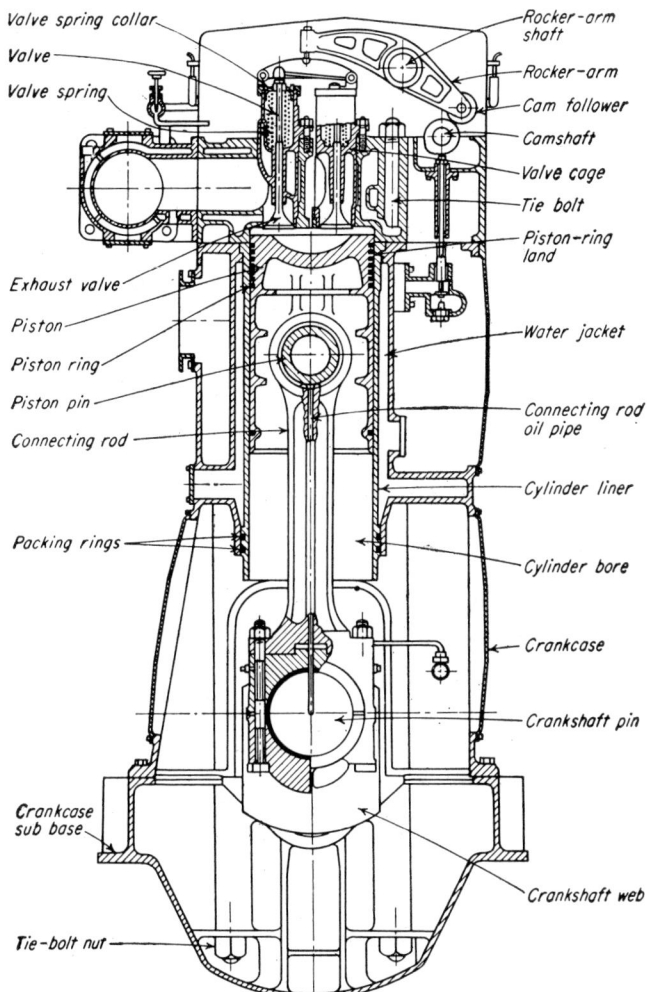

Fig. A-2. Cross section of a marine diesel engine. *Courtesy of National Supply Company.*

Acceleration The rate at which the speed of an object increases.

Adiabatic From the Greek word meaning "no pass through." Adiabatic compression or expansion of a gas is accomplished without the loss or gain of heat through the cylinder walls.

Advance Sometimes referred to as *lead,* or *angle of advance,* meaning the distance ahead of top or bottom dead center of the piston as measured in degrees of crank travel.

Air cell A small receptacle communicating with an engine cylinder into which some of the compressed air is forced, and from which air later flows back into the cylinder.

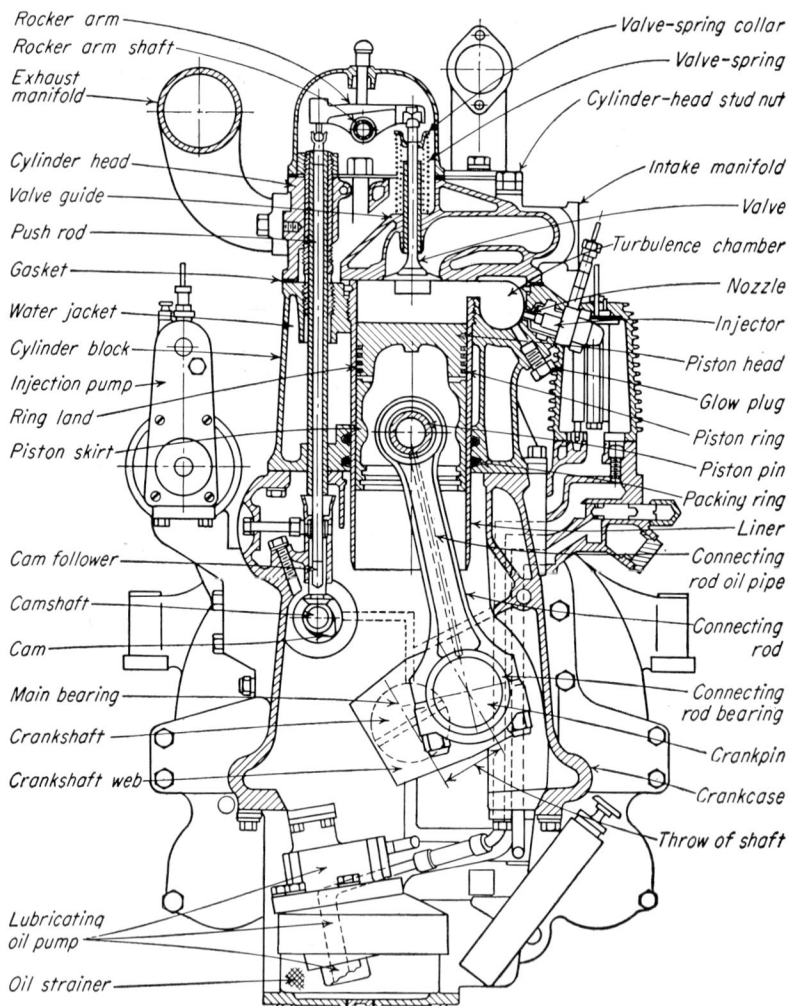

Fig. A-3. Cross section of an automotive diesel engine. *Courtesy of Hercules Engine Company.*

Air filter A device for filtering the air, before it goes into the engine, to prevent particles of dust from entering the engine.

Air injection The system of injecting fuel into the combustion chamber of a diesel engine by means of a blast of highly compressed air.

Airless injection A general term describing all methods of injecting fuel without the use of compressed air.

Air starter A system whereby an engine is turned over by admitting compressed air into the cylinders in order to initiate firing.

Aniline number The lowest temperature at which equal parts of aniline and a sample of oil are completely miscible, or the temperature at which the mixture becomes turbid or cloudy.

Antichamber Same as PRECOMBUSTION CHAMBER.

API American Petroleum Institute.

API gravity An arbitrary scale adopted by the American Petroleum Institute to designate the specific gravity of mineral oils. Diesel fuels range from 18 to 41° API.

Atmospheric pressure The pressure of the atmosphere measured from absolute zero pressure. At sea level atmospheric pressure is 14.7 psi, decreasing as the altitude increases.

Atomize To break up a liquid into extremely fine particles.

Axial Parallel to the center line of a cylinder or shaft.

Axis A center line. A line about which a body rotates or about which it is arranged.

Babbitt A soft antifriction metal used to line bearings.

Back pressure The resistance to the normal flow of gases and liquids.

Bedplate The lower part of the engine resting on the foundation.

Bore The interior diameter of an engine or compressor cylinder.

Blow-by Escape of gases from the engine cylinder into the crankcase because of unsatisfactory action of the piston rings.

Brake horsepower The useful horsepower delivered by an engine that may be found by the use of a prony brake. Abbreviated bhp.

Brake mean effective pressure The mean effective pressure, corresponding to the brake horsepower developed. Abbreviated bmep.

British thermal unit The amount of heat needed to raise the temperature of one pound of water one degree Fahrenheit, from 68 to 69 F. Abbreviated Btu.

Burning Commonly substituted for COMBUSTION, as *late burning*, meaning late or slow combustion.

By-pass A passage which permits a liquid or gas to take a course other than that normally used.

Cam A disk-like piece, attached to a shaft, a portion of which is circular, the remainder (the "nose") protruding beyond this circle. Cams are used to impart a desired motion to poppet valves.

Camshaft The shaft which carries the various cams required for the operation of inlet, exhaust, fuel, and starting-air valves.

Cam follower That part of the push rod that is in contact with the cam.

Carbon One of the chemical elements which is the main constituent of liquid and solid fuels. Also the residual substance deposited in the combustion space and exhaust system of diesel engines when combustion of fuel oil is not complete.

Carbon dioxide Gas composed of molecules made of one atom of carbon and two atoms of oxygen.

Carbon monoxide Gas composed of molecules made up of one atom of carbon and one of oxygen. It is formed when combustion is not complete because of the absence of sufficient air.

Carbon residue The carbon remaining after evaporating off the volatile portion of a fuel or lubricating oil by heating it in the absence of air under controlled test conditions. It is an indication of the amount of carbon that may be deposited in a diesel engine.

Centrifugal force The force acting on all parts of a rotating body that tends to pull them away from the axis of rotation.

Cetane A hydrocarbon used in testing the ignition quality of diesel fuels.

Cetane number A percentage indicating the ignition quality of diesel fuels.

Cetene A hydrocarbon used formerly in testing the ignition quality of diesel fuels.

Chamfer A beveled corner.

Charge efficiency The ratio of the weight of the charge actually taken in to the weight of the air at standard conditions corresponding to the piston displacement.

Check valve A valve that permits the passage of a liquid or gas in one direction only. It stops, or checks, reverse flow.

Clearance The space between a moving and a stationary part. Clearance must be provided between two surfaces to allow for lubrication and for expansion and contraction with a change of temperature.

Clearance volume The volume of air or liquid remaining in the cylinder of an air compressor or a pump when the piston is nearest to the cylinder head.

Coefficient A ratio; a factor or quantity that remains constant.

Combustion The rapid oxidation, or combination, of a combustible such as carbon, hydrogen, or sulfur, with oxygen of air.

Combustion chamber The space above the piston in which the fuel-air mixture starts to burn.

Common rail A pipe or header from which branch lines lead to each of the fuel valves in the different cylinder heads of a diesel engine and in which fuel is carried at high pressure, ready for delivery to each separate cylinder when the fuel valve is opened by a cam.

Compression The act or result of pressing a substance into a smaller space. One of the events of a combustion-engine cycle.

Compression ignition Ignition of a fuel charge by the heat of the air in a cylinder, generated by compression of the air, as in the diesel engine.

Compression pressure The pressure of the air charge at the end of the compression stroke.

Compression ratio The ratio of the volume of the charge in the engine cylinder at the beginning of the compression stroke to that at the end of the stroke.

Compression relief A device to reduce the compression in a cylinder and thus to make cranking easier.

Compression rings Piston rings placed in the upper part of a piston to seal against loss of compression pressure and against gas blowing.
Compression stroke The stroke of the piston during which the air charge in the cylinder is compressed by the piston movement.
Compressor The air "pump" which furnishes compressed air for starting the engine, or for the injection of the fuel in an air-injection diesel engine.
Concentric Having a common center.
Condensation The process by which a substance changes from vapor to the liquid state.
Connecting rod The engine part which connects the piston to the crankshaft. It changes reciprocating motion of the piston into rotary motion of the crankshaft or vice versa.
Connecting rod bearing The bearing located in the large end of the connecting rod by which it is attached to the crankshaft.
Constant A value or figure in a formula or equation which does not change, remains constant.
Constant-pressure combustion Combustion of fuel in a cylinder at so slow a rate that there is no rise in cylinder pressure. The slow-speed air-injection diesel engine is a constant-pressure combustion engine.
Constant-volume combustion Combustion in a cylinder so fast that there is no change in volume. Many high-speed diesel engines have practically constant-volume combustion.
Contraction Becoming smaller in size. In metals and fluids a result of cooling or a lowering of temperature.
Cooling water Water which is circulated through the jacket space of cylinders and cylinder heads to prevent excessive heating of these parts.
Crank That part of the crankshaft which is in the form of a crank and crankpin.
Crankcase The middle part of the engine structure surrounding the working parts.
Crankcase subbase The lower portion of the engine structure; the BEDPLATE.
Crankpin That part of the crank to which the connecting rod is attached.
Crankshaft That part of the engine which transmits the reciprocating motion of the pistons to the driven unit in the form of rotary motion. That part to which the connecting rods are attached.
Crankshaft cheek The part of the crankshaft that connects the crankpin to the main crankshaft journal.
Crankshaft journal The part of the crankshaft which rotates in the main bearings and transmits the torque developed by the engine.
Crankshaft web The crankshaft cheek.
Critical speed Speed at which the natural period of vibration of a shaft or other machine part is in synchronism with the power impulses.
Crosshead The part of an engine to which are attached the piston pin with the connecting rod and the piston rod and which is supported on guides.
Crown The top of an engine piston.
Crush The amount by which a precision bearing is compressed to ensure good contact between the back of the bearing and the bore holding it.

Cycle A series of events, operations, or movements that repeat themselves in a regular sequence.

Cylinder The cylindrical part of the engine in which the piston moves, and in which combustion takes place.

Cylinder block A number of cylinders cast in one piece.

Cylinder bore The inside diameter of an engine cylinder. Also the surface of the cylinder in which the piston slides or moves.

Cylinder head The part which covers and seals the end of the cylinder and usually contains the valves.

Cylinder-head stud Threaded round steel rod, one end of which screws into the cylinder block, the other being threaded to take a nut which holds the cylinder head in correct position.

Cylinder liner A cylindrical lining that is inserted into the cylinder jacket or cylinder block and in which the piston slides.

Delivery stroke The stroke of a pump during which the fluid in the pump is forced out of the cylinder.

Detonation A violent uncontrolled burning of a fuel in the combustion chamber.

Diesel engine A compression-ignition combustion engine first developed by Rudolf Diesel.

Distillation Separation of the more volatile parts of a liquid from those less volatile by vaporization and subsequent condensation.

Distributor A device which distributes and directs the flow of fuel or compressed air to the various cylinders of the engine in proper sequence.

Dribbling Slow seeping of fuel oil from the nozzle tip after cutoff of the fuel.

Dynamometer A device for determining the power of an engine.

Eccentric A circle not having the same center as another circle within it. A device mounted off-center for converting rotary motion into reciprocating motion.

Efficiency The ratio of output over input.

Electromotive force The potential, or voltage, developed by a dynamo, battery, or thermocouple. Abbreviated emf.

Energy Capacity for doing work.

Engine A machine which produces power to do work, particularly one that converts heat into mechanical work.

Exhaust The act of discharging gases from an engine after they have done work.

Exhaust cam The cam that controls the operation of the exhaust valve.

Exhaust gases Products of combustion which are discharged from the cylinder after doing work on the piston.

Exhaust manifold The pipe that collects the burnt gases as they are expelled from the cylinders.

Exhaust pipe Piping through which exhaust gases from an engine pass out to the atmosphere.

Exhaust pyrometer An instrument used to measure the temperature of the exhaust, mostly by the small electric current developed at the junction of two dissimilar metals when exposed to heat.

Exhaust valve The valve through which the burnt gases are allowed to pass out to the exhaust manifold.

Expansion period The portion of the power stroke during which the combustion gases expand from the movement of the piston and thus do work.

Filter A device to remove dirt and other impurities from air, oil, or water.

Fit The desired positive or negative clearance between the surfaces of two machined engine parts.

Flash point The temperature, degrees F, to which oil must be heated before the oil vapor over the oil will ignite when a small flame is passed across the surface of the oil.

Fluctuation Variation in value, such as of pressure or velocity.

Flywheel The wheel on the end of the crankshaft which gives the crankshaft momentum to carry the pistons through the compression stroke.

Foot-pound Unit in which work is measured; it is equivalent to the work of raising one pound vertically a distance of one foot or of moving an object one foot against a resistance of one pound. Abbreviated ft-lb.

Four-stroke engine An engine operating on a cycle which is completed in four strokes, or two revolutions of the crankshaft.

Framing The part of an engine between the cylinders and the bedplate; the crankcase.

Friction The resistance to relative motion between two bodies in contact.

Friction horsepower The power consumed within the engine from friction between its parts.

Fuel injector The device which sprays the fuel into the cylinder.

Fuel knock A noise produced in the cylinder of a diesel engine during combustion, usually when the fuel oil has a low ignition quality.

Fuel pump The pump that delivers the fuel to the injector.

Fulcrum The support on which a lever turns.

Gasket Packing placed between two surfaces that must have a leakproof joint.

Glow plug An electrical device used to heat fuel as it is injected in the cylinder for quick ignition and starting when the engine is cold.

Governor A mechanism used to control the speed of an engine.

Heat A form of energy.

Heat balance A tabulation showing the percentages of the heat developed by combustion in the engine cylinder that are (1) delivered in the form of power at the crankshaft, (2) lost in friction, (3) lost to the cooling water, and (4) lost in the exhaust gases.

Heat unit The unit of heat, usually British thermal unit (Btu).

Heat value The heat developed by the combustion of one pound of fuel, Btu per pound.

Helical Having the shape of a helix, or screw. Helical gears have teeth shaped like a helix.

Helix A line cut on a cylindrical surface shaped like a screw thread.
Horsepower A unit for measuring power. Rate at which work is done. One horsepower = 33,000 ft-lb per min. Abbreviated hp.
Hunting Erratic variation of the speed of the governor, also of the engine.
Hydraulic Pertaining to movement of and by water, also by other liquids, such as oils.
Idling Engine running without a load at the lowest speed possible.
Impeller The rotating part of a centrifugal pump or blower that imparts motion to a liquid or air by forcing it outward from the center of the machine.
Indicated horsepower The horsepower developed in the engine cylinder, as calculated from an indicator diagram.
Indicator Instrument used to investigate the pressures inside an engine cylinder.
Indicator diagram A diagram obtained by means of an indicator; it shows the change of pressure in the engine cylinder.
Inertia The tendency of a body to maintain its existing velocity.
Injection The forcing of fuel oil into the combustion chamber of a diesel engine by means of high pressure.
Injection pump The pump used to inject fuel oil into the combustion space of a diesel engine; the fuel pump.
Inlet cam The cam that controls the operation of the air inlet valve in a four-stroke engine.
Inlet manifold The main pipe that lies alongside the cylinder heads and from which branch pipes take the air charge to the separate cylinders.
Inlet valve The valve through which air or the air-fuel mixture is admitted to the cylinder of a four-stroke engine.
Intake stroke The suction stroke.
Integral An indivisible part of a whole, constituting a completed whole.
Intermittent Occurring at intervals.
Jack A tool to lift or move a heavy object. Also a tool to turn the flywheel to a desired position.
Jacket The outer casing forming a space around an engine cylinder that permits circulation of cooling water.
Jerk pump A fuel pump which injects fuel into the cylinder by action of a cam having a sharp nose.
Jet A small orifice used to control the flow of fuel or air. Also the stream of fuel or air coming from such an orifice.
Journal The finished part of a shaft that rotates in a bearing.
Jumper A water-pipe connection between a cylinder head and the cylinder jacket or the water-jacketed exhaust manifold.
Keeper A dowel or pin used to keep piston rings from moving from an assigned position.
Key A square or rectangular piece of steel—straight or tapering from one end to the other—used to secure a part on a shaft.
Keyway The machined slot in a shaft or hub of a wheel to take a steel key.

Kilowatt An electrical unit of measure equal to one thousand watts.

Kilowatthour A unit of energy equal to a continuous flow of one kilowatt for one hour.

Kinetic energy The energy of a moving body due to its mass and velocity.

Laminated Made of thin layers.

Laminated shim A SHIM made up of thin metal sheets soldered together but so that each layer can be easily peeled off.

Land The portion of the piston between two grooves carrying the piston rings.

Lanova cell A special combustion chamber, also called *energy cell*, for diesel engines of high rotary speeds.

Lb Abbreviation for pound.

Lean mixture A mixture in which the proportion of air to fuel is greater than that theoretically necessary for complete combustion.

Linear motion Motion in a straight line.

Liner The removable inner engine cylinder in contact with the piston.

Load The useful output of an engine at a given moment.

Lubricant A liquid or grease employed to separate two surfaces in relative motion to each other, in order to reduce friction.

Lubricating pump A pump which handles lubricating oil in an engine.

Manifold A pipe with a number of inlets to, or outlets from, the several cylinders of an engine.

Manometer A U-shaped glass tube, partly filled with a liquid, water or mercury, employed to measure pressure.

Mean effective pressure The mean or average pressure which, acting on the piston, would do the same work as does the actual variable pressure in the cylinder. Abbreviated mep.

Mechanical efficiency The ratio of brake horsepower to indicated horsepower.

Mechanical injection Injection with the fuel-valve operated mechanically from a cam. Sometimes, although wrongly, used to indicate airless injection in general.

Motor A mechanism doing work by means of a ready source of energy, such as electric current, compressed air, or oil under pressure. Incorrectly applied to the combustion engine in an automobile.

Muffler A device used to diminish noise of the intake or exhaust. Sometimes referred to as a *silencer*.

Needle valve A round steel rod with a conical or tapered point that seats against an outlet and prevents fuel oil from entering the engine cylinder except when it is lifted by a cam or oil pressure.

Nickel A metal which, when alloyed to steel and cast iron, improves their mechanical properties.

Nitrogen A rather inert gas that makes up slightly more than three-fourths of the atmospheric air by volume.

Nozzle The part of the injector or spray valve in which are located the holes through which the fuel is injected into the cylinder.

Oil-control rings The piston ring, usually located at the lower part of the piston, that prevents an excessive amount of lubricating oil from being drawn up into the combustion space during the suction stroke. Also called simply *oil ring* and *oil scraper ring*.

Oil grooves The passages cut in bearings for distributing the lubricating oil.

Opposed-piston engine An engine that has two pistons within the same cylinder, traveling in opposite directions.

Orifice A small round opening. Usually refers to the hole in the spray nozzle.

Otto cycle An engine cycle in which combustion takes place at constant volume.

Outboard bearing A bearing outside the engine proper, carrying an extension of the crankshaft.

Oxygen A gas that readily combines with other substances, such as carbon, hydrogen, sulfur, releasing heat. It makes up slightly less than one-fourth of the atmospheric air by volume.

Packing A material used to seal a joint against leakage.

Packing rings Rubber rings used to form a watertight joint at the bottom of the cylinder liner.

Pintle A small extension of the needle-valve tip projecting through the discharge nozzle. When the needle lifts, the oil passes through the opening between the circumference of the orifice and that of the pintle.

Piston A cylindrical part which reciprocates in the cylinder bore of an engine and transmits the force of the gas pressure through the connecting rod to the crankshaft.

Piston crown The top of the piston; the piston head.

Piston head The top of the piston or that part of the piston against which the gas pressure acts.

Piston pin A pin which rests in two bored holes in the piston and passes through the eye of the connecting rod, to join the two together flexibly.

Piston-pin bearing The bearing either in the eye of the connecting rod or in the bored bosses of the piston, in which the piston pin rocks.

Piston-pin boss That part of the piston on the inside, through which the hole is made to take the piston pin.

Piston-pin lock The device used to hold or lock the piston pin in the piston.

Piston ring A split ring placed in a groove of the piston to form a leakproof joint between the piston and the cylinder wall.

Piston-ring gap The space between the ends of the piston ring when it is in the cylinder bore.

Piston-ring land The part of the piston on the outside surface located between the piston-ring grooves.

Piston skirt The part of the piston below the piston-ring grooves.

Piston stroke The movement of the piston from one end to the other of the piston travel in the cylinder bore. The piston stroke is equal to twice the throw of the crankshaft.

Plunger The long piston of a single-acting pump, such as a fuel-injection pump.

Poppet valve A valve opened by the action of a cam and closed by a spring.
Port An opening, hole, or passage.
Pound per square inch The unit used to measure the pressure exerted by one body upon another. It is found by dividing the total force, pounds, acting normally upon a surface by the area of the surface, square inches. Abbreviated psi.
Pour point The lowest temperature at which fuel oil will just flow under test conditions. It is an indication as to how suitable a fuel is for cold-weather operation.
Power Rate at which work is performed.
Power factor The proportion (expressed as a decimal) which the actual power in an a-c electrical circuit bears to the apparent power indicated by instruments measuring the electrical pressure (volts) and current (amperes). Standard abbreviation pf.
Power stroke The working stroke of a piston.
Precombustion chamber A chamber in the cylinder head of some engines into which the fuel is injected, ignited, and partly burned, the rest of the fuel being thrown out into the main combustion space where combustion is completed. Sometimes also called *antechamber*.
Preignition Ignition taking place before the desired time in the operating cycle in spark-ignition engines. In a diesel engine can occur only if the fuel-injection timing is deranged.
Pressure The force due to the action of a gas or liquid in a closed vessel. Usually measured in pounds per square inch. Small pressures are measured in inches of a column of mercury or water. Also force applied to an area.
Psi Standard abbreviation for pounds per square inch.
Punk A slow-burning material inserted by means of a steel plug into the combustion chamber to provide the additional heat necessary to ignite the first fuel charge in starting some engines, especially in extremely cold weather.
Push rod The rod that transmits the action of a cam to the cam-operated valve.
Pyrometer An instrument for measuring high temperatures, as of the exhaust gases of a diesel engine.
Radial Extending from a center to the circumference, having the direction of a radius.
Reciprocating Having a back-and-forth or up-and-down linear motion, such as an engine piston.
Reclaimer An apparatus in which dirty lubricating oil, which often is discarded, is treated and made usable, reclaimed.
Relief valve A valve held closed by a spring and forced open when the pressure in the system rises above the desired height.
Resistance Mechanically, a force opposing the motion of a body, measured in pounds. Electrically, that which opposes the flow of an electric current, measured in ohms.

Rheostat A device to regulate the flow of electric current by transforming part or all of it into heat.

Ring grooves Grooves cut in the piston barrel to hold the piston rings.

Rocker arm A lever that transmits the action of the cam, usually by means of a push rod, to the stem of the intake or exhaust valve, sometimes also to the starting-air valve and fuel valve.

Rocker-arm shaft The shaft, usually at the top of the cylinder, that serves as a fulcrum for the rocker arms.

Rotary Turning on an axis.

Rotative Pertaining to rotation.

Rpm Abbreviation for revolutions per minute.

Saybolt viscosimeter The standard American instrument used to measure the viscosity of oils.

Scavenging The removing from the engine cylinder, by a stream of slightly compressed air, of the products of combustion of the preceding cycle.

Screen A wire cloth with a fine mesh used to remove dirt from oil or water.

Seal Any device to prevent leakage of gas or liquid, oil or water.

Semidiesel engine A term applied to oil engines using rather low compression pressures and requiring a hot surface for ignition of the injected fuel.

Sensitivity Change in engine speed before the governor begins to act.

Servomotor A motor operated by oil or air pressure and used for operating heavy control mechanisms.

Shaft A round bar of steel or other strong metal that is used to transmit rotary action.

Shaft horsepower The power rating of a diesel engine used for turning a propeller shaft in marine installations. Abbreviated shp.

Shell The steel or bronze backing to which the babbitt of a shaft bearing is bonded. Also the whole removable bearing.

Shim A thin sheet of metal or other material which is inserted between two machine parts to obtain their correct relative location.

Silencer A device to deaden the sound of the intake or exhaust of an engine; a muffler.

Silent chain A chain made up of small pins and steel plates that engage the teeth on sprockets resembling spur gears, and that is used to transmit power from one shaft to another and by its construction is less noisy than the ordinary roller chain.

Skirt The lower part of the piston. Also the lower part of a liner if it protrudes below the cylinder jacket.

Sludge A tarlike formation in oil resulting from the oxidation of a portion of the oil.

Solid injection A rather misleading term applied to airless injection.

Specific fuel consumption The fuel consumption per hour divided by the brake, or shaft, horsepower developed, expressed in lb per bhp or lb per shp.

Specific gravity (1) Weight of a liquid or solid compared with the weight of an equal volume of water at 60 F. (2) For a gas, its weight as compared with the weight of an equal volume of air at the same temperature and pressure.

Specific heat The amount of heat needed to raise the temperature of one pound of the substance one degree Fahrenheit.

Speed droop The difference in speed between no-load and full-load engine speed.

Spray valve The fuel injector.

Spring A coiled piece of round or square steel wire which, when compressed, exerts a force that may be used to do some work.

Stability (1) Ability of a lubricating oil to withstand physical change under severe operating conditions. (2) Ability of a governor to maintain the required engine speed without fluctuations or hunting.

Stress The internal forces set up in a body when it is subjected to forces tending to deform it by tension, compression, shear, bending, or torsion.

Stroke The distance a piston travels up or down inside the cylinder.

Suction stroke The stroke of the piston of a four-stroke engine during which a fresh charge is sucked in or forced by atmospheric pressure into the space vacated by the piston.

Supercharging Supplying of combustion air to an engine at higher than atmospheric pressure, usually 2 to 4 psig, in some engines up to 30 psig.

Surface-ignition engine The semidiesel engine.

Synchronous Occurring at the same time or in phase.

Tachometer An instrument indicating instantaneous rotary speed of a shaft in rpm.

Tangent A straight line touching a circle at one point and forming a right angle with the radius connecting this point with the centre of the circle.

Tangential Having the characteristics of a tangent.

Tangential force The component of the force applied to the piston acting at a right angle to the crank arm.

Tappet The part of the valve-actuating mechanism in contact with the cam; the cam follower.

Template A pattern used as a guide for shaping something. In engine-foundation work a wooden frame used to locate the foundation bolts.

Temperature The intensity or degree of heat.

Thermal efficiency The percentage of the total chemical energy in the fuel consumed that is converted into useful work.

Thermocouple Two strips or wires of dissimilar metals joined at one end used to measure temperature differences.

Thermodynamics The science of changing heat into mechanical work.

Thermostat A mechanism to convert the expansion of a heated metal or fluid into movement and having power sufficient to operate small devices, control electric circuits or small valves, etc. Can be set to operate at definite temperatures.

Throw of crankshaft The distance between the center of the crankpins and the center of the journals of the crankshaft. It is equal to half the stroke of the engine.

Thrust   An axial force acting on a shaft.

Thumbscrew   A screw or bolt whose head is in the shape of a flattened, vertical fin, so that the bolt can be turned by the fingers.

Timing   The angle made by the crank with its top or bottom dead-center position at which some valve opens or closes.

Timing chain   A chain that is used to connect the crankshaft and camshaft by which the camshaft is made to rotate.

Timing gears   Gears keyed to the crankshaft and camshaft, by which the camshaft is made to rotate.

Tolerance   An allowable variation in dimensions. For example: a dimension of 0.753 in. with a tolerance of $+0.000$ and -0.003 indicates that any dimension from 0.750 to 0.753 in. is acceptable.

Top dead center   The position of the crank when the piston is in its farthest position from the crankshaft, in its nearest position to the cylinder head. *Abbreviated* tdc.

Torque   The effect which rotates or tends to rotate a body. Torque is the product of force multiplied by the arm, or normal distance from the center of rotation to the force. Torque is measured in lb-ft or lb-in.

Torsion   The deformation of a body caused by a torque or twisting effort.

Torsional vibration   Oscillatory twisting vibration in a rotating shaft which tends to make a gear mounted on one end of the shaft whip back and forth with respect to a gear on the other end.

Trammel   A metal rod having pointed ends, used to mark off a span equal to its length. Also called a *tram*.

Transfer pump   A pump employed to force fuel oil from storage to the engine fuel tank.

Turbulence   A high-velocity swirling of air, fuel vapor, or a mixture of both within the combustion chamber or cylinder.

Two-stroke engine   An engine operating on a cycle that is completed in two strokes, or one revolution of the crankshaft.

U.S. gallon   Contains 231 cu in.; 1 U.S. gal of water weighs 8.33 lb.

Vacuum   A pressure below atmospheric, referring to a vessel filled with gas.

Valve   In a combustion engine, an intake or exhaust valve usually consists of a disk with a stem which is opened by a cam and closed by a spring.

Valve seat   That part of the valve mechanism upon which the valve face rests to close the port.

Valve spring   The spring which is used to close a valve.

Valve-spring retainer   The part which is held against a groove or grooves on the valve stem and in turn holds the valve spring in a state of compression.

Vanes   Baffles employed to deflect the flow of a fluid, gas or liquid. Vanes may be stationary or, as in a centrifugal-pump impeller, moving.

Velocity   The rate of motion or the speed of a body at any instant. Measured in feet per minute (fpm) or revolutions per minute (rpm).

Venturi   A tube with a narrowing throat or constriction to increase the velocity of the gas or liquid flowing through it.

Viscosity Internal resistance to flow in a liquid or gas. In practice, for oils it is measured by the number of seconds required for a definite quantity to flow through a standard orifice under stated test conditions.

Viscosity index A number given a lubricating oil to indicate its performance, particularly its change of viscosity with the temperature.

Volatility Ability of a liquid to turn into vapor.

Volumetric efficiency Ratio of the volume discharged from a pump to the piston displacement of the pump. In diesel engines a term often used instead of the correct term *charge efficiency*.

Water jacket The outer casing forming a space around an engine cylinder to permit circulation of cooling water.

Work The transference of energy by a process involving the motion of the point of application of a force. Work is done when a force moves a body through a certain distance.

Working stroke The piston stroke during which the combustion gases exert a pressure on the moving piston.

Wrist pin Piston pin.

LIST OF VISUAL AIDS

The visual aids listed below and on the following pages can be used to supplement much of the material in this book. For the convenience of users the films have been grouped under two headings: (1) films on general principles and fundamentals and (2) films on specific aspects of operation and maintenance.

It is recommended that each film be reviewed before use, in order to determine its suitability for a particular group or unit of study.

Motion pictures and filmstrips are included in the following list, the character of each being indicated by the self-explanatory abbreviations "MP" and "FS." Immediately following this identification is the name of the producer, and if different from the producer, the name of the distributor. Abbreviations are used for names of producers and distributors; these abbreviations are identified in the list of sources at the end of the bibliography. Unless otherwise indicated, the motion pictures are 16-mm sound black-and-white films and the filmstrips are 35-mm black-and-white silent films. The length of motion pictures is given in minutes (min), of filmstrips in frames (fr).

This bibliography is a selective one, and film users should examine the latest annual edition and semiannual supplements of *Educational Film Guide*, a catalog of 11,000 films, published by the H. W. Wilson Co., New York. The *Guide*, a standard reference book, is available in most college and public libraries. Readers should also write to the various diesel-engine manufacturers for copies of charts, posters, diagrams, models, and other visual aids dealing with the maintenance and repair of specific diesel engines.

Films on Principles and Fundamentals

A.B.C. of the Diesel Engine (MP, GM, 20 min, color) Animated film showing the fundamentals of operation of a diesel engine, using animated characters to personify air, fuel, and ignition. Gives examples of the use of engines as sources of power in industry and transportation.

Construction of Diesel Engines (MP, USN/UWF, 17 min) Shows the general structure of several types of diesel engines and the different frame types, cylinder parts, pistons, piston rings, connecting rods, crankshafts, bearings, camshafts, and rocker assemblies.

Diesel, the Modern Power (MP, USBM, 21 min) Demonstrates the principles, construction, and operation of diesel engines; contrasts the four-stroke and two-cycle diesel engines; explains the applications of diesel power in industry.

Diesel Engine (MP, USN/UWF, 29 min) Shows how ignition may be achieved by compression; describes basic diesel-engine types; discusses forms of air headers and fuel injectors. Introductory survey film.

Diesel Engine Fuel Systems (MP, USN/UWF, 40 min) Shows the basic structure of diesel fuel systems, parts and operation of injectors and fuel pumps, and Bosch, General Motors, and Excello equipment.

Diesel Engine, Ideal Diesel Cycle (MP, McGraw, 6 min) Explains the six processes in the diesel-engine cycle; assumes ideal conditions, and plots on a pressure-volume diagram the resulting diesel cycle; discusses the factors affecting efficiency. (College physics series produced in collaboration with American Association of Physics Teachers)

Diesel Lubrication and Cooling Systems (MP, USN/UWF, 10 min) Parts of diesel lubricating and cooling systems and how they work in relation to each other.

Diesel Story (MP, Shell, 20 min) Traces the development of the early internal-combustion engines and explains the principles and working of the modern four-stroke engine.

International Diesel Power (MP, Int Harvester, 20 min) Explains the features of the "international" diesel engine. Includes scenes showing how diesel fuel burns in a combustion chamber.

Films on Maintenance and Operation

Cooper Bessemer Diesel Engine Fuel System (MP, USN/UWF, 13 min) Covers the fuel injection system, examination of the fuel pump, fuel delivery and pressure, and functioning of the fuel injection nozzles.

Cooper Bessemer Diesel Engine Maintenance (MP series of 17 films, USN/UWF) Titles, running times, and descriptions follow:

Disassembly, Part 1 (16 min) Demonstrates methods of inspection for all possible parts with methods of removal for all lines and parts in a complete breakdown of the engine.

Disassembly, Part 2 (17 min) Step-by-step demonstration of tools used in disassembly of engine for either overhaul or maintenance, concluding with a review of the important steps of disassembly.

Bearing Disassembly and Inspection (14 min) Demonstrates how to disassemble main bearing, inspect, take readings on the bearing shell, and use care in handling the parts.

Bearing Reassembly (11 min) Demonstrates how to roll the bearing shell back into place; install cap, wedge blocks, and wedge shoe; and connect the lube oil line.

Cylinder Head and Piston (20 min) Shows how to disassemble the cylinder head and piston; inspect the cover frame; check rocker arms; and remove valves, crossheads, wedging piston, piston-pin bolts, connecting rod, piston-pin caps, and piston pin.

Inspection of Piston (10 min)   Demonstrates how to inspect piston covers and piston head, measure ring clearance, remove and inspect rings and ring grooves, measure piston pin and bearings, and record measurements.

Reassembly of Piston (11 min)   Demonstrates proper sequence of steps in reassembling pistons.

Cylinder Head (16 min)   Demonstrates how to inspect and recondition the cylinder head.

Reassembly of Cylinder Head (14 min)   Demonstrates how to reassemble and install the cylinder head; explains each operation and the use of proper tools.

Engine Reassembly, Part 1 (16 min)   How to inspect the liner, seat the seal in position in the cylinder block, lower the liner and fit the piston into the liner, put the shell into the connecting-rod bearing, and install the connecting-rod bearing cap.

Engine Reassembly, Part 2 (17 min)   How to install fire gasket, liner, cylinder nuts, motor header flange, liner drain plug, exhaust manifold gasket and coverframe, lube oil lines, injector nozzle, air-starting and fuel lines; adjust valve clearances; and install cylinder-head cover.

Engine Reassembly, Part 3 (16 min)   How to secure exhaust manifold, install indicator cock and relief valve unit, secure air-starting manifold and supply line, remove coupling gear, close indicator cocks, and tighten wrist-pin bolt.

Injector Block Removal and Disassembly (15 min)   How to put throttle in closed position, remove inspection window and injector block, and complete disassembly of the injector block and valve parts at a work bench.

Injector Block Reconditioning and Reassembly (18 min)   How to clean parts, lap valve seats, remove lapping compound, reassemble all valve parts in the injector block, and reassemble auxiliary parts in the injector block.

Injector Block Replacement and Timing (13 min)   Demonstrates how to replace the injector block and connecting lines; explains in detail how to set the injector lifts.

Fuel Pump Disassembly (15 min)   How to disconnect linkage, remove cover plate, disconnect all lines, remove fuel pump and disassemble at work bench, and remove all parts from pump housing.

Fuel Pump Reconditioning and Reassembly (23 min)   How to recondition and reassemble parts of a fuel pump system.

Diesel Engine—Scavenging and Supercharging Diesel Engines (MP, USN/UWF, 15 min)   Shows operation of two-stroke-cycle single- and double-acting engines and opposed-piston engines. Discusses method of scavenging and supercharging air.

Diesel Engine Governors. Part 1: Woodward Governors (MP, USN/UWF, 14 min)   Shows the operation of diesel-engine governors and explains the operation of overspeed, overspeed trip, and regulating governors.

Diesel Engine Governors. Part 2: GM Series 71, *Limiting Speed Mechanical Governors* (MP, USN/UWF, 12 min) Discusses three main assemblies of the governor and their functions. Reviews the operation of the manual fuel control to explain the action of the governor through low-, intermediate-, and high-speed ranges.

Diesel Engine Marquette Hydraulic Governors (MP series of 4 films, USN/UWF) Titles, running times, and descriptions follow:

Basic Hydraulic Governor (17 min) Function of Marquette hydraulic governor; basic construction; operation of principal parts of the hydraulic system. Schematic animation.

Speed Drop (5 min) Demonstrates how load equalizer is added to the governor to control speed and how the power piston and speeder spring function.

BMEP (10 min) Explains how the governor operates for normal and low speeds; how the bmep limiter functions when fuel is on and off, reduces the fuel supply to an overloaded engine, and goes out of position when the full operation is completed.

Powerhead (5 min) Shows adjustment of the speeder spring and construction and operation of the powerhead. Demonstrates the speed-setting control of the governor.

Diesel Lubrication and Cooling Systems: Lubrication of the GM-71 Series Engines (MP, USN/UWF, 12 min) Shows by the use of animation the course of the oil through the engine; describes how it lubricates each component part; explains the working principle of the ventilation system.

The Elliott-Buchi Turbocharger for Diesel Engines (MP series of 3 films, USN/UWF) Titles, running times, and descriptions follow:

Disassembly (20 min) Demonstrates how to disassemble the turbocharger, and the tools to be used.

Reassembly (25 min) Demonstrates how to reassemble the turbocharger.

Maintenance (24 min) Shows the maintenance procedure for the turbocharger, including frequent and periodic checks.

Fairbanks Morse Diesel Engine Maintenance (MP series of 11 films, USN/UWF) Titles, running times, and descriptions follow:

Inspection and Preparatory Steps (8 min) Covers the preliminary steps, including inspection, in the maintenance and repair of a Fairbanks Morse diesel engine, model 38D8 ⅛ O.P.

Removal of Injection Nozzles (5 min) Demonstrates how to remove injection nozzles, cylinder relief valve, indicator cock, and air-start check valve.

Removal of Pistons (21 min) Demonstrates how to disconnect and remove both upper and lower pistons.

Removal and Replacement of Main Bearings (22 min)   Demonstrates how to remove and replace upper and lower main bearings and clean, inspect, and measure bearing shells.

Replacement of Pistons (20 min)   Demonstrates how to replace and connect both upper and lower pistons.

Replacement of Injection Nozzles (6 min)   Demonstrates how to replace injection nozzles, cylinder relief valve and indicator cock, and air-start valve.

Air-start Check Valve, Cylinder Relief Valve, and Indicator Cock (13 min)   How to disassemble and reassemble air-start check valve, cylinder relief valve, and indicator cock and test the cylinder relief valve.

Injection Nozzle (10 min)   How to test and check injection pressure and disassemble, clean, and reassemble injection nozzle.

Pistons and Rods (17 min)   How to disassemble, clean, inspect, and reassemble pistons and connecting rods.

Removal of Cylinder Liner (27 min)   How to remove liner, including various connections and lines, timing mechanism, connecting rods, and upper crankshaft.

Replacement of Cylinder Liner (26 min)   How to replace cylinder liner, including upper crankshaft, connecting rods, timing mechanism, and various connections and lines.

Fulton Sylphon Diesel Engine Temperature Control Valve: Operation Adjustment (FS, USN/UWF, 90 fr, disk recording 12 min)   Describes the function, operation, and adjustment of the Fulton Sylphon valve for diesel engines.

General Motors Diesel Engine Unit Injectors: Disassembly and Reassembly (MP, USN/UWF, 18 min)   Shows how to disassemble and reassemble General Motors diesel-engine unit injector model 278.

General Motors Diesel Engine Unit Injectors: Maintenance (MP, USN/UWF, 18 min)   How to disassemble, handle, clean, and inspect the diesel-engine unit injector.

General Motors 16-278A Diesel Engine (MP series of 8 films, USN/UWF) Titles, running times, and descriptions follow:

Disassembly (30 min)   How to inspect blower rotors and pistons, remove head and scrape wear ridge, remove connecting rod bearings, and pull one piston and one liner in piston assembly.

Bearings (11 min)   Deals with the maintenance of bearings and demonstrates how to remove main bearings and inspect and install bearing shells.

Reassembly, Part 1 (23 min)   How to install liner, piston assembly, cylinder head, and rocker lever assembly in reassembly of the General Motors 16-278A diesel engine.

Reassembly, Part 2: Head (27 min)   How to install injector and controls, test injector, adjust injector, install valve bridges and rocker shaft, and assemble and time injector and exhaust.

Benchwork, Part 1 (14 min)   How to remove valves, overspeed trip assembly, transfer block, and check valve.

Benchwork, Part 2 (19 min)   How to clean and inspect the head, resurface valve seats, lap valves, and "mike" guides and valves.

Benchwork, Part 3 (12 min)   How to install valves, overspeed trip, and transfer block; test valve and relief-valve assembly; start air-starting check valve.

Benchwork, Part 4 (26 min)   How to disassemble the piston assembly; inspect and "mike" the piston, wrist pin, and wrist-pin bearings; check ring clearance; reassemble piston and rod; and inspect, clean, and "mike" the liner.

Installing New Rings in Tired Diesels (FS, Perf C, 175 fr, disk recording 25 min)   Explains the procedure for diagnosing and performing a complete piston ring job on a diesel engine.

Progressive Maintenance Diesel Propulsion Engine (MP series of 4 films, USN/UWF)   Titles, running times, and descriptions follow:

Disassembly of the 8-268A *Engine* (27 min)   Demonstrates how to remove air lines, manifold, rocker-arm assembly, piston, lever, injector, and cylinder head; stresses precautions to be observed.

Reassembly of the 8-268A *Engine* (36 min)   Demonstrates how to reassemble the General Motors 8-268A diesel engine.

Benchwork, 8-268A *Engine* (14 min)   Demonstrates how to check parts for cracks, use ring-expansion tool, insert wrist-pin bearings, reassemble piston, and replace needles in eye of connecting rod.

Bearing Removal and Inspection, 8-268A *Engine* (17 min)   Demonstrates how to disassemble the main bearing; inspect, clean, and replace bearing shells; and reassemble the bearing.

Progressive Maintenance on General Motors Diesel Engine 12-567A (MP series of 8 films, USN/UWF)   Titles, running times, and descriptions follow:

Cylinder Head Removal (22 min)   How to remove piston cooling tubes, cylinder-head covers, test valve, rocker-arm assemblies, fuel lines, injector linkage, injectors, and cylinder head.

Piston and Liner Removal (12 min)   How to remove connecting-rod bearing, fork-rod piston assembly, blade-rod piston assembly, and opposite liners.

Bearings (9 min)   How to remove, inspect, and install main bearing cap, lower bearing shell, and upper bearing shell.

Installation of Liner and Piston (23 min)   How to install liners, blade-rod piston assembly, fork-rod piston assembly, and pistons; and secure pistons with the bearing cap.

Installation of Cylinder Head (23 min)   How to position the cylinder head and install injector linkages, fuel lines, rocker-arm assemblies, test valves, and piston cooling lines.

Installation of Cylinder Head (23 min)   How to recondition the cylinder head, remove valve springs and valves, reface and check valve seats, check height of valve stems, and clean and inspect parts.

Reconditioning the Fuel Pump (17 min) How to install seal assembly, diaphragm, copper seal gasket, shim, setting bar, and spacer in the fuel pump.

Liner and Piston (20 min) How to inspect liner; disassemble piston-rod assembly; inspect slipper-rod assembly, piston assembly, and piston; and replace rings.

List of Film Sources

GM—General Motors Corp., 3044 W. Grand Blvd., Detroit 2, and 405 Montgomery St., San Francisco 4.

Int Harvester—International Harvester Film Library, c/o George W. Colburn Laboratory, 164 N. Wacker Dr., Chicago 6.

McGraw—McGraw-Hill Book Co., Text-Film Dept., 330 W. 42d St., New York 36.

Perf C—Perfect Circle Co., Hagerstown, Ind.

Shell—Shell Oil Co., 50 W. 50th St., New York 20, and 100 Bush St., San Francisco 6.

USBM—U.S. Bureau of Mines, 4800 Forbes St., Pittsburgh 13.

USN—U.S. Dept. of the Navy, Washington 25. (Films in this list distributed, sales basis only, by United World Films.)

UWF—United World Films, 1445 Park Ave., New York 29.

INDEX

Additive oils, 148, 187
Air cleaners, 244, 427
Air compressor, 427
Air-fuel ratio, 147
Air injection, 99
Air-intake system, 244
Air starting, 288
Aligning, main bearings, 449
   shaft extension, 335
Analysis of used oil, 200
Angularity of connecting rod, 52

Balancers, 314
Ball bearings, 71
Bearing safeguard, 415
Bearings, 65, 415
   classification, 65
   clearances, 408
   construction, 67
   fitting, 448
   loads, 196
   operating conditions, 67
   precision, 70, 450
   reciprocating motion, 70
Bedplate, 31, 440
   repairing, 466
Bore, 6, 479
Brake horsepower, 368, 382
   mean effective pressure, 368

Cam, 84
Cam follower, 86
Camshaft, 7, 86
   drive, 7, 86
   lubrication, 205
Carbon deposits, 183, 353
Centrifugal force, 311
Centrifuging, 158, 192
Cetane, 144
Cetane number, 144
Cetene, 480

Clearances, average, 407
Cold welding, 459
Color, exhaust, 349
Combustion, 19, 146, 395
   in diesel engines, 149
Combustion knock, 152, 352
Common-rail injection, 101
Compression, 18
Compression ratio, 18
Connecting rod, 6, 51, 441
Contamination, oil, 199
Cooler, soft-water, 236
Cooling (*see* Engine cooling)
Cooling towers, 238
Counterweights, 60
Crankcase, 7
Crankpin, building up, 642
Crankpin bearing, 53, 416
Crankshaft, 7, 56, 443
   balance, 312
   critical speed, 317
   dimensions, 59
   endurance failure, 58
   repairing, 470
Crosshead, 55
Crosshead guides, 71
Cylinder, 33, 412
   arrangement, 7
   honing, 35, 435
   reboring, 431
Cylinder head, 39, 411, 463
Cylinder liners, 34, 412, 430, 468
   measurements, 434

Diesel engines, characteristics, 3
   classification, 4
   cylinder arrangement, 7
   designation, 12
   fuels, 14
      properties, 142
   selection, 325

INDEX

Dilution of oil, 199
Distributor, injection system, 114
Draw cards, 362

Efficiency, charge, 371
  engine, 369
  mechanical, 369
  over-all, 371
  scavenge, 372
  thermal, 370
  volumetric, 371
Electric dynamometer, 386
Electroplating, 462
Engine cooling, 218
  air, 229
  equipment, 230
  expansion tank, 235
  liquid, 229
  recooling water, 227
  temperatures, 221
  vapor-phase, 229
  water circulation, 235
Exhaust, four-stroke engines, 261
  two-stroke engines, 278
Exhaust-gas turbine, 273
Exhaust manifold, 264
Exhaust mufflers, 267, 270
Exhaust pipe, 264
Exhaust silencer, 269
Exhaust snubber, 269
Exhaust system, 269

Failure, to come up to speed, 345
  to develop full power, 347
  to start, 343
Filters, air, 244
  fuel, 157
  lubricating oil, 188
Filtration, lubricating oil, 188, 208
Firing order, diesel engines, 56
Flash point, 144, 184, 201
Flywheel, action, 7, 74
  balancing, 79
  construction features, 80
  knock, 79
  rim markings, 78
  velocity, 78
Foundations, 327
  excavation, 329
Fuel injection, 98

Fuel injection, methods, 98
Fuel knock, 352
Fuel measurements, 387
Fuel metering, 98
Fuel nozzles, 110
  repairing, 418
  testing, 422
Fuel pumps, 418
Fuel system, 155
Fuel tanks, 165
Fuel transfer pumps, 160
Fuels, 14
  properties, 142

Glossary of engine terms, 475
Governors, 118, 426
  characteristics, 121
  classification, 121
  functions, 120
  hunting, 122, 124, 247
  hydraulic, 131
  isochronous, 122, 123
  load-limit, 135
  mechanical, 125
    pivotless, 127
  overspeed, 137
  relay-type, 131
  sensitivity, 123
  speed droop, 122
  types, 125
Greases, 187
Grouting, engine, foundation, 334
Gumming-up of piston, 353

Heat exchangers, 237
Heat transfer, 219
Heat values, fuel, 149
Honing cylinders, 35, 435
Hydraulic governor, 131
Hydraulic nozzles, 421

Ignition, 146
Ignition delay, 146
Indicated horsepower, 367
Indicator, 358
Indicator diagrams, 356
Inspection schedules, 396
Intake headers, 250
Intake silencers, 249
Irregular engine speed, 347

INDEX

Jerk-pump injection, 104
Jerk pumps, 106
Journals, crankshaft, 444, 479

Kingsbury thrust bearing, 178

Leveling, engine, 332
Limitations, performance, 372
Load-limit governor, 135
Low-temperature starting, 297
Lubricants, 179, 187
Lubricating oil (see Oil, lubricating)
Lubrication, 174, 196
    air compressor, 207
    ball bearings, 179
    bearings, 196
    diesel engines, 197
    nonrotating parts, 206
    objects of, 174
    pressure, 197
    principles, 174
    rotating parts, 204
    splash, 197
    wedge action, 176

Main bearings, 445
    checking alignment, 449
Maintenance, 395
    details, 406
    log sheets, 398
Mean effective pressure, brake, 368
Mean indicated pressure, 367
Mechanical efficiency, 369
Mechanical injection, 101
Metal spraying, 461
Mufflers, 267, 270

Neutralization number, 201
Noisy engine, 352
Nominal horsepower, 376

Offset indicator diagrams, 361
Oil, lubricating, characteristics, 179, 182
    circulation, 202
    consumption, 202
Oil coolers, 211
Oil filters, 188
Oil grooves in bearings, 177
Operation of diesel engine, 337
Opposed-piston engine, 10
    scavenging, 23
Overhauling diesel engine, 429

Overheating engine, 351
Overspeed governor, 137

Performance curves, 378
Performance limitations, 372
Piping, 239
Piston, 41, 412, 437, 469
    displacement, 14
    lubrication, 202
Piston pins, 45
Piston rings, 6, 47, 438
Piston rods, 54
Precipitation number, 201
Precombustion chamber, 115
Pressure regulator, oil, 214
Pressure-time diagrams, 366
Pumps, fuel-injection, 418
    lubricating-oil, 214
    water-circulation, 231

Radiator, engine cooling, 228
Radiator unit, 238
Rating, engine, 375
Regulator, pressure, oil, 214
    temperature, 240
Repairs, 456
    cylinder heads, 463
Reversing, 301
Running an engine, 340
Running gear, 41

Safety devices, 215, 241, 338
Safety precautions, 406
Salvage methods, 456
Scavenging methods, 21
    two-stroke engines, 256
Selection of engine, 325
Smoky exhaust, 348
Snubber, exhaust, 269
Speed factor, 26
Spring, valve, 92
Spring retainers, 93
Spring surge, 92
Starting engines, 284, 293
    failure in, 343
    low-temperature, 297
    methods, 285
    position, 296

Tail pipe, exhaust, 272
Temperature measurement, 215

Temperature regulator, 240
Testing, diesel engines, 381
 fuel nozzles, 422
Torque, engine, 376
Tramming, 452
Troubles in operating engine, 343
Two-stroke-cycle engines, 278
Two-stroke-cycle events, 19

Unit injector, 111
Unloader valve, 103

Valve cages, 92
Valve gear, 83
Valve guides, 91

Valve lash, 94
Valve seat inserts, 91
Valve springs, 92
Valves, engine, 89, 417, 469
 starting, air-operated, 294
Vapor-phase engine cooling, 229
Vibration, absorbers, 315, 318
 causes of, 307
 engine, 308, 353
 foundation, 329
 torsional, 316
Viscosity, oil, 143, 180, 201

Warming up diesel engine, 339
Waste-heat recovery, 279
Welding, repairs, 456, 461